ELEMENTARY SEISMOLOGY

By CHARLES F. RICHTER

CALIFORNIA INSTITUTE OF TECHNOLOGY

W. H. Freeman and Company

SAN FRANCISCO AND LONDON

A Series of Books in Geology

EDITORS: James Gilluly and A. O. Woodford

PREFACE

THIS BOOK developed from a lecture course organized particularly for students in geology who do not plan to specialize in seismology or geophysics. Because of the dual nature of the subject, it is necessary to strike a proper balance between instrumental seismology and field work.

When the course was given originally, mathematics was held to a low level. At present, geophysical theories and methods are permeating every branch of geology and effecting a gradual revolution of our thinking; for this reason it becomes desirable, if not mandatory, to give mathematics a more comprehensive treatment. To meet this requirement without too great a demand on the student, mathematics in the body of the book is kept at the minimum consistent with intelligent comprehension. Details and long proofs are given in appendixes. The fundamental concepts of stress, strain, and elasticity are developed in Chapter 16; the unprepared student will find it possible to use the book without working through this chapter. However, the material in Chapter 15, covering the principle of the seismograph, cannot be passed over if the student is to make intelligent use of the results of instrumental seismology. He might as well attempt to use modern petrological data without understanding the microscope.

Due emphasis on mathematics does not imply the other extreme of underrating field observation and field training. Great harm is done by poorly trained men who hasten about in the field, observe a small part of the evidence, and publish premature conclusions which are actual obstacles to serious investigation. Throughout the book, and especially in Chapters 11, 13, and 14, there are suggestions for proper seismological field work.

One reason for setting forth the methods and assumptions of seismology in detail is that geologists and engineers often accept results too literally and apply them beyond the limits of accuracy. The opposite error, of rejecting definite instrumental results because they conflict with conclusions from hasty field work, is less common now than formerly. Throughout the text, possible sources of misdirection of both kinds are pointed out for the benefit of the working geologist.

Many textbook generalizations on tectonic earthquakes are based on only a small part of the available literature. For this book the collections by Montessus de Ballore and by Davison have been extended and their interpretations revised.

Although the book is intended primarily for elementary students, it includes descriptive and reference material for instructors and research workers. Except where it is necessary to refer to original contributions, publications in the more generally accessible journals have been given preference, especially the *Bulletin of the Seismological Society of America* which is available in most large libraries.

Material omitted or given abbreviated treatment as being too advanced or too special, or needing too much space, includes:

1. General geophysics apart from seismology.
2. Discussion of the cause and nature of mountain-building.
3. Derivation of earthquake mechanism from seismograms (especially when there is dip slip).
4. Microseisms (treated briefly in Chapter 23).
5. Damage and other effects of well-investigated earthquakes where there is no direct evidence of faulting.
6. Theory of elastic waves in media not homogeneous or not isotropic, including layered media.
7. Theory of plastic deformation and of fracture.
8. Calculation of earthquake energy from seismograms.
9. Seismograph construction and testing.

Geography and statistics of earthquakes are discussed in outline only; further details are given in *Seismicity of the Earth*.

The discussion of prospecting for oil and minerals by seismic methods is limited to a short statement of general principles. The interested student should refer to special handbooks. However, techniques change with extreme rapidity; books must be supplemented by study of current periodicals, and if possible by personal contact with the work.

Of special interest to engineers are Chapters 3, 8, 11, and 24, and Appendixes II and III. Appendix II, on safe construction, is presented with apology; its subject matter is too important to pass over completely, but adequate treatment including constructional details would call for another volume, by another hand.

Chapters 4 and 5 discuss a few selected earthquakes as illustrative examples. For the sake of completeness, these descriptions include material which otherwise would have been deferred to later chapters.

I am under many obligations, notably to my colleagues Beno Gutenberg, Hugo Benioff, R. H. Jahns, C. R. Allen, and Frank Press, and to Dr. Markus Båth of Uppsala, who read the manuscript and offered valuable suggestions and references. The book as it stands would have been impossible without the extraordinary resources of Professor Gutenberg's personal library.

I am profoundly grateful for having had the opportunity of field work on earthquakes with the late and affectionately remembered John P. Buwalda. Over many years, I learned much from association and discussion with Harry O. Wood.†

The Seventh Pacific Science Congress in New Zealand (1949) afforded a splendid opportunity, not only to become acquainted with the local circumstances of that interesting region, but also to broaden my whole outlook in the geological sciences. In this book, California seismology has been discussed at length because it provides first-hand material for illustration. Comparison with New Zealand has been emphasized to avoid giving the book too parochial a character. For the necessary data and discussion I am indebted to Professors C. A. Cotton and W. N. Benson, Dr. C. A. Fleming, and Dr. A. R. Lillie. Professor Cotton has placed me under further heavy obligations by reviewing those chapters dealing with New Zealand; he has helped me to remove inaccuracies and add many paragraphs of new material.

Professor V. P. Gianella kindly reviewed the pages of Chapter 28 which deal with Owens Valley and Nevada earthquakes. He furnished many additional details and references.

During several visits to Pasadena, Professor Chuji Tsuboi contributed greatly to our understanding of Japan and its geophysical research, in a manner which it is a deep pleasure to acknowledge. He has read Chapter 30 and provided many valuable suggestions.

I am indebted to Lt.-Col. Ernest Tillotson for a large portfolio of original data on the African earthquake of 1928, and to Dr. J. B. Auden for notes on the tectonics of India and references on the earthquake of 1762.

For advice on engineering and insurance points I wish to thank Dr. G. W. Housner and Mr. H. M. Engle.

Illustration has been in charge of Mr. J. M. Nordquist. Figures have been drafted by him and by Mrs. Dorothy Hammond, Miss Phyllis Cangelosi, and Mrs. Barbara Dixon.

Maps from *Seismicity of the Earth* are reproduced with permission of the Princeton University Press. Much material on the seismology of Japan is summarized or quoted, by permission, from A. Imamura, *Theoretical and Applied Seismology,* published by Maruzen Co., Tokyo, 1937.

Except for quoted matter, I take full responsibility for statements in the text.

C. F. R.

Pasadena, California

March 1, 1957

---

† Deceased, February 1958.

CONTENTS

PART ONE

Nature and Observation of Earthquakes

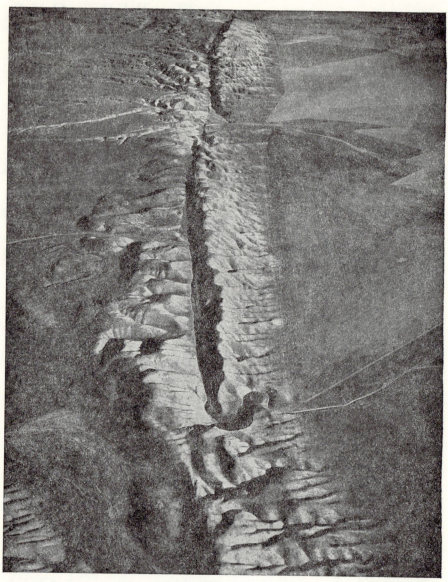

FRONTISPIECE   *View southeastward along the San Andreas Rift, San Luis Obispo County, California. Drainage descending from the highland at the left to the Carrizo Plain on the right is interrupted by the Rift and deflected in consequence of right-hand strike-slip. [Photo by William A. Garnett.]*

CHAPTER 1

Introduction to Seismology

A YOUNG SCIENCE

Investigating earthquakes is a live field of study, continually breaking down old fences and taking in new territory. It is useless to attempt exact definition; we usually say that *seismology is the science of earthquakes and related phenomena.*

Seismology developed later than most of the physical sciences. It is now as difficult to think of the seismologist without his characteristic instrument, the seismograph, as to imagine a modern astronomer without a telescope. The telescope dates from about 1600, but the first effective seismographs were built between 1879 and 1890.

In studying any earthquake one should carefully note its date in relation to the state of the science at the time. Not merely instruments, but also the whole background of information and investigation, change so rapidly that it makes an enormous difference, in terms of what we may expect to learn about it, whether an earthquake took place in 1850, 1880, 1910, 1930, or 1955.

PROGRESS OF SEISMOLOGY

As astronomy existed long before the telescope, so is seismology older than the seismograph; but our information about early earthquakes comes mainly from unscientific sources. Ancient accounts of earthquakes do not help us much; they are incomplete, and accuracy is usually sacrificed to make the most of a good story. Useful reports begin in the eighteenth century.

The following dates represent milestones in the progress of our knowledge of earthquakes and the earth to 1923.

1755, November 1. Lisbon earthquake. Some effects scientifically described.

1783, February 5 and following. Earthquakes in Calabria, Italy. Investigated by scientific commissions.

3

1819, June 16. Earthquake in Cutch, India. Earliest well-documented observations of faulting accompanying an earthquake.

1857, December 16. Earthquake east of Naples, Italy. Field investigation by Robert Mallet; first systematic attempt to apply physical principles to earthquake effects in detail.

1880. First effective seismographs developed by Gray, Milne, and Ewing in Japan.

1883–1884. Rossi-Forel scale for earthquake effects published (based on work in Italy and Switzerland).

1889, April 18. First identified seismogram of a distant earthquake. An instrument at Potsdam, Germany, recorded an earthquake in Japan.

1891, October 28. Mino-Owari earthquake, Japan. Large fault displacements; great damage. Imperial Earthquake Investigation Committee set up in consequence.

1896. Committee on Seismology formed by the British Association for the Advancement of Science. Enabled John Milne to establish seismograph stations with world-wide distribution.

1897, June 12. Great Indian earthquake. Investigated by R. D. Oldham.

1901. Geophysical Institute founded at Göttingen, Germany, by E. Wiechert.

1902. Improved intensity scale published by G. Mercalli in Italy.

1903. International Seismological Association founded. General improvement in recording and reporting earthquakes.

1906, April 18. California earthquake. Observed faulting; elastic-rebound theory of earthquakes formulated by H. F. Reid (Johns Hopkins University).

1906. Electromagnetic seismographs developed by B. Galitzin in Russia.

1909, October 8. Earthquake in Croatia. A. Mohorovičić (at Zagreb) discovered a sharp change in the speed of seismic waves at the base of the continental crust.

1913. First accurate determination of the depth of the earth's core, by B. Gutenberg (at Göttingen).

1918. First year covered by the *International Seismological Summary,* collating readings from most of the seismological stations of the world.

1922. Deep-focus earthquakes discovered by H. H. Turner in the course of editing the *International Summary* (at Oxford, England).

1923, September 1. The Kwantō earthquake, destructive at Tokyo and Yokohama. Detailed investigation; many published reports. Earthquake Research Institute established at Tokyo in consequence.

Many of these items are major themes for later chapters in this book. It should be noticed how truly international seismology is. Important contributions are made at present by hundreds of recording stations and dozens of research centers in all parts of the world.

"BELIEVE IT OR NOT?"

Reports on earthquake effects must be considered critically. Earthquakes are exciting and sometimes spectacular events, offering opportunity for mistakes and exaggeration. Accounts in the popular press, or observations by untrained persons, can be used for scientific purposes only with caution. Even scientific men without adequate training in seismology are sometimes responsible for misinformation. The serious student needs knowledge both of earthquakes and of the psychology of error. In dealing with earthquakes of old date, he also needs a knowledge of history; proper interpretation of documents may call for the help of professional historians.

Data in seismology accumulate slowly; much depends on investigating large earthquakes, which are not frequent. Actual harm has been done to the science by premature conclusions from fragmentary data, repeated from publication to publication. Few seismologists have had the admirable tact of the Russian who wrote (his own translation), "For lack of material we restrain ourselves from any kind of conclusions."

EARTHQUAKE EFFECTS

Many occurrences of different kinds have been called earthquakes (see Chapter 12). At this point we are not concerned with definition, but with the nature of earthquake observations.

A large earthquake usually attracts general attention only because it destroys houses and other works of man or breaks up and changes the ground surface. Such strong effects are limited to a relatively small area. The field geologist investigating this area finds evidence bearing on the mechanical cause of the earthquake. Faulting has been observed to occur together with twenty or thirty large earthquakes, and most important earthquakes are believed to originate in this way. Crustal blocks are raised, tilted, or displaced laterally by faulting, with corresponding effects on terrain.

The commoner consequences of earthquakes are related only indirectly to this primary geological, *tectonic,* process. When faulting breaks the rocks, waves of elastic compression or distortion spread out through the solid earth as waves spread over a pond, or as quivering spreads through jelly. These waves arriving at the surface of the earth shake it—violently near the source of a great earthquake, mildly at distant points. This shaking is responsible for most of the ordinary earthquake damage, and for nearly all minor earthquake phenomena.

With seismographs we can trace the wave disturbance far beyond the limits of destruction, and well beyond the limits of shaking felt by persons.

Sensitive instruments register major earthquakes at all distances, even near the antipodes of the heavily shaken area, thus demonstrating that such events disturb the whole earth.

SEISMOLOGY A DUAL SCIENCE

Seismology is a borderline field between geology and physics. It calls for sound physical thinking, as do all the geological sciences. Its data result from widely different techniques including especially those of field geology and (because of the seismograph) those of laboratory physics. This duality arises from the real division in nature between the visible earthquake phenomenon and the invisible waves of elastic disturbance. Few persons are well trained and practiced in both techniques; progress in seismology depends on cooperation between physicists and geologists.

THE GEOLOGIST AND SEISMOLOGY

Interpretation of seismograms depends partly on geological findings in the field. Where faulting or other geological processes occur, the geologist's knowledge is indispensable. His judgment is often called for to separate significant primary effects from such secondary occurrences as large landslides.

Observing and reporting earthquake effects in the field involve special problems, such as the behavior of objects and materials when violently shaken and the characteristics of wave motion in solid bodies. The competent geologist is no more deterred by these minor obstacles than by the special problems of volcanology, glaciology, or paleontology.

Because opportunity to investigate a strong earthquake is not common, most of the field work is done by persons without previous experience. Repetition of old and familiar errors, waste of limited time on unimportant or well-understood matters, overlooking of significant details, and incomplete reporting of unusual observations may result. Common errors are:

(1) Description of secondary effects, such as slides and fissures due to shaking, as if they were direct evidence of faulting.

(2) Overhasty identification of a conspicuous fault as the source of an earthquake, or of an erosional feature as an active fault.

(3) Failure to specify the direction of fault displacement (particularly of horizontal slip).

(4) Attribution of non-volcanic shocks to volcanic causes.

(5) Uncritical adoption of the observations and conclusions of untrained witnesses.

(6) Underestimation of the effect of unconsolidated ground increasing

the shaking, or of weak construction increasing its effects; this may lead to mismapping of intensities and mislocation of the source of the earthquake.

The geologist has a public responsibility in connection with earthquakes. He should be able properly to inform engineers, architects, property owners, and public officials without being too reassuring or needlessly alarming. He should form a clear idea of the long-term nature of earthquake risk, as well as its relation to the location of faults and the character of ground or foundation.

Because of the dispersion of seismological literature, geologists often overlook or ignore it. A recent paper on the geomorphology of a highly seismic region discusses rift valleys and faults but completely ignores well-described faulting on two historical occasions, omits study of earthquake locations made at seismological stations, and ends with an airy generality to the effect that the frequent earthquakes show that block movements are still going on. One objective of the present book is to make the results of seismology more accessible to the geologist and thus reduce the probability of such oversights.

RESULTS FROM THE SEISMOGRAPH

From the geologist's point of view, the most significant results of operating seismograph stations all over the world are:

(1) Location of all important earthquakes, in depth as well as in geographical position, even under the oceans or in remote parts of the continents.

(2) Determination of the structure of the earth, both of the upper crust and of the deep interior.

(3) Evidence as to the mechanical nature of earthquakes, including the details of the process of faulting deduced from the character of recorded elastic waves.

Fundamentally important, from the point of view of general physics, is the observation of elastic waves which have passed through the central part of the earth, where the pressure is of the order of millions of atmospheres. Since such pressures cannot be approached in the laboratory, this is valuable information about the properties of matter which partially justifies extrapolating ordinary physical principles to apply under those extreme conditions.

APPLICATIONS OF SEISMOLOGY

Prospecting by the seismic method, using explosives to generate small artificial earthquakes, is of great economic importance, particularly in the oil industry. Geologically valuable information is often obtained in this way.

The original purpose of the seismograph, the detailed recording of motion in a strong earthquake, is now carried on with special instruments of low magnification. Engineers and architects use this information in designing earthquake-resistant structures. One frequently employed type of experiment consists in applying motions, representative of those recorded in actual earthquakes, to models representative of buildings and other engineering structures, actual or proposed.

Continuous recording of earthquakes leads to compiling statistics valuable to insurance men and others interested in long-term risk. To be useful, such statistics must discriminate critically between small and large shocks.

Prediction of earthquakes in any precise sense is not now possible. Any hope of such prediction looks toward a rather distant future. Cranks and amateurs frequently claim to predict earthquakes. They deceive themselves, and to some extent the public, partly because it is not generally known how frequent earthquakes are. If small shocks are counted, 100,000 a year is a conservative estimate.

GEOPHYSICS

Geophysics is the application of general physical principles to the earth. Logically this would include all the geological sciences. Some of the recognized branches of geophysics, with their subject matter, are:

(1) Seismology: earthquakes and related phenomena
(2) Volcanology: volcanoes, hot springs, etc.
(3) Hydrology: ground and surface water
(4) Oceanography: the seas
(5) Meteorology: the atmosphere
(6) Geodesy: the size and form of the earth
(7) Terrestrial magnetism: the earth's magnetic field
(8) Tectonophysics: a newly named branch which deals with the physics of geologically significant processes in the earth

The study of the force of gravity is so closely connected with the problems of geodesy that it usually is referred to that branch. Atmospheric electricity, logically a part of meteorology, has long been associated with the study of terrestrial magnetism. The physics of the upper atmosphere, which includes the ionosphere or Heaviside layer, has expanded so much lately that when its enthusiasts write or speak of "geophysics" they mean only this one field. There has been a similar narrow use of "geophysics" to mean prospecting for oil and minerals by geophysical methods; other than the seismic method, these include measurements of gravity, the magnetic field, and electric conductivity, as well as chemical and radiological techniques—in fact, almost every procedure which might imaginably have bearing on the finding of oil has been used.

The study of the internal constitution of the earth is less a branch of geophysics than a research program, using data of many kinds. This work has a fundamental bearing on geology, since it profoundly affects theories about the origin and history of our planet. Many of the important data for this purpose are taken from seismology, but determinations of heat flow to the surface of the earth, measurements of gravity and terrestrial magnetism, and astronomical observations are all involved.

Many students in geology now wish to continue into geophysics, pure or applied; often they find their lack of mathematical preparation a serious obstacle. Some graduate students are appalled and a little offended at being confronted with differential equations, never having attended a formal course in that subject. Yet such a course is not an absolute necessity; the basic requirements are elementary algebra, trigonometry, and differential calculus. Many students do not completely grasp algebra and trigonometry during their secondary-school years, and attain only a hazy understanding of calculus.

One root of this difficulty is that many geological departments still rate field geological experience, and even paleontology, far above basic physics and mathematics. This extreme point of view is out of date; it results in drawing into geology students whose preparation or proficiency in mathematics is deficient. References below to papers by Hubbert, by Schriever, and by others, are given to illustrate the need for sounder mathematical physics in geology proper, as well as in geophysics.

General References and Reading

THE FOLLOWING books and papers are recommended for study; they or their equivalents should be available in any seismological library. Most of them are general works of reference; some are important original contributions; and a few are elementary and popularized treatments which may serve to introduce the subject. Special references follow each chapter; at this stage note particularly those for Chapter 25.

Geological classics

Lyell, C., *Principles of Geology*. (The 1st edition, 1830, is of interest chiefly as a historical monument. The material of greatest seismological importance was inserted in later editions; a good reference is the 12th edition—2 vols., 1875.)

Suess, E., *Das Antlitz der Erde*, Vols. 1–4; 1st ed. of Vol. 1, Prague, 1885; Vol. 4, 1903. (This great work has had an enormous influence on geophysics and seismology.)

————, *La Face de la terre*, Paris, Vols. 1–4, 1897–1913. (This edition, prepared and annotated under the direction of E. de Margerie, is far more than a mere translation; some students prefer it to the original.)

————, *The Face of the Earth*, Oxford University Press, Vols. 1–4, 1904–1909; Vol. 5 (index and maps), 1924.

General geophysics, and internal constitution of the earth (usually including seismology)

Gutenberg, B., ed., *Handbuch der Geophysik*, Borntraeger, Berlin, Vols. 1–4, 1929–1936. (Editing and publication of the remaining volumes was disorganized under the Nazi regime.) Vol. 1, *Die Erde als Planet*. Vol. 2, *Aufbau der Erde*. (Note especially Gutenberg, "Der physikalische Aufbau der Erde," pp. 450–564, and Born, A., "Der geologische Aufbau der Erde," pp. 565–867.) Vol. 3, *Veränderungen der Erdkruste*. Vol. 4, *Erdbeben*. (Very inclusive treatment of seismology.)

Angenheister, G., ed., *Wien-Harms Handbuch der Experimentalphysik*, Akademische Verlagsgesellschaft, Leipzig, Vol. 25, Part 1, "Geophysik 1," 1928 (concerned with the atmosphere and terrestrial magnetism); Part 2, "Geophysik 2, Physik des festen Erdkörpers und des Meeres," 1931 (contains some seismology).

Handbuch der Physik, Springer, Berlin, Vol. 47, *Geophysik 1*, 1956. (About half the contents is seismological.)

Gutenberg, B., "Geophysics as a science," *Geophysics*, vol. 2 (1937), pp. 185–187. (A brief summary and classification.)

Jeffreys, H., *The Earth*, Cambridge University Press, 3rd ed., 1952. (A research monograph, not a handbook. Often mathematically difficult, even for the advanced student. Treats most of the fundamental problems of seismology.)

Kuiper, G. P., ed., *The Earth as a Planet*, Chicago University Press, 1954.

"The planet Earth." *Scientific American*, vol. 193, No. 3 (Sept. 1955), pp. 1–211. (Good popular articles. Includes Bullen, K. E., "The interior of the earth," pp. 56–61.)

Gutenberg, B., ed., *Physics of the Earth*, Vol. VII, *Internal Constitution of the Earth*, McGraw-Hill, New York, 1939. 2nd ed., rev., Dover Publications, New York, 1951. (About half is seismological. Some chapters in the second edition were brought up to date, others almost not at all.)

Coulomb, J., *La Constitution physique de la terre*, Albin Michel, Paris, 1952.

Bullen, K. E., "Some trends in modern seismology," *Science Prog.*, vol. 43, No. 170 (April 1955), pp. 211–227. (Seismology with reference to the earth's interior.)

Gutenberg, B., "Neue Ergebnisse über den Aufbau der Erde," *Geol. Rundschau*, vol. 45 (1956), pp. 342–353.

Poldervaart, A., ed., "Crust of the Earth (A Symposium)," *Geol. Soc. Amer., Spec. Paper* No. 62 (1955). (Some important seismological data in Part I, "Nature of the Earth's Crust.")

Seismology, general

Wiechert, E., Zoeppritz, K., Geiger, L., and Gutenberg, B., "Über Erdbebenwellen," *Göttinger Nachrichten*. (See notes on periodicals and serials.) Papers I, 1907, pp. 415–529; II, pp. 529–549; III, 1909, pp. 1–30; V, 1912, pp. 121–206; VI, 1912, pp. 623–675; VIIA (Gutenberg), 1914, pp. 1–52; VIIb (Zoeppritz), 1919, pp. 66–84. (This series of fundamental papers is recommended to the instructor and the advanced student, together with other publications from the same group in the same journal. Paper IV was planned but never published.)

Sieberg, A., *Geologische, physikalische und angewandte Erdbebenkunde*, Fischer, Jena, 1923. (Exhaustive for its date. Several sections are by Gutenberg.)

Macelwane, J. B., *et al.*, *Physics of the Earth—VI, Seismology*, *Bull. Nat. Research Council*, No. 90 (1933). (Excellent in part, but incomplete, uneven, and now largely out of date.)

Imamura, Akitune, *Theoretical and Applied Seismology*, Maruzen, Tokyo, 1937. (Comprehensive except for geology. Extremely useful for studying the earthquakes of Japan.)

Milne, J., *Earthquakes and Other Earth Movements*, 7th ed., London, 1939, Kegan Paul, rev. and rewritten by A. L. Lee. (A classic of seismology, thoroughly revised up to 1939 and practically a new book.)

Byerly, P., *Seismology*, Prentice-Hall, New York, 1942. (Well balanced, but especially recommended for theory of seismometers and of seismic waves.)

Savarensky, E. F., and Kirnos, D. P., *Elementi seismologii i seismometrii*, 2nd rev. ed., Gosudarstvennoe izdatel'stvo tekhniko-teoreticheskey literaturi, Moscow, 1955. (Thorough and authoritative; the best general source for recent seismological research in the USSR.)

Seismology, smaller treatises

Rothé, J. P., *Séismes et volcans*, Presses Universitaires de France, Paris, 1946.

Jung, K., *Kleine Erdbebenkunde*, 2nd ed., Springer, Berlin, 1953. (The general treatment is excellent, but the second edition was not sufficiently revised to bring it up to date.)

Bullen, K. E., *Seismology*, Methuen and Wiley, London and New York, 1954.

Lynch, J., *Our Trembling Earth*, Dodd, Mead and Co., New York, 1940. (This and the next book offer relatively entertaining reading on the popular level.)

Leet, L. D., *Causes of Catastrophe*, Whittlesey House, New York, 1948. (Seismology in part.)

Eiby, G. A., *About earthquakes*, Harper, New York, 1957. (An excellent popular treatment; the best now in the field.)

Seismology, theory

Macelwane, J. B., and Sohon, F. W., *Introduction to theoretical seismology;* Part I, "Geodynamics," by Macelwane, McGraw-Hill, New York, 1936; Part II,

"Seismometry," by Sohon, McGraw-Hill, 1932. Both republished, St. Louis University, 1949. (Macelwane's accurate bibliography makes his references unusually valuable.)

Bullen, K. E., *An Introduction to the Theory of Seismology,* 2nd rev. ed., Cambridge University Press, 1953.

Blake, A., "Mathematical problems in seismology," Trans. Am. Geophys. Union, 1940, pp. 1094–1113.

Byerly, P., "A seismologist's difficulties with some mathematical theory or the lack of it," *ibid.,* pp. 1113–1118.

Richter, C. F., "Mathematical questions in seismology," Bull. Am. Math. Soc., vol. 49 (1943), pp. 477–493.

Other references

Davison, C., *The Founders of Seismology,* Cambridge University Press, 1927.

————, "Founders of seismology, IV," Geol. Mag., vol. 74 (1937), pp. 529–534. (Supplement to the preceding; deals with the work of Kotō, Baratta, and Oldham.)

Ballore, F. de Montessus de, *La Géologie sismologique,* Armand Colin, Paris, 1924, 488 pp.

Davison, C., *Great Earthquakes,* Murby, London, 1936. (This and the preceding book are the best available collections of earthquake data.)

————, *A Study of Recent Earthquakes,* Walter Scott, London, 1905. (Discusses several events not included in his later book; discussion is partly obsolete, and is directed largely toward Davison's idea of "twin earthquakes.")

Lawson, A. C., *et al., The California Earthquake of April 18, 1906, Report of the State Earthquake Investigation Commission,* Carnegie Institution of Washington, Vol. 1, 1908, with atlas (25 maps, 15 pls. of seismograms); Vol. 2, 1910, *The Mechanics of the Earthquake,* by Harry Fielding Reid. (Contains the fundamental data and discussion for the elastic-rebound theory of earthquakes; a source book for earthquake information of many kinds.)

Freeman, J. R., *Earthquake Damage and Earthquake Insurance,* McGraw-Hill, New York, 1932. (Reprints much valuable source material. Weak from the point of view of instrumental seismology, and to be used with caution.)

Suyehiro, K., "Engineering seismology; notes on American lectures," *Proc. Am. Soc. Civil Engs.,* vol. 58, No. 4 (May 1932). (Of much more than engineering interest, and highly informative.)

Hubbert, M. K., "Theory of scale models as applied to the study of geological structures," *Bull. Geol. Soc. Amer.,* vol. 48 (1937), pp. 1459–1520. (This paper, not bearing directly on seismology, is recommended as an example of the kind of revision needed for restoring sound physics to geology.)

Schriever, W., "Were the Carolina Bays oriented by gyroscopic action?" Trans. Am. Geophys. Union, vol. 36 (1955), pp. 465–469; discussion, *ibid.,* vol. 37 (1956), pp. 112–117.

Oakeshott, G. B., ed., "Earthquakes in Kern County, California, during 1952," *Calif. Dept. Nat. Resources, Div. Mines, Bull.* No. 171 (1955). (Contains much general information and many new results.)

Steinbrugge, K. V., and Moran, D. F., "An engineering study of the Southern California earthquake of July 21, 1952, and its aftershocks," *B.S.S.A.* (*Bull. Seismol. Soc. Amer.,* vol. 44 (1954), pp. 199–462. (A valuable adjunct to the above; reviewed in Chapter 8. Each of the two publications (this and the preceding one) includes material summarized from the other.)

Rothé, J. P., ed., *Comptes rendus des séances de la dixième conférence réunie à Rome du 14 au 25 septembre 1954,* Union Géodésique et Géophysique Internationale, Association de Séismologie et de Physique de l'Intérieur de la Terre, Strasbourg, 1955. (Those fortunate enough to have this available will find in it abstracts and discussions of many important papers,† many bibliographical data, and reports on the state of seismology in many nations. Much of this will probably be superseded by a corresponding publication for the next meeting, scheduled for Toronto in 1957.)

Periodicals and serials

Although the literature of seismology is widely dispersed, a large proportion of original material has appeared in comparatively few journals. In this book these journals have been given preference. The most frequently cited are as follows. Dates are those of the first volumes.

American Association of Petroleum Geologists, Bulletin, Tulsa, 1917.

American Geophysical Union, Transactions, Washington, 1920. At first, volumes were identified only by year; volume numbers were attached beginning with vol. 26 for 1945, and retroactively applied to earlier issues. Volumes 3 (1922) and 5 (1924) were never published.

Annales de Géophysique, Paris, 1944.

Annali di geofisica, Rome, 1948.

Bureau central séismologique international, Publications (now issued from Strasbourg).

Earthquake Research Institute, Bulletin, Tokyo, 1926. Abbreviation: Bull. E.R.I.

Geological Society of America, Bulletin, 1889; *Special Papers,* 1934; *Memoirs,* 1934.

Geophysical Magazine, Tokyo, 1926. (Contributions largely from staff of the Central Meteorological Observatory—now Japan Meteorological Agency.)

Göttinger Nachrichten. This is the usual form of reference. The full title is: *Nachrichten der königlichen Akademie der Wissenschaften zu Göttingen, mathematisch-physikalische Klasse.* The adjective "königlichen" was dropped after 1918. (Important in seismology for publications by Wiechert and his group.)

---

† The texts of papers presented at Rome appear in *Publ. Bur. central international,* vol. 19 (1956).

Gerlands Beiträge zur Geophysik, Stuttgart, 1887 (Vols. 3 and thereafter, Leipzig). Abbreviation: *G. Beitr.*

Imperial Earthquake Investigation Committee, Tokyo, Publications (especially those in foreign languages). Superseded by *Bulletin of the Earthquake Research Institute.* (Many valuable papers also appeared in *Proceedings of the Imperial Academy,* Tokyo, and in the *Japanese Journal of Astronomy and Geophysics.*)

New Zealand Journal of Science and Technology, Wellington, 1918. For recent volumes designate Series B. Abbreviation: *N. Z. Journ.* Replaced beginning 1958 by a new quarterly, *New Zealand Journal of Geology and Geophysics,* and companion periodicals in other fields.

New Zealand Institute, Transactions. Wellington, 1868. Vol. 63 (1933) is continued as vol. 64 (1934–1935) of *Transactions of the Royal Society of New Zealand.*

Royal Astronomical Society, Monthly Notices, Geophysical Supplements, London, 1922. (Each volume extends over several years. Abbreviation: *M.N.R.A.S. Geophys. Suppl.*) Replaced, 1958, by new *Geophysical Journal.*

Royal Society of New Zealand, Transactions. (Continuation of *New Zealand Institute, Transactions,* as noted above.)

Seismological Society of America, Bulletin, Stanford University Press, 1911. Beginning in 1935, University of California Press. Abbreviation: *B.S.S.A.* (Indispensable for studying California earthquakes; many other significant papers.)

United States Earthquakes, U. S. Coast and Geodetic Survey, U. S. Dept. of Commerce. First issue for the year 1928, serial number 483; annual issues thereafter, with nonconsecutive serial numbers. Government Printing Office, Washington, D. C.

Zeitschrift für Geophysik, Braunschweig, 1924.

The most important current Russian-language periodicals are those of the Academy of Sciences at Moscow (Akademiya Nauk SSSR); especially *Doklady, Izvestiya, ser. geofiz.* (this is relatively new), and *Trudy geofiz. instituta.*

Papers published in a journal are sometimes reissued separately as numbered bulletins or contributions. In such cases only the reference to the journal is given.

The following special index publications are helpful:

Bibliography of Seismology, B.S.S.A., vol. 17 (1927), pp. 149–182, 218–248; vol. 18 (1928), pp. 16–63, 110–125, 214–235, 267–283. New series, *Publ. Dominion Observatory, Ottawa,* vols. 10, 12, 13, 14.

Geophysical Abstracts, 1929. Nos. 1–86 and 112–127 issued as mimeographed *Information Circulars* by the U. S. Bureau of Mines. Nos. 87–111 and 128 ff. printed as Bulletins of the *U. S. Geol. Survey.* (Particularly valuable for notes on Russian publications.)

Geophysikalische Berichte. (Abstracts, included with *Zeitschrift für Geophysik,* 1927–1939; geophysical section of *Physikalische Berichte.*)

Montanwissenschaftliche Literaturberichte, Sonderhefte der Freiberger Fort-schritte, Freiberg (Saxony).

Zentralblatt für Geophysik, Meteorologie und Geodäsie, Berlin, 1937–1940. (Many abstracts and some good review articles.)

The well-known bibliographical publications in the fields of geology and general physics should also be consulted in searching the literature.

CHAPTER 2

Definitions for a Dual Science

AS SUGGESTED in the introductory chapter, attempts to define general terms like seismology and earthquake lead mainly to unnecessary hairsplitting. On the other hand, it is essential to be precise about the meaning of the arbitrary technical terms with which we work. These are chosen for economy of expression and thinking, and their usefulness depends on exactness.

DEFINITION

Because of the dual nature of seismology, some important terms are defined from the viewpoint of the geologist or the engineer investigating effects in the field, while others are defined by the laboratory seismologist interpreting instrumental records called seismograms. This contrast is expressed by a pair of handy terms. *Macroseismic* effects of earthquakes are those that can be observed on the large scale, in the field, without instrumental aid. *Microseismic* effects are small-scale, observable only with instruments.

Unfortunately the noun *microseism* is used in a different way. Microseisms, quite illogically, are not small earthquakes; they are more or less continuous disturbances in the ground recorded by seismographs. Most of them appear to be connected with weather; on the whole, they are among the most puzzling and provoking of the "related phenomena" skulking behind the usual definition of seismology.

A good example of a macroseismic term is *tectonic earthquake*. In geology, tectonic means structural; the subject of tectonics is structural geology. Tectonic earthquakes are those believed to be associated with faulting or other structural processes. This is taken to exclude volcanic earthquakes, as well as minor shocks due to less important causes.

Teleseism is a typical microseismic term. A teleseism is an earthquake recorded by a seismograph at a great distance. By international convention this distance is required to be over 1000 kilometers (621 miles) from the epicenter (as defined below). Earthquakes originating nearer the recording station are "near earthquakes" or "local earthquakes."

Intensity and magnitude scales describe earthquakes by using macroseismic and microseismic data respectively. *Intensity* (Chapter 11) is the older term. It refers to the degree of shaking at a specified place. This is not based on measurement but is a rating assigned by an experienced observer using a descriptive scale, with grades indicated by Roman numerals from I to XII. Repeated attempts have been made to tie accepted intensity scales to some measurable physical quantity, usually acceleration.

Magnitude (Chapter 22) is intended to be a rating of a given earthquake independent of the place of observation. Since it is calculated from measurements on seismograms, it is properly expressed in ordinary numbers and decimals. Magnitude was originally defined as the logarithm of the maximum amplitude on a seismogram written by an instrument of specified standard type at a distance of 100 kilometers (62 miles) from the epicenter. Tables were constructed empirically to reduce from any given distance to 100 kilometers. Because the scale is logarithmic, every upward step of one magnitude unit means multiplying the recorded amplitude by 10. The zero of the scale is fixed arbitrarily to fit the smallest recorded earthquakes. The largest known earthquake magnitudes are near 8¾; this is a result of observation, not an arbitrary "ceiling" like that of the intensity scales. Attempts are being made to relate magnitude to the total energy released by an earthquake in the form of elastic waves; this work has led to some revision and redefinition of magnitude.

Isoseismals, or isoseismal lines, were at first mapped as curves connecting localities where equal intensity was observed in a given earthquake. They are now more commonly mapped as boundaries between regions of successive intensity ratings, such as IV and V.

The area within the isoseismals of higher intensity is the *meizoseismal area.* This term is now out of fashion, and perhaps rightly so because it is difficult to define precisely. In this book it is used for convenience in description. Any vagueness will be consistent with the state of information, for the pattern of macroseismic observations is not tidy. In a small earthquake, the meizoseismal area is simply that where shaking is perceptible to persons; in a large one, it may be a region of great destruction and other spectacular effects. It may be elongated or irregular in form and may even consist of several detached parts.

Epicenter and *hypocenter* were originally defined before the development of seismographs and have since been redefined. The subtle change of meaning sets a trap for the student. In the 1850's, Mallet and others believed that an earthquake originated in a small volume underground, which might be represented as a point for most purposes. This subterranean point was the hypocenter, and the point vertically above it on the earth's surface was the epicenter. When investigation of such earthquakes as that in California in 1906 showed the development of faults breaking the surface for hundreds of miles, it was thought that the terms epicenter and hypocenter would have to be abandoned. However, even in large earthquakes, the first waves recorded by seismographs start practically from a single point. This point is believed to represent the position of the initial rupture of the rocks; it is taken as the hypocenter, and the point on the surface vertically above it as the epicenter.

In the absence of seismographs, the epicenter was commonly taken to be near the center of the meizoseismal area, although even then evidence was sometimes found to point elsewhere. The "instrumental epicenter" derived from seismograms often is near one end or side of the meizoseismal area, and it may even fall outside.

Focus is used as a synonym for hypocenter, for example, when writing of deep-focus earthquakes; but the word is also used in the manner familiar in optics, and we speak of the focus of a system of seismic waves.

The symbols *P, S, L* are seismological shorthand for three successive groups of waves recorded on seismograms of normal earthquakes, *L* chiefly at teleseismic distances. Leaving aside details to be found in the chapters on elasticity and seismic waves, these groups are identified as follows.

P: longitudinal, or compression-rarefactional, waves through the earth
S: transverse or shear waves through the earth
L: surface waves over the earth, generally of long period

Other seismological terms will be introduced in later chapters as they are needed. Definitions of *seismograph* and related terms will be found in Chapter 15; of *stress, strain,* and terms related to elasticity and waves, in Chapter 16; of *crust, mantle,* and *core* and of the various types of seismic waves in Chapters 17 and 18; of *shallow, intermediate,* and *deep* earthquakes, in Chapter 19.

The use of some geological terms in this book should be noted. *Basement* has various meanings; here we follow the usage of R. D. Reed. In California, New Zealand, and some other areas it is convenient to apply this name to igneous or metamorphic rocks, including partly metamorphosed sediments, but underlying the *sedimentary blanket,* which consists of Cretaceous and younger strata, generally less consolidated and less competent than the basement rocks.

As explained in Chapter 17, in this book *crust* means only that part of the earth above the Mohorovičić discontinuity; it is not necessarily to be identified with the lithosphere.

It has been necessary to use *structure* in two senses, one geological and one engineering. The meaning intended should be clear from the context.

For terminology related to faulting see any standard manual. This book prefers the terms *right-hand* and *left-hand* for horizontal displacement (see Chapter 13).

The following terms are also encountered.

Origin time is the instant at which the earthquake event (apart from foreshocks) commences at the hypocenter.

Transit time is the elapsed time between the origin time and the arrival of a given seismic wave at a specified point (usually a seismograph station). Most authors term this *travel time.*

Compression and *dilatation* (= rarefaction) are used in connection with longitudinal waves, as in acoustics. They refer to the nature of the motion at a given point, usually a recording station. When the ray emerges to the surface, displacement upward and away from the hypocenter corresponds to compression, the opposite to dilatation.

Gravity anomaly is the deviation of the observed acceleration of gravity (negative when smaller, positive when larger) from an expected value calculated from the general gravitational field of the earth, considering latitude and elevation; correction is usually added to allow for the effects of irregular topography and structure. "Isostatic" gravity anomalies are also calculated, in which the effects otherwise to be expected from the larger topographic irregularities such as high mountains or deep troughs are partially compensated according to the principle of isostasy.

As defined in building codes, *parapets,* or parapet walls, are portions of side walls extending above the roof line. They may merely serve as false fronts, but are sometimes intended as protection for firemen. When constructed of unreinforced and unsupported masonry they create a serious earthquake hazard.

Times for earthquakes recorded instrumentally are reported as Greenwich Civil Time (G.C.T.), including date. These usually differ from the local time in use in the country of origin; thus standard time† in California is 8 hours earlier than G.C.T., in Japan 9 hours later, and in New Zealand 11 hours later. In consequence, the local date may be one day earlier or later than the G.C.T. date. In dealing with historical, non-instrumental reports, the date and hour are local. On occasion both local time and G.C.T. are mentioned.

Origin times are usually given in the form 20:31:53, equivalent to 20^h 31^m 53^s. When accuracy is relatively low, the minute and tenth of minute may be given, as in 20:31.9.

It was unavoidable to give some lengths in the metric system and others in inches, feet, and miles. Equivalents are given here and there. For reference, 1 mile = 1.609 kilometers, 1 kilometer = 0.6214 mile; 1 foot = 0.3048 meter, 1 meter = 39.37 inches = 3.281 feet. Units not used here, which the reader will find in some of the literature referred to, are the *chain* (= 66 feet), the *fathom* (= 6 feet), and the *knot* (1 nautical mile per hour, 51.48 centimeters per second).

The c.g.s. unit of acceleration is the *gal* (= 1 centimeter per second per second); one thousandth of a gal is a *milligal.*

---

† International scientific usage ignores daylight-saving or summer time.

NOTATION

Without overworking the alphabet it is hard to avoid using the same letter for two different quantities. The reader will find the notation consistent in any one section. Symbols used throughout this book are given in the following list, which points out use of the same symbol in two senses.

C	phase velocity of a wave
C'	group velocity
D	straight-line distance from hypocenter
e_{xx}, etc.	components of strain
E	Young's modulus
g	acceleration of gravity
h	seismograph damping constant
H	depth of hypocenter
$\mathfrak{K}$	depth of water
i	angle of incidence of ray
j	$(1 - h^2)^{1/2}$
k	bulk modulus of elasticity (in Appendix IV, also $k = \omega/C$, the wave number)
K	seismograph spring constant (Chapter 15); also symbol for longitudinal wave in the core of the earth
m	mass (in Chapter 22, m = unified magnitude according to Gutenberg)
M	magnitude (on the original system)
p	angular frequency of ground motion (Chapter 15); parameter constant on a ray (Appendix VI); also a symbol for the direct longitudinal wave at short distance
P	general wave symbol for longitudinal wave
q	quantity calculated in determining wave velocities (Appendix VI); $\log_{10} e$ (Chapter 22)
Q	provisional damping constant (Chapter 15)
r	radius vector, usually from the center of the earth
s, S	symbols for transverse waves, corresponding to p, P for longitudinal waves
t	time
T	period of ground motion
v	velocity of S waves, or y-component of displacement

V	velocity of P waves, or static magnification of seismometer
Xx, etc.	components of stress
Δ	surface distance from epicenter
ϵ	damping ratio of seismometer
θ	dilatation (Chapter 15); central angle (Appendix VI)
λ, μ	Lamé's elastic coefficients
ρ	density
σ	Poisson's ratio
τ	period of pendulum
ω	angular frequency of free oscillation of pendulum (undamped); angular frequency of wave (Appendix VI)

CHAPTER 3

The Nature of Earthquake Motion

SOME OF THE BEST EVIDENCE on the nature of earthquakes comes from field work in the limited meizoseismal area; but this evidence is often inadequately examined or poorly reported, and vanishes with the lapse of time. Moreover, the meizoseismal area may be remote, inaccessible, or covered by the sea.

Outside the meizoseismal area, inference depends on analysis of the elastic waves generated by an earthquake. Thus, much of what is known and believed about earthquakes follows from study of the motion of the ground, which is felt as shaking and produces most of the more obvious effects. Although such study began long before the invention of the seismograph, many important points have not yet been satisfactorily cleared up. This is not due to lack of interest or effort, for the analysis of earthquake motion is of direct concern to the engineer designing structures to resist it. To the geologist it is almost equally important, although his concern is not so directly with the motion itself as with inference as to the tectonic or other processes which cause it.

This chapter summarizes chief results of the study of earthquake motion. Details and further discussion are distributed in their proper places throughout the remainder of the book.

MACROSEISMIC AND MICROSEISMIC DATA

The immediately following chapters will show how such evidence as that of displaced objects has been used by painstaking observers to analyze earthquake motion. In general, this has been supplanted by microseismic evidence; for, where seismographs exist, their description of motion is far more detailed and accurate than that derivable from macroseismic observations. Such analysis and description were the original purposes of the seismograph; they

22

remain the principal objectives of an increasing class of special instruments operated at relatively low magnification in order to record strong motion.

Our instrumental knowledge of earthquake motion thus depends on (1) a vast number of seismograms recorded at hundreds of stations during the last fifty years, showing small motion due to minor local or large distant earthquakes, and (2) a much smaller number of recordings (a few hundred at best) of locally strong motion. Generalizations depend on proper consideration of both groups.

Seismological discussion is still influenced by conclusions drawn in the pre-instrumental stage of observation. Terms defined, and theories framed, on that relatively crude basis have persisted into the present period. Ordinary observers, journalists, amateurs, and even scientific observers in regions where no seismographs are in operation often express their experiences or investigations in these partly outmoded forms, which are naturally suited to the most easily made observations. There is also a disproportionate effect of ideas developed between 1880 and 1900 with the aid of seismographs which were very imperfect by present standards.

Many correct conclusions were drawn by the earliest scientific workers in the field. Thus, from the relatively small size of the meizoseismal area, and the rapid falling off of intensity outside it, it was properly concluded that ordinary earthquakes originate at depths which are great compared with the average thickness of geological formations, but small in terms of the radius of the earth. The seismograph has confirmed this (except for the class of deep-focus earthquakes) but has shown that the falling off of motion at great distances is slow, so that a large earthquake is recorded at almost all stations.

Despite efforts to relate the intensity scale to instrumental recording, mapping of intensities and drawing of isoseismals still depend almost exclusively on macroseismic data. However, it has proved possible to confirm the known effect of varying ground instrumentally by comparing seismograms written by identical instruments at relatively near points; under equivalent conditions the ground motion on alluvium may be ten times as great as that on competent rock.

PREVALENT DIRECTIONS

Much use was once made of evidence that motion at a given locality is principally in a single direction. Too often this rested on personal impressions; but in many instances the evidence was objective, such as the falling out of walls in one direction or consistent orientation of cracks in the ground and in structures. Thus, in the Long Beach earthquake of 1933, frame houses in the most heavily shaken area were commonly shifted off their foundations northward; and the concrete caps of a row of adjacent standpipes were found, after the earthquake, rotated through about the same angle in the same sense. Seismograms show that the evidence of prevailing motion in one direction probably refers to a large group of waves of relatively short duration as compared to the entire earthquake disturbance; generally these appear to be transverse waves (S). The detail of motion may be expected to be fairly uniform over a limited area, because the wave lengths of the larger seismic waves are of the order of half a mile or more (say a kilometer or more). Nevertheless, this uniformity of motion was at one time attributed to mass motion of crustal blocks; its recognition as an effect of wave propagation is largely due to Mallet (see Chapter 4).

The total motion during an earthquake is highly irregular. Many textbooks include figures of wire models constructed by Sekiya in 1887 from seismograms at Tokyo to represent the recorded motion in three dimensions. They show a complicated tangle. Qualitatively, they give a fair representation of what goes on in an earthquake. Quantitatively, Sekiya's wires cannot be taken seriously; the precision of his data did not warrant so elaborate an undertaking. Equivalent work, using such instruments as those of the strong-motion installations operated by the U. S Coast and Geodetic Survey, has been done much more precisely in recent years.

Ordinary observation tends to exaggerate horizontal as compared to vertical motion, since a majority of obvious effects, from damage in masonry to the displacement of small loose objects, are more readily caused by horizontal disturbance. Sensations of vertical motion are most commonly reported near the epicenter. This is not easy to confirm from seismograms; those available at short epicentral distances generally refer to small shocks only and usually show no more vertical than horizontal displacement.

PERIODS AND SPECTRA

Perceptible motion includes a greater proportion of slow oscillation with long periods as either distance or the magnitude of the shock increases. This causes some trouble in the application of intensity scales, which commonly lump together effects due to long-period and short-period motion. The general phenomenon is confirmed by seismograms of all types. The increase in dominant periods with distance is in part a filtering effect; the shorter periods are attenuated more rapidly. In addition, there is at least the appearance of increasing dominant period in a given wave group; this can be explained in part as the normal spreading of an originally impulsive disturbance, but many investigators have attributed it to some more specific cause, such as viscosity or internal friction in the material of the earth's interior.

Strong-motion seismograms, especially, make it possible to set up a spectrum analysis for moderately heavy shaking and find the distribution of energy over the frequency range. The total motion is thus represented as the sum of harmonic oscillations, each with its own frequency and amplitude. Engineering discussion often hinges on the displacements and accelerations of these component oscillations. The waves with maximum displacement are, in general, not the same as those with maximum acceleration. In attempting to allow for this, the necessary elementary relation among amplitude, acceleration, and frequency must be kept in mind. Table 3-1 exhibits this

Table 3-1   Harmonic Oscillation Frequencies (cycles per second) Corresponding to Given Amplitudes A and Accelerations a

Acceleration a	A (cm)						
	0.0001	0.001	0.01	0.1	1	10	100
g	500 cps	160 cps	50 cps	**16 cps**[1]	5 cps	1.6 cps	0.5 cps
0.1g	160	50	**16**	**5**	**1.6**	0.5	0.16
0.01g	50	16	**5**	**1.6**	**0.5**	0.16	0.05

[1] Boldface numerals indicate expectation in moderately strong earthquakes.

relation.† In this table the boldface numerals indicate the combinations of amplitude and acceleration, and the corresponding frequencies, most common in the analysis of moderately strong motion; the data of a locally more violent earthquake should cover a larger area of the table. In that case it would probably remain true that the highest accelerations are associated with small amplitude, and the large amplitudes with low frequency and low acceleration.

HIGH ACCELERATION

The spectrum represented in Table 3-1 is that of motion persisting for a number of cycles and forming an appreciable fraction of the whole disturbance. Strong-motion records, as well as other evidence, indicate the occasional occurrence of short single jolts of relatively high acceleration. This may perhaps contribute to some of the startling effects which have sometimes been reported, although nearly always there is a simpler explanation. After the Long Beach earthquake of 1933, a stove was found with all four corner supports outside of the rings of grime which had surrounded them on the floor, without any break in the rings indicating sliding. However, this is not evidence for vertical acceleration in excess of g; it points more prob-

† A similar table with amplitudes in inches instead of centimeters will be found in Chapter 8.

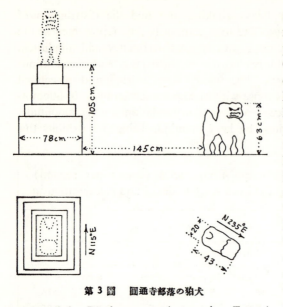

第 3 圖   圓通寺部落の狛犬

FIGURE 3-1 *Displacement of stone dog, Tottori earthquake, 1943.* [*After Matuzawa.*]

ably to rocking, tipping, and rotating, first around one support, then around another. On the same occasion a large ornamental urn at a mausoleum was found apparently lifted off a spike which had projected a few inches up through a hole in its base. Here the rational explanation is that the urn rocked laterally and slid up along the spike. This would not call for very high acceleration. After the Tottori (Japan) earthquake of 1943, ornamental stone dogs (*komainu*) associated with temples were found displaced as if they had been thrown up and to the side a meter or more; there appears to have been oscillation of the pedestals (Fig. 3-1).†

As will be shown in Chapter 5, there is good evidence that in the meizoseismal areas of the greatest earthquakes actual ground accelerations of the order of *g* or greater occur. The level of acceleration generally taken as sufficient to produce ordinary damage to weak construction is 0.1*g* (about 100 gals). The lower limit of acceleration perceptible to persons has been set by observation and experiment near 1 gal (0.001*g* or 1 cm/sec^2).

CAUSES OF COMPLEXITY

Modern seismographs, with their clear separation of the successive elements of recorded motion, show that the high complexity of the disturbance is due largely to a succession of waves of different types. The accurately determined times of these waves are used to locate epicenters and to study the interior of the earth or the structure of its crust. Interest in such work led for many years to concentration of attention on precise timing and to comparative neglect of the more laborious analysis of amplitudes.

The complexity of the total earthquake, as shown roughly by Sekiya's wires, is of engineering importance. To this complexity of a single event it is necessary, in designing for earthquake resistance, to add that due to

---

† See also the description of the Imaichi earthquake of 1949 (Chapter 30).

variable source. In its lifetime a given building may be affected by earth-quake waves coming from different directions and distances, so that the general effect is that of a random disturbance. The design problem is further aggravated by the mechanical complexity of most large engineering struc-tures.

Seismograms generally show decreasing individuality with increasing dis-tance. The earth is surprisingly symmetrical; consequently at equal distances in different directions the same succession of waves arrives at almost the same times. Two equal shocks in the same region will write almost super-posable seismograms at a distant station, particularly for instruments of long-period characteristics. Thus the complexity of a teleseismic record is due largely to the variety of paths along which an initially simple disturbance is propagated from the hypocenter to the recording station.

"TWO SHOCKS"

At short distances, especially for large earthquakes, there is clear evidence of complexity at the source; sometimes there is obvious superposition of recordings of large and small individual shocks. Certain areas characteristi-cally produce earthquakes in swarms and groups (see Chapter 6).

However, at short distances a single earthquake is often felt by persons as two sharp jolts a few seconds apart. Such observations are so common that they led Davison to set up a special theory of twin earthquakes, now obsolete. News reports of "two shocks" usually originate in this way, or in information from some station which has given recorded times; for at dis-tances less than 100 kilometers (or 60 miles) the normal seismogram con-sists principally of two sharp and short disturbances, P and S, at an interval not over about 12 seconds. Stations often report times for both P and S at greater distances, where the interval is measured in minutes. When two earthquakes occur in close succession both P and S are duplicated.

LONG-PERIOD WAVES

At distances of about 1000 kilometers (or 600 miles) the P and S groups begin to be followed by conspicuous long-period surface waves. At great distances the motion of largest amplitude, as shown by seismograms of large shallow earthquakes, consists of a long train of surface waves with periods near 20 seconds. These were highly exaggerated by the early seismographs and were called the principal waves. For the greatest earthquakes, such waves have actual ground amplitudes reaching 1 millimeter over the whole surface of the earth; but the wave lengths are great, accelerations low, and perceptibility to persons nil.

At distances of a few hundred miles great earthquakes often give rise to waves of periods of 10 to 20 seconds with amplitudes of several centimeters. These are responsible for several groups of effects, discussed in later chapters, such as:

1. Marginal effects of strong earthquakes beyond the normal distance of perceptibility; persons nauseated, chandeliers and doors swinging slowly.
2. Effects on ground and surface water; seiches; oscillations in wells (Chapter 9).
3. Earth slides, cracks, fissures.
4. Slow swaying of tall buildings and towers.

NOTE ON TIMEKEEPING

For present methods of timekeeping see Chapter 15. Many dubious or positively wrong conclusions were drawn by early seismologists who did not suspect how incredibly bad is the average untrained person's notion of determining time. Wrong velocities for seismic waves, and wrong locations of epicenters, were supported by combining times at different locations when persons looked at their watches, or when pendulum clocks were stopped.

Oldham, in his report on the Indian earthquake of 1897 (Chapter 5), writes:

A fruitful source of error, and one that is often impossible to eliminate, is the different times that are kept and used. I do not refer to isolated localities where there is neither railway station nor telegraph office, and where the local time is determined by the gastric sensations of the individual in charge of the station time-gong, or the indications of a sundial . . . but to those places where the presence of telegraphic communication should enable accurate time to be kept . . . A high degree of accuracy, according to the standards of ordinary life, may be a high degree of inaccuracy where even fractions of a second should be taken into consideration, and it is obvious from the returns that, even where accuracy is most to be expected, the time is often in error by several minutes.

The report on the 1934 Indian earthquake indicates conditions not greatly improved over those of 1897. This is no special reflection on the people of India; similar remarks would apply, especially before 1924, to rural districts the world over. Even now the general standard of timekeeping is reasonably good only in urbanized areas.†

---

† In 1918 an astronomer, who had entered military service, was stationed at an army post where the customary cannon was fired at noon. Professional interest led the astronomer to inquire of the officer in charge where he got his time; it transpired that the source was a clock in a jeweler's window in the nearby town, where the officer regularly set his watch. The logical next step was to interview the jeweler. Result: "Oh, yes; they fire a cannon at the fort every day at noon, and I set my clock by that."

With increasing experience the need for time to 1 second or better was urgently realized; in consequence seismological stations were often installed at astronomical observatories. About 1924 there was general improvement in timekeeping due to increase in the broadcasting of time signals by radio. In recent years the demand for precise time in adjusting broadcasting systems to national networks, together with the general introduction of electric clocks, has led to improved timing on the popular level.

References

READING MATERIAL appropriate to this chapter will be found among the general references appended to Chapter 1 and as special references following later chapters. The account of displaced *komainu* was published by T. Matuzawa in *Bull. E.R.I.*, vol. 22 (1944), pp. 60–65 (in Japanese, with German summary). The quotation from Oldham is taken from p. 55 of his memoir (see Chapter 5). The wire model referred to was first illustrated as plate 2 accompanying: Sekiya, S., "A model showing the motion of an earth-particle during an earthquake," *Trans. Seism. Soc. Japan,* vol. 11 (1887), pp. 175–177.

CHAPTER 4

Two Sample Earthquakes[†]— *1857 and 1929*

MALLET AND HIS EARTHQUAKE

No one man contributed more to the early organization of knowledge about earthquakes into a science than Robert Mallet. In 1846 he read a paper before the Irish Academy on the dynamics of earthquakes. In 1850 and subsequent years he contributed reports to the British Association for the Advancement of Science, attempting to bring together all that was then known on the subject. He formed definite hypotheses as to what earthquakes were, how they were caused, and how they ought to be investigated. On December 16, 1857, a destructive earthquake in Italy gave him his opportunity. The result was a large two-volume work, published in 1862: "The great Neapolitan earthquake[‡] of 1857; the first principles of observational seismology . . ." and so on through a long, old-style title page. Mallet did not reach Naples until February, 1858; he returned to England in April. In a few weeks he accumulated an enormous mass of observations, much of it by traveling through rough hill country over extremely bad roads, and climbing one by one up to the towns perched on steep hill tops. "Often," he writes, "three hours' painful toil upon our mule will but suffice to bring us—by long traverses over rough and rolling stones, and by an approach road that is often the bed of a torrent in time of rain—to the ancient gateway." Observers who visited towns damaged by the earthquake of 1930 in almost the same area traveled in comfort by automobile over good roads.

[†] This and the next chapter describe a few selected earthquakes as illustrative examples which will be referred to repeatedly in the book. For completeness some details are introduced for which the student is not prepared at this point; these will be clarified in later chapters.

[‡] Neapolitan earthquake—from Naples, ancient Neapolis. This earthquake was not very strong at Naples itself; but the meizoseismal area was in the then Kingdom of Naples, which became part of united Italy in 1861.

30

Geology in Relation to Italian Earthquakes

The principal structural feature of Italy is the arc of the Apennines, convex eastward. Arcuate structures are a common feature of regions of tectonic and volcanic activity, especially in the circum-Pacific belt, where folding and mountain-building are going on. The Apennine arc belongs to the same system as the Alps; both resulted from folding and overthrusting which began as long ago as the Jurassic, although the main compressive mountain making was mid-Tertiary,

FIGURE 4-1 *The more damaging earthquakes of Italy, 1600–   . [After Baratta, with later additions.] Earthquake of 1857: solid line, limit of destructive shaking; dotted line, limit of perceptibility.*

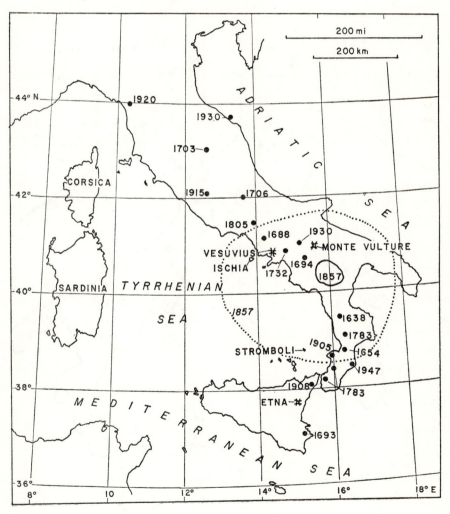

with later uplift culminating in the Pleistocene. As in many other regions, the stresses acting have shifted. It is not likely that the Alpide structures are growing; the folding and thrusting properly called Alpine is over, and the mass is being broken into blocks along comparatively new fractures. During this process the interior of the Apennine arc, to the west, has foundered under the Tyrrhenian Sea.

Most epicenters of important earthquakes in Italy are on the Apennine arc. (Figs. 4-1, 4-2) As in the Pacific, active volcanoes are within the arc, opposite its concave side. Among them are Vesuvius, Etna, and Stromboli. Volcanoes and

FIGURE 4-2   *Earthquakes felt in Italy, 1930–1939. Size of spots indicates maximum observed intensity; stars are instrumentally determined epicenters.* [*Caloi.*]

volcanic earthquakes of the Bay of Naples (including Ischia) are discussed in Chapter 9. Deep earthquakes originate almost under the volcanoes; several of those known near the coast of Italy are unusually deep for such an area, down to nearly 300 kilometers below the surface; at least one was about 470 kilometers deep. (See also Chapter 31.)

Mallet's Procedure

Mallet believed that strong earthquakes were fundamentally volcanic. He thought they were due to subterranean processes, explosive or nearly so, such as might be caused by superheated steam entering a fissure. He drew attention to the nearness of Monte Vulture (which is quite extinct) to the 1857 earthquake area. Such an explosive source, reasoned Mallet, would be essentially a point—the hypocenter. Elastic waves from the explosion should be compressional, hence longitudinal, like sound waves; transverse shear waves should not be observed. The first motion of the ground, assuming a constant wave velocity, should be radially away from the hypocenter; the horizontal component of motion at the surface should be radially away from the epicenter. Overturned objects, and loose objects thrown from high places, should usually fall away from (or toward) the epicenter. Cracks in buildings should show the direction of the wave front.

This last point, essential to Mallet's work, is at first hard to accept. In most structures, cracks are more likely to be related to the structures themselves than to the direction of vibration in elastic waves. One must first read Mallet's account of the construction usual in these hill towns, which is typical of most of southern and eastern Europe. "Lime is abundant, but the mortar often of very slender cohesion, from too great a proportion of lime and the want of a proper quality of sharp sand. Hence the general style of construction of wall, even in first-class buildings, consists of a coarse, short-bedded, ill-laid rubble masonry, with great thickness of mortar joints, very thick walls, without any attention to thorough bonding whatever." The floors are made of planks covered with a heavy layer of concrete; the roofs are heavy with thick tiles not tied down. Thus, "in the construction of the more important buildings, the mass and inertia, of walls, floors and roofs are enormous, while the bond and connection of each of these, and of all to the others, is loose and imperfect."

This makes it a little easier to understand why Mallet had some success in treating such structures as if they were merely piles of loose, brittle material in which cracks caused by a compressional wave might be expected to open along the wave front or at right angles to the ray. Mallet does not suggest that all cracks show such regularity; he is careful to say that in any one locality the first impression is one of pure chaos, and that only after long and careful examination from many points of view does one begin to see the prevailing direction of cracking. Moreover, he usually finds two principal directions of cracking, sometimes nearly at right angles. Usually one set was better developed, and the overthrow of loose objects was at right angles to that set, consequently in the direction identified as that of wave propagation.

Since Mallet was working months after the earthquake, he had to rely on interviews and reports to supplement his own observations. One of his "finds" was the cathedral at Potenza, which had developed an elaborate system of cracks. The cathedral architect, with a view to repairs, had made careful sketches of these cracks before Mallet's arrival. Moreover, the earthquake had displaced the cathedral dome, and the shift was at right angles to the main group of cracks, as Mallet would have expected.

Eccentric Epicenters

On a map (Fig. 4-3) Mallet placed at his points of observation arrows directed toward the epicenter as inferred for each place; the secondary directions were also shown. The arrows generally converged near a small community called Caggiano, not far from the northerly end of the oval meisoseismal area. Though strongly shaken, this was not the vicinity of greatest destruction, which was toward the other end of the oval, near Montemurro. Some arrows for the secondary prevailing direction point toward the Montemurro area.

Since we do not follow Mallet's idea that the earthquake was of explosive origin, we cannot accept his exclusion of transverse waves. Seismograms show that amplitudes in the transverse phase, S, are usually larger than in the longitudinal phase, P. The two directions roughly at right angles may correspond to P and S waves; but very likely Mallet's principal directions correspond to S, not P, and his secondary directions may be right.

Higher intensity near the Montemurro end does not necessarily indicate that as the epicenter. Two California earthquakes may be compared with this one. In the 1933 Long Beach earthquake, faulting probably extended from the epicenter about 20 miles northwestward. Intensity was high near the epicenter, but the most general damage was in the vicinity of Long Beach, near the northwest end. In the Kern County earthquake, faulting began near Wheeler Ridge and extended about 40 miles northeast; again, intensity was high near the epicenter, but the heaviest damage and strongest observed effects were mostly toward the northeast end.

An interesting attempt to apply Mallet's method to the 1933 earthquake was made by Professor Thomas Clements, who used the directions of overthrow of tombstones. The lines of directions converged near Compton, north of Long Beach; but, when the motion was assumed to be transverse, and perpendiculars were drawn accordingly, the lines converged near the instrumental epicenter far to the southeast.

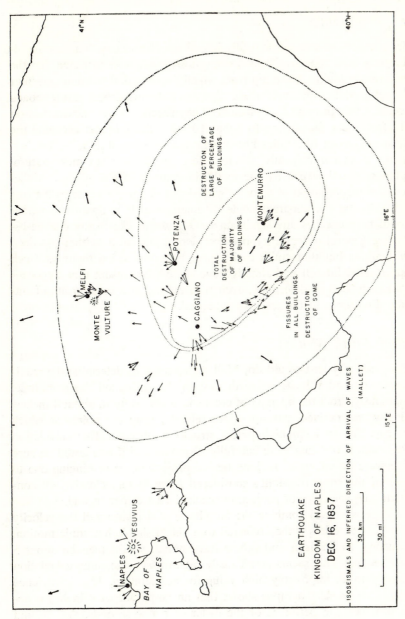

FIGURE 4-3 Kingdom of Naples, 1857. Chief isoseismals, and directions of displacement used to locate the epicenter. [After Mallet.]

Mallet's Hypocenter

Mallet pursued his method logically into three dimensions, but here he almost certainly went too far. On his assumptions it might be expected that the principal cracks would show a dip from which the direction to the hypocenter could be read. Mallet measured these dips carefully and projected them back to the vertical line through his epicenter. They reached it over a wide scatter of levels, from near the surface to about 10 miles deep. Mallet accepted the mean result of about 6½ miles but interpreted the spread as indicating extension of the source vertically, where a less enthusiastic but more natural interpretation would take it as due to the inaccuracy of the method. Mallet did not allow for variation of velocity with depth or for the effect of the free surface in altering the apparent angle of emergence. It is doubtful whether this part of his work has any significance. One would like to say that he established the general order of hypocentral depth, placing it neither too deep nor too shallow; but that could easily have been guessed without so much work, and Mallet himself had already done so. At the present time hypocentral depth is difficult to determine, even with the aid of a large group of stations recording times to the tenth of a second.

Extent of Earth Motion

As any good physicist would do, Mallet attempted to determine the maximum amplitude and velocity of earth motion. He estimated the amplitude, rather dubiously, from the opening of cracks in walls, as up to about 4 inches. The velocity was derived, perhaps more reliably, from the points at which loose objects thrown from high places struck the ground. This calculation would be exact if the data were carefully observed and if one could be sure that there were no effects of rocking before projection, or of flinging due to oscillation of the support. Mallet considered these factors closely and concluded that true velocities of projection reached 12 feet per second.

It would be wrong to combine the amplitude of 4 inches and the velocity of 12 feet per second as if they referred to a single simple harmonic motion. The frequency, as Davison and others have noted, would then be about 5 cycles per second. This seems suspiciously fast; in fact, the same calculation leads directly to an impossibly high value for acceleration, $13g$. Experience with strong-motion seismographs shows that an amplitude of 4 inches is acceptable provided the period is of the order of a few seconds. A 2-second period would imply acceleration of 2 feet per second ^{2}, which is moderate for a destructive shock. On the other hand, the velocity of 12 feet per second is difficult to reconcile with amplitudes and accelerations to be considered reasonable on the basis of later work. Probably Mallet, in spite of his care, underestimated the fling.

Intensity and Magnitude for Mallet's Earthquake

It is easy to overrate an earthquake like this, when whole towns were wrecked and the loss of life may have exceeded 12,000. In view of the extremely bad type of construction, intensity IX on the Modified Mercalli scale will probably account for the effects. In the absence of seismographs magnitude can only be estimated by comparison with a later shock, such as that on July 23, 1930, which centered just north of that of 1857. Destructive effects were comparable; the meizoseismal area was of about the same size, and the shock was felt about as far away. The magnitude of the 1930 shock was close to 6½, and the same probably applies to that of 1857.

WHITTIER EARTHQUAKE—JULY 8, 1929

The published earthquake investigations most read by students refer to large and important events; but the prospective field investigator should have a clear idea of what is involved in studying a small shock. To proceed in such a case on the model of the 1906 California earthquake, or even of Mallet's earthquake, is like using a pile driver to crack an egg.

The principal difficulty in field investigation of a great earthquake is that a single investigator or a small group must cover a large meizoseismal area in a limited time, yet make certain that no essential points are overlooked and that any really exceptional effects are studied. In a small earthquake the main difficulty is that only a fraction of the necessary data can be obtained by personal observation; there are no devastated cities, no great fractures and fissures, and much time must be used in interviewing eyewitnesses, even going from door to door.

Earlier Shocks

The Whittier earthquake of July 8, 1929, was the climax of a remarkable series. On May 4, 1929, at 5:07 P.M., there was an earthquake which slightly exceeded intensity V at Whittier and in adjacent places; that is, it was strong enough to disturb small objects and cause mild alarm, but not to cause any damage. The shock was felt by some persons as far away as Los Angeles. A smaller shock, reported felt only in the Whittier area, followed 8 minutes later, and a third, somewhat larger than the second, at 11:33 P.M. The seismograms of these three earthquakes, as written at Pasadena, were quite similar in character, showing a time interval close to 4 seconds between the two principal sharp waves (*P* and *S*). Nine more small shocks of the same sort were recorded in the next two weeks; three of these were reported felt at Los Nietos, south of Whittier.

Main Shock Field Work

The principal earthquake occurred at 8:46 A.M. on July 8. It was felt over most of the Los Angeles metropolitan area. The seismograms of this shock at Pasadena could not be read for detail because of the underexposed photographic record of large rapid motion; but numerous small shocks showed the *S-P* interval of 4 seconds and other characteristic appearance of the earlier Whittier shocks. When reports by telephone indicated strong shaking not far from Whittier, a field party set out in that direction.

Observed Intensities Approaching Whittier. The shock was not noticed by anyone then in the Seismological Laboratory building; it was felt by persons in downtown Pasadena, especially on upper floors. Successive stops en route are indicated by numbered points on the sketch map (Fig. 4-4). Abbreviated findings at some of these points are listed here (Roman numerals are intensities on the Modified Mercalli scale of 1931; they differ to some extent from intensities given in the published paper on this earthquake, which used the Rossi-Forel scale).

1. Shock of short duration felt by one person, seated; not by others (II).

2, 3. Not noticed; felt by other persons nearby (II).

4. Doors rattled (IV).

5. Heavy rattling of doors, windows, showcases, etc. (IV+).

6. Near hilltop. Rattling of toilet articles, small vases, dishes (IV).

8. Low ground. Small frame house creaked and groaned. All chairs moved. Slight alarm; ran out of house (V).

9. Low ground. Statuette fell. One block away, chimney fell (VI+).

10. Cigarette packs slid off shelves. Telephone wires whipped violently up and down as result of motion of poles (V–VI). Higher ground than No. 9.

11. Store goods off shelves. Shook building and gasoline pumps heavily (V+). Here information began to be obtained as to extensive minor damage in and near Whittier; but the first locality visited in Whittier showed comparatively low intensity.

12. Entering northwestern part of Whittier. Rolling hill ground higher than the business center. General rattling; few articles disturbed, none upset. Duration of felt shaking estimated at about 1 minute (V).

Whittier and Vicinity. Conditions in Whittier were observed, and additional information was taken from personal interviews and newspaper items. Effects decreased northward from the business center, and apparent intensity dropped rapidly on the hills north of the town. Loose objects were upset in many residences and business buildings; plate glass in a bank window was broken. At 119 North Greenleaf, just north of the business center, walls were cracked and pictures were thrown off the walls. Intensity in this area did not exceed

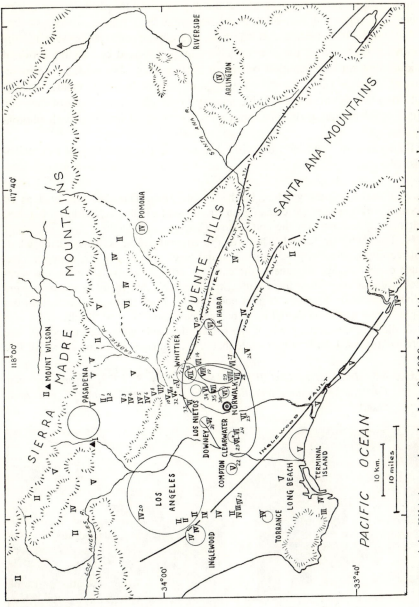

FIGURE 4-4   *Whittier earthquake, July 8, 1929. Intensities, isoseismals, instrumental epicenter.*

VI; but a short distance eastward it was generally VII. Along Painter Avenue, a north-south thoroughfare, many chimneys were damaged by loss of bricks, some were thrown down; brick and stone walls were cracked or out of line. Less than 3 hours after the earthquake, fire department trucks were rapidly removing bricks from the streets. This shows the need of prompt action if reliable firsthand information on minor earthquakes is wanted. Haste in removing debris is justifiable from the point of view of removing traffic hazards; but often there is an unmistakable anxiety to get the evidence out of the way. Misdirected public spirit sometimes goes so far as to interfere with photographers.

Points east of Whittier were visited to outline the boundary of the meizoseismal area.

13. Hacienda Country Club, in the Puente Hills. Duration estimated as 30 seconds. Rocking motion; creaking and rattling of clubhouse, rattling of dishes. Felt strongly outdoors on the golf course (V).

14. East Whittier School. This was the structure most seriously damaged. It had recently received a prize in its area and class, presumably for its appearance. Much of the damage was due to the common error of constructing sections separately without connecting them into a structural unit; but many of the parts were individually weak. The west face, a poor type of brick masonry, was thrown out of plumb and had to be shored up to prevent collapse [Fig. 4-5]. Bricks were down in all directions, especially out of the gables; the gable ends were unsecured and were overturned. J. R. Freeman wrote that ". . . damage was estimated at $20,000

FIGURE 4-5   *Damage to East Whittier School, July 8, 1929. [From Freeman, Earthquake Damage and Earthquake Insurance. Courtesy Clarke Freeman.]*

on a sound value of $80,000 . . . The foundation, basement and first floors, said to be of reinforced concrete, were not damaged. Walls above the first floor were of concrete without reinforcing except in the northwest wing, which has hollow-tile walls, stucco-clad. The roof is of tile on boards on wood joist. The 11 gables in the roof are of brick and stucco laid on top of the concrete with no other fastening than the mortar. Eight of these weak gable walls were demolished. Interior partitions are of hollow tile plastered, and were badly cracked." Intensity was certainly at least VII; in view of the poor construction a rating of VIII is excessive. This school was subsequently demolished and replaced by a building of more modern and sound construction.

Along Whittier Boulevard eastward from this point damaged brick chimneys were seen; at least one was broken off at the roof level. About 3 miles eastward, approaching La Habra, such evidences became rarer.

15. La Habra. Only one chimney was seen with a loose brick. To the west, in the one residence which was visited, bottles and other articles were upset, a crack opened between the brick fireplace and the wall, and a clock stopped at 8:45 (VI).

18. The Murphy Ranch packing house, south of the center of Whittier. Steel frame, with hollow tile walls, which were conspicuously cracked and had to be rebuilt. Ground here was exceptionally bad, being artificial fill (VII).

19. Corner of Gunn Avenue and Mulberry Drive. Two frame stucco homes diagonally across from each other were shifted on their foundations. In one the damage (estimated at $2500) was chiefly due to the brick chimney and fireplace, which collapsed inward, filling part of the living room with brick. The other house had rotated on its foundation; this effect, as in most cases, is probably due to differential friction and to successive blows in different directions, rather than to actual twisting. The brick chimney broke at the roof line; the upper part did not fall, but remained in place, rotated about 30 degrees with respect to the lower part. A cellar window was distorted from a rectangle into an oblique parallelogram. Near this location a gas and a water main were broken (VIII).

Further Observed Intensities.   Subsequent numbered points are from another field tour 2 days later.

20. The author's home, in western Los Angeles near Wilshire Boulevard. Shaking felt by one of two persons, duration not over 2 seconds. Clock weights on chains swung; clock did not stop. Creaking of walls and frame (IV). Intensity III to IV was found at points southward from this.

21. Gardena. Intensity still IV.

22. Compton. Newspaper office: heavy rattling of windows and loose objects, one small object upset (V). Residence, corner of Main St. and Long Beach Boulevard: rattling of house, loud roaring sound, vases upset, plaster down (VI).

23. Hynes. Two or three chimneys fell. Walls cracked, plaster down, goods off shelves (VI, approaching VII).

24. Bellflower. Walls cracked, plaster down, bottles off shelves (VI).

25. Artesia. Plaster down, goods off shelves (VI).

26. Buena Park. A few bottles off shelves in one store. No walls cracked. Not much earthquake noise. Rattling and creaking in frame structures (V).

27. La Mirada station. Plaster down, a few tiles off roof, glasses and dishes off shelves, bottles out of medicine chest. Roar like a concussion. Clock affected by shock (VI).

28. Carmenita. Goods off shelves. Heavy noise. Chimney down. "Stop the clock? Why, it ripped the pendulum right out of it!" (VII).

29. Sewage disposal plant, north of Carmenita. Standpipe broken (VIII).

30. Norwalk. Brick out of at least one chimney. Grocery goods off north-south shelves, but not off east-west shelves. Roaring sound. Clocks stopped (VI).

31. Downey. Cracks at top of walls. Plaster down. Pictures off walls. General alarm (VI). North and west of Downey, intensity appeared to decrease; upset objects were rarer.

32. Pico. Only a few objects off shelves. Several localities visited, all fitting intensity V.

33. Rivera. Bottle down. Vases upset. Loud roar. Plaster cracked. Furniture moved. One chimney damaged in the vicinity (VI).

34. Between Los Nietos and Santa Fe Springs. Articles off shelves (including a flashlight and a small clock), but plates remained on plate rail. Pickled beets splashed out of a bowl inside the refrigerator, to the housewife's dismay (VI or perhaps slightly less).

35. Norwalk State Hospital. Buildings used for public institutions of this type have performed discreditably in several California earthquakes. Damage was estimated at $6500 on a total value of about a million dollars. Concrete arches were cracked and bricks were dislodged; three chimneys had to be taken down. Supports on porches were broken. Tile roofs were damaged. Plaster was extensively cracked, and some fell. Pictures were off walls, and many dishes and bottles were broken (VII). Here many of the small aftershocks were felt, and shaking in each case was preceded by an audible roar. Since most of these aftershocks were very small, the epicenter could hardly have been more than a few miles distant.

Further information was taken from reports of other investigators, from questionnaire cards circulated by the U. S. Coast and Geodetic Survey, and from newspapers. In Fig. 4-4 isoseismals have been drawn enclosing areas of fairly consistent intensity VI, VII, or VIII. These are drawn with relatively little allowance for the expected increase of intensity on soft ground.

Epicenter and Time of Origin

The Whittier earthquake occurred just at the beginning of adequate timing at the southern California stations. Only times at the three nearest stations are valuable for locations. For the principal shock these were:

Pasadena	8:46:11.7 A.M.
Mt. Wilson	:12.6
Riverside	:17.5

Because of large motion following P, the times of S could not be read; but the time intervals from P to S in aftershocks were usually clear and fairly constant. Representative of these are:

Pasadena	4.0 sec
Mt. Wilson	4.7
Riverside	7.8

On the simplest assumptions (Chapter 20) these should be multiplied by 1.37 to give the corresponding transit times of P from the source, which come out as 5.5, 6.5, and 10.7 seconds. Subtracting these from the times of P for the main shock, one finds 8:46:06.2, 8:46:06.1, 8:46:06.8. The last result (from Riverside) is a little less reliable; but, giving it some weight, we find for the time of origin of the shock 8:46:06.3. The present standard assumption for southern California shocks is a hypocentral depth of 16 kilometers and an average velocity for P waves of 6.34 kilometers per second. This would give distances from the three stations of 30, 37, and 69 kilometers. The three arcs do not quite intersect in a point; but, if we give a little more weight to Riverside and take a slightly later time of origin, a good intersection is found near Norwalk, at 33° 54′ N, 118° 06′ W. This epicenter, which is marked on Figure 4-4, differs from that originally published because of the different velocities used. Both, however, indicate association with the Norwalk fault and not with the Whittier fault.

In 1929 the Whittier fault was the only one generally known to exist in that vicinity; evidence establishing the Norwalk fault was then brought forward, chiefly from commercial oil exploration work. This fault is of great importance in estimating earthquake risk in the Los Angeles areas. The central deep of the Los Angeles Basin, in simplest terms, represents a block dropped down thousands of feet between the Inglewood fault on the south and the Norwalk fault on the north. The destructive Long Beach earthquake of 1933 originated on the Inglewood fault; but before that date the only noteworthy shock known to have been associated with that fault was one which damaged Inglewood in 1920 and could hardly have been greater in magnitude than the Whittier earthquake of 1929.

Since 1929 smaller earthquakes have continued to occur on the Norwalk fault, or at least with epicenters not far from it. The activity is fairly well established. There is good reason to suppose that the Norwalk fault is capable of producing an earthquake of the magnitude of the Long Beach earthquake (6¼). Such an event would be heavily damaging in the metropolitan area.

Magnitude of the Whittier earthquake was placed at 4.7. The maxima of most of the seismograms were photographically underexposed; but the probable maxima were estimated by comparison with records of aftershocks, and the results proved consistent. The maximum observed intensity and the extent of the meizoseismal area and of the area of perceptible shaking are all comparable to those for other shocks of the same magnitude.

Aftershocks

During the first 24 hours following the main earthquake at 8:46 A.M. on July 8, 155 aftershocks were counted on the seismograms at Pasadena. They became less frequent very rapidly; between 9 and 10 A.M. 26 were noted, while between 9 and 10 P.M. there were only 4. The total for July was 203. Considering that these were recorded by torsion seismographs, and not by the high-magnification Benioff instruments later developed at Pasadena, this is unusually high aftershock activity for a shock of magnitude 4.7.

The Reflected Transverse Wave

All the small shocks of the series showed a remarkable additional impulse on the seismograms (Fig. 4-6). At Pasadena this was regularly about 4.4 seconds after *S*, and much stronger in the N-S component than in the E-W component. Similar waves arrived at Mt. Wilson, and less definite ones at Riverside. The directional effect suggested a transverse wave polarized by reflection at the base of the continental crust (the Mohorovičić discontinuity). This identification has since been confirmed; these waves, and corresponding reflected longitudinal waves, are regularly recorded in southern California at distances from about 25 to about 80 kilometers from the epicenter.

The existence of such sharp reflections indicates that the continental base is marked by a very sharp break in physical properties. This has an important bearing on geophysical discussions of the history of the earth and the continents. Workers in other areas have often overlooked these observations or questioned them. Naturally only a small fraction of the available evidence can be published, but skeptical visitors to Pasadena are usually shown a selection from the hundreds of clear seismograms of this type which are on file (Fig. 4-7). Lately a number of observers have been getting sharp reflections of the same type while recording artificial explosions with the equipment used in geophysical prospecting and operating it at a longer time after the detonation than is usual in such work. In Europe such reflections have been reported by Reich in Germany and by Rothé and others in France. In southern

FIGURE 4-6   *Pasadena seismograms, Nov. 3, 1930. Aftershock of the 1929 Whittier earthquake, showing P, S, and S reflected from the Moho.*

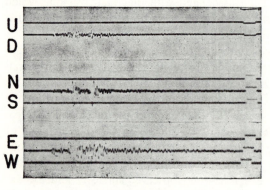

30 sec

California Dr. G. G. Shor has observed waves of this type, particularly from a large quarry blast at Monolith. The principal reflected P arrives about 10 seconds after the blast, and there is probably another about 1 second earlier; similar duplications can be found on earthquake records.†

Note for Students

The Whittier earthquake took place just as a well-equipped research program was commencing effective activity. In consequence, this minor event contributed an exceptional amount of new data. Equally high dividends cannot be expected from every little shock; but in no active science, and certainly not in seismology, can we be so sure of our ground that we know beforehand whether or not an investigation will produce new results. A geologist or an interested student will find it well worth while to undertake field study of any earthquake that comes to his notice; even though the results may not be new, the experience will be of much value to him.

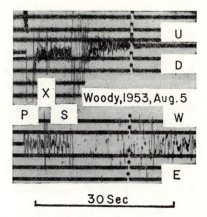

FIGURE 4-7 *Seismograms at Woody, California, showing P, S, and P reflected from the Moho.*

References

1857 earthquake

Mallet, Robert, *Great Neapolitan Earthquake of 1857. The first principles of observational seismology,* Chapman and Hall, London, 1862, 2 vols. Quotations in our text are from Vol. 1, pp. 29, 26, 28. Mallet's discussion and conclusions are in his Part III, beginning on p. 233 of Vol. 2. Mallet's work is summarized in: Davison, C., *A Study of Recent Earthquakes,* Chapter II.)

1929 earthquake

Freeman, John R., *Earthquake Damage and Earthquake Insurance,* McGraw-Hill, New York, 1932. Whittier earthquake of 1929, pp. 435–440; Italian earthquake of July 23, 1930, pp. 513–544.

Wood, H. O., and Richter, C. F., "Recent earthquakes near Whittier, California," *B.S.S.A.,* vol. 21 (1931), pp. 183–203.

---

† In 1955 J. Galfi and L. Stegena reported a reflection arriving after about 9 seconds for explosions in the Hungarian plain. Dr. Båth informs the writer that reflections arriving after about 10 seconds have been observed for explosions near Kiruna (northern Sweden). E. W. Janczewski in 1956 identified S reflected from a depth of about 33 kilometers on records of small earthquakes in Poland. Kamitsuki finds earthquake waves in central Japan reflected from a depth of 30 kilometers.

Reflected *S*

Gutenberg, B., "Reflected and minor phases in records of nearby earthquakes in southern California," *B.S.S.A.*, vol. 34 (1944), pp. 137–159 (especially p. 146).

Reich, Hermann, "Uber seismische Beobachtungen der PRAKLA von Reflexionen aus grossen Tiefen . . ." *Geol. Jahrbuch*, vol. 68 (1953), pp. 225–240.

Shor, George G., Jr., "Deep reflections from Southern California blasts," *Trans. Am. Geophys. Union*, vol. 36 (1955), pp. 133–138.

Galfi, J. and Stegena, L., "Deep reflections in the region of Hadjúszoboszló," (in Magyar, with Russian and English summaries) Geofizikai Közlemenyek (Eötvös Geophysical Institute, Budapest) vol. 4 (1955), pp. 37–40.

Janczewski, E. W. Correspondence, 1956.

Båth, M. Personal communication, 1957.

Geneslay, R., Labrouste, Y., and Rothé, J.-P., "Réflexions à grande profondeur dans les grosses explosions (Champagne, October 1952)," Publ. Bur. centr. séism. internat. (A), vol. 19 (1956), pp. 331–334.

Kamitsuki, A., "On the seismic waves reflected at the Mohorovičić discontinuity (I)," Mem. College Sci. Univ. Kyoto, series A, vol. 28 (1956), pp. 143–159.

CHAPTER 5

Some Great Indian Earthquakes

IT IS EASY to underestimate India, geographically and otherwise. That is why Figure 5-1 has an inset showing California to the same scale. The geology of India is a huge subject; to summarize it at all is like attempting a thumbnail sketch of the geology of North America, but something of the sort must be done if the background of the great Indian earthquakes is to be understood.

FIGURE 5-1 *India, with principal isoseismals of 1897 [Oldham] and epicenters of other large shocks. Inset: California to same scale, with isoseismals of 1906.*

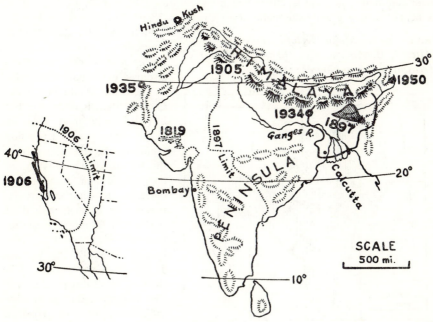

GEOLOGY OF INDIAN EARTHQUAKES

The three chief subregions, the Himalaya, the plain of the Ganges and other great rivers, and the Peninsula, are very different in structure and in geological history. In North America these compare with the Pacific Cordillera, the lower Mississippi Plain, and the Canadian Shield. (See Fig. 5-1A.)

The Himalayan arc, convex southward and fronting on the alluviated depression of the great plain, has often been compared to the island arcs of the Pacific, many of which are convex toward and front upon oceanic troughs or foredeeps. Like many great ranges, the Himalaya is made primarily of sediments accumulated over long geological time in a shallow sea. This particular sea, which Eduard Suess named Tethys, stretched across what is now Eurasia; the Mediterranean is a remnant of it, and the Alps and Apennines arose from it at about the same time and in the same way as the Himalaya. In India, the main collapse and folding into mountains began during the passage from Cretaceous to Eocene, at about the time when the Rocky Mountains were rising. Folding and thrusting continued, with a climax in the mid-Tertiary; Eocene marine sediments are found as high as 20,000 feet. The higher parts of the present Himalaya consist of igneous and metamorphic rocks from which the sedimentary cover has been eroded. In front of the range are foothills, the Siwaliks and others, composed of Tertiary sediments. Although the great thrusts of the Himalaya are now apparently quiescent, the foothills show evidence of geologically very recent faulting and thrusting on a large scale. As in Europe, there is suggestion that the stresses in action have changed materially in the last few tens of thousands of years, and the displacement continues along old fractures because they are established lines of weakness. A belt of strong gravity anomalies (see Chapters 2 and 25), indicating lack of equilibrium, runs along the hill front; in or near this belt are the epicenters of many of the stronger earthquakes of India.

FIGURE 5-1A *Major tectonic units of India.* [*After J. B. Auden, with slight modifications.*] *Areas of sedimentation are stippled.*

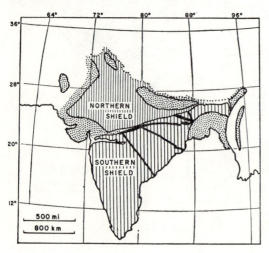

The depth of alluvium in the Ganges Plain is unknown, but it is certainly great. Like an ocean, this great depression separates the Himalaya from the Peninsula, which is an ancient stable area, a continental oldland. Archean rocks are exposed over more than half of the Peninsula; much of the remainder is covered by the basaltic flows of the Deccan Trap, which were extruded in the Cretaceous-Eocene interval. The Peninsula has no marine sediments of consequence younger than the Cambrian, except near the coast and in one long narrow belt where shallow waters entered at the height of the Cretaceous floods.

The Himalayan arc appears to be pressing southward toward the Peninsula. To the west and east are the arcuate structures of Baluchistan and Burma, also convex toward the Peninsula, as if the latter were the center for pressures converging from three sides.

Between the Peninsula and the Himalaya at the east is the mainly igneous and metamorphic mass of the Assam hills. Here was the meizoseismal area of the earthquake of 1897.

OLDHAM AND THE 1897 EARTHQUAKE

The name of R. D. Oldham is associated with much pioneer work during the years when seismology was passing from the pre-instrumental period into the era of the seismograph. As head of the Geological Survey of India, he directed, and personally carried out most of, the investigation of the great earthquake of June 12, 1897. His monograph is one of the most valuable source books in seismology. Its contents fall principally into five categories: (1) determination of intensities and drawing of isoseismals; (2) estimation of displacement, velocity, and acceleration; (3) investigation of the meizoseismal area (Fig. 5-2); (4) study of seismograms; (5) hypotheses as to the cause of the earthquake.

Isoseismals and Magnitude

Figure 5-1 shows Oldham's three most important isoseismals, which bound the area of perceptible shaking, the region of significant damage to masonry, and the meizoseismal area. The inset shows the first two corresponding isoseismals for the California earthquake of 1906; on that occasion effects comparable with those in the meizoseismal area of 1897 were observed only close to the San Andreas fault. The greatest linear extent of the 1906 isoseismals is nearly the same as in 1897, but the area included is much narrower. The magnitude of the 1906 earthquake was 8¼. Seismographs which registered the 1897 earthquake were not of modern type; it is difficult to use their records for deter-

FIGURE 5-2 *Indian earthquake, 1897. Meizoseismal area.* [Oldham.] *Solid outline, region of violent shaking; dashed outline, hill area.*

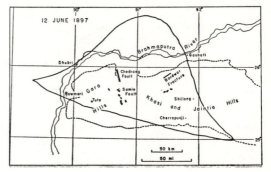

mining magnitude. In a detailed study of the large earthquakes of 1896–1903, Gutenberg assigned this one a magnitude of 8.7±, placing it among the dozen or so largest known (Chapter 22). We may compare the macroseismic data with those for the shock of 1934, which has a well-determined magnitude of 8.4. With due consideration for incompleteness of information and for the irregularity of figure of the isoseismals, we may take the following figures.

	1897	1934
Mean radius of area of perceptibility	900 miles	800 miles
Mean radius of area of serious damage	300	200
Longest dimension of meizoseismal area	160	65

These figures establish the 1897 event as the greater.

Amplitudes and Acceleration

Like Mallet, Oldham estimated amplitudes from cracks in the ground and in buildings; but he was dissatisfied with the results and searched for better data. His best evidence he considered to be that of a pair of damaged brick tombs at Cherrapunji (Fig. 5-3), which had impinged against each other and against the walls of the depression in which they stood. He inferred an amplitude of 10 to 18 inches, probably near the mean of 14 inches. His observations were minutely careful and his reasoning ingenious.

Seismograms of large earthquakes often indicate quite large amplitudes at short distances. The Long Beach earthquake of 1933, of magnitude 6.3, recorded at Pasadena (distant about 30 miles) with a ground amplitude of about half an inch and a period of 7 seconds. Simple application of the magnitude scale, without consideration of the many possible modifying factors, indicates that if that earthquake had been of magnitude 8.3 the amplitude at Pasadena should have reached 50 inches. This is probably two or three times too large. Near the epicenter of a great earthquake the amplitudes of slow elastic wave motion may be comparable with the displacements observed in faulting, which in the 1897 earthquake reached 35 feet.

Oldham concluded that the maximum acceleration exceeded that of gravity. Unlike less careful field workers, who sometimes report startling observations without convincing detail or proper documentation, Oldham presents

FIGURE 5-3

Indian earthquake, 1897. Damaged tombs at Cherrapunji. "The edge of the western depression has the grass growing undisturbed up to the edge of it, and along the edge small fragments of lime and plaster show that this was originally in contact with the edge of the tomb, which has now moved away to a distance of 18 inches." [Oldham.]

FIGURE 5-4 *Indian earthquake, 1897. Dislodged boulders.* [*Oldham.*]

his result fully and with circumstantial care. Not merely did eyewitnesses report seeing pebbles bouncing on the ground "like peas on a drumhead," but numerous instances were observed, photographed, and figured in detail, of posts shot out of their holes and of boulders lifted out of the ground without cutting the edges of their former seats (Fig. 5-4). This high acceleration is consistent with evidence in the granitic rock of the Assam hills, of widespread surface distortion, and of complex fracturing best characterized as shattering. The few features Oldham was able to find in a limited time in difficult jungle country can be no more than a representative fraction of those formed.

Faults and Fractures

Two true faults were found. The Chedrang fault was the greater, extending over 12 miles with throws up to 35 feet, in crystalline rock. It followed the general line of a stream, which suggests an old line of weakness. However, the winding course of the stream took it back and forth across the fault. Result: a series of waterfalls alternating with pools, as the stream dropped down over the fault scarp or flowed against it. Ponding was also observed along the stream where the former grade had reversed; and in the jungle, out of line with the Chedrang and Samin faults, similar pools indicated extensive warping.

There was other evidence that the surface had been distorted. Oldham's most picturesque instance comes from heliograph signaling, a means of communication widely used in nineteenth-century India. A station at Tura, in the hills, was regularly exchanging signals with Rowmari, on the Brahmaputra. "Before the earthquake it was just possible to do this from a certain spot by a grazing ray over an intervening hill. Now there is no difficulty at all, and instead of Rowmari being just visible over the hilltops, a broad stretch of the plains east of the Brahmaputra is visible." Resurveys after the earthquake confirmed extensive change; but, since the monuments were all in the disturbed area, no points could be assumed to be fixed and no details of the warping could be derived.

Oldham's investigation of what he named the Bordwar fracture is a good story that deserves retelling. He visited the Bordwar tea plantation to check a report that a hill had been "rent from top to bottom." That sort of language naturally made him suspicious. As he expected, the report was based on a misinterpretation. "A huge mass of rock, dislodged from near the crest of the hills has rolled down the slope, scoring the side of the hill. On the opposite side an equally large block has been dislodged, and in its downward course cleared a straight track down the hill; and on the summit a gap has been cleared by the overthrow of trees . . ." This was all that had been seen by anyone but Oldham. He writes: "At first, on seeing what was the true character of the appearances on which the report was based, the natural inclination was to reject it as one of the fables which are narrated of earthquake effects; but the band of trees, killed as they stood . . . showed that something unusual had happened." Where a less scientific worker might have gone no further, Oldham investigated thoroughly. "The actual fracture is only a few inches wide, it has rent the solid rock and in its immediate neighbourhood the violence of the shock was extreme. Trees have been overthrown or killed . . . it was found that great slabs of weathered gneiss had been rent in two, and on the crest there was a well-marked depression, like a small ditch, running away to southwest." Oldham was able to trace the feature for about 7 miles through the jungle, following it by broken trees and cracks in the ground. As there was little or no evidence of relative displacement of the two sides, the ordinary signs of faulting were absent. It is natural to suggest that there may have been displacement at the time of the earthquake followed by return to the original position.

After this experience Oldham noticed evidence of other, smaller fractures of the kind, and he remarked that similar features may have been seen and not correctly interpreted before the Bordwar fracture was found and studied.

Here we meet one of the most tantalizing difficulties of field seismology. Everyone recognizes the value of experience in all geological field work; one has to learn, by repeatedly going over the ground, what to look for and how to interpret it. (One also has to beware of falling into routine, losing one's freshness of viewpoint, and overlooking or misinterpreting exceptional conditions, or even points new to one's individual experience.) However, earthquake effects on the ground do not remain; many of them are erased by the weather or by human activity in a single season. The ground has to be gone over in a hurry, and investigation simply cannot be thorough. Unfortunately there is little real chance to accumulate sound experience. Earthquakes differ, and few workers have opportunity to investigate strong earthquakes in the field. The only source of help is enough acquaintance with the literature to profit by what has already been written into the record.

Shaking and Damage

Secondary effects of shaking in the earthquake of 1897 were extreme in many respects. In the alluviated plain of the Brahmaputra there was much flooding by ground water coming to the surface, sometimes as fountains during the earthquake. There were spectacular slides in the sedimentary rocks at the edge of the Assam hills. Oldham describes, and figures, hillsides so denuded of soil that the bedrock stratification was exposed.

Effects on artificial structures came as near total devastation as was ever described. The following is an eyewitness account reported by Oldham for Shillong, the administrative center of Assam:

I was out for a walk at the time. At 5–15 . . . a deep rumbling sound, like near thunder, commenced . . . followed immediately by the shock . . . The ground began to rock violently, and in a few seconds it was impossible to stand upright, and I had to sit down suddenly on the road . . . The feeling was as if the ground was being violently jerked backwards and forwards very rapidly, every third or fourth jerk being of greater scope than the intermediate ones.

The surface of the ground vibrated visibly in every direction, as if it was made of soft jelly; and long cracks appeared at once along the road. The sloping earth-bank round the water tank, which was some 10 feet high, began to shake down, and at one point cracked and opened out bodily. The road is bounded here and there by low banks of earth, about 2 feet high, and these were all shaken down quite flat. The school building, which was in sight, began to shake at the first shock, and large slabs of plaster fell from the walls at once. A few moments afterwards the whole building was lying bent and broken on the ground. A pink cloud of plaster and dust was seen hanging over every house in Shillong at the end of the shock.

My impression at the end of the shock was that its duration was certainly under one minute . . . subsequent tremors lasted some time . . . The whole of the damage done was completed in the first 10 or 15 seconds . . .

The buildings of Shillong might, before the earthquake, have been grouped into three classes, which correspond now to three degrees of ruin:

1. Stone buildings—It is not too much to say that every bit of solid stone work in the neighbourhood of Shillong, including most of the bridges, is absolutely levelled to the ground. The stone houses, and conspicuously the church, are now reduced to flat heaps of single loose stones, covered with torn and burst sheets of corrugated iron—the remains of the roofs.

The walls do not show the slightest partiality in their direction of falling. The stones have in every case been shaken loose, and have collapsed equally on both sides of the wall.

Heaps of stones along the roads, broken for mending purposes, which stood 1 foot high before the shock, are now flat, roughly circular patches 3 or 4 inches in thickness.

Two tall monuments of excellent cut stone work, about 20 or 30 feet in height, are in ruins; though in each case some feet of the masonry at the base still remain in an upright position—the individual stones being shaken from each other. The ruins are scattered most impartially on all sides in a rough circle . . .

2. Ekra-built buildings—A wooden framework, with walls of *san* grass covered with plaster. About half the buildings of this description are ruined in the same way as the stone buildings. All the large ekra buildings are utterly ruined inside, the chimneys in all cases being of stone work, the whole of which has fallen with the plaster from the walls, and in many cases the roofs also.

Small outhouses and villages of ekra-work have in some cases escaped with the loss of the plaster. Some of the new large buildings would also have escaped but for the stone chimneys, which have in every case wrecked the house.

3. Plank buildings—Built on the "log hut" principle, a wooden framework covered with planks, resting unattached on the ground. The only buildings of this description were stables or outhouses. In every case they have escaped untouched, except where the supporting stone work has been shaken away, when they have been slightly displaced. . . .

Shillong was rebuilt after the earthquake. Since 1950 its population has been reported as 27,000 or even larger. In 1953 it became the headquaters of the seismological service in India and now has an excellent station.

The whole group of effects in the meizoseismal area of 1897 furnished the principal model for the highest grade, XII, of the Modified Mercalli intensity scale.

Mechanism of the Earthquake

Some earthquakes, like the California earthquake of 1906, have been associated with extensive displacements on a single great fault. This is not the pattern of the 1897 earthquake. Although the 35-foot throw on the Chedrang fault ranks among the largest known, even this feature appears only as the most remarkable detail in the deformation of a large area. It is natural to seek explanation in terms of a great deep-seated disturbance which expressed itself in a complex way at the surface. Oldham at first followed this line, with the reasonable suggestion that there might have been thrusting on some fault surface beneath the Assam hills, presumably dipping north toward the Himalaya. The upper plate of the thrust might well be broken up and distorted.

The chief trouble here is that there was little geological evidence for any such thrusting in that area. Oldham later abandoned the thrust as an explanation, but he was guided further by evidence which we must regard with great suspicion. Since many aftershocks were felt widely in Assam, he arranged to have telegraph operators along the railway communicate at such times to discover where the shock was felt first. Many of these shocks appeared to be felt simultaneously at all points, within the limit of accuracy. Oldham concluded that this required a very deep source, say 200 miles below the surface. Here the remarks in Chapter 3 on the unreliability of timekeeping by ordinary observers apply with much force. The authors of the report on the 1934 Indian earthquake, perhaps with Oldham's conclusions in mind, observe:

It is interesting to note the accuracy of local records as compared with those of the seismograph stations. As was the experience in previous earthquakes, no reliance can be placed on the majority of time estimates made in the earthquake area. In spite of the fact that railway stations and telegraph offices are supposed to receive Indian Standard time every day, estimates given by the observers varied greatly.

Even when alert observers promptly communicate with one another, it is doubtful whether determinations of relative times for small shocks, in which the first motion noticed might be much later than the arrival of the first true seismic wave, have much meaning. We need not reject Oldham's conclusions altogether, although a depth of 200 miles would bring the main earthquake into the class of deep-focus earthquakes, which conflicts with most of what is known of such shocks (see Chapter 19). There can have been no great fundamental difference between the earthquakes of 1897 and 1934, and the seismograms leave no doubt that the latter was an ordinary "shallow" shock. The comparatively crude seismograms of 1897 indicate normal depth. However, shallow shocks originate at depths of the order of 15 miles. A thrust or other fault at such a depth would not necessarily be associated in any obvious way with geological structures at the surface above it. The true interpretation of the earthquake, therefore, may lie somewhere between Oldham's first and second hypotheses.

In correspondence, Dr. J. B. Auden has kindly supplied a contemporary interpretation. He considers the Shillong-Garo plateau as clearly an inlier of peninsular India, raised up between the southward Himalayan overthrusting and the northward Naga overthrusting. Its rocks differ from the foundation rocks of the Himalaya south of the main range; those, although originally of peninsular character, have been modified by tectonic processes and, in Dr. Auden's opinion, by granitization. The Shillong plateau was elevated mainly in the Pliocene, along a fault overthrust southward with a large east-west strike slip component. It is an area of large gravity anomalies. (For the large tectonic units see Fig. 5-1A.)

The student having access to Oldham's original memoir will find it very instructive reading, for Oldham's thinking in terms of fundamental physical principles was sounder than that of many later investigators. The preface to the memoir is a short treatise on the general principles used; it is outdated only by the progress of knowledge, particularly as contributed by seismograms. The derivation of displacement from the tombs at Cherrapunji is a model of scientific observation and inference. The discussion of projection of objects out of the ground, where Oldham was handling an almost unique set of data, is full of interesting detail. Thus Oldham found an explanation for the projected stones being all "much of a size"—mostly about 1 foot to 3 feet in diameter. (For small stones, the square-cube rule applies. Area and friction go up as the square of linear dimensions; volume, mass, and hence force under a given acceleration, as the cube. For very large stones, the force is enough to crush the surrounding earth, and the stone sinks in instead of being projected.) There is a thought-provoking chapter on the rotation of pillars and monuments in which Oldham refutes all the more obvious explanations of what is undoubtedly a complicated phenomenon which may have a different mechanical basis in different instances. His conclusion, certainly acceptable, is that waves must have arrived successively from many different directions during the shaking. Actual twisting, or vortical motion, is possible, if at all, only on very soft ground where conditions approach those of a liquid.

THE BIHAR-NEPAL EARTHQUAKE OF 1934

Turning from Oldham's memoir to the later number in the same series dealing with the earthquake of January 15, 1934, one sees the importance of date in studying seismology. The investigators of 1934 were a well-trained team, all of whom made significant contributions to the result. They had competent acquaintance with the progress of seismology up to 1934, including familiarity with Oldham's work. Seismograms at stations in India as well as in all distant parts of the world made it possible to locate the epicenter and to fix the magnitude at 8.4.

The extent of the isoseismals (Fig. 5-5) places this earthquake only a little below that of 1897. Intensity X on the Mercalli scale was assigned to a belt about 80 miles long by 20 miles wide, and to two spots almost 100 miles distant from the main belt, at Monghyr to the south and in the Nepal Valley to the north. The isoseismal of intensity IX was drawn to include an area which the authors of the report name "the slump belt," about 190 miles long and of irregular width exceeding 40 miles in places. The main belt of intensity X lies entirely within the slump belt.

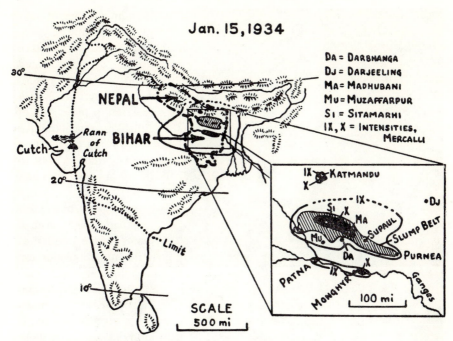

FIGURE 5-5   *Bihar-Nepal earthquake, 1934. Location map, showing outer limit of perceptibility. Inset: isoseismals and slump belt. [Data from Geological Survey of India, Memoir 73.]*

The Slump Belt

The chief criterion adopted in the demarcation of this belt was the behaviour of buildings and other structures. These tilted and slumped bodily into the alluvium, but seldom tumbled brick from brick. Sinking was often differential, in proportion to the relative pressures of the parts of buildings per unit area.

Subsidence of road causeways and railway embankments were marked, in some cases embankments originally 6 feet high sank down level with the surrounding country. Tanks, lakes, borrow pits and other depressions became noticeably shallower as a result of uplift of their bottoms—the tendency on the whole was for elevations and depressions to approach a common level. Fissuring and emission of sand and water reached their maximum development along this belt . . . The damage to buildings along this belt is in contrast to that of the area between Muzaffarpur and Darbhanga, where houses were razed to the ground. [Fig. 5-6.] . . . The epicentral tract and the slump belt, lying vertically above the focal region, received vertical shocks rather than oblique. Muzaffarpur and Darbhanga, on the other hand, being further removed from the focal region, received an oblique shock, and suffered less slumping but greater shattering.

Effects of this type had never before been reported in so much detail and on such a scale. At Sitamarhi "only one house of any weight . . . escaped

FIGURE 5-6   *Bihar, 1934. Judge's house, Muzaffarpur, before and after the earth-quake.* [*Government report.*]

tilting and sinking into the ground . . . and none was inhabitable. Many of the walls of buildings sank as much as two feet and foundations and floors were broken up completely. Sand covered the floors of sunken buildings to a depth of up to three feet. Concentric fissures formed in the ground around several buildings." At Purnea "the east wall of Darbhanga house . . . sank between three and four feet." At Supaul "the sinking continued for many days . . . doors which could be opened at first became jammed later." Similar reports are given by the dozen, and are supported by good photographs. (Fig. 5-7.)

FIGURE 5-7 *Slump belt, Bihar, 1934. Central part of house sank through floor halfway up to the lintels. [Government report.]*

Fissures and Fountains

Large fissures were opened in alluvium, mainly by lurching (Fig. 5-8), but were usually partly filled by emerging sand. One at Champaran (Fig. 5-9), 15 feet deep, 30 feet wide, and 300 yards long, remained open. At Sitamarhi

FIGURE 5-8 *Bihar, 1934. Lurching of formerly level ground; bank of lake at Motihari. [Government report.]*

"a typical fissure was 80 yards long, eight feet wide and infilled with sand to within three feet of the top." Near Purnea "there was a hole five feet wide and twelve feet deep."

As in 1897, ground water was disturbed and emerged at the surface, often in spectacular fountains. (Eyewitness accounts are quoted in Chapter 9.) So much sand was brought up that it was feared agriculture would suffer; but the sand was distributed by wind and heavy rains, and much of the land was actually made more fertile.

FIGURE 5-9   *Bihar, 1934. Fissure in Champaran; 30 feet wide, 15 feet deep, 300 yards long. [Government report.]*

Intensity and Ground

There were no effects suggesting great violence like that of 1897. This is not hard to understand in terms of surface geology. In 1897 the seismic waves arrived at the surface after traveling through solid basement rock, and probably with little loss of energy; in 1934 the corresponding waves emerged through a great thickness of sedimentary rock and alluvium. There must have been much absorption of energy in unconsolidated material, with compaction, subsidence at the surface, and a tendency for heavy structures during the shaking to press downward into the disturbed ground.

Normally we expect earthquake intensity to be higher on loose material than on basement rock; but this holds only when there is no great difference in conditions along the path. Then the absorption of energy takes place chiefly near the surface, but the amplitude of seismic waves is actually increased there and the effects on structures are greater.

Epicenter and Depth

The instrumentally determined epicenter is near the eastern end of the slump belt; as in many large earthquakes, it is at one end of the meizoseismal area, suggesting that faulting began at that end and proceeded westward.

Only one aftershock was large enough to be located instrumentally; its epi-center was near that of the main earthquake.

There are no data for determining the depth of origin with any precision. However, the times of arrival of first recorded waves at distant stations, to-gether with the registration of large long-period surface waves, establish that the Bihar earthquake originated at the usual depth for great "shallow" earth-quakes, say 20 to 30 kilometers (or 12 to 20 miles) below the surface.

Visible Waves

A tantalizing deficiency in this otherwise excellent report is the perfunctory treatment of accounts of waves seen on the ground during the earthquake. This vexed matter will be discussed in Chapter 10. There is undoubtedly a real and interesting phenomenon, obscured by optical, physiological, and psychological illusion, and often further colored by exaggeration. Although the authors of the 1934 memoir admit that "many seismologists are disin-clined to believe in the existence of such undulations," they assert that "their occurrence in an earthquake of this magnitude must be accepted as authen-tic." One might fairly expect careful and detailed documentation; instead there is a half page of very condensed information, repeating accounts of exactly the same character as had given rise to question when presented for earlier earthquakes, including that of 1897. India is so traditionally the home of marvelous stories that more care might be expected. Moreover, seismology will not progress with this problem unless the details of observa-tion are available for discussion and analysis.

Monghyr and Nepal

By contrast, we have a wealth of excellent detailed observation on the two puzzling outlying spots of high intensity. "The devastation at Monghyr was greater than in any other part of Bihar. The entire town was reduced to ruins, scarcely a house or hut escaped destruction or damage." "The damage . . . was entirely due to the severe shaking which the town received; neither fissures nor slumping of the ground were noticeable except near the edge of the river to the north." The loss of life here (1260) was greater than in any other one town, and about one-sixth of the total for Bihar. At Monghyr a ridge of Archean quartzite, an outlier of the Peninsular rocks, emerges through the alluvium. "The quartzite forming the higher ground appeared definitely to have resisted the severe shaking and the greatest damage was confined to the alluvium immediately surrounding it." In harmony with the suggested explanation for the effects in the slump belt, one is led to think that the seismic waves were transmitted from a source miles deep under the Ganges alluvium and sediments and, passing through competent rock with-out appreciable absorption, emerged with considerable violence into the al-

FIGURE 5-10   *Nepal, 1934. Tri-Chandra College, Katmandu.* [*From Geological Survey of India, Memoir 73.*]

luvium about Monghyr, which was accordingly more disturbed than the similar ground to the north, where the shock waves traveled upward through a greater thickness of absorbing material.

A similar explanation may be applied to the other exceptional spot, the Nepal Valley. This area includes Katmandu, the capital of Nepal, which in 1934 was officially an independent state, not part of British India (Fig. 5-10). The valley is a basin containing Pleistocene and Recent unconsolidated sediments resting on metamorphics and on partially metamorphosed pre-Tertiary series. As at Monghyr, the heavy damage was on the unconsolidated ground. A broad band of hill country to the south intervenes between the Nepal Valley and the main area of high intensity in the Ganges Plain. Because this hill area is thinly settled, the lower intensity assigned to it, while acceptable, rests on relatively few observations. Katmandu and Monghyr are nearly in opposite directions from the epicenter.

The known loss of life in India is given as 7253. In the Nepal Valley it is estimated as 3400. This is not high for so great an earthquake, especially in view of the widespread devastation. Fortunately the event occurred in the early afternoon, when most people were awake and many were outdoors.

OTHER EARTHQUAKES OF INDIA

In view of the size of India, great earthquakes there are relatively no more frequent than in California or in New Zealand. They are not nearly so frequent as in Japan. Moderate earthquakes, damaging a small area, appear to be relatively uncommon. Some of the historically important events are:

1819, June 16. Cutch. This great earthquake provides the earliest well-documented instance of faulting during an earthquake. See Chapter 31.

1905, April 4. Kangra. The earliest large Indian earthquake for which a well-documented instrumental magnitude (8.6±) can be assigned. This was a great disaster; the loss of life is stated as 19,000. Instrumental data are not adequate to fix the epicenter. The meizoseismal area, including Kangra, was on the Tertiary rocks of the foothills of the Himalaya. An isolated area of high intensity, lower than that at Kangra but not approached elsewhere, included Dehra Dun, also in the foothills; this was separated from the Kangra meizoseismal area by about 100 miles. The available evidence does not support the idea of two separate earthquakes; it is more likely that there was a great linear extent of faulting.

1935, May 30. Quetta. This earthquake laid waste the city of Quetta, the capital of Baluchistan (now part of Pakistan), with a loss of about 30,000 lives. While its magnitude (7.6) was less than those of the others discussed in this chapter, the epicenter was close to the city, so that the intensity there was relatively high.

1950, August 15. Assam and Tibet. Strictly this was not an Indian earthquake; the epicenter was near Rima, in a region claimed by both China and Tibet. It is one of the few earthquakes to which the instrumentally determined magnitude, 8.7,[†] is assigned. This shock was more damaging in Assam, in terms of property loss, than the earthquake of 1897. To the effects of shaking were added those of flood; the rivers rose high after the earthquake, bringing down sand, mud, trees, and all kinds of debris. Pilots flying over the meizoseismal area reported great changes in topography; this was largely due to enormous slides, some of which were photographed. The only available on-the-spot account is that of F. Kingdon-Ward, a botanical explorer who was at Rima. However, he had little opportunity for observations; he confirms violent shaking at Rima, extensive slides, and the rise of the streams, but his attention was perforce directed to the difficulties of getting out and back to India. Aftershocks were numerous; many of them were of magnitude 6 and over and well enough recorded at distant stations for reasonably good epicenter location. From such data Dr. Tandon, of the Indian seismological service, established an enormous geographical spread of this activity, from about 90° to 97° east longitude, with the epicenter of the great earthquake near the eastern margin. One of the more westerly aftershocks, a few days later, was felt more extensively in Assam than the main shock; this led certain journalists to the absurd conclusion that the later shock was "bigger" and must be the greatest earthquake of all time! This is a typical example of

---

† Until lately only this and one other earthquake were rated as high as 8.6; revision gives a longer list, with at least two larger than the 1950 shock. See Chapter 22 and Appendix XIV.

confusion between the essential concepts of magnitude and intensity. The extraordinary sounds heard by Kingdon-Ward and many others at the time of the main earthquake have been specially investigated. (See the discussion in Chapter 10.) Seiches (Chapter 9) were observed as far away as Norway and England.

References

Oldham, R. D., "Report on the great earthquake of 12th June 1897," *Mem. Geol. Survey India,* vol. 29 (1899), pp. 1–379. (Summarized by Davison in *Great Earthquakes,* Chapter X; in *Recent Earthquakes,* Chapter IX.)

————, "The Cutch (Kachh) earthquake of 16th June 1819, with a revision of the great earthquake of 12th June 1897," *ibid.,* vol. 46 (1928), pp. 71–147.

Middlemiss, C. S., "The Kangra earthquake of 4th April 1905," *ibid.,* vol. 38 (1910), pp. 1–409.

Officers of the Geological Survey of India, and Roy, S. C., "The Bihar-Nepal earthquake of 1934," *ibid.,* vol. 73 (1939), pp. 1–391. (The officers of the Survey were J. B. Auden, J. A. Dunn, A. M. N. Ghosh, and D. N. Wadia.)

The Central Board of Geophysics, "A compilation of papers on the Assam earthquake of August 15, 1950" (compiled by M. B. Ramachandra Rao), Government of India, Calcutta, 1953. (Includes 11 papers presented at a symposium on April 24, 1951, and 7 papers reprinted from other sources, among them the next:

Kingdon-Ward, F., "Notes on the Assam earthquake," *Nature,* vol. 167 (1951), pp. 130–131.

See also:

Kingdon-Ward, F., "Caught in the Assam-Tibet earthquake," *Natl. Geographic Mag.,* vol. 103 (1953), pp. 403–415.

————, "The Assam earthquake of 1950," *Geographical Journ.,* vol. 119 (1953), pp. 150–182.

Tandon, A. N., "Study of the great Assam earthquake of August, 1950 and its aftershocks," *Indian Journ. Meteorology Geophysics,* vol. 5 (1954), pp. 95–137.

Kvale, Anders, "Seismic seiches in Norway and England during the Assam earthquake of August 15, 1950," *B.S.S.A.,* vol. 45 (1955), pp. 93–113.

Brett, W. B., *A Report on the Bihar Earthquake and on the Measures Taken in Consequence Thereof up to the 31st December, 1934,* Patna, Govt. Printing Office (Bihar and Orissa), 1935. (Text deals almost wholly with relief work, but maps are good and photographs are excellent.)

Krishnan, M. S., "The structure and tectonics of India," *Mem. Geol. Survey India,* vol. 81 (1953), pp. 1–109.

Auden, J. B., "A geological discussion on the Satpura hypothesis and Garo-Rajmahal gap," *Proc. Nat. Inst. of Sciences of India,* vol. 15 (1949), pp. 315–340. (Alignment of the Assam hills with the northern edge of the Peninsular shield. Sketch map of suggested tectonic units; our Fig. 5–1A.)

CHAPTER 6

Foreshocks, Aftershocks, Earthquake Swarms

AN EARTHQUAKE of consequence is never an isolated event. It is likely to be preceded by foreshocks, normally few in number and small in magnitude; and there are almost certain to be many aftershocks, gradually decreasing in frequency and magnitude. However, some earthquake sequences depart widely from this pattern.

Question often arises whether a preceding or following event was actually a foreshock or aftershock. In any active region small earthquakes are common; occurrence of a small shock before or after a larger one is not necessarily significant, unless both appear to have the same epicenter or to be associated with the same fault system.

FORESHOCKS

Examples of Foreshocks

The earthquakes studied in Chapters 4 and 5 afford few good examples of foreshocks. In the absence of seismographs, evidence for foreshocks is hard to judge. Mallet cites a few small shocks felt at various times and places before his 1857 earthquake; but the reports scarcely indicate anything other than the usual minor activity of Italy. For so small an event, the Whittier earthquake was preceded by an unusual number of preliminary shocks.

Oldham reports no foreshocks for the great Indian earthquake of 1897. Nor were any noted for that of 1934; there were certainly no large ones, or seismographs then operating would have recorded them clearly. For the Assam-Tibet earthquake of August 15, 1950, negative evidence is even better. Moderate earthquakes are not rare with epicenters in the same vicinity, but the last preceding was probably one on November 17, 1949. A small shock, possibly in Assam, was recorded at Indian stations on August 12, 1950; its epicenter must have been far from that of August 15.

The great earthquakes of California (Chapter 28) furnish only doubtful examples. A shock felt at San Francisco a few hours before that of 1857 may have been a foreshock. The documents mention none for that of 1872. That of 1906 was preceded by many small shocks in its large area, but there was no identifiable immediate foreshock. By contrast, the large earthquake in Nevada in 1915 was preceded by many small foreshocks and by two strong enough to be felt throughout northern Nevada and recorded at distant stations.

Foreshocks as Forewarnings?

Case Histories. In 28 years of seismograph recording from 1929 through 1956 there was only one earthquake in southern California of magnitude 5, or over, which was not preceded by others from the same or an adjacent epicenter, although weeks or months might intervene between the last previous shock and the large one. The exception was the Manix earthquake (magnitude 6.4) on April 10, 1947, on the Mojave Desert 10 to 15 miles from the nearest previously established epicenters.

Events preceding the Long Beach earthquake of 1933 are well documented. A minor damaging shock at Inglewood in 1920 led to the identification of the active Inglewood fault; small recorded earthquakes from 1927 on were attributed to it. During 1932 there was a slight general increase in activity over much of southern California. So many small shocks were felt at Huntington Beach, near the Inglewood fault, that Mr. Martin Murray of that city set up a homemade seismograph to record them. After a major earthquake in Nevada on December 20, 1932, the general activity declined. Early in March, 1933, there was a long series of small shocks which originated near Old Woman Springs in the southern Mojave Desert. On March 9 at 1:13 A.M. a true foreshock (magnitude 4) was sharply felt at Huntington Beach. The slacking in activity had led Mr. Murray to take his instrument out of operation. After this foreshock he reconditioned it and recorded the destructive earthquake (magnitude 6.3) at 5:54 P.M. on March 10. The epicenter of the foreshock was not over a few miles from that of the main earthquake.

The area of the major Kern County earthquake of 1952 had been affected by many small shocks during the preceding 25 years of recording; but there was one true foreshock, with epicenter near that of the main shock, a little more than 2 hours preceding it. The magnitudes were 3.1 and 7.7, respectively.

As these examples indicate, foreshocks seldom afford any opportunity for warning or prediction of major earthquakes, since there is nothing to distinguish foreshocks from ordinary small shocks. "Suspicious" series of small earthquakes may be followed by no event of consequence or by a considerable event in some other part of the area, as the Old Woman Springs shocks were followed by the Long Beach earthquake. Increase in general activity before an important shock is not uncommon; a more strikingly localized example than that of Long Beach is the Helena, Montana, earthquake sequence of 1935, discussed later in this chapter. However, definite anticipations are in general unjustified.

Forewarnings According to Davison. In 1897 a most misleading paper, including discussion of foreshocks of the great Japanese earthquake of 1891, was published by Charles Davison. The conclusions were often reiterated by Davison and by other writers, Japanese as well as occidental. They aroused largely unfounded expectations of foreknowledge of earthquake activity. On the whole, seismology has benefited because seismological stations, established for purposes of prediction, have gone on operating after proving their value in other ways.

Davison states, in short, that before the 1891 earthquake there was a distinct increase in minor activity in what was to be the meizoseismal area, which was ultimately almost completely outlined by the epicenters of such foreshocks. Probably very little of this is true. The pertinent considerations are:

(1) The epicenters were not located instrumentally; the seismographs of 1891 were inadequate for that purpose. Epicenters were determined at Tokyo from macroseismic reports sent in by officials and by volunteers.

(2) While the number of such reports was large at first, it increased enormously year by year. From 1885 to 1892, more than 8300 earthquakes throughout Japan were catalogued, and by the latter year 968 observers were reporting regularly. Increase in located earthquakes in any given area does not necessarily mean increase in frequency of occurrence.

(3) During this period seismographs were installed in the critical area at Nagoya and Gifu. Earthquakes recorded at these stations were added to the count.

(4) Intensity tended to be higher, and consequently macroseismic epicenters were located, in the alluviated valleys about Nagoya rather than in the surrounding hill country. The intensity of the great earthquake was also increased on the alluvial ground.

(5) Because of the denser population in the valleys, the number of reports from the area and the number of located epicenters there increased.

None of these points was adequately considered by Davison; it remains unsettled whether there was any prelude to the 1891 earthquake even as definite as that for the comparatively small Long Beach earthquake, although there probably was a preceding increase in the general seismic activity of Japan.†

AFTERSHOCKS

Where means of observation are adequate, aftershocks are almost always noted after a sizable earthquake. Their number and magnitude, and the

---

† See also material quoted from Kotō in Chapter 30.

duration of aftershock activity before it drops below a given magnitude level, tend to increase with the magnitude of the principal shock.† Persons without much experience in earthquakes are often perturbed by the continued shaking, reacting to each new tremor as if it were a sure sign of a devastating convulsion to follow. Less excitable individuals become adjusted to these repetitions. One Nevada housewife, filling out a questionnaire form during an aftershock period, responded to the printed query as to whether anyone was frightened: "I am not alarmed, but sometimes the children complain."

In Japan Omori expressed the rate of aftershock activity by $N(1 + kt) = A$, where N is the number of aftershocks in a specified time interval such as an hour or a day, centered at time t after the main shock, while k and A are constants chosen to fit the data. The graph is a hyperbola; by analogy with other decay curves in physics one would rather have expected an exponential. The procedure is incompletely defined, and the meaning somewhat doubtful, because of the simple counting of aftershocks without regard to their rapidly diminishing magnitudes. Physically one should prefer a formula more clearly related to release of strain or to the radiation of energy; this has been accomplished by Dr. Benioff's procedure (outlined later in this chapter).

Second-Order Aftershocks

Aftershocks of the second order exist; that is, a large aftershock may be closely followed by a train of small ones falling off more rapidly than the general aftershock activity. There were such secondary sequences among the aftershocks of the 1929 Whittier earthquake and the 1933 Long Beach earthquake which, in both series, were abnormally numerous. Such observations can be made only in exceptional cases, since otherwise the aftershocks of the second order cannot be separated from the activity directly related to the large shock.

The well-investigated aftershocks of the 1952 Kern County series (Chapter 28) presented a wide variety of patterns of this kind. Those relating to the largest of the series can be summarized as follows (times are G.C.T.).

July 21, 11:52. Main shock, magnitude 7.7. Extremely high activity for several hours.

July 21, 12:05. Magnitude 6.4. Any secondary shocks lost in the general aftershock background.

July 23, 00:38. Magnitude 6.1. Identified as beginning a new phase of strain release. Epicenter differs from that of any preceding shock of the series. Six others of magnitude 4 or over within 7 hours following.

---

† "Båth's law." Dr. Båth has lately noted that in many instances the magnitude of the largest aftershock is about 1.2 less than that of the main shock.

July 25, 19:09 and 19:43. Both of magnitude 5.7. Epicenter near the terminal point of displacement along the White Wolf fault. Shocks from this point identified beginning on July 23, more numerous on July 25. Aftershocks with this epicenter frequent; until the end of 1953 they were more numerous than those from near the epicenter of the main shock.

July 29, 07:03. Magnitude 6.1; new epicenter near Bakersfield. Foreshock with this epicenter at 05:56; magnitude 3.9. Aftershock with slightly different epicenter at 08:01; magnitude 5.1. Additional small shocks continued from this vicinity.

August 22, 22:41. Magnitude 5.8. Epicenter nearer Bakersfield than on July 29; more obvious damage there than was caused by the main shock of July 21. Followed closely by smaller shocks classifiable as aftershocks of the second order, although the first-order aftershock activity of the entire Kern County disturbed region was still going on.

The occurrence of second-order aftershocks suggests that the larger aftershock with which they are associated is to some extent a new event, dynamically independent of the main earthquake at the head of the series. Such reasoning can apply only partly to the Kern County shocks, among which second-order activity was insignificant in comparison with that of first order. This was established by setting up a scatter plot comparing the locations of immediately successive shocks of the series; it appeared that a given shock was in general much more commonly followed by one in another part of the active area than by one with the same or a nearby epicenter. Over a two-year period the aftershocks originated mostly either near the epicenter of the main shock of July 21 or near that at the other end of the disturbed area (active on July 25). However, other points were well represented, and epicenters filled a roughly quadrilateral area with sharp boundaries on three sides, presumably indicating the boundaries of the crustal blocks most affected.

For many important earthquakes the data at least suggest the geographic pattern of: epicenter of main shock at one end of the active fault segment, epicenter of a large aftershock at the other end, aftershocks most numerous near the two ends, and foreshock epicenter close to that of the main shock but in the opposite direction to that of subsequent faulting.

Late Large Aftershocks?

Not uncommonly, when aftershocks have apparently subsided, what seems to be an unusually large one occurs without warning. Such a shock usually has its own aftershock series, strongly suggesting that it is dynamically independent. The Santa Barbara earthquake of 1925 was followed by a strong shock a year later, to the day. The Long Beach earthquake of March 10, 1933, was followed on October 2 by a shock of magnitude 5.2, with epicenter near the probable terminus of faulting in the main earthquake. In the Kern County series there were no shocks of magnitude as large as 5.0 from August 22, 1952, to January 12, 1954, when one of magnitude 5.9, with numerous aftershocks, took place almost precisely at the epicenter of the main earthquake of July 21, 1952. A similar occurrence was included in the aftershock series of the Hawke's Bay (New Zealand) earthquake of February 2, 1931 (magnitude 7.9); the activity had largely subsided when on September 15, 1932, the Wairoa earthquake, of magnitude 6.8, with a somewhat different epicenter, took place.

Aftershock Risk

Usually neither foreshocks nor aftershocks are comparable in magnitude with the main shock of a series. When a strong earthquake has occurred, the normal expectation is that serious danger is over, and the public is often reassured on this basis. However justifiable on humanitarian grounds, this is not a scientifically sound procedure, because it overlooks at least three well-recognized possibilities:

(1) A strong damaging shock may be a foreshock of a great and destructive earthquake.

(2) Aftershocks of a major earthquake may be as large as ordinary locally damaging shocks. Moreover, the epicenters are likely to be distributed over a wide area; a large aftershock may originate closer to a center of population, and cause more damage there, than the main earthquake. This is what happened at Bakersfield in 1952.

(3) In certain areas a strong shock may be habitually followed within a comparatively few hours by an equal or a larger one.

EARTHQUAKE SWARMS

Swarm Types

Certain localities are frequently visited by *earthquake swarms,* long series of large and small shocks with no one outstanding principal event. Such swarms are common in volcanic regions; they often occur before and during eruptions (Chapter 12). They are also observed in areas of geologically recent, but not current, volcanism. For example, long series of small earthquakes, sometimes as many as 100 in a day, are occasionally recorded at the Haiwee seismological station in Owens Valley. This station is founded on Pleistocene tuff; cinder cones, basaltic flows, hot springs, and other evidences of volcanism are only a few miles distant.

Swarm earthquakes are observed in some definitely non-volcanic areas, as in the Vogtland on the southern border of Germany and at Comrie in

Scotland. Swarms of a rather special pattern in time are usual in Imperial Valley, California, where residents are well aware that a sharp earthquake is often soon followed by an equal or a larger one. Schools are closed, and persons remain outdoors; such action has often saved lives. Repeatedly in that area the pattern has begun with two or more sharp shocks of nearly equal magnitude, with smaller shocks between, followed by a few hours or even a full day of quiet; then another large shock with a normal aftershock sequence. This pattern was followed on December 30 and 31, 1934, when the two largest shocks were of magnitude 6.5 and 7.1; the epicenters were in the thinly populated region south of the international boundary. After the large earthquake of 1940, and its immediate aftershocks, the seismicity of Imperial Valley was very low until 1950, when activity resumed on July 27 with a complicated swarm, including two shocks of magnitude 5.4 on July 28 and 29 and one of magnitude 4.8 on August 1. A swarm with shocks up to magnitude 5.4 originated close to the Imperial Valley town of Brawley on December 16–17, 1955, and caused some minor damage there.

A Foreshock Swarm

Imamura picturesquely describes the foreshocks of the Riku-Ugo earthquake of 1896; the series began with 15 perceptible small shocks on August 23 and others on following days. The principal shock took place on August 31.

Seismically speaking, the morning of the eventful day was unusually active. Then there was a short lull which was broken by a strong shock . . . in the afternoon. After that it appeared to the anxious people as though the earth had been trembling without intermission for almost an hour when the climax was reached, with the arrival of the violent shock that caused such widespread destruction.

Imamura details a similar but smaller series in the same region in 1914, and continues:

The foregoing would seem to suggest a tendency of foreshocks to precede earthquakes that originate from certain seismic zones rather than others. If so, it may turn out to be a matter of considerable importance in the study of earthquake prediction. At all events it behooves us to be on our guard lest we confound other earthquakes with foreshocks, as certain localities are peculiarly subject to development within a short interval of several groups of small shocks that are in no way related to the principal shock.

OTHER ABNORMAL SEQUENCES

Such a foreshock swarm is an extreme example of abnormal sequence, in which the earliest noticeable disturbance is not the principal event. Such sequences, if only they can be identified, offer a chance of forewarning. An abnormal pattern of much interest when there is temptation to minimize risk after a sharp earthquake is that of the disastrous earthquake series at Helena, Montana, in 1935:

October   3. Sharp shock; no damage, but some alarm.

October 12. Damaging shock ($50,000).

October 18. Destructive earthquake; damage $3,000,000; 2 killed.

October 31. Second large earthquake; additional damage $1,000,000; 2 more killed.

The two who lost their lives in the second large earthquake at Helena were workmen on repairs which had been started in the hope that aftershock activity would continue to be small.

Where the normal sequence of small foreshocks and aftershocks is expected, abnormal patterns are dangerous to the public. Since they characterize certain regions particularly, continuous recording of earthquakes in an active area is important in order to discover where such behavior is to be expected. In most of southern California excluding Imperial Valley, foreshock and aftershock patterns have followed the normal pattern. Among the exceptions are shocks along faults in the Transverse Ranges which may or may not be related to the San Andreas fault. Abnormal patterns are apparently more frequent in the Basin and Range province of Nevada and eastern California; an example is the 1954 series in Nevada:

July 6. Earthquake of magnitude 6.6, damaging at Fallon, perceptible over a wide area. Minor faulting in the Rainbow Mountains east of Fallon. Moderate aftershock activity.

August 24. Earthquake of magnitude 6.8; further minor faulting. Epicenter near that of July 6. Numerous aftershocks.

December 16. Major earthquake of magnitude 7.1. Epicenter and major faulting about 30 miles farther east. Foreshock of magnitude 4.0, 4 hours 37 minutes earlier. Large aftershock of magnitude nearly 7 about 4 minutes after the main shock, with faulting on another zone. Aftershock activity then at first falling off rapidly, but continuing normally in the following months.

UNUSUAL ACTIVITY

Sometimes a single large shock occurs in an area where earthquakes are comparatively rare, as at Charleston, South Carolina, in 1886; near the Grand Banks in the north Atlantic in 1929; in Baffin Bay in 1933; in Matto Grosso, southwestern Brazil, in 1746 and 1955. Or an unusual series of strong earthquakes may occur in a region which afterwards becomes nearly quiet, as in Korea about three centuries ago.

During the years 1925–1930, in a limited area of the Indian Ocean southeast of Madagascar, there were at least 42 shocks large enough to be clearly recorded at distant stations; about 30 of these were of magnitude 6 or over, and 3 reached 7.0. In 1933 there was another of magnitude 7.0, followed by a long pause which was broken in 1949 by one more of about that magnitude, and in 1951 by one of magnitude 7.8, the largest known for the region.

Known seismic areas occasionally show periods of unusual activity. A series of earthquake disasters in the Near East in the early centuries of our era materially contributed to the decline of civilization in Palestine and Syria. Beginning with an earthquake of magnitude 7.9 in 1939, Turkey has had a long series of damaging shocks; those in 1942, 1943, 1944, 1953 and 1957 exceeded magnitude 7. In November, 1938, off the east coast of Japan, there were 6 shocks of magnitude 7.0 to 7.7 (several of which were followed by small tsunamis) and many smaller ones—an earthquake swarm on the grand scale.

NATURE OF AFTERSHOCKS

Aftershocks as Afterworking

The mechanics of aftershocks can be understood properly only in relation to that of the principal shocks. This is discussed chiefly in Chapters 14 and 22; the student may find it best to read those pages before proceeding here. A large earthquake is held to be due to the fracture of rocks under strain. The energy released in seismic waves and otherwise is held to be derived from the potential energy in the strained blocks as they snap back toward equilibrium—*elastic rebound*. Aftershocks indicate that the readjustment to equilibrium is not completed at once.

Aftershocks are often journalistically called "settling shocks"; this phrase suggests one simple but inadequate theory of their cause. The displaced masses are not perfectly coherent; they have many internal breaks and irregularities. Internal readjustments may be expected to go on for some time with decreasing frequency and magnitude. Minor subsidences may occur under the action of gravity.

Observation both of earthquake sequences and of the behavior of specimens fractured in the testing laboratory strongly suggests that in most instances the aftershock process is not so passive. Rather it constitutes an active continuation of the major strain release involving phenomena such as have been described under the names *elastic afterworking* and *creep*. A few days after the Imperial Valley earthquake of 1940 the writer visited the fault trace just north of the international boundary, where strike-slip amounting to about 10 feet was distributed among several fractures. Several bicycle tracks crossing one of these were offset a few inches in the same sense (right-hand) as the main displacement, indicating that the original movement had continued on a minor scale, presumably accompanying the numerous aftershocks.†

Benioff's Theory

Dr. Hugo Benioff has set up an exact analysis of this process, both theoretical and observational. His theory involves careful definition of such terms as creep, which have often been used too vaguely. The essential consideration is that the chief process of a major earthquake runs its course in a time not likely to exceed 2 minutes; yet there is readjustment of strain involving blocks whose linear dimensions are of the order of 100 kilometers or more. Even if these blocks were perfectly coherent internally, their elastic recovery could hardly be completed in so short a time. This recovery will be delayed by the lack of perfect elasticity, and its continuance after the main event will produce creep. In this way further strain will accumulate along the line of original fracture, until it overcomes the resistance of the temporary blocking which brought the first event to a close, and so produces further faulting. The process will be repeated, in general on a smaller scale at each recurrence, until there is insufficient residual strain to break through the remaining resistance. Further motion will depend on the continuing tectonic forces which caused the original major strain; with lapse of time these may again raise the strain to the point of breaking through resistance. On the small scale, this may account for the sudden large apparent aftershocks, already noted, which occur after the general activity has subsided. On the large scale, this primary process determines the interval between major earthquakes on a given fault system.

Dr. Benioff has applied the magnitude scale to express these ideas quantitatively and to exhibit the results in graphical form. The contribution of each individual shock to the release of strain is taken as proportional to the square root of the energy radiated in elastic waves. The energy is calculated from the magnitude relationship presented in Chapter 22. As set forth in that chapter, the magnitude-energy relationship is subject to various qualifications and uncertainties, but its exact form fortunately has only a minor influence on the result. The calculated square roots of the energies are closely

---

† Refer also to the account of the Tango earthquake of 1927 in Chapter 30. From observations on that and other occasions, Professor Chuji Tsuboi concludes that there are two distinct types of aftershock event: (1) that involving chiefly vertical displacements, which gradually decrease from their initial values, and (2) displacements either vertical or horizontal, but continuing in the original sense. The latter would need another subdivision according to Dr. Benioff's analysis now to be described.

but not exactly proportional to the measured amplitudes from which the magnitudes are derived.

The calculated strain increments are accumulated after each successive shock, and the sums plotted as ordinates, with the logarithm of elapsed time from the main shock as abscissa. Two types of plot result: (Figs. 6-1, 6-1A, 6-1B) one, nearly straight, agrees with the behavior of laboratory specimens under compression; the

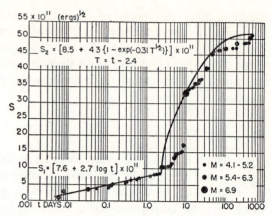

FIGURE 6-1   *Strain release curve, Hawke's Bay, New Zealand, aftershocks, 1931.* [Benioff.]

other, a curve becoming nearly level, was originally identified as due to shear, on the basis of experiments by A. A. Michelson many years ago. Research in 1955 by Dr. Cinna Lomnitz gave the same type of strain-release characteristic for shear as for compression. The interpretation of the second type of aftershock curve thus becomes questionable, although it agrees satisfactorily with the known circumstances of the earthquakes involved. Dr. Benioff now inclines to think that this type of curve may represent conditions where there is lateral flow or creep, perhaps subcrustal. Aftershock sequence plots from some earthquakes, like the Long Beach earthquake of 1933 and the Hawke's Bay (New Zealand) earthquake of 1931, begin nearly straight, as for compression, and then break abruptly into a curve of the supposed shear type. This behavior was found for the Kern County aftershocks of 1952, but it then appeared that the epicenters were separable geographically into two groups, one of which is responsible for the shear-type curve, while the other,

FIGURE 6-1A   *Strain release curve, Imperial Valley aftershocks, 1940.* [Benioff.]

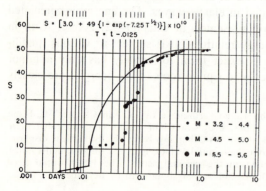

plotted alone, gives the straight compression-type plot. Compare the results on displacements following the Tango earthquake of 1927 (Fig. 30-13).

The occurrence of aftershocks of the second order, which in theory might suggest modification of the plotting or its interpretation, in practice hardly affects either. Strain release in such shocks is an insignificant fraction of

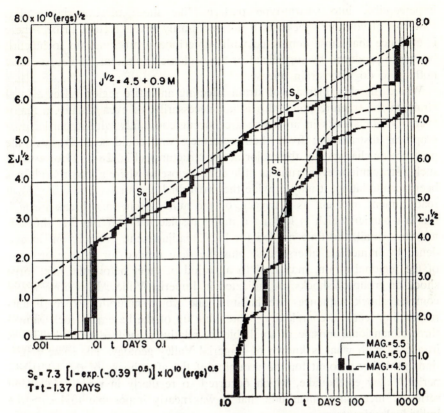

FIGURE 6-1B *Kern County aftershock sequence, 1952–1954. Strain release curves for the two sides of the White Wolf fault plotted separately [Benioff.]*

the general release going forward; it is at least in part legitimately lumped together with it, since it means at most a passing complexity in the process. Similarly, although some of the aftershocks may represent minor internal readjustments of the disturbed blocks taking place at a distance from the principal fracture surface, their contribution is small and, in part, is legitimately included.

Dr. Benioff has applied the same form of plot to the large shocks of limited regions and to the largest shocks of the world. The implications of the linear addition of strain bear on the problem of whether minor shocks can act as a "safety valve" to delay the occurrence of large ones. These points are taken up in Chapter 22.

NATURE OF FORESHOCKS AND SWARMS

Dr. Benioff's procedure does not explicitly include foreshocks, but the ideas involved are readily applicable to them. The occurrence of foreshocks grades imperceptibly into swarm-type patterns like those of Imperial Valley. (Swarms of volcanic shocks need not necessarily fit this type of theory; there is evidence that those connected with subterranean processes do so in part, but those produced by superficial or atmospheric explosions, of course, have nothing to do with strain release.)

Where the slow increase of strain affecting an entire region raises a local strain to the fracture point, this is not necessarily, or even usually, a major event. Minor shocks in an active area occur constantly with impartial geographical distribution; from their epicenters one could scarcely infer the presence or character of the major active structures (Fig. 28-3). However, if such a stray minor event runs its little course close to where a major strain has been accumulating, it may provide the proverbial last straw. The minor local readjustment may act as a trigger, or the minor event may break through the last barrier which is locking a major displacement. If the first shock is not very small, the time needed for local readjustment, possibly put in evidence by immediate aftershocks, plus the time needed for further progress of the regional strain, may lead to a period of quiet before the next large shock or group of shocks, as indicated in Imperial Valley. In regions where complex shock sequences are habitual one would naturally expect a corresponding tectonic complexity, a breaking up of the basement rocks into a three-dimensional mosaic of small blocks relatively easily displaced. This is in accord with what is known of Imperial Valley geology. The same conditions probably account for the grouping of the aftershocks of the 1940 Imperial Valley earthquake, which occurred so regularly in rapidly following series of increasing magnitude that it was nearly impossible to locate the individual epicenters.

After the damming of the Colorado River and the impounding of Lake Mead, small earthquakes became frequent in that area. Three local stations were established by the U. S. Coast and Geodetic Survey in cooperation with the Bureau of Reclamation. It was found that the numbers of the shocks could be correlated with changes in reservoir loading. The weight of the water in the new lake, of the order of 10^{10} tons, is small compared to the weight of the crustal blocks displaced. The tectonic equilibrium was sufficiently delicate to be disturbed by this comparatively minor additional load. This result represents a local condition; similar shocks were not observed in tests at other large reservoirs.

Near Chatra, India, Guha and collaborators report correlation of the frequency of small local earthquakes with variations in flood intensity of streams.

References

1891 foreshocks

Davison, Charles, "On the distribution in space of the accessory shocks of the great Japanese earthquake of 1891," *Quart. Journ. Geol. Soc. (London)*, vol. 53 (1897), pp. 1–15.

Swarm earthquakes

Etzold, Franz, "Die sächsischen Erdbeben während der Jahre 1907–1915," *Abhandl. math.-phys. Kl. sächs. Ges. Wiss. (Leipzig)*, vol. 36, No. 3, pp. 16–429. (Vogtland earthquakes.)

Davison, Charles, *A History of British Earthquakes,* Cambridge University Press, 1924. (Comrie earthquakes, Chapter IV.)

Imamura, Akitune, *Theoretical and Applied Seismology,* Tokyo, Maruzen, 1937. (Riku-Ugo earthquakes, pp. 82–83.)

Helena, Montana, earthquakes of 1935

Engle, H. M., "The Montana earthquakes of October, 1935; structural lessons," *B.S.S.A.,* vol. 26 (1936), pp. 99–109.

Ulrich, F. P., "Helena earthquakes," *ibid.,* pp. 323–339.

Neumann, Frank, "The Helena earthquakes of October and November, 1935," in *United States Earthquakes, 1935,* pp. 42–56, U. S. Coast and Geodetic Survey, Ser. No. 600, U. S. Dept. of Commerce, Washington, D. C., Govt. Printing Office, 1937. (See also pp. 18–28.)

Strain release

Benioff, Hugo, "Earthquakes and rock creep, Part 1," *B.S.S.A.,* vol. 41 (1951), pp. 31–62.

Lomnitz, C., "Creep measurements in igneous rocks," *Journ. Geology,* vol. 64 (1956), pp. 473–479.

Michelson, A. A., "Elastic viscous flow, Part I," *Journ. Geology,* vol. 25 (1917), pp. 405–410; Part II, vol. 28 (1920), pp. 18–24.

Carder, D. S., "Seismic investigations in the Boulder Dam area, 1940–1941, and the influence of reservoir loading on earthquake activity," *B.S.S.A.,* vol. 35 (1945), pp. 175–192.

Guha, S. K., Ram, G., and Rao, G. V., "Trigger causes in earth movements," *Publ. Bur. central international, ser. A,* vol. 19 (1956), pp. 345–355.

CHAPTER 7

Effects of Earthquakes— General

FOREGOING CHAPTERS have offered samples of earthquake effects, common and uncommon. At this point a broader view is needed.

CLASSIFICATION

Earthquake effects may be classified as (1) *primary,* due to the causative process, such as faulting or volcanic action, or (2) *secondary,* due to shaking, or more generally to the passage of elastic waves generated by the primary process. Secondary effects therefore include the registration of seismograms.

Table 7-1 includes primary effects only for tectonic earthquakes. When the cause is volcanic, primary effects may be of an entirely different sort. For description of the phenomena accompanying eruptions and subterranean movements of magma the student should refer to works on volcanology.

Elastic waves being similar no matter how generated, there are no essential differences between the secondary effects of tectonic and volcanic earthquakes; the more obvious observed distinctions are attributable to the generally shallower hypocentral depths of volcanic shocks (Chapter 12).

Table 7-1 separates secondary macroseismic effects into two groups: permanent and transient. This division is practical rather than fundamental. Certain effects, such as damaged buildings, the investigator may expect to observe in the field after an earthquake; these are here classed as "permanent," although the evidence always disappears eventually. Other effects, like the swinging of suspended objects and the rattling of windows, can only be observed on the spot while the earthquake is going on; they are "transient," and the investigator must depend for his information on personal interviews or written reports.

80

Table 7-1 Principal Macroseismic Effects of Tectonic Earthquakes

Effects on	Primary	Secondary Permanent	Secondary Transient
Terrain	Regional warping, etc. Scarps Offsets Fissures, mole tracks, other trace phenomena Elevation or depression of coasts; changes in coast line	Landslides (slumps, flows, avalanches, lurches)[1,2,3,4] Secondary fissures[3] Sand craters[5] Raising of posts and piles	Visible waves? Perceptible shaking
Water (ground and surface)	Damming; waterfalls; diversion Sag ponds Changes in wells, springs		Changes in well levels Earthquake fountains[5] Water over stream banks Seiches Tsunamis Seaquakes
Works of construction	Offsets, and destruction or damage by rending or crushing; buildings, bridges, pipe lines, railways, fences, roads, ditches	Most ordinary damage to buildings, chimneys, windows, plaster	Creaking of frame Swaying of bridges and tall structures
Loose objects		Displacement (including apparent rotation) Overturning, fall, projection (horizontal or vertical)	Rocking Swinging Shaking Rattling
Miscellaneous		Clocks stop, change rate, etc. Glaciers affected Fishes killed Cable breaks	Nausea Fright, panic Sleepers wakened Animals disturbed Birds disturbed Trees shaken Bells rung Automobiles, standing or in motion, disturbed Audible sound Flashes of light?

[1] Earth flows properly belong with water phenomena.
[2] Landslides may produce damage to works of construction.
[3] Landslides and secondary fissures may produce the effects on terrain and on surface water listed as primary.
[4] Classification of landslides from California Earthquake Commission, vol. 1, pt. 2, p. 385.
[5] Production of sand craters and fountains is a single phenomenon.

PRIMARY AND SECONDARY EVIDENCE

For the physicist, every earthquake phenomenon, primary or secondary, presents a problem of explanation and interpretation. Some effects are as yet incompletely explained, while others (such as tsunamis or seismic sea waves; see Chapter 9) have been interpreted by different authors in widely divergent ways. Some, to which an offhand simple explanation is commonly applied, on close study show much more complexity.

To the engineer, the earthquake phenomena of greatest interest are those which affect the stability and safety of buildings and other construction. Such are the damaging effects of shaking, of subsidence, and (in rarer instances) of actual fault displacement. They are discussed in the following chapter.

Because of the special principles involved, effects on ground and surface water are also given a separate chapter (Chapter 9).

To the geologist, earthquake effects are of interest principally as evidence bearing on the nature and mechanism of faulting or of other tectonic processes, and on the relation of earthquakes to geological structures. Attempts like Mallet's to use macroseismic effects to find an epicenter and otherwise physically analyze an earthquake occurrence are now obsolete, particularly in areas where modern seismographs are operated. However, these effects remain the basis for assigning intensities and drawing isoseismals; even where good seismographs are available, careful mapping of intensities may shed much light on local earthquake problems.

Primary effects, on the other hand, are of fundamental importance. A fresh fault scarp or a new cinder cone is in the nature of direct evidence, compared to which even that derived from good seismograms is circumstantial.†

DISTINCTIONS AND DIFFICULTIES

Distinction between primary and secondary effects has to be made by the geologist in the field. His decision is often of crucial importance in the

---

† We all are best impressed by evidence of the type with which we are most familiar. I am indebted to Mr. H. O. Wood for an anecdote concerning Hugo De Vries, the distinguished Dutch botanist responsible for the biological concept of mutation, who was in residence at the University of California at Berkeley in 1906. He was taken into the field to see some of the effects of the great earthquake of that year. Being told in advance of displacements ranging from 15 to 20 feet, he listened with polite interest but obvious scientific skepticism. The first few localities to which he was conducted appeared not to convince him; he was clearly not familiar with geological matters. The party finally arrived at the often photographed locality between Point Reyes Station and Inverness, where the road was offset 21 feet. The fault trace here was close to Papermill Creek; small trees and bushes were uprooted, some were torn and some killed. "Ah, here is *botanical evidence!*" cried De Vries, and he fell to examining these details with close attention, until his companions insisted on going on.

interpretation of an earthquake; but it is not always easy. Questions often arise in evaluating cracks and fissures, and more frequently in separating primary effects from those of landslides.

Cracks and fissures, especially in firm rock, may be primary features directly related to faulting; even in loose materials they may be produced directly by displacement of the underlying bedrock. Much more commonly they are due to shaking, further modified by differential settling and occasionally by the ejection of ground water. Even along an active fault, cracks found after an earthquake may be secondary effects of this sort due to disturbance of the broken material in the fault zone.

Large landslides set off by an earthquake often imitate all the more spectacular effects on terrain and ground water produced by faulting, including the formation of extensive scarps. Later chapters will present instances of landslides or slumps described as faulting, and of faulting described as landslides. Extremely difficult problems may arise in this way. Thus, the principal scarps of the Kern County earthquake in 1952 developed at the base of Bear Mountain in an obvious landslide area; nevertheless, the weight of evidence indicates that these scarps represent a real tectonic displacement in the basement rock, although the surface expression was certainly modified by the presence of unconsolidated material.

UNCLASSIFIED EFFECTS

In the general discussion of primary effects (Chapter 14), and in the description of individual earthquakes with faulting (Part II), it will appear that there are at least two additional groups of effects on terrain which are not satisfactorily classified as primary or secondary. One such group may be illustrated by a hypothetical case in which there has been an upthrust of basement rock underlying a great thickness of alluvium or unconsolidated sediments. The surface expression will then not take the form of a clean rupture related to the edge of the upthrust block; instead there will be an area of surface warping and cracking from which the primary disturbance can be inferred only roughly and indirectly.

The remaining group of effects is exemplified by observations in the zone of the Garlock fault after the Kern County earthquake of July 21, 1952. A belt of cracking about 4 feet wide crossed a minor thoroughfare, and the disturbance extended a few hundred yards as a narrow trace along the fault zone. There was no sign of strike-slip. The very broken terrain included a meadow of sag-pond type. This occurrence was taken in some quarters as evidence that the major earthquake originated on the Garlock fault—a natural supposition, made by the writer and many others during the first few hours after the earthquake when information was scanty and partly distorted, but definitely contradicted by instrumental evidence as well as by the strong development of trace features along the White Wolf fault. Presumably there was very minor tectonic activity in the Garlock fault zone—to the extent that a small and relatively superficial block dropped a short distance, either as a result of shaking or because of the general regional readjustment of strain following the major earthquake. Such effects as this might grade insensibly, with increasing scale, up to the displacement of a large crustal block, giving rise to a new earthquake precipitated by the first event. Compare the Nevada earthquakes of December 16, 1954 (Chapter 28).

Problems of interpretation may thus become serious even when field observation is competent and thorough, and when all the accessible facts are in hand. When dealing only with historical (and sometimes quite unscientific) accounts of an earthquake, it may be impossible for a seismologist or a historical scholar to tell what occurred.

CHAPTER 8

Earthquake Effects on Buildings and Other Construction—California, 1952

VALUE OF THE OBSERVATIONS

Effects of earthquakes on artificial structures are of prime importance to engineers and architects. In the general study of earthquakes they have no special intrinsic importance; but because of their human interest they attract more attention, and are likely to be better observed, than any but the most spectacular effects on terrain or water. The older historical accounts of earthquakes rarely include much useful information other than reports of damage and of casualties.

The middle and higher grades of most intensity scales depend largely on observations of structural damage. In this way even reports by untrained observers may be of scientific value, but good judgment is required for interpretation. Knowledge of the prevailing styles of construction at the given time and locality is necessary, and details of the individual structures are highly desirable. The field seismologist should work, in so far as possible, in cooperation with qualified architects and engineers, making the maximum use of their evaluation of the original weakness or strength of structures showing damage.

USUAL TYPES OF DAMAGE

Most descriptions of earthquake damage report effects of shaking on masonry (brick or stone) or wood-frame structures, of small to moderate size, and of the general styles originally European but now introduced all over the world with the diffusion of occidental culture.

In some earthquake regions (including southern and southeastern Europe) effects on structures of other types are reported. Such structures generally belong to one of the following groups.

(1) Huts or small dwellings made of light materials, flexible under moderate shaking. Usually these are not tied down, so that violent earthquakes may merely slide them about without serious injury (compare Shillong in the 1897 earthquake, Chapter 5). Even total collapse is not necessarily a serious menace to persons, and damage is soon repaired. In some countries, as in Japan, a heavy roof is added to light construction; the inertia of this roof in violent shaking frequently leads to serious damage or dangerous collapse.

(2) Substitutes for masonry using weak materials, such as the adobe of the southwestern United States and Latin America, the tamped earth or pisé of the Near East, the similar tapia of Spanish-speaking countries, or the mud-block houses of Formosa. Unless such materials are carefully faced and supported (as by a stout wood frame), these structures are highly dangerous; they have contributed heavily to earthquake death lists.

(3) Large and heavy cut-stone structures, such as temples or palaces. The solidity required to pile these up to their imposing heights usually is sufficient to withstand moderate earthquakes that wreck weak structures in the same localities; but violent shaking may cause collapse of such an edifice that has stood for many years.

Some damage is caused by the effect of shaking, not directly on the structure, but on its foundation. Warping and fissuring of its base can be withstood only by an exceptionally strong and structurally consolidated building.

Settling may affect the structure differentially; or the whole may sink into loose ground, as in the slump belt in 1934 in Bihar. Large lurches and other slides may be destructive to structures located on the moving ground; these risks, with the allied ones of subsidence, also arise independently of earthquakes. Relative motion of the two sides of an active fault may destroy a structure of any kind by rending or crushing. Finally, the effects of shaking, slumping, settling, and faulting have to be considered in relation to railways, bridges, ditches, canals, and other works of engineering.

DAMAGE TO MODERN STRUCTURES

Information about the behavior of large and modern structures under strong shaking is as yet limited largely to experience at San Francisco in 1906 and at Tokyo and Yokohama in 1923. In other well-described twentieth-century earthquakes, either no such structures were in the meizoseismal area, or the intensity in their vicinity was lower (as at Long Beach and Los Angeles in 1933), or defects of design, materials, and execution were so grave as to add little to our data for sound construction (as at Fukui, Japan, in 1948, or at Mexico City in 1957). The performance of smaller structures of steel-frame type or of concrete (with and without reinforcing) has been well established and studied in many instances.

Proper earthquake-resistant design for large and tall structures is still vigorously debated in the engineering profession. At present the questions are only of minor concern to the general seismologist, and the debate tends to confuse more important issues on which there should be no serious differences of opinion. The engineering student will find a short discussion, with some references, in Appendix II. Proper design for simpler structures is considered, together with questions of risk, in Chapter 24.

DAMAGE BY SHAKING

Horizontal Forces

Most of the notable secondary effects of earthquakes are produced by the horizontal rather than by the vertical part of the motion—not because the horizontal motion is larger, although it is likely to be so everywhere except in the immediate vicinity of the epicenter. Vertical motion operates against gravity; horizontal motion meets with no similar resistance. Thus, objects falling from a shelf are usually not thrown off by vertical whipping, which would call for very violent motion; instead, a small vase, for example, begins to rock during the horizontal shaking, "walks" about on its rim, and without ever having been upset may pitch off the shelf to smash on the floor.

This principle is of importance in judging earthquake damage to buildings and other construction. Most ordinary structures are engineered to withstand a considerable vertical load, with a liberal factor of safety to allow for overload. Only in an earthquake with vertical oscillation approaching that of 1897 in India is there likely to be failure due to vertical overloading in any but the most contemptible construction. But much "ordinary substantial" construction in which no special provision has been made against horizontal forces is incredibly weak laterally. Official standards in many localities still permit the erection of structures which would hardly stand in a good strong wind, let alone an earthquake. This applies still more to parapets, fire walls, cornices, and architectural ornaments in communities where earthquake risk is not considered.

Duration and Frequency

Earthquake effects are often evaluated exclusively in terms of acceleration. Grades on the intensity scale have repeatedly been correlated with maximum acceleration (see Chapter 11). Building-code provisions, for practical reasons, are usually formulated in terms equivalent to requiring resistance to accelerations of one-tenth of gravity, or some other specified fraction. However, even simple harmonic motion is not completely described, nor its effects properly accounted for, without also taking into consideration (1) the duration of the oscillation and (2) another quantity which may be displacement or velocity of the oscillating particle, or its frequency of oscillation (any one of these together with acceleration determines the others). In relation to damage it is probable that the velocity is more significant than either acceleration or displacement taken alone.

Table 8-1   Harmonic Oscillation Frequencies (cycles per second)
Corresponding to Given Amplitudes *A* and Accelerations *a*

	A (In.)						
Acceleration a	0.0001	0.001	0.01	0.1	1	10	100
g	300 cps	100 cps	30 cps	10 cps[1]	3 cps	1 cps	0.3 cps
0.1g	100	30	**10**	3	1	0.3	0.1
0.01g	30	10	3	1	**0.3**	**0.1**	0.03

[1] Boldface numerals indicate expectation in moderately strong earthquakes.

Table 8-1 repeats Table 3-1, except that for engineering convenience the amplitudes *A* are given in inches. Although earthquake vibration is not at all simple harmonic, it can be represented by a superposition of harmonic oscillations (Fourier analysis; compare Chapters 3 and 15). The boldface numerals indicate the combinations of amplitude and acceleration and the corresponding frequencies most common in the analysis of strong shaking; the data of a locally more violent earthquake ought to cover a larger part of the table.

The effect on structures of oscillations represented in the parts of even this small tabulation would differ widely. At the upper left, accelerations up to gravity correspond to very rapid oscillations of small amplitude. Such oscillations are almost never observed in earthquakes, but they may originate in other ways. Even if of unlimited duration, they should not directly produce the normal types of structural damage; the effect to be expected would be that of slow disintegration of materials.

With an amplitude of 1 inch and a frequency of 1 per second, we have acceleration of 0.1g, which is that ordinarily considered damaging; but here we may apply the duration factor. If such motion continues only a few seconds, it will not damage the ordinary sound structure. In repeated events of this kind such a structure may be gradually weakened; or if motion continues for, say, 15 or 20 seconds, as it may in a great earthquake, damage may be very much greater. Most strong-motion seismograms, as well as most ordinary seismograms of smaller or more distant earthquakes, show individual waves corresponding to acceleration significantly higher than the maximum of the more continuous motion; it is the latter which largely governs the damaging effects.

At the lower right of Table 8-1 it is shown that an amplitude of 10 inches with a period of 10 seconds gives acceleration only of the order 0.01g. A great earthquake may send out surface waves which a thousand miles away will show an amplitude of an inch with a period of 20 seconds; if these cause any damage it can only be by setting up seiches in water (Chapter 9) or producing some extraordinary long-period resonance effect.

Table 8-1 and the accompanying discussion originated from suggestions by Mr. H. M. Engle, who writes:

Duration of shaking is possibly the single most important factor in producing excessive damage. It takes time to break buildings up once damage starts. Long duration, reasonably high acceleration, and considerable amplitudes, are the combination which does most damage in buildings. Too many engineers take from seismograms accelerations of 0.2g, 0.3g or more, without any consideration or knowledge of what duration and amplitude go with them.

Here it should be noted that the duration factor, as Mr. Engle intends it, is taken care of in the complete Fourier analysis of motion by proper distribution of amplitudes among the various components.

Effect of Magnitude

Not only does the range of frequency in strong motion increase with intensity, but it also increases with the magnitude of the earthquake. The long-period effects at great distance are observed almost exclusively in large earthquakes (see Chapters 3 and 11). It follows that the earth motion may have different characteristics and produce different prevailing types of damage at localities assigned the same intensity in earthquakes of large and small magnitude; see, for example, the difference in effects at Bakersfield and Los Angeles in the large and small shocks of 1952, noted in the engineering report quoted from in this chapter.

TYPICAL DAMAGE—CALIFORNIA, 1952

The Engineering Report—Losses

One of the best recent engineering reports on earthquake effects is that on the California earthquake of July 21, 1952, and its aftershocks.† This

---

† Steinbrugge, K. V., and Moran, D. F., "An engineering study of the Southern California earthquake of July 21, 1952, and its aftershocks," *B.S.S.A.*, vol. 44 (1954), pp. 199–462. Reviewed by permission of the authors.

earthquake is described as a tectonic event in Chapter 28 and is referred to, for illustrative purposes, throughout this book.

The engineering report is based on data collected and compiled by the earthquake department of the Pacific Fire Rating Bureau. It is one of a long series of such publications sponsored by this and other insurance organizations, generally characterized by accurate and disinterested presentation of the facts of damage and the lessons to be derived from them. Apart from introductory historical and geological material, the report includes:

(1) A general section on damage to structures, grouping them by class of construction and by location.

(2) Special sections on public schools, sprinkler systems, oil wells and production, refineries and pumping plants, railroad properties, agriculture, dams, roads and bridges, electrical equipment, and on the action of building and fire departments.

(3) Appendixes giving details, with photographs, sketches, and plans, of damage to 14 individual buildings.

(4) Appendixes on the performance of multistory buildings, old brick buildings, elevated water tanks, and steel ground tanks.

(5) Conclusions and recommendations.

The report presents the following estimates of property damage.

Building Damage	
Bakersfield	$23,000,000
Kern County (excluding Bakersfield)	4,250,000
Los Angeles, Long Beach, Pasadena	10,000,000
Santa Barbara	400,000
Other Damage	
Oil wells and refineries	2,000,000
Agriculture	6,000,000
Public utilities (power, water)	600,000
Dams, roads, bridges	100,000
Railroads	2,300,000

Each item has its individual story. Bakersfield was the only large city in the meizoseismal area. The figure quoted includes damage in both the main earthquake of July 21 and the aftershock of August 22 (which caused little damage elsewhere). Damage in Kern County outside Bakersfield includes that at Tehachapi, which attracted so much popular attention that the press still refers to the July 21 event as the "Tehachapi earthquake." This distorted impression was due to spectacular damage to old and weak construction. To quote the report: "Tehachapi was essentially an older town, the construction of the business district being primarily unreinforced brick with sand and lime mortar. Brick and adobe buildings suffered extensively." It is then remarked that, although structural damage in Tehachapi was severe, engineering judgment concluded that relatively it did not exceed that in Long Beach in 1933. "The fact that 10 of the 12 deaths occurred in this community may have created an unwarranted assumption regarding building damage." (These deaths also focused the attention of the press, the public, and even of officials who should have been better informed, on the losses at Tehachapi disproportionately to those elsewhere.)

The items for Los Angeles, Long Beach, and Santa Barbara principally represent renewed injury to structures damaged and inadequately repaired after the earthquakes of 1925 and 1933.

No details are given for Pasadena. At the time of the earthquake the press reported damage of $20,000; 30 feet of a brick parapet on a two-story building collapsed.

Of the damage to oil wells and refineries, the authors of the report estimate that about 75 per cent is accounted for by fire at the Paloma refinery.

The figure for agricultural loss is severely reduced from semi-official values which are quoted in the report but are considered to be overestimated. Crops in this area depend largely on irrigation; damage figures are divided between actual loss of crops and cost of repairs to the water supply system.

Of the damage to railroad properties, about $1,400,000 represents the cost of repairs to tunnels and track put out of service by displacement on the White Wolf fault.

Wood Frame

Construction and Support.   The ordinary wood-frame structure, especially if only of one story, is damaged in a few simple ways. The extent of damage depends on the soundness of construction; presence or absence of diagonal bracing in walls and ceiling makes a great difference in the amount of distortion suffered. Very strong shaking may seriously rack the frame, either directly or by warping and distorting the foundation. A common omission occurs when the building is not bolted or otherwise well anchored on its foundations; in comparatively moderate shaking it will then slide across and even off the foundation, usually damaging the whole frame. The 1952 report notes very few occurrences of this sort and comments on their rarity as compared with the Long Beach earthquake of 1933. This difference has a simple explanation: in the Long Beach area at that time many frame structures rested, not directly on the foundation, but on posts known as "cripples." The horizontal displacement tipped these posts, leaving the structure free to slide.

Attached Masonry.   Brick chimneys and fireplaces are responsible for some damage to otherwise uninjured frame structures, as in the Whittier earthquake (Chapter 4). Fire regulations often demand that a chimney shall pass through the roof without contact. Without further provision this results in a clear space where the chimney can oscillate as an inverted pendulum, eventually striking the roof edge hard enough to break off the upper part of the chimney at the roof level. Building codes now generally call for vertical steel reinforcing rods in chimneys and fireplaces.

Carelessness in using masonry in connection with frame construction is still found, especially in small communities. The 1952 report figures a frame structure with a stone veneer not anchored to the wood; the stone collapsed over parked automobiles. Practically all chimneys at Tehachapi were broken or damaged; this represents construction of old date, not conforming to present building standards.

Dangerous Innovations. The report also draws attention to certain practices, becoming frequent in frame dwellings of supposedly modern architecture, which tend to decrease resistance to earthquakes; for instance, the introduction of large openings for picture windows and garage entrances without any compensating bracing. Replacing of wood sheathed walls by plastered walls may give increased stiffness up to the limit of strength of the plaster, but may result in great weakening in successive shocks and possible failure in violent shaking which a wood frame would withstand. Disregard for earthquake and other risks exists where homes are being set up on steep hillsides. "Construction in some instances amounts to no more than a building on stilts."

Masonry

The Dead Hand of Tradition.   A large part of the loss of life and property in earthquakes is due to the failure of masonry. That this continues to be so after centuries of experience is an outstanding example of social inertia, which is particularly obvious in the older centers of civilization. No condemnation is too strong for the kind of masonry Mallet found prevalent in Italy in 1857; yet observers of the destructive effects of the earthquake of 1930 in the same region remarked how well Mallet's description still applied. In the seismic regions of Latin America one still finds dwellings and even public buildings constructed of adobe. In Japan tradition has made it difficult to alter indigenous Japanese styles of construction; but much additional unnecessary loss has resulted from copying too closely masonry and other structures of northern Europe, where earthquake risk ordinarily is not considered.

California—Experience and Reform.   In California tradition is not strong. Since failure of brick masonry has been conspicuously responsible for most California earthquake losses since 1906, a prejudice has developed which sometimes even results in public statements that brick is not a suitable building material for earthquake regions. When it is pointed out that the fault is not so much in the material as in the manner in which it is used, the reply is that such misuse continues and meanwhile the people perish. This attitude tends to underrate the equal failure of other types of structures when they are constructed as badly as the old brick buildings which go to ruin in earthquakes year after year. Poor construction of all kinds persists in communities which have not adopted adequate building codes, or where enforcement of existing codes is lax.

In California such discussion has led to increasing emphasis on the necessity for lateral reinforcement in structures of all types, including brick. Another good effect has been the gradual disappearance of adobe; this material is too weak and dangerous for use in any but the most carefully designed and supervised construction. The authors of the 1952 report remark that all adobe structures in the meizoseismal area were seriously damaged or destroyed; practically all those left standing were torn down afterwards. Many of these were at Tehachapi. At Grapevine, adobe cabins for the accommodation of motorists collapsed.

When the materials are not unusually weak, failure of masonry in earthquakes is due to imperfection in the mortar, in workmanship, or in design. (Omitting reinforcement is a flaw in design.)

Defective Mortar. In California mortar has often been so bad that after several destructive earthquakes it has been profitable to collect bricks from damaged buildings, wash off the remains of mortar with a hose, and sell secondhand bricks practically as good as new. At present, even when good mortar is used, it is commonly applied so rapidly that the full bonding effect is not reached. The bond between bricks does not approach in strength that of the bricks themselves, so that when masonry is cracked during an earthquake the cracks take a zigzag course between the bricks. Such cracks in walls are often diagonal, crossing in the form of an X (Fig. 8-1); these are tension cracks due to shearing by lateral motion.

FIGURE 8-1 *Diagonal shear cracks in masonry. [Diagram based on photographs.]*

Defective Workmanship. Inspection of damaged buildings often shows that nothing was done to anchor brick facings to the rest of the structure, or that bricks were not laid so as to tie together. The Santa Barbara earthquake of 1925 called public attention to these defects. Many unsecured facings fell out; in some buildings there was not even the normal tie at the corners by crossing bricks in alter-

nate courses. These are deficiencies of workmanship rather than of design; often such omissions are perpetrated in defiance of plans and specifications.†

Defective Design. Failure to anchor firewalls and parapets is due to faulty design; so is the construction of projecting cornices, which are usually unnecessary as well as unsafe. The 1952 report has a section on damage by exposure, which includes damage of adjacent structures by fall of unsupported parapets. There were numerous instances of this kind at Bakersfield; in the July 21 shock parapets as far away as Santa Barbara and Pasadena fell. Parapets are intended as a fire precaution; but, as the authors of the report remark, "parapets can often be strengthened at a nominal cost, and in a few instances can be removed without endangering the appearance, fire insurance rate, or safety of firemen." Official action to require such procedure has been taken (and enforced) in Los Angeles and some other California cities, but only abortively in San Francisco. Exposure also includes the effect of pounding between adjoining structures without sufficient intervening free space.

Unreinforced Masonry—1952. The major fault of design in masonry is lack of reinforcement. An unreinforced brick wall, unless abnormally thick, will not withstand a strong horizontal thrust at right angles to the wall. If it is a bearing wall supporting a considerable part of the load of a structure, its failure may bring about total collapse. Unreinforced non-bearing walls, acting as partitions or fillers, may be badly damaged and may be costly to repair. Internal reinforcing of the wall increases its resistance; but the soundest design also uses lateral bracing extending through the structure.

In 1952 most of the obvious damage was to structures of one or two stories with unreinforced brick bearing walls and wood interiors, as at Arvin and Tehachapi on July 21 and at Bakersfield on August 22. In several instances the supposedly non-structural wood partitions held up the structure after the exterior brick fell out; at Tehachapi such a case on the main thoroughfare afforded photographers opportunity for the typical California earthquake picture: hotel bedroom furniture exposed to public view (Fig. 8-2).

The behavior of buildings with steel frames and unreinforced brick bearing walls was similar. The report details damage to the Kern County General Hospital at Bakersfield. Three older wings, constructed 1924–1929, had not been designed to resist earthquakes; they had interior steel frames but unreinforced brick bearing walls with poor mortar. The damage on July 21 was serious, more so than in Bakersfield generally on that date. These wings

---

† After the 1933 earthquake an engineer, noticing an unsecured gable which had fallen out flat on the ground, was at first puzzled to see what were apparently the brick ends of a correct tie-in. On closer inspection he found that the thrifty but dishonest contractor had used half-bricks to cheaply counterfeit the appearance of a tie-in, which undoubtedly had been specified.

FIGURE 8-2   *Juanita Hotel, Tehachapi, California, July 21, 1952. [Photo by F. E. Lehner.]*

were then braced with outside steel columns and steel tie rods throughout. They survived the shock on August 22 but were later torn down. The new wing, built in 1938 of reinforced concrete, had only negligible damage.

Reinforced Masonry—1952.   Examples of structures with reinforced brick walls are taken from Arvin, where the central business district consisted largely of buildings nearly as old as those at Tehachapi, of the same type, and showing similar damage. The Arvin high school had a large group of buildings constructed in 1949–1951 under the earthquake resistance provisions of the Field Act. Damage amounted to about 1 per cent; the most conspicuous item, a cracked wall, was traced to faulty workmanship. Also at Arvin was a recently constructed brick supermarket, reinforced and designed to be resistant; damage was negligible.

Other Materials—1952.   The report also refers to masonry using materials other than brick. The weakness of adobe has already been noted. Hollow clay tile was found in use chiefly for partitions; even there it performed badly and appeared undesirable as a structural material. Many examples of hollow concrete-block construction were inspected; the performance is in general similar to that of brick masonry, and similar damage occurs when there is no reinforcement. Most of the structures were reinforced, because concrete block came into use since it became customary to reinforce masonry; this has tended to create a prejudice in favor of hollow concrete over brick, by comparison with old and unreinforced brick structures. A badly constructed church school of hollow concrete at San Bernardino, over 110 miles from the July 21 epicenter, is reported to have had 10 per cent damage.

Steel Frame

Unless there were evident faults of design or workmanship, steel-frame structures have suffered relatively little damage in most earthquakes where their performance could be studied. Where steel frame is used to support masonry, the resistance depends largely on the latter. (Note the account of the Kern County Hospital above.)

In the area of the 1952 earthquakes, most of the all-steel structures were small, such as gasoline service stations; these were undamaged. On July 21 some larger steel structures suffered minor damage at Maricopa Seed Farms in the valley southwest of Bakersfield; this was very near the instrumentally determined epicenter. Similar minor damage occurred to the same structures in the strong aftershock on January 12, 1954, which had almost the same epicenter.

Concrete—Reinforced and Unreinforced

Deficiencies—Construction Joints. Most concrete construction now includes steel reinforcing, at least nominally. To judge by performance, this is potentially the strongest and most earthquake-resistant type of construction, but even such structures may be badly designed or badly executed. What passes for concrete may be inferior material carelessly poured together. To

FIGURE 8-3 *Cummings Valley School. Inferior "reinforced" concrete, wrecked by the earthquake of July 21, 1952. [Photo by Edwin A. Verner, civil and structural engineer, San Francisco.]*

quote the 1952 report: "Perhaps the most noticeable workmanship deficiency . . . was poor construction joints. Concrete is poured in sections . . . The junctures between these pours is called the construction joint." To ensure proper bonding, these joints should be kept clean during construction; but this is often overlooked. "In every major instance where reinforced concrete structures were damaged, movement at construction joints was also noted."

Nominal reinforcing is illustrated by the Cummings Valley School. The report calls this "a classic in poor design, poor material and poor workmanship." The concrete was weak, and the reinforcing bars did not lap at the construction joints; consequently during the earthquake the building separated into blocks and collapsed (Fig. 8-3).

Reinforced Concrete—1952. Not far from Cummings Valley was the State Prison for Women. The frames of the several buildings were of reinforced concrete, which withstood the earthquake well. Numerous hollow clay tile partitions fractured badly. Grotesque was the fate of several false ornamental chimneys, some of which were thrown to the ground while others were overturned on the roof.

A large department store at Bakersfield, with negligible injury on July 21, was badly damaged on August 22; rehabilitation was a long process, costing 40 per cent of the value of the building. This was the only monolithic reinforced-concrete structure in Bakersfield seriously damaged in these shocks; there was evidence of somewhat inferior material and workmanship, and of imperfect design which allowed internal torsional strains to develop.

A conspicuous success was a two-story structure with reinforced-concrete walls at Tehachapi, a former lodge building in use as a youth center; it was undamaged and provided shelter for the injured and homeless (Fig. 8-4).

Non-Structural Damage

Plaster. The behavior of plaster is often misjudged. Plaster of the best quality, bonded with hair or fibrous material, cracks significantly only under strong shaking and usually does not fall until there is other structural damage. Plaster of mediocre quality without bonding soon develops conspicuous cracks due to shrinkage and will fall in quantity in a shaking far below the level of other damage. Moreover, wood-frame structures are often constructed of unseasoned timbers which shrink or expand in dry or wet weather; the resulting strains in the walls produce plaster cracks which open and close regularly with the seasons but are often first noticed and reported after an earthquake.

The 1952 report remarks: "Plastered ceilings represent a life hazard not often considered. Such a ceiling may weigh up to 8 pounds per square foot.

FIGURE 8-4   *Undamaged concrete structure, Tehachapi, California, July 21, 1952.* [*Photo by F. E. Lehner.*]

For a small room, say 15 feet square, the ceiling will weigh close to one ton."

Windows, Lighting Fixtures, Sprinklers.   Windows, especially large plate-glass ones, are broken during earthquakes much more easily than might be supposed. The breaking is not due to a sudden blow but to distortion of the frame, which puts a strain on the glass that will develop a fracture along any weakness.

A freakish but costly result of the July 21 earthquake was the destruction of "literally miles" of fluorescent fixtures in the newly occupied Prudential Insurance Building in Los Angeles; the supports were broken and the tubes dropped to the floor. The swinging of chandeliers, and the detachment from their suspensions or supports of inverted bowls for indirect lighting, have sometimes caused extensive minor damage.

Experience in the Long Beach earthquake of 1933 showed the need for bracing interior sprinkler pipes against shaking, to preserve fire protection and prevent unnecessary flooding. Such measures had been adopted generally by 1952, and there were no significant failures of sprinkler systems.

Lack of Coherent Construction

Serious damage often results when a complex structure is not designed or connected so as to behave as a mechanical unit. This is very common, and difficult to forestall, when an annex is constructed in contact with a building.

Frequently earthquake motion affects the two parts differently, so that strains develop at the junction. If the separate units are individually well consolidated, such strains may act to distort both, subjecting them to effects which would have not occurred if they had been separated, or had been so connected as to respond coherently.

The damage to the East Whittier School, described in Chapter 4, was in part due to lack of coherence in construction. More striking examples were provided by large structures destroyed at Santa Barbara in 1925, such as the Arlington Hotel and the San Marcos Building.

Tall Structures at a Distance

The characteristic difference between the effects of long-period and short-period motion, to which we shall return in the discussion of intensity scales (Chapter 11), is clearly set forth in the 1952 report:

Los Angeles. Damage in Los Angeles as a result of the July 21, 1952 shock was generally confined to fire resistive structures over 5 or 6 stories high. A few isolated instances of minor damage to one- and two-story non-fire resistive buildings were noted, but they are not significant.

This pattern of damage is opposite to that which was experienced in Kern County on July 21st and in Bakersfield on August 22, 1952, in that the one- and two-story brick bearing wall buildings were most affected as compared to the taller fire resistive type such as the Hotel Padre and the Haberfelde buildings. One explanation for this difference is that the earth motion in the Los Angeles area was generally of longer periods, which adversely affect taller buildings with correspondingly longer natural periods. In other words, the motion some 70–80 miles from the epicenter was such as to excite vibrations of crack-producing magnitudes in tall structures while not affecting the lower more rigid buildings.

Another contributing factor is the previous damage to these tall buildings in past shocks, particularly the Long Beach shock of 1933. It is known that effective repairs were generally not made after these shocks or even after the July 21, 1952 shock for that matter.

No cases of structural damage were noted, and principal damage was to partitions, masonry filler walls, ceilings, marble trim, veneer, and exterior facing. Considering the relative value of these items as compared to the structural frame and floors, it can be seen and has been proven in past shocks that non-structural damage can amount to 50% of the value of the building.

Long Beach. Behavior of tall buildings in Long Beach was similar to that in Los Angeles. However, it is disquieting to note rather extensive damage to major structures in some cases when one considers that they were located some 100 miles south of the epicenter. Again damage was confined to partitions, unreinforced masonry panel walls, and other non-structural items.

In the 1933 shock these buildings, in general, suffered more extensive damage than those in Los Angeles, and the methods of repair were often equally ineffective. The Pacific Fire Rating Bureau's files contain damage reports on the taller buildings which are practically identical for 1933 and 1952, and there is no reason to believe that a future shock would produce any different results.

Santa Barbara.   The damage pattern in Santa Barbara was similar . . . except that somewhat more damage was suffered by several one- and two-story masonry structures. Three taller buildings suffered varying degrees of damage . . . and again this could be attributed to previous poorly repaired earthquake damage in 1925, 1926, and 1941 . . .

Public Schools

In its effects on public schools, the 1952 earthquake provided an excellent controlled scientific test of the merits of proper building regulation. The California law known as the Field Act, passed as a consequence of the appalling damage to schools in the 1933 earthquake, brings all new public school construction under standards set by Title 21, California Administrative Code. Because the law is not retroactive, many school buildings constructed prior to 1933 are still in use although they do not conform to the provisions of the code. In 1952, as on some former occasions, many of these older structures were seriously damaged, and some rendered unfit for use, while those conforming to the code showed little or no damage. The Vineland School had an original building which was seriously damaged on the pattern of the Long Beach schools in 1933, and a new annex which was not damaged at all.

Other Damage

Rending by Faulting.   Relative displacement of the two sides of a fault involves forces which are practically irresistible so far as man-made structures are concerned. Comparatively few buildings have been so unfortunately located as to be damaged in this way. In the 1906 California earthquake a wood-frame house at Wright's in the Santa Cruz Mountains was torn in two, but the parts remained standing. The barn at the Skinner ranch, which was pulled off one corner of its foundations and shifted over 15 feet, remained intact. In the 1940 earthquake an adobe structure at Cocopah (Lower California) astride the fault trace was demolished.

Railway Lines.   Railway lines have been badly damaged in many earthquakes. Change of level in soft ground, usually due to slumping and subsidence, can put a line out of service. In similar situations rails are often badly bent; Oldham reports and figures such effects for the 1897 earthquake. The breaking of a railroad in Baluchistan in 1892 was definitely due to strike-slip faulting; the same thing happened at Cocopah in 1940. Bridges, whether on railways or highways, are often seriously damaged by lateral shaking; but, if a bridge crosses a fault line where there is displacement, it may fail completely.

Tunnels—1952.   The 1952 report contains details of damage to tunnels on the route of the Southern Pacific Railroad (here also used by the Santa Fe Railroad) west of Tehachapi. The track winds up a steep grade with 15 tunnels, crossing the zone of the White Wolf fault twice. Four tunnels were located in these crossings; all were greatly damaged, and to restore service large-scale excavation and reconstruction were required. One tunnel was dug out, or "daylighted," completely, another for 200 feet of a total of 700 feet. A third tunnel was bypassed by an open cut. The fourth tunnel was at first bypassed by a temporary track, so that through traffic interrupted on July 21 could be resumed on August 16; reconstruction of this tunnel was completed on December 16. All this damage was due to thrusting on the White Wolf fault, which broke and offset the tunnel structures. The tunnels were shortened by the displacement; rails in the tunnels and at the entrances were sharply bent.

Irrigation Systems.   Damage to irrigation systems may be due to slumping, or to the emergence of ground water and sand, or to the direct breaking and offset of canals and ditches by faulting. These effects were well illustrated in the Imperial Valley earthquake of 1940 (Chapters 9 and 28). The Kern County earthquake of July 21, 1952, affected the water system indirectly by throwing down 846 transformers from poles, thus cutting off electric power from pumps used in irrigation. Emergency action restored service in time to prevent much loss of crops.

Dams.   Large earthquakes occasionally result in the failure of dams. In the comparatively moderate shaking of the 1925 earthquake at Santa Barbara, an earth fill dam retaining the Sheffield reservoir gave way. Probably this was not different from other effects of intensity VIII in soft material; however, it suggested the possibility of the water being forced against a dam by earthquake motion with the effect of a soft-nosed hammer. Calculation for large dams indicated a serious danger of the kind under horizontal acceleration of the order of $0.1g$; this has since been taken into account in design.†

In the 1906 California earthquake the principal dams in the fault zone did not fail. At the San Andreas reservoir the fault trace passed through a knoll forming a common abutment between two embankments. The Crystal Springs dam, a first-class concrete structure, was located completely off the rift line, although it was less than 300 yards away; it was undamaged. Minor dams

---

† Mr. Engle notes that the Sheffield reservoir was retained by a hydraulic fill dam; water often remains in such a fill in plastic lenses. For this and other reasons, hydraulic fill dams are poor earthquake risks.

and auxiliary structures were sheared by faulting. The pipeline carrying water to San Francisco followed the fault, crossing the trace repeatedly; large pipes were completely ruined by rending or compression.

Underground Pipes.   Even in small earthquakes, underground pipes are put out of service by slumping and subsidence in soft ground; the initial failure need only be a small crack. Old and weakened pipes are commonly reported as out of service after earthquakes in which the local shaking was hardly perceptible to persons; usually it is impossible to decide whether the failure occurred at the time of the shock.

References

Observed earthquake effects

Freeman, J. R., *Earthquake Damage and Earthquake Insurance,* McGraw-Hill, New York, 1932. (The largest readily available collection of data on earthquake damage, with many details for San Francisco and Tokyo; for the former, some of Freeman's conclusions were based on incomplete data; see Chapter 28.)

Suyehiro, K., "Engineering seismology; notes on American lectures, *Proc. Am. Soc. Civil Engrs.,* vol. 28, No. 4 (1932), pp. 1–110. (Valuable general discussion, with many specific illustrations from Japan.)

Steinbrugge, K. V., and Moran, D. F., "An engineering study of the Southern California earthquake of July 21, 1952, and its aftershocks," *B.S.S.A.,* vol. 44 (1954), pp. 199–462. (An abbreviated version of this report appears in: *Calif. Dept. Nat. Resources, Div. Mines, Bull.* 171, "Earthquakes in Kern County, California, during 1952," pp. 249–270, with several reports on damage by other authors.)

Derleth, Charles, Jr., "The destructive extent of the California earthquake. Its effect upon structures and structural materials within the earthquake belt," in: Jordan, D. S., ed., *The California Earthquake of 1906,* Robertson, San Francisco, 1907, pp. 81–212.

Dewell, H. D., *et al.* "The Santa Barbara earthquake," *B.S.S.A.,* vol. 15, No. 4 (1925), pp. 251–333.

Kirkbride, W. H., "The earthquake at Santa Barbara, June 29, 1925, as it affected the railroad of the Southern Pacific Company," *ibid.,* vol. 27 (1927), pp. 1–7, pl. 1–7.

Chick, A. C., "The Long Beach earthquake of March 10, 1933 and its effect on industrial structures," *Trans. Am. Geophys. Union* (1933), pp. 273–284.

Engle, H. M., "The Montana earthquakes of October, 1935; structural lessons," *B.S.S.A.,* vol. 26 (1936), pp. 99–109.

The Great Earthquake of 1923 in Japan, Bureau of Social Affairs, Tokyo, 1926.

Eckart, N. A., "Development of San Francisco's water supply to care for emergencies," *B.S.S.A.,* vol. 27 (1937), pp. 185–204. (Contains many details of damage to the water system in 1906.)

Louderback, G. D., "Characteristics of active faults in the central Coast Ranges of California, with application to the safety of dams," *ibid.,* pp. 1–27.

These references are representative, and bear directly on discussion in this chapter. Data on earthquake damage appear in nearly every monograph on a large shock, such as the papers listed in the chronological bibliography, Appendix XVI. Much valuable source material appeared in the U. S. Coast and Geodetic Survey annual, *United States Earthquakes.*

Inertial effect on dams

Morris, S. B., and Pearce, C. E., "Earthquake forces on dams," *B.S.S.A.,* vol. 21 (1931), pp. 204–215.

Nature of damaging motion

Benioff, H., "The physical evaluation of seismic destructiveness," *ibid.,* vol. 24 (1934), pp. 398–403.

Neumann, Frank, *Earthquake Intensity and Related Ground Motion,* University of Washington Press, Seattle, 1954.

Housner, G. W., "Characteristics of strong-motion earthquakes," *B.S.S.A.,* vol. 37 (1947), pp. 19–31.

——, "Properties of strong ground motion earthquakes," *ibid.,* vol. 45 (1955), pp. 197–218.

Housner, G. W., Martel, R. R., and Alford, J. L., "Spectrum analysis of strong-motion earthquakes," *ibid.,* vol. 43 (1953), pp. 97–119.

Consult also the references for Chapters 11 and 24, and Appendix II.

CHAPTER 9

Effects on Ground and Surface Water

LISBON EARTHQUAKE—1755

The Lisbon earthquake of November 1, 1755, was one of the greatest seismic events of which we have scientific description. In western Europe such an earthquake catastrophe is rare; this one aroused tremendous popular and scientific interest, and many reports have come down to us. The student may read and compare the summary accounts by Davison and by Reid. The former is handy but relatively uncritical; the latter is a careful study, making full use of sound critical methods and of physical principles. The new book by Kendrick can be recommended.

The Lisbon earthquake is frequently mentioned in general literature. Rousseau seized upon it as the text for a pamphlet against the artificialities of civilization: if we lived out-of-doors, earthquakes would not kill us. This was too much for Voltaire, who introduced the Lisbon earthquake as an episode of his brilliant satirical tale, *Candide,* in which Rousseau's ideas were handled rather mercilessly.

By coincidence, on November 18, 1755, one of the strongest earthquakes ever felt in New England took place; chimneys fell, notably at Salem. The news of the Lisbon earthquake did not reach America until mid-December, and the two events were often popularly connected and confused. Certainly it is the Lisbon earthquake, with its correct date, to which Oliver Wendell Holmes refers in *The Deacon's Masterpiece, or the Wonderful "One-Hoss Shay."*

November 1 is All Saints' Day (this is why Halloween is October 31). In Lisbon, as in other Catholic cities, most of the population was in the churches when those medieval structures collapsed. Loss of life in Lisbon alone is commonly stated as at least 60,000; this presumably includes those who perished by fire or were drowned in the great sea waves.

The geographical extent of macroseismic effects was vast (Fig. 9-1). Destructive intensity affected all of Portugal and most of Spain. Violent shaking occurred in Morocco, and at Algiers there was so much damage that—even allowing for masonry of the weakest sort—most investigators believe there must have been a separate earthquake about the same time. Intensity II, actual shaking of the ground felt by persons, applied as far away as northwest Germany. The evidence has to be examined closely, because the marginal effects classed under intensity I were developed to

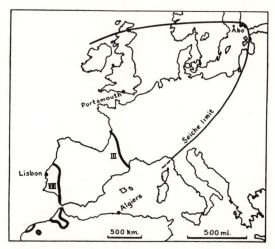

FIGURE 9-1 *Area affected by the Lisbon earth-quake, 1755, in Europe and Africa.*

almost unparalleled extent; chandeliers, loose objects, and water were set in motion far beyond the limits of direct perceptibility.

In the absence of seismograms, magnitude can be assigned only roughly, but conclusions can be drawn by comparing the areas affected with those for earthquakes of known magnitude. Taking an epicenter rather arbitrarily about 100 kilometers (or 60 miles) west of Lisbon, we find for the mean radius of the outer limit of damage 600 kilometers; for that of the outer limit of perceptibility, 2000 kilometers; for oscillations in surface water (seiches), 3500 kilometers. Oldham, comparing the 1897 Indian earthquake with this one, concluded that the effects of the two were about equally extensive. The magnitude of the Lisbon shock could scarcely have been less than 8½ and may have approached 8¾.

The effects of this earthquake on ground and surface water, including the ocean, were spectacular. They will be described in the proper places in the general discussion which follows.

PRIMARY EFFECTS ON WATERS

Primary effects have to do mainly with the diversion of drainage, on the surface or underground, by the displacement of crustal blocks. Instances have been given in preceding chapters. Unless the initiation of a seismic sea wave is a primary effect (discussion later in this chapter), the Lisbon earthquake presents no known effects under this heading.

OSCILLATIONS IN WELLS

Many wells are now equipped with floats, connected with recorders which give continuous graphs of the water level. These recorders sometimes show oscillations of several feet during strong earthquakes, even during the passage of the large long-period surface waves of a teleseism. Oscillations up to about 1 inch were reported to have been caused in wells near Lodi in the San Joaquin Valley, California, by an earthquake which caused some damage at Bishop on the other side of the Sierra Nevada on September 17, 1927 (no one felt this shock at Lodi). On June 3, 1932, a great earthquake (magnitude 8.1) took place near Colima, Mexico. Seismographs at Pasadena recorded very large amplitudes; some instruments were put out of action. These large slow waves were not felt, nor were there any reports of the marginal effects of intensity I; but there were large oscillations in several wells in the adjacent San Gabriel Valley.

Reports of this kind became more frequent about 1932, and several papers dealing with the effect were published. Since 1943 the U. S. Coast and Geodetic Survey has been systematically publishing such data.

In 1935 Blanchard and Byerly attached a seismograph recorder to a float in a well near Lodi. A number of earthquakes were recorded, and the seismograms were compared with those written by the regular instruments at Berkeley. The records were similar, except that, as was expected, the well level recorder responded less to S waves and to surface waves involving shear.

The theory that most easily accounts for such observations holds that when a passing seismic wave compresses the aquifer it squeezes additional water into the well, raising its level; the opposite half of the wave dilates the aquifer and draws water out of the well. A similar response would be expected from a pressure gauge attached to a buried tank.

EARTHQUAKE FOUNTAINS

Where there is plenty of ground water, a strong earthquake often produces fountains, spouts, "geysers," which play during the strong shaking and for some time afterward.

An Eyewitness Description

The following account is from the Bihar-Nepal earthquake of 1934. A witness writes that, while driving:

. . . my car suddenly began to rock in a most dangerous fashion . . . Owing to the sound of the engine I noticed no noise, but was told such was heard from the west, a deep, terrifying rumble. As the rocking ceased, mud huts in the village, on either side of the road, began to fall. To my right a lone dried palm trunk without a top was vigorously shaken, as an irate man might shake his stick, then water spouts, hundreds of them throwing up water and sand were to be observed on the whole face of the country, the sand forming miniature volcanoes, whilst the water spouted out of the craters; some of the spouts were quite six feet high.

In a few minutes, on both sides of the road as far as the eye could see, was vast expanse of sand and water, water and sand. The road spouted water, and wide openings were to be seen across it ahead of me, then under me, and my car sank, while the water and sand bubbled, and spat, and sucked, till my axles were covered. "Abandon ship" was quickly obeyed, and my man and I stepped into knee deep hot water and sand and made for shore. It was a particularly cold afternoon, and to step into water of such temperature was surprising.

It was distressing to see the villagers, running some east some west, others to, others from their fallen houses, wailing and beating their chests.

In less than half an hour I should say, the water spouts ceased to play, though water oozed out of the land and trickled from the mouth of the lesser sand heaps.†

This occurred in the heavily alluviated Ganges Plain; there are many reports of such effects during large Indian earthquakes.

Heights and Duration

Some observers in 1934 in India reported heights of 30 feet; the authors of the memoir believe this is exaggerated, and accept heights of 6 to 8 feet as more probable. In many accounts it appears that the water did not begin to be ejected until after the strong shaking was over; in some places it continued to emerge for at least 3 hours. Spouts were seen during the California earthquake of 1906; some reports describe them as reaching heights of 20 feet, which also seems excessive.

Pulsation

There are at least two reported instances in which the ejection of water pulsated. One refers to the Nevada earthquake on October 2, 1915: "with every lurch of the ground the spring, which ordinarily has but a feeble flow, spurted water into the air to a height of two or three feet." (Jones.) The other is reported by Imamura for the great earthquake of 1923, at the town of Hōjō on the Bōsō peninsula, in the grounds of a primary school where there had formerly been a marshy paddy field:

---

† From "The Bihar-Nepal earthquake of 1934," *Mem. Geol. Survey India,* vol. 73 (1939), pp. 33–34.

The earthquake having wrecked the school building, the teachers who had fled into the open witnessed the interesting phenomenon . . . Two fissures, each nearly 40 metres long, traversed the middle of the former paddy field . . . spurting muddy water to a height of about 10 ft. Geyser-like it would stop and then resume action a few seconds later. This was repeated some six times . . . the fissures as well as a few sand cones remained intact for more than a month.

Ejection of Sand

The Imperial Valley earthquake of 1940, centering under the delta of the Colorado River, produced many large spouts. These were very damaging to the Yuma Valley irrigation district in the southwestern corner of Arizona. The spouts brought up great quantities of sand which covered fields and choked canals and ditches, necessitating repairs and reconstruction.

Sand brought up in this way is deposited around the spout in the form of a miniature volcanic crater, which remains as a "permanent" effect. The crater may be small or large, depending on the local intensity and on the availability of ground water. In 1940 such craters 6 or 8 inches in diameter were found in a channel west of the town of Brawley. South of the Mexican boundary larger ones were formed; the writer saw a crater which measured about 4 feet in diameter and 18 inches high at the rim.

Vertical Drag

The emerging water and sand may drag up posts or other light material. Near Tokyo the earthquake of 1923 brought up posts in a river flat; they had never been seen before by peasants farming the area for rice, and old records showed that they belonged to a bridge built in 1182. In the Yuma Valley in 1940 a light bridge crossing a canal had one end raised about 3 feet by the emergence of posts; a flume crossing another canal had its back broken by raising of posts in the middle.

Causes of Fountains

There is no single explanation for all occurrences of ejection of water and sand. The passage of large seismic waves, which produces the oscillation of well levels, might be expected sometimes to bring water out at the surface, but scarcely so violently or continuously. Few earthquake fountains have been reported to pulsate or flow intermittently; rather there is a continuous flow which gradually falls off. Earthquake fountains might be produced in the same way as artesian wells, which exist in many areas where the fountains have been observed. Shaking could break up local resistance in the aquifer, overload the substratum with water, and build up pressure which might lead to ejection at the surface. Lawson attributed the fountains in

1906 to settling and compaction of the ground due to shaking, which would also create a subsurface pressure. Such explanations are in harmony with the increased flow in streams, as in Kern County in 1952.

The field investigator, if he is lucky enough to see earthquake fountains in action, or if he is collecting reports from good witnesses, should make sure of the time, in relation to strong shaking, of the beginning and end of flow; the pulsating or continuous nature of emission; the height reached in ejection; and artesian water conditions in the vicinity.

SEICHES

The word *seiche* comes from Switzerland, whence Forel introduced it to general use. A typical seiche is a standing wave set up on the surface of an enclosed body of water such as a lake, pond, or tank.

Mechanism of Seiches

Figure 9-2 represents a vertical section of a square tank in which water has been made to "slosh" from side to side. Two positions of the water surface are shown by a full and a dotted line. If this motion is maintained it is a standing wave, with antinodes at the side and a node in the middle—a *uninodal* seiche. The time for one complete oscillation, the period T of the seiche, is determined by gravity and the width and depth of the tank:

$$T = 2L/(g \mathfrak{z} c)^{1/2} \tag{1}$$

Seiches may be multinodal, just as standing waves on a string or in a pipe may have several nodes. They occur not only in closed bodies of water but also in partially closed bodies such as harbors or channels,† and as lateral oscillations in rivers, canals, or ditches. It is only necessary that the geometry of the water boundary define a natural period of oscillation. Where there are currents, or where the enclosure is incomplete, part of a seiche may be transformed into a progressing wave. Earthquakes are a comparatively rare cause of seiches, which are usually set up by winds, currents, or tides. A seiche in a harbor may be started by the arrival of a tsunami (see the following section).

FIGURE 9-2 *Schematic section of seiche in a tank.*

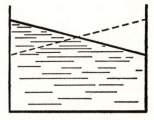

Long-Period Seiches

The classical studies of seiches were made about 1890 by F. A. Forel (of the Rossi-Forel

---

† In a square enclosure open to the ocean on one side there may be a uninodal seiche with the node across the opening and period double that given by equation (1).

intensity scale) in connection with a monograph on the Lake of Geneva. There the fundamental seiche period is about 72 minutes. Seiches in other lakes, many of them with similarly long periods, have been studied by hydrologists. Certain seiches on the coast of southern California are of public importance; they arise in the channels between the mainland and the islands of Santa Catalina and San Clemente. They appear to be set up by long-period waves originating in storm centers far out in the Pacific; often they produce high and destructive surf along the beaches, even during otherwise calm weather.

River Seiches

Seiches are responsible for river water being splashed out over the banks during earthquakes (if extensive, this indicates intensity X). The report on the 1934 Indian earthquake includes a hair-raising and probably exaggerated eyewitness account of such a seiche in the Ganges, which temporarily drew the water completely off part of the river bed.

Lisbon Earthquake Seiches

The Lisbon earthquake set up seiches over much of western Europe; the most distant ones reported were in Scandinavia and Finland. (Reports that the American Great Lakes were disturbed are in error.) Some of the best descriptions, referring to English harbors and ponds, were published in the *Proceedings of the Royal Society*. The following is part of the account of the seiche at Portsmouth:

In the north dock, whose length is about 229 feet, breadth 74 feet, and at that time about 16½ feet of water, shut in by a pair of strong gates, well secured, his Majesty's ship the Gosport, of forty guns, was just let in to be dock'd and well stay'd by guys and hawsers (certain large ropes, so called). On a sudden the ship ran backwards near three feet, and then forwards as much, and at the same time she alternately pitch'd with her stern and head to the depth of near three feet; and, by the liberation of the water, the gates alternately opened and shut, receding from one another near four inches . . .

The account describes similar disturbance to other vessels. Using $T = 2L/(g\mathfrak{Z})^{1/2}$ with the given length and depth, Reid finds $T = 19\frac{1}{2}$ seconds. This is almost too good, since the prevailing period of the large surface waves of teleseisms is usually close to 20 seconds. The Portsmouth seiche was uninodal; but in some other localities more complex oscillation was set up. Reid discusses the motion in the rectangular moat at Shirburn Castle, where the water rose and fell simultaneously at two opposite corners, and was undisturbed at the other two corners, which were nodes.

Kvale has reported in detail on seiches in Norway and England caused by the great Assam earthquake of 1950. He also refers to those of the Lisbon earthquake but repeats much doubtful material from Davison without noting the critical comments by Reid. Kvale finds seiches especially well developed in directions radial from the epicenters, which indicates resonance to Rayleigh waves (Chapter 16).

TSUNAMIS

Nomenclature

Many populated coasts, notably those of Chile, Peru, Japan, and Hawaii, have been swept again and again by great waves surging as high as 60 feet above normal sea level. These are commonly called "tidal waves," although they have nothing to do with the tides. Most of them follow strong earthquakes and may properly be called *seismic sea waves;* but waves of the same character have been started by violent volcanic explosions such as that of Krakatoa in 1883, and others have accompanied great storms or typhoons. In South America the name is *maremoto,* neatly paired with *terremoto;* but the term which has come into international use is the Japanese *tsunami* (sometimes spelled tunami by Japanese authors).

This word is another local term like seiche; its exact origin is uncertain, although it may originally have meant a large wave entering a bay or inlet. The Japanese coast has many deep inlets which narrow rapidly; and in these the tsunami surges to extreme heights.

Tsunamis are not completely understood. Some writers have oversimplified the matter; this is understandable because there is no one good work of reference on the subject, and observational material is scattered through many books and periodicals.†

Places of Origin

At least two great earthquakes with epicenters on land have been followed by tsunamis on the nearest coast (Chile, 1922; Turkey, 1939). A tsunami also followed the great Mexican earthquake of June 3, 1932, for which the best available epicenter is inland.

Most large tsunamis, especially in the Pacific, originate at one or another of the great submarine troughs or trenches, such as the Atacama Deep off Chile, the Tuscarora Deep (Nippon Trench) off Japan, the Mindanao Trench, and the Aleutian Trench. The corresponding earthquakes usually have their epicenters near the trench, on its landward side. The tsunami may reach the nearest coast 15 minutes to some hours after the earthquake. At

---

† The author has not seen the book by A. E. Svyatlovsky, *Tsunami,* Akademiya Nauk, Moscow, 1955.

Lisbon the great wave came in just after the second strong shock, about 20 minutes after the first.

Speed of Travel

Reliable observations of time, including records written by automatic tide gauges (mareograms), show that tsunami waves travel approximately with the velocity calculated for gravity-controlled waves on shallow water:

$$C = (gz)^{1/2} \tag{2}$$

(For small ripples with wave lengths of a few centimeters the principal controlling force is surface tension, not gravity, and the theory is different.) "Shallow" here means that the ocean depth z is smaller than the wave length. The period of oscillation in these waves is of the order of an hour, and their wave lengths accordingly are measurable in hundreds of miles. In crossing the Pacific C ranges up to 300 meters per second (more than 600 miles per hour) over the deepest waters.

Height

The crest-to-trough height of these waves in the open ocean is no more than a few feet. This and their great wave lengths render them imperceptible by ordinary means; consequently the large waves sometimes suddenly encountered by vessels in the open sea must be explained otherwise. Vessels in Sunda Straits in 1883 were not seriously disturbed by the wave started by Krakatoa; and in 1896 a Japanese fishing fleet easily rode the waves of a tsunami which surged to destructive heights on the coast of Honshu. This increase in height is only partly due to the confining effect of inlets; the main cause appears to be the decrease in velocity with depth. With decreasing speed, the kinetic energy in the wave is maintained by increasing amplitude. Physically this effect is the same as that suggested in another chapter for the increase in earthquake intensity where seismic waves emerge from basement rock into unconsolidated material.

Approaching the Hawaiian Islands, the water becomes shallower very suddenly. Tsunamis from almost all directions—from Japan, from the Aleutian Trench, from South America—pile up here and constitute a serious danger. So much damage was caused by a wave from the Aleutian Trench in 1946 that a warning service under the auspices of the U. S. Coast and Geodetic Survey has been established with headquarters at Honolulu. A 24-hour watch is maintained with the aid of an alarm system triggered by a seismograph. With cooperation from the armed services and other government agencies, facilities are provided for rapid communication with seismological observatories and tide-gauge stations to forewarn Hawaii and other coasts of the Pacific against destructive waves. At least 4 hours can be counted on between the occurrence of an earthquake and the arrival of the following tsunami at Hawaii. Great tsunamis have caused damage after crossing the entire Pacific; between Japan and South America this takes almost a full day.

Withdrawal of the Sea

Arrival of a tsunami at the near coast is often preceded by a withdrawal of the sea, sometimes well below normal low tide. If the bottom descends gradually, this may lay it bare far out from the coast. On the South American and Japanese coasts such withdrawal is a known danger signal. In Japan the people take to high ground after any strong earthquake, whether the water retires or not, because the withdrawal is not invariable; the tsunami may arrive without it, or there may be withdrawal on some parts of the coast and not on others.

California Tsunamis

Minor tsunamis have originated off many coasts usually unaffected by major ones. Tsunamis on the California coast are rare, but on December 21, 1812, an earthquake which damaged several of the missions in southern California was followed by a wave that entered Refugio harbor, west of Santa Barbara; it is believed to have risen there to 30 feet or over, and possibly to 50 feet at Gaviota about 20 miles west. On November 4, 1927, a shock of magnitude 7.4 sent in a wave 5 to 7 feet high along the coast north of Point Arguello.

Tsunami disturbance in a harbor may be prolonged by seiches.

Important Tsunamis

Really great tsunamis are uncommon.† Of those which are well known, only one, that of the Lisbon earthquake, occurred in the Atlantic Ocean. All the others were in the Pacific area. However, large and locally destructive tsunamis are known from nearly all the seas. Many have occurred in the West Indies; one tsunami in 1945 affected the northern Arabian Sea; and they have been reported almost everywhere in the East Indies. The extraordinary wave of April 2, 1868, which originated off the coast of Hawaii, will be discussed in Chapter 12.

The best known tsunamis are probably the following five:

November 1, 1755. The tsunami of the Lisbon earthquake reached the city at about 10 A.M. local time, about 20 minutes after the first destructive shock. On

---

† Bobillier lists 12 for Chile, 1562–1922. For A.D. 599–1949 Kawasumi's catalogue of 342 large earthquakes in Japan includes 69 accompanied by tsunamis.

the Portuguese coast it rose at many points to heights of 20 feet, and, in places, up to 50 feet. From its offshore source the disturbance spread in every direction. It was damaging on the coast of the island of Madeira, and reached reported heights of 15 feet in the West Indies. In the Mediterranean it died out rapidly; but northward it entered the English Channel, disturbing British and French harbors in the afternoon, and was observable as far away as Holland. In that area the difference of several hours makes the tsunami clearly separable from the seiches which occurred in the morning.

August 8, 1868. This tsunami followed an earthquake which devastated Arica— then in Peru, but now, after settlement of a long-standing boundary dispute, in Chile. Practically the whole Pacific Ocean was disturbed. The waves were damaging at Hawaii, and high in the region of New Zealand; on the Japanese coast they caused considerable alarm. At Arica the ocean was greatly disturbed and retired, coming back in a succession of waves. The greatest wave came in about 4 hours after the earthquake, wrecking most of the ships in the harbor, overwhelming the city, and rising to a height of at least 47 feet. The U. S. gunboat *Wateree* was carried 3 miles up the coast and 2 miles inland, to within 200 feet of a sheer cliff; here it rested safely on its flat bottom, but it had to be abandoned. An officer of the *Wateree,* L. G. Billings, who later became a rear admiral, in 1915 published a horrifying account of this experience.

May 9, 1877. Much of what has been said of the 1868 tsunami applies to this one. The associated earthquake was very destructive at Iquique (then in Peru, now in Chile). Heights reached by the waves on distant coasts are given as: Samoa, 6 to 12 feet; Japan, 5 to 20 feet; New Zealand and Australia, 3 to 20 feet.

March 2, 1933 (March 3, Japanese time). This wave followed a very great earthquake off the eastern, or Sanriku, coast of the main Japanese island of Honshu. The earthquake damage was moderate, but the tsunami was devastating. In places the waves rose as high as 75 feet. As far as local effects are concerned, this is perhaps the most thoroughly studied of all tsunamis; a voluminous report was published by the Earthquake Research Institute at Tokyo.

Miyabe investigated the times of arrival of the waves as recorded on mareograms. At such distant points as San Francisco, Honolulu, Iquique, times of arrival check well enough for a wave starting from the earthquake epicenter at about the time calculated as origin time from seismograms (which is good to within a very few seconds). However, the times recorded at Japanese tide stations are in part anomalous; at some of the more distant stations the tsunami was actually recorded as beginning 20 minutes earlier than at the nearest ones. This discrepancy may be due in part to seiches started by the seismic waves in the ground; Miyabe prefers to conclude that the tsunami probably did not originate from a point. In so great an earthquake, which almost certainly involved extensive faulting or block movement, this is to be expected. Omote proposes that there was an earlier wave originating around the margin of an area several hundred kilometers wide and a later wave from a more limited source of the order of 10 kilometers in diameter. An interesting detail was the recording at tide stations in Australia and New Zealand of waves, reflected from the American coast, traveling twice across the Pacific in about 47 hours.

April 1, 1946. This was a great tsunami, but it was set off by an earthquake of only moderate magnitude, 7.4. Earthquake and tsunami both originated close to the Aleutian Trench. At Scotch Cap, the west point of Unimak Island (Aleutians), the waves demolished a large and well-built lighthouse 45 feet above sea level and surged to a height of over 100 feet. The highest wave levels on the Hawaiian islands were (in feet): Oahu 37, Maui 31, Kauai 45, Molokai 54, Hawaii 55. In the vicinity of Hilo, on Hawaii, the largest populated center affected, the level was 26 feet. In the islands 159 lives were lost, 96 of them at Hilo; 488 homes were demolished. The greatest single loss in damage to structures was $375,000 at a sugar mill near Hilo.

This is the only tsunami from a distant source known to have amounted to much on the California coast; it rose to 11 feet at Halfmoon Bay and 12 feet at Santa Cruz, where one man was drowned. There were reports of large waves on the South American coast. Studies of timing again indicated a line source rather than a point source. (This was also true of a tsunami on the Japanese coast following the Nankai earthquake of December 21, 1946.)

A tsunami was started by the great earthquake (magnitude 8.3) on March 9, 1957 in the Aleutian Islands, but the wave was much smaller than that of 1946. Nevertheless, damage on the coasts of Oahu and Kauai amounted to about 3 million dollars, and two villages were destroyed. Loss of life, and greater property loss, were forestalled by the warning service set up after the 1946 disaster; Hawaiian authorities were alerted more than three hours before the arrival of the wave. By misunderstanding, warnings were also issued by local centers in California, where the wave was detected only by tide gauges.

Causes and Controversy

The cause of tsunamis is much debated. Not all of them are started by earthquakes. Nor does an explanation for minor and local waves necessarily apply to the great tsunamis which sweep the oceans.

Block Movements? The most commonly offered explanation calls for displacement of submarine blocks of the earth's crust. Since we know of a number of great earthquakes on land which have been accompanied by vertical movements up to 30 feet—and some of these have affected crustal blocks hundreds of miles long and at least a hundred miles wide—it is easy to think of a similar displacement under the sea as starting a wave. Like many simple explanations, this is convincing as long as one does not work out the implications or study the observations in detail. It seems so obvious that seismologists are sometimes told dogmatically that the epicenter must be wrong because it fell on land when a tsunami was observed, or even because it fell offshore and there was no tsunami. There are well-documented instances of both kinds; but such occurrences neither prove nor disprove the hypothesis. As to the former, the instrumental epicenter represents merely the point of initial rupture; there is no general reason why fracturing should not start under a continent and then extend out under the sea, or why a crustal block, displaced and fractured along one edge, should not readjust its whole width. As to the latter, we need not expect every large submarine earthquake to start a wave.

Slides. It has often been suggested that tsunamis are started by submarine slides. Large tsunamis originate at the great oceanic troughs; on their slopes there may be great masses of unconsolidated material which, disturbed by an earthquake, may slide into the depths, displacing a great deal of water. Such a hypothesis helps to explain a tsunami which follows an earthquake with epicenter on land; it also explains long time delays, such as the 4 hours between the earthquake and the greatest wave at Arica. Naturally there is little positive evidence to support the idea, although slides entering the sea or occurring just off shore are known to have caused large waves locally; this may account for some of the minor tsunamis in regions where large ones are unknown. Recent investigation of cable breaks bears on this problem (Chapter 10). Sliding may well be a contributing factor in starting great tsunamis; but the author regretfully differs from the eminent seismologists who have insisted that it is the principal cause.

Surface Waves? Resonance? Some years ago Dr. H. Benioff suggested possible correlation between tsunamis and the great surface waves with periods over a minute, which are especially well developed on the seismograms of certain large earthquakes.† Usually other phases of the same seismograms show striking development of long periods, and aftershocks are relatively numerous. This evidence indicates displacement of large crustal blocks, or perhaps great linear extent of faulting. Even at distant stations the large surface waves may have amplitudes of one or more centimeters, suggesting amplitudes at relatively short distances of the order of a meter. The resulting deformation of the sea bottom, especially in a trough or trench, might readily produce great displacement of the water surface. This type of explanation fits the evidence that the tsunami does not originate at a point, but the association between great tsunamis and great earthquakes which it implies is seriously called in question by the circumstances of the 1946 Aleutian occurrence. The earthquake was not a great one (magnitude 7.4), and the seismograms show no unusually large long-period waves, although an unusually long and numerous sequence of aftershocks was recorded at many distant stations. On the other hand, the great Aleutian shock of 1957, with an extraordinary number of aftershocks, many of them large, started only a moderate tsunami.

---

† See Chapters 16 and 17. An inconclusive effort was made to correlate tsunamis with T waves (Chapter 17).

The author has often speculatively entertained the possibility that a tsunami might result from resonance in the water of a trough or trench, with a mechanism perhaps distantly related to that of seiches. A mass of water of such form and dimensions might have a long fundamental period measurable in hours which would agree with the slow oscillations of observed tsunamis.

Minor Events. As suggested earlier, various explanations may apply to minor tsunamis, particularly to those not associated with great earthquakes. Some of these occur where major tsunamis are unknown or rare, as in California. However, they are fairly frequent on the coasts of Japan. On November 5–6, 1938, four major earthquakes originated off the east coast of Honshu, each followed by its individual minor tsunami (as shown on mareograms).

SEAQUAKES

The word seaquake at first was used rather generally for any seismic disturbance of the ocean, including tsunamis. Indeed, the word is a possible literal translation of *maremoto*. In more recent literature the term is restricted to mean shaking caused by an earthquake, but felt on board a vessel at sea. Such occurrences are not rare, but they were imperfectly understood until the publication of two monumental papers by Rudolph in 1887 and 1895. He did his work so well that very little has been added to the subject since.

Information comes mostly from ships' logs and from the reports of ships' masters to nautical offices. Rudolph collated hundreds of such accounts, most of them German or British. Instrumental seismology was then in its infancy, and these reports were the principal source of information about submarine earthquakes. Now, when even small shocks under the ocean can often be located as accurately as those under the continents, Rudolph's work has lost in geographical importance. However, seaquakes are curious and interesting; it is a pity that no one has carried on Rudolph's work.

Descriptions of seaquakes are surprisingly similar. Here are two British reports used by Rudolph.†

[March 11, 1855, in the equatorial Atlantic.] While standing near the wheel I heard a sound as of distant thunder; on walking over to port side to look to the southward, experienced a tremendous and grating motion of ship, as if grazing over a coral reef; it caused everything to shake for about a minute after the sound had ceased. The whole lasted two minutes. I tried for soundings, but had no bottom with 120 fathoms line. There was not the least ripple on the surface of the water, but the sound seemed to come from ship's bottom, and the motion

---

† In "Über submarine Erdbeben und Eruptionen," Part II, vol. 2 (1895).

was not unlike letting go the anchor in deep water when the chain runs out quickly.

[August 10, 1884, in the far south Pacific, on the sailing route between New Zealand and Cape Horn.] I had just gone down from the deck, when I was startled by hearing a noise, as if the ship were scraping over something hard. I rushed on deck and found that everyone in the ship had felt the same; the watch below turning out, feeling sure the ship had struck. I had a cast of the deep sea lead taken—40 fathoms, no bottom—within 3 minutes of the occurrence, and also went right round the ship with the hand lead-line, but found no bottom. I may mention that the man forward felt sure that it was aft that the ship had struck; while the mate, the man at the wheel, and myself were as positive that it was forward. The force of the shock was sufficient to keep the glasses in the swinging tray in the cabin, and the lamp jingling for fully half a minute, although the shock itself did not last more than 10 seconds. The mate stated that a rumbling noise was heard for a few seconds before the actual shock to the ship . . .

Many such reports note a sensation as of the ship running aground. Navigational maps carry numerous indications of points at which a reef or shoal has been reported but not verified by soundings, at the time or later. Some of these are certainly due to seaquakes, particularly in the rather seismic area of the equatorial Atlantic. The two accounts just quoted are from sailing vessels; but observations on modern steamers are similar. We have two independent reports of one occurrence. The first is from the British *Marine Observer:*

The following report has been received from SS *Magician,* Captain P. O. Nicholas: 5th November, 1926 . . . in lat. 10° 25′ N, long. 88° 10′ W, a very severe submarine disturbance was experienced. Two distinct shocks, lasting about 10 or 15 seconds, with an interval of about 1½ minutes, were felt. The vessel shook violently, a rumbling, grating sensation was experienced. Masts, funnel, and superstructures vibrated and rattled alarmingly, giving the impression that the ship was running aground on to hard bottom and buckling fore and aft . . . The chart shows 1800 to 1900 fathoms water in the vicinity. The American SS *Eagle,* then some 15 miles NW of our position, was later communicated with by wireless and her master replied that the shocks had been felt on board his vessel with such severity that the engines were stopped in the belief that the ship was running over something.

The second report is from American newspapers:

November 5, 1926. Aboard SS *Eagle* in lat. 10° N long. 88° W, two severe shocks were felt. They were of about one minute duration, with an interval of one minute between each. The *Eagle* listed four or five degrees, and the masts, booms, rigging, and stock vibrated considerably. Capt. P. O. Nicholas, of the British steamer *Magician,* when about 17 miles distant from the *Eagle* and at the same time, experienced two severe earthquake shocks with a similar effect on the vessel.

These ships were not far off the coast of Nicaragua, where this earthquake caused much damage; at Managua about half the houses were damaged and there was some loss of life. The instrumentally located epicenter was on land, near Managua; but the hypocenter was at the great depth of 135 kilometers (85 miles); these data eliminate such causes as submarine explosions or displacement of the ocean bottom. Clearly we are dealing with elastic waves transmitted first through the rock to the sea bottom, then up through the water, so that it is not surprising that an earthquake originating under the land was felt on vessels off the coast. The wave velocity is much less in water than in rock, so that the rays are refracted steeply upward; this agrees with the many signs that the disturbance arrives as a series of vertical blows, and accounts for the difficulty of establishing direction. The two successive disturbances in the last reports quoted must be due to the two principal types of seismic waves (P and S) arriving successively at the sea bottom and starting successive trains of compressional waves in the water.

All this was clearly set forth by Rudolph, who refuted the idea that seaquakes are due mostly to submarine explosions, presumably volcanic. However, there are some exceptional reports which point to a volcanic cause; they include such features as disturbance of the water surface, rise in temperature, even jets of hot water out of the sea, and not uncommonly an erupting volcano in sight.

Because Rudolph correctly recognized that most seaquakes are due to tectonic earthquakes, it is surprising to find he was wrong about the causes of tsunamis, which he ascribed chiefly to volcanic action. Apparently he was fascinated by the conspicuous example of the great wave started by the explosion of Krakatoa, not realizing how exceptional it was.

At least one quite different cause may produce an apparent seaquake. Rudolph quotes the following:

We were greatly surprised with a rumbling noise like an earthquake, and the ship trembling, the watch below coming on deck, being alarmed with the shock. It was repeated 12 times at intervals of one or two minutes . . . After the lapse of 20 minutes, to our great astonishment, two enormous whales appeared from under the bottom of the ship, going round the ship and blowing several times, and then going under the ship and repeating the same shock several times. I am of the opinion that the shells on the iron bottom is what they must have been rubbing themselves against, and the shock more felt than in a wood ship, but the hollow booming sound was the strongest . . . One of them was the largest whale I ever saw . . .

Underwater recording equipment now exists which would make possible a systematic instrumental study of seaquakes, but none has been made as yet. Although, insofar as we know, there are no unsolved mysteries about seaquakes, the matter has been neglected so long that it is a promising field for new investigators with new ideas.

References

Lisbon earthquake, 1755

Kendrick, T. D., *The Lisbon Earthquake,* Lippincott, Philadelphia and New York, 1955.

Davison, Charles, *Great Earthquakes,* Murby, London, 1936, Chapter 1.

Reid, Harry F., "The Lisbon earthquake of November 1, 1955," *B.S.S.A.,* vol. 4 (1914), pp. 53–80.

Trans. Royal Soc. (London), vol. 49 (1755).

Sousa, F. L. Pereira de, "Sur les effets en Portugal du megaséisme du 1er novembre 1755," *Comptes rendus,* vol. 158 (1914), pp. 2033–2035. (See also Appendix XVI.)

Oscillations in wells

Stearns, Harold T., "Record of earthquake made by automatic well recorders in California," *B.S.S.A.,* vol. 18 (1928), pp. 9–15.

Blanchard, F. B., and Byerly, Perry, "A study of a well gauge as a seismograph," *ibid.,* vol. 25 (1935), pp. 313–321.

Earthquake fountains, etc.

Jones, J C., "The Pleasant Valley, Nevada, earthquake of October 2, 1915," *B.S.S.A.,* vol. 5 (1915). (Quotation is from p. 197.)

Imamura, Akitune, *Theoretical and Applied Seismology,* Tokyo, 1937. (The account of pulsating fountains is quoted from p. 74.)

Yamasaki, N., *Journ. Faculty Science, Imp. Univ. Tokyo,* vol. 2 (1926), p. 106. (Bridge pillars out of the ground; photographs reproduced in: Davison, C., *The Japanese Earthquake of 1923,* London, 1931, pls. V and VI.)

Officers of the Geological Survey of India, "The Bihar-Nepal earthquake of 1934," *Mem. Geol. Survey India,* vol. 73 (1939), pp. 33–34, 185–187.

Seiches

Trans. Roy. Soc., vol. 49 (1755).

Forel, F. A., *Le Leman,* Lausanne, 3 vols., 1892–1904.

Kvale, Anders, "Seismic seiches in Norway and England during the Assam earthquake of August 15, 1950," *B.S.S.A.,* vol. 45 (1955), pp. 93–113.

Tsunamis, general

Heck, N. H., "List of seismic sea waves," *B.S.S.A.,* vol. 37 (1947), pp. 269–286.

Gutenberg, B., "Tsunamis and earthquakes," *ibid.,* vol. 29 (1939), pp. 517–526.

(States the case for submarine slides as a cause of tsunamis, with special reference to the Chilean occurrence of 1922.)

Bobillier, C., "Resumen histórico de los principales maremotos acaecidos en Chile," *Bol. servicio sismológico,* Univ. Chile (Santiago) (1933), No. 23, pp. 34–41.

Kawasumi, H., "Measures of earthquake expectancy. . . .", *Bull. E.R.I.* vol. 29 (1951), pp. 469–482.

Darbyshire, J., and Ishiguro, S., "Tsunamis," *Nature,* vol. 180 (1957), p. 150. (Meaning of the word tsunami.)

Tsunamis, chronological

1755

See references for the Lisbon earthquake, and:

Sousa, F. L. Pereira de, "La raz de marée du grand tremblement de terre de 1755 en Portugal," *Comptes rendus,* vol. 152 (1911), pp. 1129–1131.

1812 (California)

Professor Louderback's findings are noted on p. 6 of: Wood, H. O., and Heck, N. H., *Earthquake History of the United States,* Part II. U. S. Coast and Geodetic Survey, U. S. Dept. of Commerce, Ser. No. 609, rev. 1951.

1868

von Hochstetter, F., "Über das Erdbeben in Peru am 13. August 1868 . . ." *Sitzber. Akad. Wiss. Wien,* math.-naturw. Kl. vol. 58 (1868), p. 837.

————, "Die Erdbebenfluth im Pazifischen Ozean vom 13. bis 16. August 1868 . . ." *ibid.,* vol. 59 (1869), p. 109.

Billings, L. G., "Some personal experiences with earthquakes," *Natl. Geographic Mag.,* vol. 27 (1915), pp. 57–71.

1877

Geinitz, E., "Das Erdbeben von Iquique am 9. Mai 1877 und die durch dasselbe verursachte Erdbebenfluth im Grossen Ozean," *Akad. Halle, Nova Acta,* vol. 40 (1878), p. 385.

1883

Symons, G. J., ed., *The Eruption of Krakatoa and Subsequent Phenomena,* Royal Society, London, 1888.

1896

Davison, Charles, *Great Earthquakes,* Murby, London, 1936, Chapter IX.

Imamura, Akitune, *Theoretical and Applied Seismology,* Maruzen, Tokyo, 1937.

1927

Byerly, P., "The California earthquake of November 4, 1927," *B.S.S.A.,* vol. 20 (1930), pp. 53–66.

1933

Papers and Reports on the Tunami of 1933 on the Sanriku Coast, Japan, Bull. E.R.I., Supplementary Volume 1, Tokyo, 1934. (Includes: Miyabe, N., "An investigation of the Sanriku tunami based on mareogram data," pp. 112–126. Summarized by Davison, in *Great Earthquakes,* Chapter IX.)

Omote, S., "On the central area of seismic sea waves, *Bull. E.R.I.,* vol. 25 (1948), pp. 15–19.

1938

Otuka, Y., "On the earthquakes that occurred in November, 1938, on the Pacific coast of northeastern Japan," *Bull. E.R.I.,* vol. 17 (1939), pp. 168–178. (In Japanese, with English summary. Group of major earthquakes, several followed by tsunamis.)

1945

"Earthquake in the Arabian Sea," *Nature,* vol. 156 (1945), pp. 712–713.

Beer, A., and Stagg, J. M., "Seismic sea wave of November 27, 1945," *ibid.,* vol. 158 (1946), p. 63.

1946

Green, C. K., "Seismic sea wave of April 1, 1946, as recorded on tide gages," *Trans. Am. Geophys. Union,* vol. 27 (1946), pp. 490–500.

Shepard, F. P., Macdonald, G. A., and Cox, D. C., "The tsunami of April 1, 1946," *Bull. Scripps Inst. Oceanography,* vol. 5 (1950), pp. 391–528, pls. 6–33.

Macdonald, G. A., Shepard, F. P., and Cox, D. C., "The tsunami of April 1, 1946, in the Hawaiian Islands," *Pacific Science,* vol. 1 (1947), pp. 21–37.

Seaquakes

Rudolph, E., "Über submarine Erdbeben und Eruptionen, Part I," *G. Beitr.,* vol. 1 (1887), pp. 133–373; Part II, vol. 2 (1895), pp. 537–666; Part III, vol. 3 (1898), pp. 273–336. [The quoted reports of ships' masters are from Part II, pp. 545, 557, 581. Quotations for the seaquake of November 5, 1926, are taken from the *International Seismological Summary,* 1926, pp. 315–316 (which repeats them from the *Marine Observer* and from mimeographed press reports circulated by the seismological station at Georgetown, D. C.).]

CHAPTER 10

Other Secondary Effects of Earthquakes

LANDSLIDES

Among the larger secondary effects of earthquakes are landslides, which often imitate or obscure primary effects such as faulting. In the report on the earthquake of 1906 Lawson classified its associated slides into four groups: earth slumps, earth flows, earth avalanches, and earth lurches.

Earth slumps in this sense are the result of earthquake motion accelerating processes which go on at all times wherever there is unconsolidated material on steep slopes. Later authors have separated these processes into several subclasses; one is soil creep, which is a general downgrade motion of soil and weathered rock over underlying bedrock. The typical slump (Fig. 10-1) is distinguished from soil creep by having definite fractures as boundaries. As described by Lawson, such a slump is recognizable by a scarp at its head which is crescent-shaped in plan, with the horns pointing downgrade. The slumped material may pile into a considerable toe below, the effect being partly that of rotating the slumped mass about a horizontal axis. In section the slump fracture is steepest upgrade, rising into the nearly vertical scarp; downgrade it becomes level and may even rise under the toe of the slide. Some authors refer to slow-moving slumps of this kind as

FIGURE 10-1 *Idealized plan and section of earth slump.*

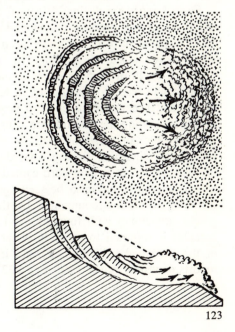

earth flows, reserving the term slump for similar but faster-moving slides; Lawson's earth flows are then called mud flows. Very large slumps have been precipitated by many earthquakes, and some of them have led to casualties. On April 1, 1955 (local date), an earthquake in the Philippines caused part of a village to slump under the waters of Lake Lanao, with loss of 174 lives; at least this is the account in the official publication, which unfortunately lacks detail, although it has many pages of generalities and material copied from other sources. The toe of a great slump at Whitecliffs on the New Zealand coast, precipitated by the 1929 earthquake, raised a strip of sea floor about a mile long and averaging about 500 feet wide to heights reaching 100 feet above sea level (Chapter 29 and Fig. 29-5).

Earth flows, as defined by Lawson, might be included with ground-water phenomena. The earthquake is accompanied or followed by a sudden burst of water from a locality where it normally appears in springs; this water carries sand and mud with it downgrade in a flow which may be destructive to buildings or fences in its path.

Earth avalanches, on the other hand, are flows of relatively dry material. The majority of those started by the 1906 earthquake consisted of rock and soil cast down from cliffs or bluffs along the seacoast to add to the accumulating talus. However, large avalanches also started from the steep sides of inland canyons, some of which were buried under debris for lengths up to a half mile. Lawson points out that, although such a slide often carries with it a certain amount of soil and vegetation, the bulk of it consists of loose rock. This kind of slide often occurs without an earthquake.

Earth lurches are distinctively earthquake effects. Minor or incipient action of this kind may produce cracks and fissures parallel to streams or gulches; some of these have occasionally been misinterpreted as primary effects and taken as evidence of faulting. In lurching, earthquake motion at right angles to a cliff or bluff or, more commonly, to a stream bank or an artificial embankment leads to yielding of material in the direction in which it is unsupported. The initial effect is to produce a series of more or less parallel cracks separating the ground into rough blocks. With harder or longer shaking, the outer of these, adjacent to the bluff or bank, slides down, usually holding together and tipping toward the bank. Others may follow, giving the appearance of a backward-canted flight of stairs (Fig. 5-8). This stepwise appearance may also be seen against the scarps of large earth slumps. Lurching is often accompanied by the ejection of ground water from the cracks, sometimes with fountains and the formation of sand craters; similar cracking is reported to occur in water-soaked level ground with the passage of the earthquake disturbance, and it may have some connection with the reports of visible waves (see page 132).

CABLE BREAKS

An occurrence allied to earth flows has been cited lately as an explanation for breaks in submarine cables consequent to large earthquakes (but sometimes occurring independently). Milne became interested in cable breaks, partly because of their economic importance, partly because of their suggestion of submarine faulting. The tendency of such breaks to occur repeatedly at points along a given line was tentatively attributed by Milne to displacement of fault blocks. In 1952 Heezen and Ewing offered a different explanation.† Rapidly moving currents carrying dense material, amounting almost to mud flows, have been described in lakes and reservoirs. Heezen and Ewing, on the assumption that such currents may also occur under the sea, explained cable breaks in this way, with particular application to the 1929 earthquake in the region of the Grand Banks, south of Newfoundland. In their words,

. . . a relatively severe earthquake on the continental slope set in motion slides and slumps which, with the incorporation of water, were transformed into turbidity currents of high density. These converged to form a gigantic turbidity current which swept across the sea floor for well over 350 nautical miles, breaking each succeeding submarine cable.

All the main submarine cables across the North Atlantic traverse this area; there are more than in any other area of similar size in the world. Six cables were broken almost at once at the time of the earthquake; then

. . . for 13 hours and 17 minutes following the earthquake there occurred an orderly sequence of breaks of each succeeding cable lying in increasingly deeper water for over 300 miles south of the epicentral area . . . The instants of the cable interruptions were accurately recorded by automatic machines which record the telegraphic messages, and the locations of the breaks were determined by resistance measurements from the shore ends of the cables.

Five cables broke after the earthquake, each at two or three locations. The velocity with which the current must be assumed to have traveled ranges from 18 to 80 miles per hour (12 to 50 knots). This high apparent velocity was taken by other authors as evidence that the breaks were due to faulting. However, Heezen and Ewing have lately reported on similar breaks following the Algerian earthquake of September 9, 1954; this had an epicenter on land, and the cable breaks did not begin until hours after the earthquake.

---

† In "Turbidity currents and submarine slumps, and the 1929 Grand Banks earthquake," *Am. Journ. Science,* vol. 250 (1952).

MISCELLANEOUS EFFECTS

Clocks

The stopping of pendulum clocks is an earthquake effect which has long been known; in the absence of seismographs it may even be used to fix the time of occurrence of an earthquake. One must be sure that a report actually refers to a pendulum clock. Stoppage of electric clocks merely means that power has been cut off—sometimes accidentally, but usually as a deliberate precaution against fire. Pendulum clocks are stopped chiefly by motion at right angles to the face, which causes friction between the pendulum and the escapement.

Automobiles

As civilization develops new techniques, new earthquake effects come to attention. Thus, effects involving automobiles have now taken regular places in intensity scales. A standing automobile with the motor not running is an excellent simple seismoscope, since the heavy body is free to oscillate up and down on the springs. The author, like many others, had opportunity to observe this effect in Kern County while that area was still being shaken frequently by aftershocks of the major earthquake of 1952. A shock barely strong enough to be noticed by the observer standing on the ground would cause plainly visible motion, with much creaking of springs and rattling of Ford parts. It is said that a car parked on Bear Mountain during the main earthquake had its springs broken in this way. An observer seated in a car will notice very small earthquakes, which may not be perceptible to others standing nearby.

An observer driving an automobile, on the contrary, is not likely to notice an earthquake until the shaking is strong enough to cause damage in his vicinity. Then steering is affected, and there is a sensation as of a tire suddenly losing air and going slightly soft; at higher intensity observers report that it feels like all four tires going flat at once.

Glaciers

Effects on glaciers are chiefly known from the Alaskan earthquakes of 1899 (Chapter 31).

Trees and Bushes

Effects on trees and bushes are of two types. The short-period part of the motion rustles leaves and twigs like a sudden gust of wind; the longer periods sway large branches and limbs. Palm trunks have snapped; and the 1906 earthquake cut a swath of devastation through the coastal redwood forest of northern California, littering the ground with broken limbs. Small motion of trees probably accounts for the many reports of birds taking flight an instant before a shock is felt.

Fishes and Domestic Animals

Seaquakes often kill fishes in great numbers. The shock wave transmitted through the water has the same lethal results as dynamiting a pond. The legendary material of seismology includes many stories of horses and other domestic animals being uneasy during the hours preceding a large earthquake. It is impossible to judge such evidence scientifically. If there is any explanation beyond coincidence and incomplete reporting, it probably rests on the occurrence of small foreshocks, not noticed by persons but disturbing to sensitive animals. During earthquakes animals are seen to react as they do to almost any sudden and unexpected event. Horses snort and bolt; cattle stampede; dogs bark and whine; cats spit. After the Long Beach earthquake the press reported that many cats failed to return to their homes for several days until the aftershock activity had died down.

Human Reactions

Human beings begin to notice shaking when the acceleration approaches 1 gal (1 cm/sec²). Some persons are nauseated when no shaking is noticed by anyone. Sleepers are wakened either by strange noises or by actual shaking. There are often ridiculous reports of persons being "thrown out of bed" by comparatively small earthquakes. When sincere, these accounts probably mean that the person, disturbed by the earthquake, gave a nervous start which threw him to the floor; being fully awakened by that event, he is convinced that the earthquake threw him down. Lying awake in bed is a favorable condition for noticing a small earthquake; the springs act as seismoscopes.

Fright and panic are such regular effects of strong shaking that they form an established part of all intensity scales. Persons and populations differ according to their previous experience; but only an abnormally cold-blooded person can remain calm when the structures over his head are being damaged and the ground under his feet is shaking so as to destroy the basic feelings of security. The most universal impulse is to run, even when already outdoors. Unless an obvious danger is avoided, this is an unwise thing to do. Many persons have dashed out of comparative safety to meet their death under debris falling on the outside of a building. It is stated that in 1933 the chief of police at Huntington Beach, attempting merely to cross his office during the earthquake, was thrown off balance and suffered a broken leg. All the normal effects of fright and shock are seen; there are usually a number of "earthquake babies."

SOUND

Quite naturally, some of the most mysterious earthquake effects are among the transient group. Perception of audible sound is now fairly well understood, although some seismologists are not quite satisfied on certain points. Especially indoors, an earthquake produces miscellaneous noises by rattling and shaking loose objects and straining the structure. These noises may obscure the true earthquake sound, which is heard more clearly out of doors; it usually is a heavy sound, pitched near the lower limit of hearing, and most commonly compared to thunder, gunfire, or heavy traffic at a distance. It is produced directly by the transfer of elastic wave energy from the ground to the air; there is considerable loss of energy in this, but the human ear is quite sensitive, so that a small amplitude of vibration in the air is heard as a loud sound. Problems of interpretation arise only from the numerous reports of sound heard appreciably before any shaking is felt. In most such cases it is nearly certain that the sound is produced by the P group of longitudinal waves, while the shaking begins to be felt only with the arrival of the transverse S waves, which usually have a larger amplitude. Dr. Pierre St. Amand had opportunity to verify this repeatedly at the University of Alaska (at College, near Fairbanks) in 1947, when a long series of earthquakes was felt and heard there. On several occasions he was in the recording vault, where he could observe the seismograph and verify that the sound arrived with the first recorded motion while felt shaking was often delayed until the arrival of S. Seismograms of local earthquakes often show that the first motion is followed a second or more later by a much larger wave which still belongs to the longitudinal series; the first of these may be heard and only the second felt. Some earthquakes of magnitude 6 and over have transferred enough energy to the air to record on sensitive barographs at Pasadena, even at distances of several hundred miles. The first such instance was identified by Dr. Benioff in 1939; many similar records have been written, especially since 1950. In great earthquakes the energy may be still greater; thus Kingdon-Ward, near the epicenter of the great Tibet earthquake of 1950 (Chapter 5), heard heavy explosive sounds following the shock, coming apparently from high in the air. These sounds were heard at many points in India and Burma, to distances of over 750 miles. Investigators in India have considered and rejected the alternative that these sounds were due to landslides. Thomson described similar sounds for the New Zealand earthquake of 1929,

VISIBLE WAVES

A much more controversial matter is the reported seeing of waves moving over the ground during an earthquake. This is a fascinating subject for anyone who likes unsolved mysteries. The best and most circumstantial accounts, taken alone, are believable and no more extraordinary than some well-established earthquake effects; many competent investigators have been convinced by such evidence that there is a real phenomenon of the kind. On the other hand, in collecting more instances one finds such descriptions shading off into others which simply cannot be believed. Some observations almost prove the existence of an illusion, which may be physical, physiological, or psychological. If to this are added excitement, bad observation, unconscious exaggeration, and deliberate lying, the result is such a tangle that some writers have rejected all the observations, usually offering one simple explanation to cover the whole complex.

Whatever the facts, the observations manifestly cannot refer to the seismic waves recorded by instruments; these travel at speeds measurable in miles per second and cannot possibly be followed by an observer's eye. Waves actually seen would have to be of another physical type with much lower velocities. They might, however, be a modification of standing waves; that is, an interference pattern of nodes and loops may be set up which shifts over the ground as the exciting disturbance changes. A high school instructor described to the author what he saw in the streets at Long Beach in 1933; it could easily be such a pattern of nodes and loops. There was no suggestion of large motion of the surface of the ground, but the loops or antinodes were put in evidence by dust thrown into the air, while the nodes appeared quiet. The account was rather convincing because the observer was not a physicist and did not have the technical vocabulary used here.

Quoted Reports

Visible waves are most commonly reported from the meizoseismal areas of great earthquakes, particularly on soft or alluvial ground. Consequently many of the clearer and more accessible accounts come from India. Oldham includes a number of these for the 1897 earthquake. To quote only one:

The ground rocked violently and we were both thrown down . . . The shock appeared to come from the southwest, and on looking in the direction we saw a series of earth waves approaching over the surface of the ground exactly like rollers on the sea, as these passed us we had some difficulty in standing, but none of these waves reached the intensity of the first, which had overthrown us . . . As the waves above subsided the ground began to crack at our feet . . .

This was immediately followed by the emergence on the spot of earthquake fountains, which the observer unsuccessfully tried to photograph. This occurred in an alluviated region, south of the hill country where the earthquake reached its highest intensity. The appearance is evidently not the same as that mentioned by the observer at Shillong (Chapter 5) who saw the ground in every direction shaking like soft jelly; another observer at Shillong compares the effect to the surface of a storm-tossed sea.

Many reports of this kind were collected by Dutton for the earthquake of 1886 at Charleston, South Carolina. The main earthquake occurred at night.

. . . The vibrations increased rapidly and the ground began to undulate like a sea. The street was well lighted, having three gas lamps within a distance of 200 feet, and I could see the earth waves . . . distinctly . . . The first wave came from the southwest . . . The waves seemed then to come from both the southwest and northwest and crossed the street diagonally, intersecting each other, and lifting me up and letting me down . . . I could see perfectly and could make careful observations, and I estimate that the waves were at least two feet in height.

Another observer reports that he was deliberately watching for such waves in aftershocks during the night but did not see any until the following afternoon.

After the first vertical tremor had passed . . . I distinctly saw four or five separate waves pass across Tradd street from the northeast to the southwest . . . they were about as wide as the roadway between the sidewalks; as to their height, I would not like to venture an estimate, but each seemed to be at least a foot high, and was certainly high enough to be plainly seen.

If such waves are real, their heights must be overestimated. Reliable observation in such a case would be exceedingly rare. Certainly there was no such devastation at Charleston as might be expected from waves 2 feet high; and the same remark applies with more force to the much smaller aftershock in the following afternoon.

A similar overestimation must be involved in the following observation, which almost amounts to positive proof that some part of the waves seen is illusory.

. . . On the occasion of the San Jacinto earthquake . . . on April 21, 1918, Dr. J. A. Anderson of the staff of the physical laboratory of the Mount Wilson Observatory was in the laboratory in Pasadena. When the earthquake motion became strong he decided to leave the building, a one-story structure with a cement-concrete floor . . . As he moved toward the door this floor appeared to him to be thrown into waves with vertical amplitudes (trough to crest) of not less than four to six inches and wave-lengths of six to ten feet . . . Immediately after the cessation of the shock he returned and inspected the floor which showed no new cracks of even minute dimensions. [The search for cracks was made with the aid of a hand lens.] Moreover, many relatively unstable objects on the tables and shelves in the laboratory remained in position, apparently undisturbed.[†]

Dr. R. H. Jahns relates that, at the time of the Long Beach earthquake in 1933, he seemed to see large waves along the top of the wall of a dry swimming pool, although no effect was apparent on the floor of the pool, and Venetian blinds on a far wall were only slightly disturbed.

Possible Illusions

Illusion certainly exists. Some of it may be psychological. The most obvious type could hardly apply to Dr. Anderson's observation, as anyone acquainted with him would be sure; but it is well known that many reasonable and sane persons, when exposed to a sudden and terrifying event, seem to see and experience things which bear little relation to fact, and they report them later with perfect honesty and conviction.[‡] Some writers have referred to the illusion of relative motion, familiar in railway stations when one is uncertain which train is moving, but this hardly goes far enough. A possible physiological factor is disturbance of the liquid in the semicircular canals of the ear, which control the sense of equilibrium.

Possible Realities

Two types of optical effects have been suggested. The more familiar is that of the "road mirage" seen on highways in warm weather, when light is reflected from the top of a layer of warm air, producing an illusion of water. The reflecting layer is easily disturbed, as by wind; its surface may be thrown into waves during an earthquake.

Mr. Frank Perret and Dr. Hugo Benioff independently suggested that elastic waves emerging into air, as they do in producing audible sound, may locally change the refractive index of the air sufficiently to deflect the light ray reaching an observer at some distance; rapid changes in the ray path would give the ground an appearance of motion.

Some other observations are more easily accounted for. Sometimes waves of disturbance are reported seen traveling across grain fields; the individual stalks of grain behave as inverted pendulums, providing the observer with a multitude of natural seismoscopes. Waves seen traveling over metal roofs or along fences are to be understood in much the same way.

[†] From *Bull. Nat. Research Council,* No. 90 (1933).

[‡] Compare Terada's remarks in connection with the Izu earthquake of 1930 (Chapter 30).

Lurching?

There is often reference to cracking open of the ground at the crests of the waves, sometimes even with emission of sand and water. Earth lurching in action must look very much like this, as individual blocks of earth sway back and forth with cracks developing between them; such oscillation might readily produce the impression of a progressing wave. Earth lurching may also account for the so-called "frozen waves" which excite public interest after an earthquake, when walls, embankments, and the like are left in a wave form.

Conclusions

This is a field in which more (and more carefully reported) observations are needed, if the matter is ever to be cleared up. On present evidence the writer concludes:

(1) There is almost certainly a real phenomenon of progressing or standing waves seen on soft ground in the meizoseismal areas of great earthquakes. The effect is possibly connected with earth lurching, with which it may be confused.

(2) Especially where waves are reported in areas of relatively low intensity, illusions of several kinds affect the observations. If visible waves occur under such circumstances, they are probably rapidly shifting systems of standing waves; and estimates of their amplitude are almost always excessive.

LIGHTS

A still more elusive earthquake effect or illusion is that of lights, flashing or persistent, in the sky during an earthquake. They are usually compared to summer "heat lightning." Such reports are noted by Mallet for the 1857 earthquake, and rarely have they been missing from reports of any large earthquake in a populated area. In the nineteenth century the favored explanation was in terms of local thunderstorms. Since the general use of electric power, arcing along transmission lines is an available explanation; much arcing occurred in the Los Angeles area and in other heavily shaken parts of California during the Kern County earthquake of 1952, and many reports of light seen then are certainly due to that cause.

A systematic investigation by Terada of lights reported during the Izu earthquake of 1930 led to the conclusion that there probably is a real phenomenon independent of thunderstorms and electrical engineering (see Chapter 30); but the direction and nature of the reported lights fitted into no recognizable pattern. Some authors have suggested that relative displace-ment of large crustal blocks may perhaps manifest itself to some extent in the form of static electricity.

WEATHER

"Earthquake weather," as commonly described, is merely a popular fable. The idea that earthquakes take place chiefly when the weather is hot and sultry can be traced to classical writings. The Greeks attributed earthquakes to imprisonment of the winds underground; if they were so imprisoned, the weather above ground should be windless. In California, minor earthquake activity often shows a perceptible increase toward the end of the year, about the beginning of the rainy season, when large air masses are being shifted and the load on the earth's surface is changing.

Some further discussion of certain secondary effects will be found in the following chapter.

References

Slides and slumps

Lawson, A. C., "Landslides" in: *The California Earthquake of April 18, 1906, Report of the State Earthquake Investigation Commission*, Vol. 1, Part 2, pp. 384–401, Carnegie Institution of Washington, Washington, D. C., 1908.

Gilluly, J., Waters, A. C., and Woodford, A. O., *Principles of Geology*, Chapter 11, "Downslope Movements of Soil and Rock," W. H. Freeman and Co., San Francisco, 1951.

Sharpe, C. F. S., *Landslides and Related Phenomena*, Columbia University Press, New York, 1938.

Kintanar, R. L., Quema, J. C., and Alcaraz, A. P., *The Lanao Earthquake, Philippines, April 1, 1955*, Weather Bureau, Manila, 1955; mimeographed.

Cable breaks

Milne, John, "Suboceanic changes," *Geographical Journ.*, vol. 10 (1897), pp. 129–146, 259–289.

Heezen, B. C., and Ewing, M., "Turbidity currents and submarine slumps, and the 1929 Grand Banks earthquake," *Am. Journ. Science*, vol. 250 (1952), pp. 849–873.

————, "Orleansville earthquake and turbidity currents," *Bull. Am. Assoc. Petroleum Geologists*, vol. 39 (1955), pp. 2505–2514.

Heezen, Bruce C., "The origin of submarine canyons," *Scientific American,* vol. 195, No. 2 (Aug. 1956), pp. 36–41.

Air waves

Thomson, A., "Earthquake sounds heard at great distances," *Nature,* vol. 124 (1929), pp. 686–688.

————, "Abnormal audibility of sound of Murchison earthquake and Tarawera eruption," *N. Z. Journ.* vol. 12 B (1930), pp. 16–18.

Benioff, H., and Gutenberg, B., "Waves and currents recorded by electromagnetic barographs," *Bull. Am. Meteorol. Soc.,* vol. 20 (1939), pp. 421–426.

Mukherjee, S. M., "Landslides and sounds due to earthquakes in relation to the upper atmosphere, Part 1," *Indian Journ. Meteorology Geophysics,* vol. 3 (1952), pp. 240–257.

Visible waves

Dutton, C. E., "The Charleston earthquake of August 31, 1886," *U. S. Geol. Survey,* 9th Annual Report, 1887–1888, pp. 203–528, pls. VII-XXXI. (Visible waves are described on pp. 265–267.)

Wood, Harry O., in *Bull. Nat. Research Council,* No. 90 (1933), *Physics of the Earth,* vol. VI, *Seismology,* Chapter 6, p. 61; also *B.S.S.A.,* vol. 23 (1933), p. 171. (Reports Dr. Anderson's observation.)

Earthquake lights

Terada, T., "On luminous phenomena accompanying earthquakes," *Bull. E.R.I.,* vol. 9 (1931), pp. 225–255.

CHAPTER 11

Intensities and Isoseismals

INTENSITY SCALES

In Mallet's day it was becoming generally known that the distribution of the macroseismic effects of earthquakes could be represented by the drawing of isoseismals, or lines of equal apparent intensity of shaking.

Special Scales

At first each earthquake was quite properly investigated independently; even at the present time this is considered good practice. Especially when a large earthquake is being investigated and many observations are being correlated, it is scientifically preferable to begin by setting up isoseismals with reference to local conditions of terrain and construction. After this is done, the isoseismals may be correlated with those of some more generally applicable scale.

Local conditions sometimes almost force a special scale on the investigator. Thus, workers who took the field after the Turkish earthquake of 1939 found that conventional intensity scales failed to describe the damage to the tamped-earth construction common in that region, and they fell back on estimates of the percentage of damage in the various localities.

The Rossi-Forel Scale

Intensity scales intended for general application developed gradually, as the comparison of individual investigations led toward a common pattern. De Rossi in Italy and Forel in Switzerland, who had been working in this direction more or less independently, joined forces in 1883 to set up the Rossi-Forel scale. It was widely adopted. In seismological and engineering literature, when no particular scale is specified, earthquake intensity is usually expressed in terms of this scale; it is commonly indicated by the abbreviation R.F. followed by the Roman numeral of the scale degree. The scale is reproduced in Appendix III.

With the general advance of technology the R.F. scale grew progressively more out of date. An enormous range of intensity was lumped together at its highest level, X. Moreover, the descriptions of effects both on construction and on natural objects proved to be too specifically European.

The Mercalli Scale

These defects were largely removed in an improved scale put forward by Mercalli in 1902 at first with ten grades of intensity, later with twelve following a suggestion by Cancani who attempted to express these grades in terms of acceleration. An elaboration of the Mercalli scale, including earthquake effects of many kinds and ostensibly correlated with Cancani's scheme, was published by Sieberg in 1923. This form was in turn used as the basis for the Modified Mercalli Scale of 1931 (commonly abbreviated M.M.) by H. O. Wood and Frank Neumann.†

Modified Mercalli Scale Restated

The original publication gives the M.M. scale in two forms: one a lengthy statement modeled on that of Sieberg, with additions and modifications suggested by later experience; the other an abridgment meant for rough-and-ready use. The abridged form was prepared chiefly by one author, and at a few points is in conflict with the main scale. At the risk of putting a third version into circulation, this chapter presents an expansion of the shorter form, including most of the items in the complete form. Some items are omitted for definite reason, and a few additional notes are included, with initials (CFR) to separate them from the scale proper. Additional whys and wherefores of a technical nature will be found in Appendix III.

To eliminate many verbal repetitions in the original scale, the following convention has been adopted. Each effect is named at that level of intensity at which it first appears frequently and characteristically. Each effect may be found less strongly, or in fewer instances, at the next lower grade of intensity; more strongly or more often at the next higher grade. A few effects are named at two successive levels to indicate a more gradual increase.

Masonry A, B, C, D.   To avoid ambiguity of language, the quality of masonry, brick or otherwise, is specified by the following lettering (which has no connection with the conventional Class A, B, C construction).

Masonry A.   Good workmanship, mortar, and design; reinforced, especially laterally, and bound together by using steel, concrete, etc.; designed to resist lateral forces.

Masonry B.   Good workmanship and mortar; reinforced, but not designed in detail to resist lateral forces.

---

† A modification suited to conditions in the USSR has been worked out by Medvedev,

Masonry C. Ordinary workmanship and mortar; no extreme weaknesses like failing to tie in at corners, but neither reinforced nor designed against horizontal forces.

Masonry D. Weak materials, such as adobe; poor mortar; low standards of workmanship; weak horizontally.

Modified Mercalli Intensity Scale of 1931 (Abridged and rewritten†)

I. Not felt. Marginal and long-period effects of large earthquakes (for details see text).

II. Felt by persons at rest, on upper floors, or favorably placed.

III. Felt indoors. Hanging objects swing. Vibration like passing of light trucks. Duration estimated. May not be recognized as an earthquake.

IV. Hanging objects swing. Vibration like passing of heavy trucks; or sensation of a jolt like a heavy ball striking the walls. Standing motor cars rock. Windows, dishes, doors rattle. Glasses clink. Crockery clashes. In the upper range of IV wooden walls and frame creak.

V. Felt outdoors; direction estimated. Sleepers wakened. Liquids disturbed, some spilled. Small unstable objects displaced or upset. Doors swing, close, open. Shutters, pictures move. Pendulum clocks stop, start, change rate.

VI. Felt by all. Many frightened and run outdoors. Persons walk unsteadily. Windows, dishes, glassware broken. Knickknacks, books, etc., off shelves. Pictures off walls. Furniture moved or overturned. Weak plaster and masonry D cracked. Small bells ring (church, school). Trees, bushes shaken (visibly, or heard to rustle—CFR).

VII. Difficult to stand. Noticed by drivers of motor cars. Hanging objects quiver. Furniture broken. Damage to masonry D, including cracks. Weak chimneys broken at roof line. Fall of plaster, loose bricks, stones, tiles, cornices (also unbraced parapets and architectural ornaments—CFR). Some cracks in masonry C. Waves on ponds; water turbid with mud. Small slides and caving in along sand or gravel banks. Large bells ring. Concrete irrigation ditches damaged.

VIII. Steering of motor cars affected. Damage to masonry C; partial collapse. Some damage to masonry B; none to masonry A. Fall of stucco and some masonry walls. Twisting, fall of chimneys, factory stacks, monuments, towers, elevated tanks. Frame houses moved on foundations if not bolted down; loose panel walls thrown out. Decayed piling broken off. Branches broken from trees. Changes in flow or temperature of springs and wells. Cracks in wet ground and on steep slopes.

IX. General panic. Masonry D destroyed; masonry C heavily damaged, sometimes with complete collapse; masonry B seriously damaged. (General damage to foundations—CFR.) Frame structures, if not bolted, shifted off

---

† The author takes full responsibility for this version, which, he believes, conforms closely to the original intention. He requests that, should it be necessary to specify it explicitly, the reference be "Modified Mercalli Scale, 1956 version," without attaching his name. The expression "Richter scale" is popularly attached to the magnitude scale; it is desirable to forestall confusion between magnitude and intensity.

foundations. Frames racked. Serious damage to reservoirs. Underground pipes broken. Conspicuous cracks in ground. In alluviated areas sand and mud ejected, earthquake fountains, sand craters.

X. Most masonry and frame structures destroyed with their foundations. Some well-built wooden structures and bridges destroyed. Serious damage to dams, dikes, embankments. Large landslides. Water thrown on banks of canals, rivers, lakes, etc. Sand and mud shifted horizontally on beaches and flat land. Rails bent slightly.

XI. Rails bent greatly. Underground pipelines completely out of service.

XII. Damage nearly total. Large rock masses displaced. Lines of sight and level distorted. Objects thrown into the air.

Long-Period Effects. The most important general consideration in applying such a scale is that it brings together long-period and short-period effects. The latter are in the majority and may be roughly correlated with acceleration. The long-period effects represent large displacement, which often goes with comparatively moderate acceleration. With increasing magnitude the proportion of long-period to short-period phenomena tends to increase at all distances from the epicenter. Since the scale in general places the long-period effects where they appear during earthquakes of moderate magnitude, serious confusion has sometimes arisen in dealing with large shocks.

Large landslides, particularly those of the earth-slump type, are typical long-period effects; they are triggered more readily by large slow motion than by rapid shaking. This is the effect referred to in assigning large slides to X. Smaller slides, many of them of the earth-avalanche type, are common, as indicated, at intensity VII. However, great earthquakes sometimes precipitate large slumps in distant areas where the intensity is otherwise indicated as low as VI. Cracks and fissures, especially those due to earth lurches, behave similarly, so that intensity from such evidence has to be assigned with some reference to magnitude. The same applies to effects on works of construction where a long-period resonance is involved, as in the swaying and distortion of tall buildings or towers and in the overturning of elevated tanks.

A special group of long-period effects is that referred to under I. The complete scale lists them as: dizziness or nausea; birds or animals uneasy or disturbed; swaying of trees, structures, liquids, bodies of water; doors swing slowly. The swinging of chandeliers may be added. All these may be observed when no actual shaking is perceptible. Many of them are pendulum effects; chandeliers and large branches of trees may act as long-period seismoscopes. The oscillation of bodies of water is analogous; these effects are seiches (Chapter 9). The increased number of such observations with higher magnitude depends in part on the greater proportion of long-period motion. There is another factor of importance: intensity measured by any reasonable criterion falls off with increasing distance at first rapidly and then more and more slowly. For relatively small magnitude, the limiting distance for perceptibility is short, and the range of distance over which intensity is close to the limiting level is narrow. For large magnitude, intensity decreases gradually near the limiting distance, and the critical zone of marginal effects expands into a broad band surrounding the area of intensity II.

Other Notes on the M.M. Scale. Intensity II is often characterized as "felt by few," since usually a small proportion of a large group will notice shaking at this level. The increased perceptibility on upper floors, some years ago still regarded with doubt, has been put beyond question by recordings from strong-motion instruments operated by the U. S. Coast and Geodetic Survey simultaneously in the basements and on the upper floors of tall buildings in California. In practically every instance the instrument on the upper floor shows notably larger recorded motion.

Intensity III is an "in-between" intensity with no characteristic effect of its own; observations which seem too much for II and too little for IV are assigned to it. Intensity IV, on the other hand, is marked by a number of characteristic effects; in a well-investigated earthquake it can usually be separated into an upper and a lower level.

Intensity IX is another "in-between" level. General damage to ordinary foundations, however, is characteristic and has been added here. Ejection of sand and water, particularly in the form of earthquake fountains (Chapter 9), beginning on a small scale at VIII, becomes notable at this level provided that the necessary subsurface conditions exist; large and spectacular phenomena of this sort belong to X.

At intensities X, XI, and XII the 1931 M.M. scale describes some effects which are primary rather than secondary and hence are doubtful indicators of the degree of shaking. Disturbance in the immediate vicinity of a moving fault appears to depend largely on the nature of the ground. Ordinary structures located close to the fault trace in 1906 were not shaken with extreme violence. One barn was actually dragged off its foundations and shifted 15 feet by the faulting without being destroyed, and at the same place a brick chimney remained standing. On the other hand, great damage was done to redwood trees close to the fault trace, indicating large amplitudes in the long-period motion which snapped off their branches. In Imperial Valley in 1940 weak adobe structures close to the fault trace were injured no more than those several miles distant. A different description applies to Oldham's observations in 1897, where the fault traces broke through sound igneous rock, and there was every evidence of extreme violence in the immediate vicinity. For this reason the M.M. scale under XII specifies "Fault slips in firm rock."

Intensity and Acceleration

The author has participated in an attempt to correlate the degrees of the M.M. scale with ground acceleration in the manner attempted by Cancani. Many excellent seismograms written by the U. S. Coast and Geodetic Survey instruments in California and elsewhere are available for such study. A passable empirical relation is

$$\log a = \frac{I}{3} - \frac{1}{2} \tag{1}$$

where a is the acceleration in cm/sec^2 and I is the M.M. intensity. This is similar to Cancani's result, although it differs somewhat numerically. Here, of course, the intensity grades must be treated as true numerical quantities, which they are not. If one lets I = 1½ represent the limit of perceptibility between intensities I and II, $\log a = 0$ or $a = 1$ cm/sec^2. Various lines of evidence point to this as the level of shaking ordinarily perceptible to persons. If one lets I = 7½, $\log a = 2$ or $a = 100$ cm/sec^2 = 0.1g approximately. This is the acceleration commonly accepted by engineers as that which damages ordinary structures not designed to be resistant. One gets acceleration equal to g for I = 10½, which is rather low.

Mr. Frank Neumann has engaged in an elaborate effort using the same data to correlate intensity with acceleration, and eventually to complete Cancani's project by redefining intensity in quantitative physical terms. The chief difficulties are: (1) extreme variations introduced by differing types of ground; (2) effect of increasing magnitude in altering the proportion between long-period and short-period vibrations, and consequently between the corresponding groups of phenomena; (3) crudity of the non-instrumental data used to assign intensities, which often leads to legitimate debate as to their significance in relation to actual earth motion.

Professor G. W. Housner has obtained fairly satisfactory results by analysis of the integrated spectrum of ground motion, a procedure proposed by Benioff long before.

Intensity Meters

From time to time it is proposed to construct a cheap instrumental intensity indicator which could be widely distributed as are household thermometers. This is much less simple than it sounds. It used to be common to set up groups of blocks with different base and height, requiring a different overturning moment; but the results in actual earthquakes proved to have no simple meaning. Frequently the supposedly most stable block would be overturned while others remained standing; this is attributable to rocking, continuous motion, and repeated impulses. Trigger and relay systems might partly eliminate such difficulties, but the installation then rapidly approaches the complexity of a seismograph. A more effective proposal would be the wide distribution of low-magnification seismographs, not to the general public, but to responsible persons and institutions who will keep them in operating condition, either continuously or subject to triggering. Strong-motion equipment like that installed by the U. S. Coast and Geodetic Survey would serve the purpose, but their stations are not yet numerous enough to be used in drawing isoseismals. A special seismometer for use as an intensity meter has been developed and applied by Medvedev in the Soviet Union.

DRAWING ISOSEISMALS

Assigning Intensity

In assigning intensity on the basis of a given report it is important not to cling exclusively to one criterion. Rattling of windows is characteristic of IV; but it is often reported when the other evidence indicates no more than III or even II. Intensity VI characteristically causes alarm; but some observers may not be alarmed, or insist that they are not, when heavy objects are displaced and other evidence indicates VI or even VII.

Combining Observations. It is never enough to assign an intensity to each individual observation or report and then proceed directly to mapping. Effects on a given spot of ground, or damage to a particular structure, may be greatly increased or decreased by some special or local circumstance. Ground may crack and settle badly over an old well, long since abandoned and filled; or the failure of a single timber that has been weakened by termites may cause great damage to an otherwise sound dwelling. The field worker cannot hope to discover in a limited time all the specific causes of such individual variations; he brings together and compares the varied information bearing on intensity in a particular locality and eliminates those indications which diverge widely. He is on sounder basis if he can assign a cause to such divergences, but he is not always so lucky. Naturally, when his information comes from interviews or from written reports he must remember that most persons are not trained in scientific observation and that their accounts, framed under exciting circumstances like an earthquake, are to be regarded critically; but, since exceptional local effects do occur, he must not be too hastily skeptical.

The intensity assigned at a given place is often loosely called an "average"; some misguided workers even add up and average their individual intensity ratings as if they were measured numbers. In statistical terms, what is wanted is the mode, not the median or the mean; the intensity to be selected is that which represents the largest number of observations, after special circumstances and obviously divergent instances are allowed for. How small a local subdivision is undertaken will depend on the extent of detailed information available. Thus, for earthquakes in South Germany in 1911 and 1913, the investigators carried out a door-to-door canvass of certain city areas so that intensities could be assigned to individual blocks. After the earthquake of 1906, Mr. H. O. Wood examined every built-up block in the city of San Francisco. Intensity is usually assigned to a whole town, or to an extensive section of a city.

Subdivisions.   Often there is a clear subdivision which should not be ignored or averaged out. After the Kern County earthquake of 1952, the town of Arvin was separable into an eastern section, including the old business center, and a western section built up during a newer expansion. The character of damage might easily have led to intensity rating of IX to the east and VII to the west; but the masonry involved was respectively C to D and A to B. Among the masonry A was a large new reinforced brick supermarket, which was practically undamaged. With all due allowances, the writer believes that the actual intensity was higher in the eastern section of Arvin, because there is a possibility that the ground differs, perhaps in degree of consolidation or in the level of the water table.

Near the Epicenter.   There is evidence, which should be regarded less as established fact than as working hypothesis, that in the neighborhood of the epicenter the vertical component of motion is larger relative to the horizontal components than elsewhere.† Near the epicenter this effect would decrease the ordinary manifestations of intensity, for reasons already discussed (Chapter 8), and cause an underestimate of the actual shaking.

Effect of Ground and Path

How to allow for or represent the effect of ground is the chief problem in drawing isoseismals. Intensities and isoseismals are used for different purposes by the engineer and the geologist; their preferences on this point differ. The engineer's interest is primarily in what actually took place; if Arvin is assigned intensity IX, he will accept that the locality was heavily shaken and judge the various types of construction by their performance under such conditions. If the effect of ground in modifying the intensity at adjacent points is mapped with attention to detail, an isoseismal chart becomes an index to danger spots to be avoided in future construction or calling for special safety measures; and general study of the relation of intensity to different types of ground makes it possible to estimate the relative safety of foundations in places not yet tested by a severe earthquake.

---

† An instance of this was the Imaichi earthquake of 1949 (Chapter 30).

The geologist looks at an isoseismal map as evidence of what went on in the basement rock, especially the nature and extent of faulting. Effects of sedimentary and alluvial cover on intensity distribution are, from this point of view, merely disturbing and distorting. In drafting isoseismals, the geologist accordingly tends to disregard or compensate for the raising of intensity by local areas of bad ground and to retain only the ground effects of larger units like the Los Angeles Basin or the San Joaquin Valley. Some workers, optimistically confident that all apparent irregularities are due to local effects of this kind, have drawn their isoseismals as perfect circles; even for small earthquakes this is certainly going too far. Where large corrections are made for ground effects, two maps should be constructed—one showing the actual observed intensities, the other those inferred for basement or average ground.

Path. Isoseismals drawn from adequate data are rarely circular and often show elliptical elongation in the direction of the major structural trends. Thus the isoseismals of the San Francisco earthquake (Fig. 28-4) are elongated parallel to the Coast Range structures; of course, this is due in part to the great linear extent of faulting. However, a similar elongation appears for the 1952 earthquake (Fig. 28-24), although the strike of the active fault was then in a different direction. It has often been suggested that waves are transmitted with less loss of energy along the structures; it is not easy to see how this can come about except by better conduction through sound rock. There is often a longer continuous extent of competent rocks along a structural trend than in the transverse direction; when the waves emerge from such rocks into alluvium or unconsolidated sediments there is considerable absorption, accompanied by increase of local intensity. Frank Neumann notes that "a 22-fold increase in acceleration is indicated when earthquake vibrations passing through the granitic basement force the most responsive types of surface formations into vibration." In alluviated valleys the parts close to basement rock are extreme danger spots, exposed to local

FIGURE 11-1

Earthquakes on east flank of the Sierra Nevada, Nov. 28, 1929, and Sept. 14, 1941. Outer isoseismals indicate limits of perceptibility. [U. S. Coast and Geodetic Survey.] Instrumental epicenter, nearly the same for both, indicated by star.

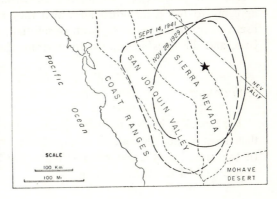

high intensity. (See the discussion in Chapter 5 of the effects at Monghyr in the Indian earthquake of 1934.)

This increase of intensity on emergence from rock into alluvium is followed by relatively rapid falling off of intensity as the disturbance proceeds further in alluvium. In California, earthquakes originating east of the Sierra Nevada often manifest higher intensity on alluvium near the west base of these mountains than at points equally distant eastward from the epicenter (Fig. 11-1); but intensity then falls off rapidly westward across the San Joaquin Valley. There is similar rapid decrease of intensity southward in Imperial Valley for earthquakes originating north of that area (see Fig. 28-13A). Such effects are similar to those described in some of the older literature as "earthquake shadows."

Ground and Long-Term Risk.   The examples from India and California show that interpretation of isoseismals of a particular earthquake calls for studying the effect of path, as well as of ground at the point of arrival. When estimating long-term risk at a locality exposed to earthquakes having many different origins, the general rule is to expect higher intensity on unconsolidated ground. The effect of emergence from basement rock accounts for this in part; but, even at a distance from basement rock, soft ground is especially subject to heavy shaking. This is due partly to slumping and local compaction; partly to lurching and perhaps to oscillations like those supposed to give rise to visible waves; partly to the disturbance of ground water. By comparison, the cushioning effect of absorption in unconsolidated material hardly needs to be considered. The Biblical comparison between the man who founded his house upon a rock and the man who built his on the sand —"and great was the fall thereof"—is still good.

Numbering Conventions

The old practice of numbering isoseismals with intensities supposedly applying to the points through which they are drawn is now not much in use. Generally the areas between isoseismals are marked with intensity numbers, and the lines are drawn to separate these areas.

Epicenters

The practice, in the absence of seismographs, of drawing isoseismals and then locating an "epicenter" at the center of figure should be discontinued. In the majority of cases the instrumentally located epicenter proves to be at one side of the meizoseismal area.

Interviews and Press Reports

The best method of obtaining field intensities is to examine effects on the spot as soon as possible after the earthquake; for the lower intensities, to interview several persons in each locality, being careful not to put questions so as to suggest particular answers. Newspaper reports are useful, if one becomes accustomed to journalistic phrases and to the habit of many reporters of inserting details which "ought to" be there whether they correspond to fact or not. "Buildings here were shaken" means only that persons felt an earthquake and does not imply that any structures vibrated visibly. "The shock traveled from east to west" is someone's estimate of direction of shaking; it means little unless the observer was out-of-doors, since impressions of direction indoors are too much affected by the orientation of buildings. "The shaking lasted for thirty seconds," if true, suggests either intensity VI or an earthquake of large magnitude with a distant epicenter; but it is likely to be an overestimate. Or, the duration reported for one locality may be given as if it applied to an entire area; or the general duration of a seismograph record, perhaps far outside the meizoseismal area, may be reported as if it were the duration felt by persons. This last misstatement is so common that the Pasadena laboratory long ago ceased to give any reply to the reporters' frequent question over the telephone—"How long did it last?"—since it appeared that the information, of little value in any case, was almost sure to be misapplied. "Two shocks" in a newspaper report usually refers to P and S waves, either as felt successively by persons or as reported by a seismological station. Magnitude numbers are often reported as intensities, and vice versa. "The strongest shock since . . ." usually refers to a particular locality, but it is often reported as if it were a general statement applying to magnitude. It has been a standing joke among seismologists that for fifty years almost every earthquake which reached noteworthy intensity at Eureka, California, was reported as the strongest there since 1906. Needless to say, there was no such steady increase as a literal interpretation would suggest.

For lack of better information, seismologists have sometimes been reduced to assigning intensities and even drawing isoseismals on the basis of such vague remarks as that the shaking was "slight," "sharp," "severe." Individuals differ widely in their use of such terms.†

---

† Townley's catalogue of California earthquakes includes the following under date of August 30, 1912: "The station agent at Cisco, Placer Co., reported: 'Had a hell of an earthquake.' It is a little difficult to translate this description into terms of the Rossi-Forel scale, but it would seem that even a mountain station agent would hardly be tempted to use such picturesque language for any intensity less than V."

Questionnaires

Information from carefully designed questionnaires is much more useful. For example, the U. S. Coast and Geodetic Survey circulates post cards carrying a questionnaire based on the M.M. scale. (See Appendix III.) These are mailed after almost every large earthquake; in addition, a considerable number of volunteer observers send in the cards regularly. These valuable observations are published, together with isoseismal maps, at first in detailed mimeographed form, then more condensed and summarized in the series *United States Earthquakes*.

THE GROUND FACTOR IN DETAIL

One of the U. S. Coast and Geodetic Survey strong-motion recording installations is on the principal campus of the California Institute of Technology, in the city of Pasadena about 3 miles from the Seismological Laboratory,

FIGURE 11-2   *Seismograms written simultaneously by standard torsion seismometers in the city of Pasadena* (A) *on alluvium, and at the Pasadena laboratory* (P) *on granitic rock. Clock marks at intervals of 1 minute. February 13 and March 2, California shocks. February 27, teleseism (southwest Pacific).* A + P, *superposition of* A *and* P.

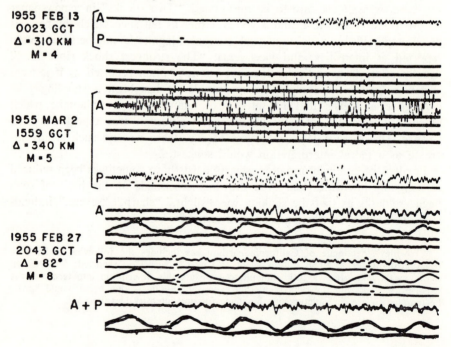

1955 FEB 13
0023 GCT
Δ • 310 KM
M • 4

1955 MAR 2
1559 GCT
Δ • 340 KM
M • 5

1955 FEB 27
2043 GCT
Δ • 82°
M • 8

and founded on alluvium at least a thousand feet thick; the Laboratory is on granitic rock. Studies in connection with the magnitude scale indicated that ground motion at the campus installation in local earthquakes was several times larger than that at the Laboratory. Two identical torsion seismometers were operated for several weeks, one at each location. Amplitudes recorded at the campus for short-period motion, both in local earthquakes and in teleseisms, averaged nearly 4 times larger than at the Laboratory; while the long-period waves from teleseisms, with wave lengths large compared to the distance between the two stations, were recorded with equal amplitudes and with motion almost exactly superposable (Fig. 11-2).

Dr. Gutenberg has compared these and other records wave for wave and has constructed plots of amplitude ratio as a function of period. His published results are the first fruits of an extensive program planned to compare ground motion in detail on sites of different character, at first in the vicinity of Pasadena, later in other parts of southern California.

INTENSITY UNDERGROUND

Scattered through the literature are statements that the effect of earthquake shaking is less in mines or caverns than at the surface. There were many reports of this nature for the Japanese earthquake of 1923. A thoughtful description of similar observations in Arizona for the Sonora earthquake of 1887 has been given by Staunton. On August 22, 1952, the strong aftershock of the Kern County earthquake series, destructive at Bakersfield, was felt generally in the Giant Forest area of Sequoia National Park, and caused some anxiety for the safety of a party being conducted through Crystal Cave; however, no one in the party noticed the earthquake.

Mines are usually in competent rock, while surface structures may be on alluvium or unconsolidated foundation. Thus the ordinary surface ground effect may account for some of the data. However, the ground in the Giant Forest region is largely granitic, while Crystal Cave is in limestone.

These are observations of shaking, not of faulting. On at least two occasions fault displacement in tunnels has been larger than at the surface above.†

D. S. Carder operated seismographs at the surface and at the 5000-foot level of the Homestake Mine, South Dakota. Records at 5000 feet showed no significant difference from those at the surface, except for disappearance of minor local and superficial disturbances. Kanai and Tanaka operated instruments simultaneously at the surface and at depths of 150, 300, and 450 meters in the Hitachi mine and compared the seismograms in detail; differences were not large. The macroseismic data for decreased intensity underground remain incompletely explained.

---

† Japan, 1930: Chapter 30; California, 1952: Chapter 28.

References

Intensity scales

de Rossi, M. S., "Programma dell'osservatorio ed archivo centrale geodinamico," *Boll. del vulcanismo italiano,* vol. 10 (1883), pp. 3–124. (Rossi-Forel scale, pp. 67–68.)

Forel, F. A., "Les tremblements de terre étudiés par la commission sismologique suisse pendant l'année 1881; 2me rapport," *Arch. sciences phys. et nat.,* vol. 11 (1884), pp. 147–182. (Rossi-Forel scale, pp. 148–149).

Mercalli, G., *Boll. soc. sismologica italiana,* vol. 8 (1902), pp. 184–191. (Improved Mercalli scale, still with 10 degrees.)

Cancani, A., "Sur l'emploi d'une double échelle séismique des intensités, empirique et absolue," *G. Beitr.,* Ergänzungsband 2 (1904), pp. 281–283. (Scale in terms of acceleration; modification of Mercalli scale to 12 degrees.)

Sieberg, A., *Erdbebenkunde,* Fischer, Jena, 1923. (Sieberg's scale, pp. 102–104.)

————, *Die Erdbeben,* Part V *in Handbuch der Geophysik,* Vol. 4, Borntraeger, Berlin, 1930, pp. 527–686. (Sieberg's scale, pp. 552–554.)

Wood, Harry O., and Neumann, Frank, "Modified Mercalli intensity scale of 1931," *B.S.S.A.,* vol. 21 (1931), pp. 277–283.

de Ballore, Montessus, "Sur la non-existence des courbes isoséistes," *Comptes rendus,* vol. 154 (1912), p. 1461.

Davison, C., "On scales of seismic intensity and on the construction and use of isoseismal lines," *B.S.S.A.,* vol. 11 (1921), pp. 95–129. (History and references.) Supplementary paper, *ibid.,* vol. 23 (1933), pp. 158–165.

Voigt, Dorothy S., and Byerly, P., "The intensity of earthquakes as rated from questionnaires," *ibid.,* vol. 39 (1949), pp. 21–26.

Medvedev, S. V., "Novaya seysmicheskaya shkala," *Trudy Geofiz. Inst. Akad. Nauk SSSR,* No. 21 (148), 1953.

Intensity meter

Medvedev, S. V., "Seysmometr dlya opredeleniya ball'nosti zemletryaseniy," *ibid.,* No. 36 (163), 1956, pp. 127–133.

Intensity and ground motion

Gutenberg, B., and Richter, C. F., "Earthquake magnitude, intensity, energy and acceleration," *B.S.S.A.,* vol. 32 (1942), pp. 163–191. (Includes the formula log $a = I/3 - 0.5$.) Second paper, *ibid.,* vol. 46 (1956), pp. 105–145 (especially p. 123).

Benioff, Hugo, "The physical evaluation of seismic destructiveness," *ibid.,* vol. 24 (1934), pp. 398–403.

Neumann, F., *Earthquake Intensity and Related Ground Motion,* University of Washington Press, Seattle, 1954.

Housner, G. W., *Intensity of Ground Motion during Strong Earthquakes,* California Institute of Technology, Pasadena, 1952.

United States Earthquakes, annual serial, U. S. Coast and Geodetic Survey, Washington, D. C. (Includes summarized data from strong-motion records used by Neumann and by Housner.)

Intensity near a moving fault

Louderback, G. D., "Faults and earthquakes," *B.S.S.A.,* vol. 32 (1942), pp. 305–330.

The ground factor in detail

Gutenberg, B., and Richter, C. F. (1956 paper as above, especially p. 123.)

Gutenberg, B., "Effects of ground on shaking in earthquakes," *Trans. Am. Geophys. Union.,* vol. 37 (1956), pp. 757–760.

Lais, R., "Die Wirkungen des Erdbebens vom 20. Juli 1913 in der Stadt Freiburg i. Br.," *Mitt. Badische Geol. Landesanstalt,* vol. 7 (1914), pp. 671–698.

Gutenberg, B., "Effects of ground on earthquake motion," *B.S.S.A.,* vol. 47 (1957), pp. 221–250.

Intensity underground

Staunton, W. F., "Effects of an earthquake in a mine at Tombstone, Arizona," *B.S.S.A.,* vol. 8 (1918), pp. 25–27.

Carder, D. S., "Seismic investigations on the 5000 foot level, Homestake Mine, Lead, S. D.," *Earthquake Notes,* vol. 21 (1950), pp. 13–14.

Kanai, K., and Tanaka, T., "Observations of the earthquake-motion at the different depths of the earth," *Bull. E.R.I.,* vol. 29 (1951), pp. 107–113.

Kanai, K., "The result of observation of wave-velocity in the ground," *ibid.,* pp. 503–509.

CHAPTER 12

Types of Earthquakes;
Volcanic Earthquakes

EARTH DISTURBANCES

Complete classification of changes going on in the earth would involve the whole of geology and geophysics. It would include the very slow changes over geological periods, the motions like those of earth slumps and slides continuing over periods of years, seasonal changes, the daily tides, and so on down to the relatively rapid and short-lived phenomena of earthquakes. Recorded along with earthquakes on seismograms, and sometimes becoming a great nuisance to the seismologist, is a miscellany of occurrences. (See Table 12-1.)

Continuous Disturbances

To the seismologist interested primarily in earthquakes, the continuous disturbances are chiefly a nuisance. At best, they are responsible for the background level of "noise" which ultimately fixes the highest useful level of magnification for seismographs. At worst, they make it practically impossible to record earthquakes at certain localities. The effects of traffic and machinery make it advisable to set up first-class installations at a distance from centers of population. The usual need for electric power adds a difficulty, since powerhouse machinery may put enough vibration in the ground to disturb seismograms several miles away. Powerhouse disturbances may also include those due to water going over a dam spillway or to waves produced by wind on a reservoir.

The disturbances just named are usually rather short-period in character, ranging from 3 to 10 cycles per second. They interfere with recording of local earthquakes, but affect long-period instruments for teleseismic work less seriously. However, traffic, construction, and other activity may produce local settling and tilting, which may throw long-period recorders out of adjustment.

Natural microseisms are allotted a separate chapter (Chapter 23). Their prevalence in a given location may affect instrumental work even more than

150

Table 12-1 Earth Disturbances Recorded by Seismographs*

A. Continuous disturbances
  1. Artificial
     Traffic
     Machinery
  2. Natural (microseisms)
     Meteorological: storms, wind, frost
     Water in motion: surf, streams, waterfalls
     Volcanic tremor
B. Single disturbances
  1. Artificial (chiefly explosions)
     Blasting: quarry or road work, geophysical exploration
     Explosives tests
     Demolitions
     Bombing and bomb tests
     Gunfire
     Accidental large detonations
  2. Natural (including earthquakes)
     I. Minor causes
        Collapse of caves
        Large slides and slumps
        Rockbursts in mines
        Meteorites
     II. Volcanic shocks
        Superficial, explosive
        Magmatic or eruptive
     III. Tectonic shocks
        Shallow or normal (depths not over 60 kilometers)
        Intermediate (depths 70 to 300 kilometers)
        Deep (depths 300 to 720 kilometers)

* This table, and the following discussion, show the problems involved in defining the term "earthquake." Authors have differed extremely. Some have practically restricted the meaning to tectonic shocks, others accept only volcanic shocks in addition; some include artificial shocks, and a few have framed definitions so broad as to include microseisms.

artificial disturbances. Because both artificial and natural continuous disturbances are accentuated on soft ground, seismological stations are founded preferably on well-consolidated rock, best on granite or Paleozoic sediments.

Single Disturbances

Gunfire. Gunfire has been listed under single disturbances, since usually separate detonations can be distinguished. Seismograms in California have often shown disturbances due to air waves from gunfire. Sound waves passing up toward the stratosphere are often refracted downward again; at distances of 100 miles or more they may arrive with considerable energy, rattling windows and giving a popular impression of an earthquake. The blow of such a sharp acoustic wave against an exposed rock surface is often sufficient to record on a nearby sensitive seismograph; or there may be dynamic coupling between rock and air. In this way air waves from large explosions are often recorded; at least one such wave from a detonating meteor has been reported, as well as some due to aircraft exceeding the speed of sound. Conversely, Dr. Benioff has noted several instances in which air waves generated by strong earthquakes in or near the meizoseismal area have been recorded by sensitive barographs at distances of several hundred miles (Chapter 10).

Explosions. Explosions, accidental or intentional, are important in seismology. Their place of origin being known, they provide a control for the methods used to locate earthquakes. Still more information can be derived when the time can be fixed precisely, as by recording the closing of the firing switch for a quarry blast.

In central and northern Europe, where large earthquakes are rare and there are many seismological stations with good timing, explosions have played an exceptionally important part. After World War I a large supply of surplus nitrates was stored at Oppau, in the Rhine Valley; these were ammoniated and intended for use as fertilizer. Ammonium nitrate has the property of sudden chemical change. On September 21, 1921, the site of Oppau became a large crater; windows were shattered at Mannheim, 5 kilometers away, and the shock was recorded at stations over 300 kilometers distant. From seismograms Jeffreys calculated the energy radiated in elastic waves and found it to be of the order of one thousandth of the chemical energy computed from the known quantity and heat of reaction. Even smaller fractional energies have been found entering the earth from other superficial detonations. A similar explosion of nitrates occurred in 1946 in the harbor at Texas City, but no seismographs were near enough to give useful records.

European seismology was advanced by a large accidental explosion at Burton-on-Trent in 1944, and even more by a deliberate one at Helgoland in 1947, when a large quantity of explosives was used to demolish fortifications and other installations. The 1947 explosion was carefully timed, and was recorded at numerous specially occupied locations as well as at permanent stations in Europe. A smaller explosion at Haslach in south Germany also was timed and observed carefully; it gave results valuable for correlation with the earthquakes which are fairly frequent in that area and southward. The most important accidental explosion outside Europe was probably that of a large quantity of high explosive at Port Chicago, near San Francisco, in 1944; this was recorded at the California stations and has been thoroughly studied by Professor Byerly.

Many of the best data of this type have come from quarry operations. It is economical to set enough explosive to break up a large quantity of rock at once; for this purpose the charge is buried so that relatively little energy goes into the air. In southern California many of the large blasts timed have been at limestone quarries operating in connection with cement plants; but some of the largest and most important have been fired near Corona, California, where porphyritic dacite is quarried to be ground up for use in roofing.

Atomic Bombs. Seismic waves from atomic bombs have been recorded at great distances. With careful planning and preparation, such observations could materially advance seismology and general geophysics. Arbitrary security restrictions have usually prevented such planning and have interfered with the publication of such data as have been obtained. The bomb tests in New Mexico and Nevada provided seismograms at the California stations which were consistent with established conclusions as to wave velocities and crustal structures in the region between, but could not be used for more precise purposes. Of the two Bikini bombs in 1946, the first, fired in the air, is not known to have recorded at distant stations; but the second, Test Baker, fired under water on July 24, was recorded (*P* waves only) by almost all the short-period Benioff seismographs in the California area. Since the detonation was timed very exactly, the time of arrival of *P* at these distant stations provided a fundamentally important check on the established travel times for earthquake waves. The actual arrivals were about 3 seconds earlier than the times then calculated from standard tables for a surface source. About 1 second of this difference is probably accounted for by higher mean velocity in upper levels in the Pacific basin than in the continental areas where most of the earthquakes used for constructing the tables originated. The remaining 2 seconds are removed by later revision of standard times and by reinterpretation of seismograms at short distances (Chapter 18). The results have been confirmed and extended by data from later tests; unfortunately it has been possible to publish the latter only in a very incomplete form.

As in other surface detonations, only a small fraction of the energy of an atomic bomb enters the earth as seismic waves. *P* waves recorded for fission bombs at distant stations are of about the same size as those of an earthquake of magnitude 5.5. However, even the total energy released is small compared to that of large earthquakes. A recent official publication gave the energy of the largest earthquakes as a million times that of the nominal bomb of Hiroshima type; however, this was based on a magnitude-energy relation which has since been revised (Chapter 22), and the factor of ? million should be reduced to ten thousand.

MINOR CAUSES OF EARTHQUAKES

Collapse of Caves

General. In the mid-nineteenth century a number of writers, most notably Volger, attributed the majority of large earthquakes to the collapse of caverns or underground cavities (German *Einsturzbeben,* sometimes rather dubiously translated "impact earthquake"). This view was slowly abandoned as more and more evidence appeared that important earthquakes were connected with faulting and related processes, as well as originating at depths of 10, 20, or more miles. That small shocks may originate by collapse is an obvious speculative possibility, but there is little in the way of direct evidence. In limestone regions, where caverns are numerous, occasionally the roof of a cave falls in, or a large block drops out of place far underground. Such an event must certainly cause a small earthquake; however, shaking would be perceptible only nearby, and only the most sensitive seismographs could register it at a distance of 10 miles or so. Similar collapse of cavities may take place in volcanic areas, but it is at least as possible that earthquakes in these areas attributed to such a cause are due more directly to volcanic action.

Thuringia—1926. One fairly well-documented instance of a possible collapse earthquake is a small shock in Thuringia (Germany) on January 28, 1926. This caused slight damage (intensity VI) at the town of Stadtroda and was felt to distances of about 40 kilometers. It was recorded at four stations, the most distant being Göttingen (146 kilometers). It thus had a rather high epicentral intensity for so small a shock, indicating relatively shallow origin. However, later work by Hiller and others at Stuttgart has shown that in southern Germany, along with ordinary earthquakes with depths of 30 to 40 kilometers there occur others, usually small, at depths near 5 kilometers. This would fit the 1926 data within their limits of accuracy. Sieberg discusses the local geology and makes a fairly good case for a collapse earthquake due to outwashing of a salt deposit by underground water. In his writings Sieberg had previously championed the reality of small *Einsturzbeben* (Stadtroda is not far from Jena, where Sieberg was a professor).

Slides and Slumps

Somewhat similar questions arise when large earthquakes are attributed to great slumps or landslides. In 1911 there was a great earthquake under the Pamir Plateau in Asia which was accompanied by an enormous and devastating slide. Galitzin, and others after him, calculated that the gravitational energy set free in the descent of the slide was of the same order as that radiated in the elastic waves recorded by seismographs. Later work suggests that this involved perhaps as much as a hundred-fold underestimate of the elastic wave energy. Moreover, large earthquakes are not infrequent in that region. The efficiency with which the energy of the slide was communicated to the ground must have been low; the process took time, and generally we find that only a small fraction of energy from superficial sources of shocks penetrates the earth to great depths and is recorded at distant stations.

Terminal Island, California. The writer's earlier attitude toward such hypotheses has been modified somewhat by a remarkable series of disturbances under Terminal Island, in Long Beach harbor, about 45 kilometers from Pasadena—perhaps another good instance of one's being most impressed by events near home. What goes on is describable as slumping on an enormous scale, incidental to subsidence. A Tertiary sedimentary series is being faulted, or rather fractured along the bedding planes, at a depth of about 1700 feet. As the fracture approaches the surface, it departs from the bedding planes and becomes steeper, so that the whole has the characteristic profile of a slump. This motion is attributed to the removal of support by oil operations. Oil is produced from a lower level, so that the well casings pass through the fracture zone. In 1947 a number of wells were damaged by fracturing involving motion up to about a foot which sheared some casings and sanded some wells. The date of occurrence is not known, but it may well have been December 14; on that day seismographs at Pasadena and auxiliary stations recorded an apparent earthquake of peculiar characteristics; the epicenter could not be located closely but was placed in the vicinity of Long Beach.

On November 17, 1949, the instruments recorded a disturbance of similar character, but much larger. There were reports that a slight earthquake had been felt in Long Beach by a few persons. The proper authorities were notified, and investigation revealed a major economic disaster. Nearly 200 oil wells went off production, many of them permanently; damage was at least 9 million dollars. There were similar occurrences on August 15, 1951, and on January 25, 1955; less damage was done on these dates, partly because of a shift in the center of the disturbance.

Some of the seismograms are reproduced in Figure 12-1. They may be compared with those of ordinary earthquakes originating at roughly the same distance from the recording stations (see also Fig. 18-6). The outstanding feature is the relatively large development of long-period motion, which presumably represents surface waves; this is theoretically to be expected as a result of the shallow depth of origin. There is also a lack of sharpness in the beginning of the motion which may be correlated with a more gradual

APRIL 6, 1933
TERMINAL ISLAND   NORMAL EARTHQUAKE

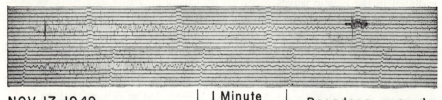

NOV. 17, 1949 | I Minute | Pasadena records
TERMINAL ISLAND   torsion seismometer

FIGURE 12-1   *Pasadena seismograms. Slump earthquakes, Terminal Island, Long Beach harbor, California. The 1933 seismogram also shows a normal earthquake in the same area.*

event than the breaking of rocks in normal tectonic earthquakes. For this reason it is difficult to read exact times, and the epicenter cannot be placed precisely from seismometric data.

Seismograms of the same kind were written on April 6, 1933, some years before the oil development in the harbor area. This disturbance occurred during the aftershock activity following the Long Beach earthquake of that year, but the other small shocks of that series wrote records of normal appearance. Subsidences and large slumps are constantly going on in similar sedimentary rock along the coast.

Triggered Slumps—Inglewood Fault Zone.   At least two instances are known of a subsurface displacement of this kind following or accompanying an earthquake of normal character. They occurred in 1941 and 1944 in oil fields along the Inglewood fault zone. The corresponding seismograms do not show the peculiar features described, and the earthquakes appear to have originated at the usual depth of about 16 kilometers. The damaging displacements must have been triggered, either by the direct shaking of the earthquake or by the readjustment of the local strain pattern.

Caloi and collaborators have attributed a series of earthquakes in the Po Valley to the commercial extraction of methane gas.

Rockbursts

Rockbursts in mines are due to failure of the rock under stresses resulting from removal of material. In many mines such failures occur gradually as mining proceeds, but, where the rock is very competent, or structurally resistant, the stresses may accumulate to high values before failure. The occurrence is then sudden, comparable to an explosion, and may be disastrous in the mine. The energy thus suddenly released produces a true earthquake, which may be recorded by distant seismographs. Since the locality of origin is known, rockbursts have afforded opportunity for special seismological investigations, notably in Canada and South Africa. A permanent station was established at the mine at Kirkland Lake, Ontario; the records at this and at the other fixed stations in eastern Canada and New England were supplemented by others at fifteen locations which were occupied temporarily a few at a time. In South Africa an instrument at the mine was triggered by a rockburst and sent out a radio impulse which started the field instruments recording at high speed in advance of the arrival of the earth wave; these records were telemetered back to a home station. Valuable data on seismic wave velocities and crustal structure have been obtained in this way.

Meteorite Earthquake—1908

The only documented instance of an earthquake caused by the impact of a meteorite refers to the great fall in Siberia on June 30, 1908 (Gregorian date; the Julian calendar was then in use in the Russian empire). For this extraordinary type of event E. Tams proposed the German term *Aufsturz-beben*. It is this term, rather than *Einsturzbeben,* which is properly translated as "impact earthquake."

The point of impact was about 61° N, 101.3° E, in the tundra region inhabited by the chiefly nomadic Tungus peoples. Russian authors use the identification *Podkamennaya Tunguska* (Stony Tunguska) for the meteorite, since the fall was near the river of that name. The huge air waves shook structures a hundred miles away like an earthquake; they were recorded on barographs at distant points, notably in England. There is no report that the earthquake as such was felt; it was large enough to write, in central Europe, legible records which consisted chiefly of surface waves, as might be expected.

The central region was first reached by a scientific expedition in 1927. Impact craters were found, and the forest was devastated over an area 15 to 20 kilometers in diameter. Trees were felled to distances of about 20 kilometers, and occasionally as far as 30 kilometers. Popular writers have reveled in horrifying speculation on the consequences if these meteorites had descended on a great city.†

VOLCANIC EARTHQUAKES

If one looks at two small-scale world maps, one showing active volcanoes and the other epicenters of large earthquakes, the distribution is seen to be similar. The circum-Pacific belt of activity (Chapters 25, 26) stands out

---

† On February 12, 1947, there was another great Siberian meteorite fall, in the Sikhota Alin at 46° 10′ N, 134° 39′ E. It was not felt as an earthquake, even at points only 15–20 kilometers distant. No registration was found on searching the seismograms at Vladivostok (the nearest station, 400 kilometers distant), and no other station reported a recording at the time.

particularly well, so that with reference to volcanism it is often called the "circle of fire."

Large-scale maps modify this first impression. There is usually a clear geographical separation of a hundred miles or more between the belts of active volcanoes and of major tectonic activity. The Aleutian Islands bear a chain of active volcanoes which continues into the Alaska peninsula, while the principal earthquake epicenters are well offshore to the south. In Mexico the major earthquakes originate near the coast, and the active volcanoes are inland. In California the principal volcanic line extends from Mt. Lassen and Mt. Shasta northward through the Cascade Mountains, while important earthquake activity is associated either with the Coast Range or with the Basin and Range structural province which extends from the Sierra Nevada eastward. In Chapter 4 we noted that the active volcanoes of Italy are well within the tectonic active arc marked by the Apennines.

Earthquake Types in Volcanic Regions

It is not astonishing that in Mallet's time large earthquakes were commonly attributed to volcanic action, perhaps at a late stage and not manifesting itself by eruptions. The notion still persists popularly, and the issue is often confused by attributing all earthquakes in volcanic areas to volcanic causes; this is almost certainly incorrect. There are at least four groups of such shocks: (1) superficial explosions; (2) shocks at shallow depths of the order of a few kilometers, probably associated with magmatic movements or other volcanic processes; (3) tectonic earthquakes at their usual depths ranging in a given region from 15 or 25 to 60 kilometers; (4) earthquakes in the intermediate depth class, usually near 100 to 150 kilometers.

Explosions.   During eruptions seismographs near volcanoes usually record numerous small sharp shocks which can sometimes be correlated with actual explosions seen or heard. These may occur in the volcanic vent considerably above the ground level, or even among the gases already emitted from the crater. In the latter case the recording is of the same type as the seismograph registrations due to air waves from other explosions.

Shallow Shocks.   The second group of shocks is of far greater interest; they afford some of the few opportunities to use seismograph data for valid prediction. Such shocks normally increase in number before an eruption; if more than one station is available, it may even be possible to localize them and so to forecast the place of the approaching outbreak. For this purpose seismographs of low magnification are operated in connection with many volcano observatories. Since shocks of this type are of relatively shallow origin, they may produce effects of surprisingly high intensity in a small area.

Volcanic Tremor. Successive small shocks of the first two groups are probably partly responsible for the continuous vibrations called volcanic tremor. An alternative explanation is that continuous irregular vibrations due to volcanic processes may set up and maintain a natural oscillation of a local structure or subdivision of the crust. This type of interpretation has also been applied to microseisms associated with weather conditions (Chapter 23).

Tectonic Shocks in Volcanic Areas. The occurrence of the third class, tectonic shocks, is responsible for much of the confusion about volcanic earthquakes in the literature, since often such earthquakes have occurred so obviously in association with an eruption as to leave little doubt of causal connection. The fourth class, shocks of intermediate depth, do not present this kind of problem, and their discussion will be deferred to Chapter 19.

An Example—Mauna Loa, 1916

The following description of volcanic shocks is quoted from H. O. Wood:

In the early morning of the 18th of May, 1916 . . . a very considerable . . . burst of fumes rushed rapidly upward from a new vent (a line of fissures) high up on the south flank of Mauna Loa . . . Some thirty very feeble earthquakes were registered during an interval of about twelve hours before, during, and after the . . . voluminous fuming. Some twenty-four hours then elapsed without further observed seismic or volcanic event. Then . . . earthquakes began to be registered at short intervals. Most of these were stronger than any of the thirty that accompanied the outbreak of fuming . . . Shocks now continued to occur in greater and greater number, and the stronger ones reached greater and greater intensity, for several days; and this held true even after flow had broken out, which occurred about thirty-six hours later (about sixty hours after the beginning of the fume eruption) . . . about thirteen miles further down the mountain than the source of the first outrush of fumes . . . As the flow declined the earthquakes fell off in numbers, but those which did occur continued to be stronger than those of the early group. One of the last of the series was stronger than any previous one.

Ischian Earthquakes

The Naples Area. Many similar descriptions have been published. For a striking series of earthquakes not followed by an eruption we may consider the famous shocks of Ischia. The volcanic area of the Bay of Naples (map, Fig. 12-2) is one of the best known and most studied in the world. Many of its features have provided type examples for volcanologists. The most conspicuous is Vesuvius, east of Naples; but to the west is the complex area

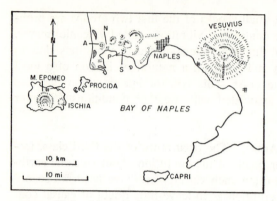

FIGURE 12-2   *Map of the Bay of Naples, show-ing Ischia and other features. A, Lake Avernus (Averno); C, Casamicciola; N, Monte Nuovo; P, Pozzuoli; S, Solfatara.*

of the Phlegrean Fields, which includes: (1) Lake Avernus, the mythical gateway to the infernal regions; (2) Monte Nuovo, which broke out as a new volcano in 1538; (3) Pozzuoli, the location of the "temple of Jupiter Serapis" (actually a market place; the columns of this structure, with their marks of boring mollusks showing former submergence, have figured in geological texts since Lyell's day as proof of changes of level in historical time); (4) The Solfatara, whose emission of fumes and gases has made it the type for similar features elsewhere.

To the southwest is the small island of Procida, and beyond that is Ischia —larger, but still only 6 miles across—rising to the volcanic crater of Epomeo. There were eruptions on Ischia in classical antiquity; but there appear to have been about a thousand years of quiet preceding 1302, when a new crater opened on the east flank of Epomeo and lava flowed down to the sea. Since then there has been no further eruption.

Shocks at Casamicciola.   The principal town of Ischia is Casamicciola on the north slope, a resort dependent on hot springs. Beginning in 1762 earthquakes occurred on Ischia which were most intense at and near Casamicciola, notably in 1796, 1828, 1841, 1867; there was a destructive shock in 1881, and finally a disastrous one on July 28, 1883, which destroyed most of Casamicciola and took over 2300 lives. The earthquakes of 1881 and 1883 were investigated by several workers who published their findings.

The most remarkable feature of these occurrences, aside from their repetition from nearly the same source, is their low intensity at comparatively small distance, considering their destructiveness at Casamicciola. To be sure, we are dealing with masonry C or D of the sort described by Mallet for the 1857 earthquake; but the contrasts in intensity are extreme. In 1881 there was a small area in which masonry was completely destroyed; yet there were places on Ischia where the earthquake was not noticed; shaking was slight on Procida, and perceptible only to a few persons at the nearest points of the mainland. In 1883 the area of destruction was 2 miles wide, but at the more distant points on the island there was no damage; while the shock was felt on the mainland, it was perceptible to few persons at Naples, only 20 miles away.

Abortive Eruption? These circumstances strongly suggest a comparatively shallow hypocenter. In both 1881 and 1883 field workers identified a narrow band near Casamicciola running a little west of north and radial to the crater of Epomeo; in this band apparent intensity was extreme, and there was evidence that the motion had been largely vertical.

It is easy to see why these repeated earthquakes were interpreted as signs of an approaching eruption like that of 1302, as if a mass of magma were being forced toward the surface and breaking the rock on its way. It is of interest, therefore, that since 1883 there have been only minor earthquakes on Ischia. The conclusion might be drawn that the earthquake of that year relieved the accumulated strain and inhibited or at least postponed an eruption.

Hawaii—1929

Some of the well-observed earthquakes of Hawaii (map, Fig. 12-3) are of interest in this connection. Near the west coast of the large island Hawaii is the volcano Hualalai, which erupted last in 1802. In the month beginning September 19, 1929, thousands of small earthquakes originated in the vicinity of Hualalai, in the direction of Mauna Loa. An eruption, of either Hualalai or Mauna Loa, was expected but did not occur. There were two larger shocks in the group, one on September 25 of magnitude near 5½, and one on October 5 of magnitude 6½. These, and especially the latter, registered at distant stations. Seismograms for these larger shocks are of normal appearance and do not suggest especially shallow origin. They appear to have been tectonic shocks associated with and perhaps modifying a volcanic process. Their occurrence sheds a little more light on the events of March and April, 1868.

Hawaii—1868

On the morning of March 27, 1868, an eruption began near the summit of Mauna Loa. Later in the same day persons on Hawaii began to notice frequent earthquakes; one on March 28 at 1:28 P.M. was strong enough in some areas to throw down stone walls. The largest shock of the series took place on April 2 at 4 P.M.; on Hawaii it was most violent in the southern half of the island, but even at Hilo stone walls were thrown flat. Farther south, "nearly every wooden house at Keiawa, Punaluu, Ninole and beyond were shot off their foundations or tumbled over, and straw houses with posts in the ground were torn to shreds. At Kapapala the vault of the cistern . . . was shot off like a quoit, and the cistern itself smashed in and shut together so that not a vestige was to be seen of it." Intensity X on Hawaii was accompanied by perceptibility on all the main islands of the group; at Honolulu, roughly 150 miles distant, the intensity was IV or V. There was an

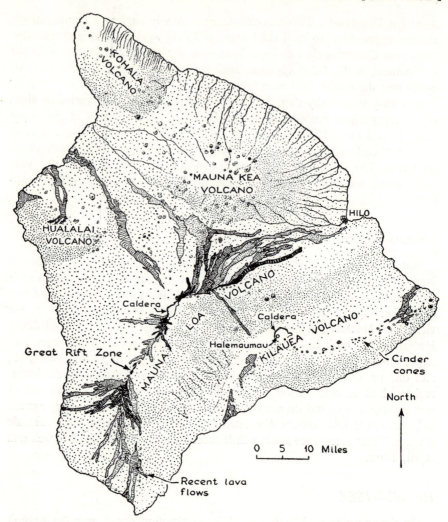

FIGURE 12-3   *Generalized map of Hawaii.* [*From* Principles of Geology *by James Gilluly, Aaron C. Waters, and A. O. Woodford. San Francisco: W. H. Freeman and Company, 1951.*]

accompanying sea wave; independent accounts state that it came in on the southern coast of Hawaii over the tops of cocoanut trees, at a height of 50 or 60 feet. At Honolulu the water rose and fell 5 feet.

This large earthquake was clearly no shallow disturbance associated merely with volcanic action. It must have been a tectonic shock like those of non-volcanic regions, or like those under Hawaii in 1929. The sea wave was no typical tsunami, but must have been due to block displacement or local subsidence. Since 1929 several strong shocks have occurred in the Hawaiian area, some of which show no special correlation with eruptions; the largest of these was on January 22, 1938 (Hawaiian local date), of magnitude 6¾.

Coincident Tectonic Shocks

Coincidence of tectonic earthquakes with volcanic eruptions has often been reported, but such reports should be studied critically. Three comparatively well-documented instances, from Japan, Alaska, and Mexico, follow.

In January, 1914, there was a large eruption of Sakurajima, in an inlet of Kyushu. After the eruption had started, an earthquake occurred which was destructive in the city of Kagoshima, not far from the volcano. This was a major shock (magnitude 7), with an estimated depth of 50 kilometers. The epicenter cannot be fixed closely.

On June 7, 1912, after the great eruption of Katmai had started, an earthquake of magnitude 6.4 occurred off the adjacent Alaskan coast; another of magnitude 7.0 followed on June 10.

On February 22, 1943, about 3 days after the outbreak of the new volcano Parícutin, a destructive earthquake of magnitude 7.5 occurred near the coast about 100 miles to the southeast of the volcano.

Critical objection to these few instances may well be made, in view of the much more numerous occasions when a major eruption is not accompanied or followed by a large earthquake in the same region. It is difficult, however, to dismiss them as mere coincidences. The geographical separation between volcano and epicenter in the Alaskan and Mexican instances makes it clear that these are not volcanic earthquakes in any useful meaning of the term.

Occurrence of repeated earthquakes in areas of presumably extinct volcanism has been noted in Chapter 6. Such are those of the Kaiserstuhl, a volcanic mass in the Rhine Valley; yet Lais, who studied their history closely, considered them tectonic.

On August 20–23, 1954, a swarm of earthquakes originated near the Arctic island Jan Mayen. About 30 of these were of magnitude ranging from 5 to 5¾ and were recorded at many European and American stations. Although Jan Mayen is an active volcano, no eruption was reported at the time.

References

Gunfire, etc.

Gutenberg, B., and Richter, C., "Pseudoseisms caused by abnormal audibility of gunfire in California," *G. Beitr.*, vol. 31 (1931), pp. 155–157.

MacCarthy, G., "Earth tremors produced by a large fireball," *Earthquake Notes,* vol. 26 (1955), p. 20.

Artificial explosions (see also references to Chapter 18)

Wrinch, Dorothy, and Jeffreys, H., "On the seismic waves from the Oppau explosion of 1921 Sept. 21," *M.N.R.A.S. Geophys. Suppl.,* vol. 1, pp. 1–22, (1923).

Jeffreys, H., "On the Burton-on-Trent explosion of 1944. November 27," *ibid.,* vol. 5 (1947), pp. 99–104.

Willmore, P. L., "Seismic experiments on the North German explosions, 1946 to 1947," *Phil. Trans. Royal Soc. (London),* Ser. A, vol. 242 (1949), pp. 123–151.

Reich, H., Schulze, G. A., and Förtsch, O., "Das geophysikalische Ergebnis der Sprengung von Haslach im südlichen Schwarzwald," *Geol. Rundschau,* vol. 36 (1948), pp. 85–96.

Rothé, J. P., and Peterschmitt, E., "Etude séismique des explosions d'Haslach," *Ann. inst. physique globe (Strasbourg),* part 3, vol. 5 (1950), pp. 13–38. (Extensive bibliography of artificial explosions.)

Byerly, P., "The seismic waves from the Port Chicago explosion," *B.S.S.A.,* vol. 36 (1946), pp. 331–348.

Wood, H. O., and Richter, C. F., "A second study of blasting recorded in southern California, *ibid.,* vol. 23 (1933), pp. 95–110.

Gutenberg, B., "Travel times from blasts in southern California," *ibid.,* vol. 41 (1951), pp. 5–12.

————, "Waves from blasts recorded in southern California," *Trans. Am. Geophys. Union,* vol. 33 (1952), pp. 427–431.

Byerly, P., "Subcontinental structure in the light of seismological evidence," in *Advances in Geophysics,* vol. 3, 1956, pp. 105–152.

Atomic bombs

Compare last preceding reference, and:

Gutenberg, B., "Interpretation of records obtained from the New Mexico atomic bomb test, July 16, 1945," *B.S.S.A.,* Vol. 36 (1946), pp. 327–330.

Gutenberg, B., and Richter, C. F., "Seismic waves from atomic bomb tests," *Trans. Am. Geophys. Union,* vol. 27 (1946), p. 776.

Gutenberg, B., "Travel times of longitudinal waves for surface foci," *Proc. Nat. Acad. Sciences,* vol. 39 (1953), pp. 849–853.

"The effects of atomic weapons," Govt. Printing Office, Washington, D. C., 1950. (Cf. especially pp. 13 and 111.)

Einsturzbeben

Volger, O., *Untersuchungen über das Phänomen der Erdbeben in der Schweiz,* Gotha, 1857–1858. (An obsolete classic of seismology.)

Sieberg, A., and Krumbach, G., "Das Einsturzbeben in Thüringen vom 28. Januar 1926," *Reichsanstalt für Erdbebenforschung*, Jena, No. 6, (1927).

Grant, U. S., "Subsidence of the Wilmington oil field, California," Calif. Dept. Nat. Resources, *Div. Mines, Bull.* 170 (1954), chapter X, pp. 19–24.

"Seismological notes," *B.S.S.A.*, vol. 40 (1950), pp. 69–70; vol. 45 (1955), pp. 161–162. (Terminal Island disturbances.)

Benioff, H., Gutenberg, B., and Richter, C. F., "Progress report, Seismological Laboratory, California Institute of Technology, 1949," *Trans. Am. Geophys. Union*, vol. 31 (1950), pp. 463–467. (Terminal Island, p. 467.)

Bravinder, K. M., "The Los Angeles Basin earthquake of October 21, 1941, and its effect on certain producing wells in Dominguez field, Los Angeles County, California," *Bull. Am. Assoc. Petroleum Geologists*, vol. 26 (1942), pp. 338–339.

Martner, S. T., "The Dominguez Hills, California, earthquake of June 18, 1944," *B.S.S.A.*, vol. 38 (1948), pp. 105–119.

Caloi, P., *et al.*, "Terremoti della Val Padana del 15–16 maggio 1951," *Ann. geofisica (Rome)*, vol. 9 (1956), pp. 63–105.

The Pamir earthquake and landslide

Galitzin, B., "Sur le tremblement de terre du 18 février, 1911," *Comptes rendus*, vol. 160 (1915), pp. 810–813.

Klotz, O., "Earthquake of February 18, 1911," *B.S.S.A.*, vol. 5 (1915), pp. 206–213.

Jeffreys, H., "The Pamir earthquake of 1911 February 18 in relation to the depths of earthquake foci," *M.N.R.A.S. Geophys. Suppl.*, vol. 1, pp. 22–31 (1923).

————, "On the materials and density of the earth's crust," *ibid.*, vol. 4, pp. 50–61 (1937); cf. especially p. 61.

Rockbursts

Hodgson, J. H., "Analysis of travel times from rockbursts at Kirkland Lake, Ontario," *B.S.S.A.*, vol. 37 (1947), pp. 5–17.

————, "A seismic survey in the Canadian shield, Part I: Refraction studies based on rockbursts at Kirkland Lake, Ont.," *Publ.* Dominion Observatory, vol. 16, No. 5 (1952), pp. 109–163. Ottawa, 1952.

Gane, P. G., Hales, A. L., and Oliver, H. A., "A seismic investigation of the Witwatersrand earth tremors," *B.S.S.A.*, vol. 36 (1946), pp. 49–80.

Logie, H. J., "The velocity of seismic waves on the Witwatersrand," *ibid.*, vol. 41 (1951), pp. 109–121.

Willmore, P. L., Hales, A. L., and Gane, P. G., "A seismic investigation of crustal structure in the western Transvaal," *ibid.*, vol. 42 (1952), pp. 53–80.

Gane, P. G., Seligman, P., and Stephen, J. H., "Focal depths of the Witwatersrand tremors," *ibid.*, vol. 42 (1952), pp. 239–250.

Siberian meteorites

Tams, E., "Das grosse sibirische Meteor vom 30. Juni 1908 und die bei seinem Niedergang hervorgerufenen Erde- und Luftwellen," *Zeitschr. Geophysik*, vol. 7 (1931), pp. 34–37.

Whipple, F. W., "The great Siberian meteor and the waves, seismic and aerial, which it produced," *Quart. Journ. Royal Meteorol. Soc.*, vol. 54 (1930), pp. 287–304.

Crowther, J. G., "More about the great Siberian meteorite," *Scientific American*, May 1931, pp. 314–317.

Brown, H., Kullerud, G., and Nichiporuk, W., *A Bibliography of Meteorites*, Chicago University Press, 1953. (Gives references to many publications by L. A. Kulik on this subject, among which are the following.)

Kulik, L. A., "On the history of the bolide of 1908 June 30," *Doklady Akad. Nauk SSSR* (A) 1927, pp. 393–398 (in Russian); *Popular Astronomy*, vol. 43 (1935), pp. 499–504 (in English).

———, "On the fall of the Podkamennaya Tunguska meteorite in 1908, *ibid.*, (A) 1927, pp. 399–402 (in Russian); *Popular Astronomy*, vol. 43 (1935), pp. 596–599 (in English).

———, "Auffindung des Tunguskischen Riesenmeteors vom 30 Juni 1908," *Petermanns geographische Mitt.*, vol. 74 (1928), pp. 338–341.

———, "Preliminary results of the meteorite expeditions 1921–1931," *Trudy Lomonosov. Inst. Akad. Nauk SSSR*, 1933, pp. 73–81 (in Russian); *Popular Astronomy*, vol. 44 (1936), pp. 215–220 (in English).

———, "La météorite de Sibérie," *La Nature*, vol. 67 (1939), pp. 129–131.

———, "The meteorite expedition to Podkamennaya Tunguska in 1939," *Doklady Akad. Nauk SSSR*, vol. 28 (1940), pp. 596–600 (in Russian).

Fesenkov, V. G., "Sikhote-Alinskiy meteorit," *Astr. Zhurnal*, vol. 24 (1947), pp. 302–317.

Volcanic earthquakes

Williams, Howel, "Problems and progress in volcanology," *Quart. Journ., Geol. Soc. London*, vol. 109 (1954), pp. 311–332. (An excellent summary of current research.)

Wood, H. O., *Nat. Research Council, Bull. 90, Seismology*, "Volcanic earthquakes," Chapter 3, pp. 9–31, Washington, D. C., 1933. (Quotation is from pp. 26–27; includes rediscussion of the 1868 earthquakes; see next reference.)

———, "On the earthquakes of 1868 in Hawaii," *B.S.S.A.*, vol. 4 (1914), pp. 169–203.

Johnston-Lavis, H. J., *Monograph of the Earthquakes of Ischia, with Some Calculations by S. Haughton*, London and Naples, 1886.

Mercalli, G., *L'Isola d'Ischia ed il terremoto del 28 luglio 1883*, Milan, 1884.

Davison, C., "The Ischian Earthquakes," Chapter III, pp. 55–74, in: *A Study of Recent Earthquakes*, Walter Scott, London, 1905.

Lais, R., "Die Erdbeben des Kaiserstuhls," *G. Beitr.*, vol. 12 (1913), pp. 45–88.

Lyell, C., *Principles of Geology*, 12th ed. 1875, "The volcanic district of Naples," Chapters 24–25, vol. 1, pp. 559–655.

CHAPTER 13

Primary Effects of Tectonic Earthquakes; Faulting

APART FROM regional warping, tilting, and general change of level, most of the primary tectonic effects to be discussed are fault trace phenomena. The discussion is prefaced by a few words on fault zones

FAULT ZONES AND RIFTS

The field investigator looking for faulting due to a particular earthquake will, of course, first search where stratigraphic or geomorphic evidence indicates the presence of an active fault.

Stratigraphic Evidence

On a major fault stratigraphic evidence may be extremely impressive. An observant traveler crossing the San Andreas fault generally finds that he has abruptly entered an area of different character, with different geological history; localities facing each other across the fault zone along a great part of its extent show almost no stratigraphic resemblance. Nevertheless, this great fault sometimes cuts between masses of apparently similar rock.

Geologic maps, especially generalized small-scale maps, are sometimes distorted by a tendency to draw faults along contacts rather than through formations; in this way a broad, straight fault zone with divergent individual fractures may appear as a single winding fault.† This occurs especially in mapping large strike-slip faults, where the principal surface breaks are often not in one line, but in a series of segments offset *en echelon* (stepwise) in the manner described later in connection with the tension cracks accompanying strike-slip faulting. (See Fig. 13-6.)

---

† An example is the course of the San Jacinto fault in southern California, as shown on the geologic map of the United States published by the U. S. Geological Survey in 1932.

168

Complex surface expression and complex stratigraphic relations are characteristic of many active faults. To the seismologist this is readily comprehensible; his data indicate that faulting in major earthquakes is initiated at a depth of 10 miles or more. Geologically young fracturing may establish itself at such depths without direct relation to surface geology (probable instances are given in Chapter 14). If such fracturing then breaks its way up to the surface, a single break between different exposed formations may become the exception rather than the rule.

Geomorphic Evidence

Geomorphic evidence is of the greatest importance. The expert in this field may deduce the presence of active faulting from the major land forms where the fault itself is buried under alluvial or fan material. The ordinary worker will depend more on the obvious and usually relatively small features which point to geologically young displacements by the very fact that the evidence has not yet been removed.

False Leads. Any long straight feature in the relief is naturally suspected of being due to a fault; but the fault may be inactive, and the straight feature due to differential erosion in the formations on the two sides. The straight feature may indicate a joint rather than a fault. Especially in desert areas, erosional scarps may be surprisingly straight; or a fan surface may intersect the general level in a nearly straight line.

Grabens and Rifts. Where there is much vertical block motion, the terrain may be broken into a series of elevated blocks, or *horsts,* separated by depressed blocks, or *grabens,* as in the basin-and-range topography of Nevada, Utah, and adjacent regions. Long narrow grabens are often called *rifts,* the type examples being those of equatorial Africa, which contain the great lakes Nyasa and Tanganyika. The Owens Valley of California is a graben, with faults on both sides, on which the adjacent mountain blocks rise more than 10,000 feet above the valley floor. It is a good example of a rule laid down by Montessus de Ballore: Where there is high relief expect active faults and high seismicity. (The great Owens Valley earthquake of 1872 will be described briefly in Part Two.)

Strike-Slip Rifts. Lawson introduced the term "San Andreas Rift" for the geomorphic feature associated with the San Andreas fault. (Figs. 13-1, 13-2, 13-3.) This usage has been objected to, since it assigns the term "rift" a meaning essentially different from its use in connection with the African rifts; however, there is no good alternative, and in this book we shall recognize the existence of two types of rifts. This second type appears to be associated primarily with faults in which the principal displacement is horizon-

FIGURE 13-1   *View west-northwest along San Andreas fault zone near Palmdale, California. Palmdale reservoir in foreground; most active fault trace lies along its right side. Older trace, 0.5 miles to right, bounds a block of Pliocene formation underlying low hills in center foreground. Note bend in distant drainage lines crossing the fault zone. Tehachapi Mountains on right skyline contain the Garlock fault. [Photo by Fairchild Aerial Surveys, Inc.]*

tal, or strike-slip. The following is a typical description of such a feature (quoted from McMahon†)

. . . we found that a well-marked line of depression or indentation in the ground was traceable at the edge of the plain . . . Following this line, or, as I may call it, this earthquake crack, we found it to run some 18 miles in a well-defined line to the very place where the earthquake fissure had damaged the railroad in 1892. Thence it ran on, gradually ascending diagonally the slopes of the Khwaja Amran range until it actually cut the crest of the main range near its highest peak. Descending again into the Spintizhe Valley, it began again to ascend diagonally the slopes of a continuation of the Khwaja Amran range. Cutting this range in a similar manner, it descended to the Lora river, and crossing that river, ran along the whole length of the foot of the Sarlaz range to Nushki . . . The total length . . . was not less than 120 miles. It is a well-defined broad line of deep indentation, in places as clearly defined as a deep railway cutting. Along the whole course of it are to be found springs of water, cropping up here and there. Both from the presence of water and from its forming a short cut across the mountainspurs, this crack is largely used as a thoroughfare. We found that

† "The southern borderlands of Afghanistan," *Geographical Journ.*, vol. 9 (1897).

FIGURE 13-2 *View east-southeast along San Andreas fault. Palmdale reservoir in foreground; most active trace extends along its left side, past small sag pond just beyond highway, along the evident trough, and into the San Gabriel Mountains, which it crosses at an elevation of 7000 feet (note the broad notch on the skyline). [Photo by Fairchild Aerial Surveys, Inc.]*

the old greybeards of the tribes residing in the neighborhood all knew of its existence. They told us that during their lifetime, on some three occasions after severe earthquake shocks, deep fissures had appeared along this line, and that they had had similar accounts handed down to them by their fathers. After one of these occurrences, the water supply of the springs along the crack had, they said, largely increased.

It it were not for the Asiatic place names, this could well pass for a description of the San Andreas Rift in southern California. Even the phrase "earthquake crack" is familiar in that connection. The feature here described is actually in Baluchistan (now part of Pakistan) near the Afghan border; the earthquake of 1892 was accompanied by strike-slip that broke and offset a railroad line where it crossed this fault zone. (See Fig. 31-10.)

Rifts in this sense are best known in California, where they are well developed along the San Andreas fault and its principal associated faults or branches (the Haywards, Garlock, and San Jacinto faults), and along the Elsinore fault which parallels the San Jacinto fault some 30 miles to the west. They are found in association with active strike-slip faults in Baluchistan, Turkey, the Philippines, and Sumatra (where jungle and volcanic deposits obscure the feature). There is a rift in southwestern China which is

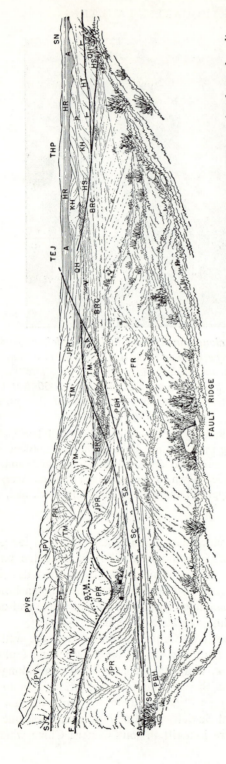

FIGURE 13-3 *San Andreas fault zone; view west across Valyermo Ranch. At right are Holcomb Ridge (HR) and exhumed pediment (P), with Antelope Valley (A), Tehachapi Mountains (THP), and Sierra Nevada (SN) beyond. The Holcomb thrust fault (HT) and Hidden Springs fault (HS) bound the San Andreas fault zone on the north. Beyond and to the left of the fault ridge (FR) in foreground is the San Andreas fault (SA), which defines Shoemaker Canyon (SC), crosses Big Rock Creek (BRC), and extends west-northwestward toward Tejon Pass (TEJ). In the distance at left are Pinyon Ridge (PR), the Punchbowl trough (PT), the San Jacinto fault zone (SJZ), and Pleasant View Ridge in the San Gabriel Mountains (PVR). Major rock types, from left to right, include Pleasant View complex (JPV) of the San Gabriel Mountains, Martinez formation (TM) with its unconformable contact (BTM) on the Pinyon Ridge granodiorite (JPR), Harold formation (QH), and Holcomb quartz monzonite (KH). Valyermo Post Office (V) and Pearblossom Highway (PBH) lie in trough of San Andreas fault zone. [Sketch by Philip B. King.]*

172

similar in appearance and accompanies an active fault to which several strong earthquakes are attributed; but direct evidence of strike-slip is lacking. In New Zealand, on the South Island, such a rift marks the major Alpine fault along which there is ample evidence of postglacial strike-slip, but it has not shown much seismicity in the short time for which documents are adequate. On the same island a perfect miniature San Andreas Rift, on which strike-slip took place in 1888, follows a narrow zone in the valleys of the Waiau and Hope rivers (Chapter 29).

A great fault runs through the Alaska Range, following its curving course convex to the north. Sainsbury and Twenhofel, and independently St. Amand (who calls it the Denali fault), have identified it as a strike-slip fault. It determines a rift feature of San Andreas type, with the characteristic independence of general topography and scissoring reversal of vertical displacements.

If W. Q. Kennedy is correct, the Great Glen fault of Scotland is an ancient strike-slip feature. It is associated with rift topography, including Loch Ness, and with some of the strongest earthquakes of the British Isles.

FAULT ZONE TOPOGRAPHY

Dip-Slip

Scarps and Scarplets. Features of smaller scale than those just discussed indicate the geologically recent activity of a fault; in some cases they may be the only geomorphic evidence of its existence. Vertical or dip-slip displacement naturally is expressed in the formation of scarps, which may be indicated as relatively fresh by slickensides (unless the fresh appearance has been preserved underground and only the exposure is recent). Low scarps are often discovered by their disturbance of drainage or ground water. With proper lighting they may be more evident from the air than on the ground.† In New Zealand many such features have been discovered by photographing from the air, some of them in forested areas where ground work is difficult. A scarplet is often small enough in throw and in linear extent to have originated in a single earthquake. In New Zealand, such scarplets often bear no close relation to the larger topographic features, but in California and some other regions small scarps are most commonly seen at the base of and parallel to larger ones; a whole mountainside there may exhibit a steplike profile attributable to successive elevations with intervals of erosion. Scarps cutting alluvial fans are frequent; they usually face away from the mountains.

---

† Where the fault in southern California now called the Banning fault by Allen, formerly misidentified as the direct continuation of the San Andreas Rift, crosses Whitewater Creek east of San Gorgonio Pass, its location is rendered conspicuous from any elevated point of view by the luxuriant growth of willows and other vegetation on the upstream side and the barrenness on the downstream side of the fault line.

They are generally taken as good evidence of faulting. Near Cucamonga, in southern California, such scarps paralleling the mountain face south of San Antonio and Cucamonga peaks reach a height of 250 feet in interstream areas; this is taken to record the sum of repeated movements, while lower scarps nearer the streams measure only the latest displacement. Interpretation of such features in terms of faulting requires caution. Professor James Noble has described scarps in the alluvium near Rosamond on the Mojave Desert as high as 25 feet which fade out on approaching exposed bedrock, within which there is no evidence of recent faulting. Similar straight scarps, not continued into adjacent bedrock, cut fans on the west side of Owens Valley; at least some of these are attributed to the earthquake of 1872. They may have been due to faulting under the valley, the breaks having extended not vertically but westward and upward, more or less following the dip of the fan deposits.

Reversed Scarplets.   Scarplets are sometimes found facing the larger and older mountain scarps, with a minor graben feature between. In most instances these probably do not mean reversal of displacement; evidence is strong that on most faults displacement continues in the same sense for geologically significant times. Such a minor graben may be produced by sinking of the broken rock of a fault zone relative to more consolidated rock on both sides. This may occur on strike-slip faults, since shearing displacement results in both raised and depressed slices within the fault zone. Strike-slip may also bring different formations opposing each other along the fault, in such a way that differential erosion produces a back-facing scarp; dip-slip may do the same, if easily eroded material is faulted upward to be exposed against more resistant rock. Strike-slip may also lead to reversed scarps by scissoring, as discussed later in this chapter.

In New Zealand, where evidence frequently suggests that motions now going forward differ from those of the comparatively late geologic past, Cotton once identified certain back-facing scarps with their intervening troughs as indicating actual reversal; he now believes that most of these may be due to large strike-slip. The troughs have a sharpness in relief which makes their cross sections differ from those of the side-hill valleys which are common along the San Andreas Rift.

Strike-Slip

Side-Hill Furrows.   Side-hill furrows are among the more conspicuous of the permanent smaller features characteristic of strike-slip fault zones. They are in large part attributable to slicing; the shearing distortion, with accompanying compression affecting the whole width of the fault zone, has altered the relative level of many long, narrow blocks, so that some of them are forced up as long, narrow ridges, or miniature horsts, while others are depressed into comparable grabens. The latter tend to collect water, which evaporates and leaves alkali flats during dry seasons; they range in size from small sag ponds to minor lakes, many of which have been raised by damming and used as reservoirs. Two in the San Andreas Rift, Crystal Springs and San Andreas reservoirs (whence the name of the fault and features), were in 1906 the principal water storage for San Francisco. The almost complete destruction of the mains which ran from these reservoirs along the Rift toward the city cut off the water supply, while storage and distribution within the city was put out of service by damage due to shaking, so that no water was available to fight the great fire which broke out after the earthquake.

Scissoring. While the prevailing sense of relative horizontal displacement along the San Andreas fault is uniform (right-hand, as defined on a later page), the vertical component of relative displacement reverses along the fault, so that now one side and now the other side is upthrown. This applies to both large-scale and small-scale features; naturally the reversals of throw on the small scale are more frequent, and they apply with individuality to the separate small slices and blocks in the Rift. Such reversal along a fault is called *scissoring,* and the points of zero throw where the reversals begin are *scissor points*. Scissoring is a common characteristic of strike-slip faulting, but no proof of it; where it is observed there is good reason to look for other evidence of strike-slip. Strike-slip may produce apparent reversal of throw by relatively displacing blocks which stand at different levels on opposite sides, as already noticed for scarps facing back against grade. Irregularities in vertical displacement, or apparent irregularities due to differential erosion, may have the same effect. It is never sufficient to deal simply with the surface topography; the dip of the formations involved must be observed, and the whole considered in three dimensions.

Strike-Slip Stratigraphy. The best geological evidence of continuing strike-slip is, of course, stratigraphic—when it can be obtained—as when marker beds, contacts, or unique structures, which match across the fault zone and appear to have once been continuous, are seen to be offset relatively on the two sides. Here again the third dimension cannot be neglected; thus, if a formation dips parallel to the strike of the fault, purely vertical displacement will produce apparent relative horizontal displacement of exposed marker beds, giving the surface effect of strike-slip faulting (Fig. 13-4).

Stratigraphic evidence becomes rarer and less convincing as the relative displacement increases. It has been applied often to show displacements of a few feet in a single earthquake. Along the San Andreas Rift there are many points where evidence of this sort establishes accumulated strike-slip up to a mile or so; near Tejon Pass Tertiary volcanics appear to have been separated and shifted several miles.

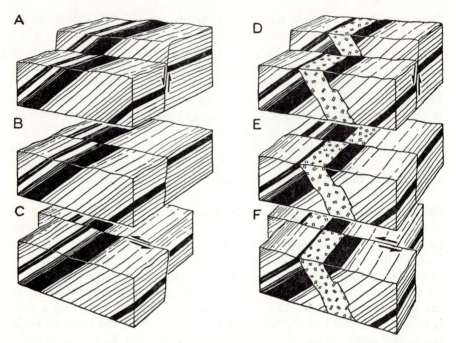

FIGURE 13-4   *Diagrams showing how normal and strike-slip faults can produce identical outcrop patterns* (A, B, C), *and how these can be differentiated under exceptionally ideal conditions* (D, E, F; *note the outcrops of the intersecting dikes*). [*From* Principles of Geology *by James Gilluly, Aaron C. Waters, and A. O. Woodford. San Francisco: W. H. Freeman and Company, 1951.*]

Offset Streams.   Strike-slip is often plainly evidenced by offset streams and minor watercourses. There are hundreds of such features along the San Andreas Rift (see Frontispiece); in some localities the whole drainage is offset, so that one stream after another flows down to the fault zone, makes a sharp right turn, and then makes an equally sharp left turn to depart from the Rift through a gap as much as a mile distant from its upper course. An individual feature of this kind might be due to the fault zone acting as a channel, plus possible local tilting; the weight of evidence lies in the repetition. As would be expected, the larger streams in general show large accumulated deflections. Apparent exceptions arise when a stream which was probably at first deflected later finds an escape through a new gap which the progressive displacement has brought close to its original outlet position. The stream may even run a short distance "backward" to find such an escape, producing an appearance of reversed deflection.

The description of the Baluchistan rift, quoted earlier, reports no actual strike-slip features, although strike-slip offset the railway in 1892. This illustrates how readily such features are overlooked by layman and geologist alike.

Springs

Emergence of water as springs through the crushed rock of the fault zone is common among many faults, active and inactive. Where the water is hot, it raises special problems (see the end of this chapter).

INDIVIDUAL FAULT TRACES

The permanent features of fault zones have been discussed here rather summarily. To the field investigator of an earthquake they are evidence to be observed and correlated with his findings; but he should not fall into the error, which has trapped many competent men, of assuming that a fault with conspicuous geological evidence of recent activity is necessarily the one responsible for a strong shock in its vicinity. The San Andreas fault has been blamed repeatedly for minor earthquakes which instrumental records proved to have originated elsewhere. Even the great Owens Valley earthquake of 1872 did not originate at the base of the huge escarpment of the Sierra Nevada which bounds the graben on the west, but along the edge of the comparatively low internal block of the Alabama Hills. Petrushevsky and other writers have lately emphasized the strong discordance between conspicuous young tectonic features and present seismicity, in the Tian-Shan and elsewhere in the USSR.

Fault trace phenomena include the development of minor features, such as scarplets and horizontal offsets, of the same type as the larger permanent features. They also include other effects, such as cracks, fissures, and mole tracks, which are less lasting in character and are removed by normal agencies in a few seasons. When such evidence is small or obscure, the only means to distinguish it from secondary effects or pure accidents may be its continuation along a narrow more or less linear trace, especially if the course of this trace is independent of topography.

Scarps

Small scarps appearing near the base of a slope should be examined critically to distinguish them from local sliding. If a scarp has a crescent form with the horns downhill it almost certainly represents an earth slump. A series of slides may occur along the contact between two formations, especially if one is relatively unconsolidated. Scarps curving around ridges and tending to follow contours are likely to be due to slides. Fresh scarps may result from differential settling due to shaking; this may happen along the surface exposure of a fault not otherwise related to the earthquake under investigation. Misleading scarps may be produced in other ways. Scarps of very fresh appearance may actually be of old date, particularly in arid regions. Even there, if field investigation does not immediately follow an earthquake, minor effects of erosion must be expected which may alter the appearance of scarps and perhaps produce new ones.

Offsets

The formation of scarplets makes the recognition of dip-slip relatively easy. In open country without man-made features strike-slip is difficult to establish, and extremely hard to measure, unless there is one of those rare stratigraphic accidents by which a conspicuous marker is offset. Horizontal offset of a dry watercourse may be roughly measurable if the fault makes a sharp and narrow break; flowing streams are unlikely to leave the terrain in a condition favorable for such detailed observations. Paths beaten by animals may show an offset, but practically all our reliable data on horizontal displacement in earthquakes come from culturally developed areas and refer to railroads, highways, lanes, footpaths, fences, ditches, or canals.

Drag

Effects of elastic rebound (Chapter 14) should not be confused with those of drag, which are often more conspicuous in the field and are actually in the opposite sense. Thus in the Imperial Valley earthquake of 1940 (Chapter 28) the appearance of strike-slip in heavy alluvial ground was almost invariably modified by drag; the offset was often distributed over a width of a hundred feet or more. Outside this belt of drag, fence lines across the fault remained straight to ordinary observation, the two lines on opposite sides being mutually displaced by the full offset. Within the drag belt, fence posts progressively nearer the center were progressively less shifted, so that the fence lines curved (Fig. 13-5). In the firmer terrain of the 1906 earthquake drag

FIGURE 13-5 *Drag in strike-slip faulting. Circles indicate posts of originally straight fence.*

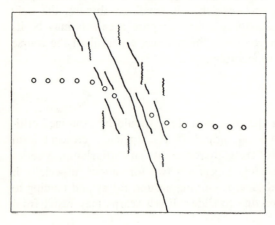

effects were generally less evident; several photographs in the Commission report show fences sharply offset at the fault trace with no evidence of curvature.

Cracks and Fissures

Cracks and fissures are among the most useful and also the most mislead-ing types of evidence by which a fault trace may be followed. In firm rock any conspicuous line of cracking is likely to be a primary effect, even when there is no clear evidence of relative displacement of the two sides, as with Oldham's Bordwar fracture (Chapter 5). Most earthquake cracks, however, are in unconsolidated ground, and their evaluation demands good judgment and some experience. Large and impressive systems of cracks and fissures may develop from shaking, settling, and lurching, as in Bihar in 1934. Long straight belts of cracking are always of interest; but, when they parallel a watercourse, a bluff, or a contact between very different formations, the cause may be sought in wave propagation and consequent lurching rather than in displacement of the underlying bedrock.

Diagonal tension cracks associated with a main fault trace may indicate strike-slip. They are often associated with pressure ridges. Directions relative to the two opposite types of strike-slip are drawn in Fig. 13-6 (compare Fig. 13-7). The visible trace may consist merely of a succession of such cracks arranged *en echelon,* like those drawn in the figure on one or other side of the fault line. Even a large fault trace with conspicuous scarps may break into a series of segments similarly offset, so that the investigator fol-lowing a scarp finds it dying out and replaced by another rising to his right or left; the sense of offset is usually that of tension cracks. Parts of the fault trace of the 1954 Nevada earthquakes were of this type.[†] The same echelon feature appears on a still larger scale in fault zones and accounts in part for the complexity, already noted, which distorts oversimplified small-scale maps.

FIGURE 13-6 *Diagram illustrating right-hand and left-hand strike-slip faulting, with associated tension cracks and pressure ridges.*

Mole Tracks

A spectacular trace phe-nomenon which appears to develop chiefly when large strike-slip traverses a heavily alluviated terrain was termed a mole track by B. Kotō. In

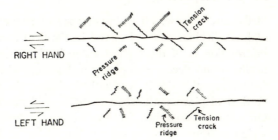

RIGHT HAND

LEFT HAND

Pressure ridge

Tension crack

[†] See E. R. Larson, *B.S.S.A.,* vol. 47 (1957), pp. 377–386.

FIGURE 13-7   *Anatolian earthquake, 1953. Sketch showing strike-slip offsetting a road, and tension fissures.* [*Ketin and Roesli.*]

this he followed the traditional usage applied by the Japanese farmers of 1891, who found a line of disturbed earth crossing their fields that looked like the track of a gigantic mole, or as if the earth had been turned by a great plough-share. Near the California-Mexico boundary the Imperial Valley earthquake of 1940 thus piled up earth in great clods to a height of 4 feet.† The effect is apparently due to the same transverse pressure that brings about pressure ridges.

Proper Reporting

Strike and Throw.   The importance of proper description of the geometry of fault displacements observed in the field cannot be overemphasized. Such information is of the highest value in determining the forces at work. Naturally, as no decent mapping can be done without reading the strike of the fault trace, this detail is usually included. As a rule, scarps indicate nearly vertical displacement; when there is evidence of dip it is usually reported. On uneven topography faulting without a steep dip results in a curved trace which is apparent in mapping. Surprising as it may seem, observers have

---

† Such effects were much rarer in the California earthquake of 1906. See Chapter 28 for quotation of a detailed description of strike-slip evidence in a grassy area.

sometimes failed to state which side was downthrown. Sometimes it is vaguely stated that the throw reversed in places; the value of this fact as an indication of scissoring and probable strike-slip, and the need for specifying the scissor points, may be missed entirely. Scissoring on a fresh fault trace often repeats a reversal of throw evident in the larger topography. In the 1940 Imperial Valley earthquake one scissor point was at the crossing of the Alamo River, and it was natural to suppose that the slight permanent depression which determined the river course was connected with previous scissoring at that point. However, other causes of reversal, real or apparent, must be considered also.

Direction of Strike-Slip. Inadequate reporting of strike-slip is common. The strike of the trace is usually given, but the sense of the relative displacement of the two sides is often not stated clearly, and sometimes not at all. The more natural forms of statement are cumbersome; for example, at points where the San Andreas fault strikes northwest the southwest side is displaced to the northwest relative to the northeast side. One can express this briefly by describing strike-slip as right-hand (or dextral) or left-hand (or sinistral). These terms correspond respectively to A and B of Fig. 13-6. The relative displacement named is that of either block as seen by an observer on the other. The distinction of right-hand and left-hand is independent of the strike. Some writers use "clockwise" and "counterclockwise" for A and B.

Survey Data

Fault displacements can be confirmed and investigated in more detail by surveying—releveling for vertical motions, retriangulation for horizontal. This procedure implies a previous survey based on well-placed monuments, with lines of surveying extending out of the disturbed area.

Complex Traces

The field worker must remember that faulting is not necessarily confined to a single linear trace. Airplane reconnaissance or, better still, photographic survey is desirable. Many large earthquakes have shown such multiplicity; it is usually not possible to say whether the minor traces originated in the main earthquake or during an aftershock. The California earthquake of 1906, usually cited as the type case of a single fault trace, probably involved at least one auxiliary trace (Chapter 28). One of the best-documented instances of multiple faulting is the Tango earthquake (Japan, 1927) when faulting took place on two sides of a crustal block. The Nevada earthquakes of 1954 provide an excellent and thoroughly investigated example.

Thrusting and Regional Distortion

The foregoing discussion of fault traces applies when the dip is nearly vertical, and the trace accordingly is nearly straight and independent of topography. Except for Tsuya's report on the Japanese earthquake of 1945 (Chapter 30), there is no good account of a large earthquake producing a surface trace along a low-angle thrust. The best-known earthquake during which a fault with dip much off the vertical broke the surface is the 1952 earthquake in Kern County, California. Some of the effects of the New Zealand earthquake of 1931 (Hawke's Bay) were strikingly similar. Both of these are discussed in later chapters.

In 1933 and thereafter continuous thrusting unaccompanied by earthquakes of consequence was reported from the Buena Vista Hills oil field in the meizoseismal area of the Kern County earthquake of July 21, 1952. The faulting was discovered by its repeated shearing of well casings, but the distortion as it progressed warped pipes up out of the ground into arches over the fault. There is survey evidence that the displacements slowed down somewhat before the 1952 earthquake, and thereafter resumed their former rate. There was no visible surface effect on July 21, 1952. Faulting was still going on vigorously in 1956.†

Many strong earthquakes have produced regional distortion, often with displacement on a number of short faults. The Indian earthquake of 1897 (Chapter 5) and the Japanese earthquake of 1923 are examples. Either or both of these may have been due to motion on a thrust which did not break the surface; in the Japanese instance, the surface break may have been under the waters of Sagami Bay (and may have been primarily strike-slip). The problem of investigating and identifying the primary effects of such earthquakes is complicated by the number of these effects, the wide area over which they occur, and their lack of apparent connection. If difficulties in getting over the terrain are added, as in 1897, complete investigation in limited time and with limited facilities is impossible. In dealing with the individual minor traces in such cases all the same field problems already discussed arise, including those of establishing strike-slip and dip-slip.

Inland changes of level are not easily established or studied, unless lines of precise leveling were previously carried far outside the area; only the general fact of distortion can be established, as in 1897. Changes at a seacoast are more positive in nature, but they must be analyzed carefully for effects of changes in tides and currents.

---

† The writer is indebted for information to Mr. J. W. Wilt of the Honolulu Oil Corporation.

OTHER PRIMARY EFFECTS

Direct effects on surface water of the movement or warping of crustal blocks are well known and obvious. Good instances have been cited in Chapter 5—the alternate waterfalls and ponding along the Chedrang fault, the stray ponds due to general disturbance of drainage.

Similar disturbance of the flow of underground water may account for some of the changes in wells, streams, and springs reported for almost every large earthquake. The great flooding of the streams entering India after the Tibet earthquake of 1950 constituted a major disaster. Immediately after the Kern County earthquake of 1952, springs at Clear Creek Café, on the highway above Caliente Creek, ran dry, and water had to be tanked in; simultaneously new springs appeared in the railroad cuts below, to the inconvenience of repair workmen. About 4 days later, during strong aftershocks, water returned to the Clear Creek springs and ceased emerging in the railroad cuts. Caliente and Tehachapi creeks showed greatly increased flow and continued to run through the summer, a season when their beds are normally dry.

Permanent changes in apparent sea level after strong earthquakes have been reported from many coasts, notably those of Chile and Japan. Some of these changes indicate actual vertical displacement on a fault at the coast line, but most of them are evidently incidental to large-scale regional warping like that of 1897. Seismic sea waves (Chapter 9) have sometimes been attributed to the direct action of displaced crustal blocks beneath the ocean. This would make them primary effects, but the cause is more probably of secondary character.

VOLCANISM?

Tectonic earthquakes, and certain of their effects, have sometimes been attributed to volcanic causes. In the early period of seismology this was common; we have noted how Mallet associated his earthquake of 1857 with the nearby extinct Monte Vulture, although similar earthquakes have occurred along the whole Apennine chain. In later years there was a tendency to err in the other direction; fissures and scarps in volcanic areas were described as faulting. An instance is the "earthquake crack," popularly so called, in northern Owens Valley, California, not far from Mammoth. This fissure system is in an area of Pleistocene volcanism; it trends directly toward the Inyo craters. Although it is sometimes attributed to the Owens Valley earthquake of 1872, there is record that it was in existence before that time.

The emergence of hot water along fault lines, especially immediately after earthquakes, is particularly difficult to account for. Many active faults and rifts are marked by springs. At least a dozen hot-spring localities are scattered along the San Jacinto fault in southern California; most of them have been developed as resorts. There is no doubt that the water reaches the surface through the crushed material of the fault zone, but the high temperatures are not readily explained except on the supposition of residual heat from otherwise extinct volcanism or other igneous sources. However, relative movement at depth must liberate great quantities of heat, so that the cause of the high temperatures may after all be tectonic.

References

Faulting and geomorphology

Sharp, R. P., "Physiographic features of faulting in southern California," *Calif. Dept. Nat. Resources, Div. Mines, Bull.* 170, 1954, Chapter V, "Geomorphology," pp. 21–28.

Lahee, F. H., *Field Geology*. McGraw-Hill, New York, 5th ed., 1952.

Cotton, C. A., "Tectonic scarps and fault valleys," *Bull. Geol. Soc. Amer.,* vol. 61 (1950), pp. 717–757.

———, "Tectonic relief; with illustrations from New Zealand," *Geographical Journ.,* vol. 119 (1953), pp. 213–222.

———, "Revival of major faulting in New Zealand," *Geol. Mag.,* vol. 84 (1947), pp. 79–88. (Possible reversals.)

———, *Landscape as Developed by the Processes of Normal Erosion*, Cambridge University Press, 1941; 2nd ed., 1948. (Refer especially to chapters xx, xxi, xxii. See also other volumes by Cotton on geomorphology, and references for Chapter 27.)

Lawson, A. C., "The San Andreas Rift as a Geomorphic Feature" in: *The California Earthquake of 1906, Report of the State Earthquake Investigation Commission,* Carnegie Institution of Washington, 1908, Vol. 1, pp. 25–52. (Details also in other divisions of the same publication.)

Rifts and grabens

The literature is voluminous. The student should approach the matter by way of the general handbooks, which will guide him to special and detailed presentations. Standard references on the African rifts are the following:

Gregory, J. W., *The Rift Valleys and Geology of East Africa,* London, 1921, 479 pp.

Krenkel, E., *Die Bruchzonen Ostafrikas,* Berlin, 1922.

Willis, B., *East African Plateaus and Rift Valleys,* Carnegie Institution of Washington, Washington, D. C., 1936.

Basin and range structure, and Owens Valley

Davis, W. M., "The mountain ranges of the Great Basin," *Bull. Museum Comparative Zoology,* vol. 42 (1903), pp. 129–177; reprinted in: Davis, W. M., *Geographical Essays* (republished, Dover, 1954).

Nolan, T. B., "The Basin and Range province in Utah, Nevada, and California," *U. S. Geol. Survey, Prof. Paper* 197–D, pp. 141–196, Washington, Gov. Printing Office, 1943.

Longwell, C., "Tectonic theory viewed from the Basin Ranges," *Bull. Geol. Soc. Amer.,* vol. 61 (1950), pp. 413–434.

Knopf, A., "A geological reconnaissance of the Inyo Ranges and the eastern slope of the southern Sierra Nevada," *U. S. Geol. Survey, Prof. Paper* 110, 1918.

Known or suspected strike-slip rifts (consult also references in Part Two)

San Andreas Rift

See preceding reference to Lawson, and:

Noble, L. F., "The San Andreas Rift and some other active faults in the desert region of southern California," *Carnegie Institution of Washington Yearbook,* vol. 25 (1926), pp. 415–428; reprinted, *B.S.S.A.,* vol. 17 (1927), pp. 25–39.

————, "The San Andreas Rift in the desert region of southern California," *Carnegie Institution of Washington Yearbook,* vol. 31 (1932), pp. 355–363.

————, "Excursion to the San Andreas fault and Cajon Pass," *Guidebook* No. 15, 16th International Geological Congress, 1933, pp. 10–21.

————, "Geology of the Pearland quadrangle," *U. S. Geol. Survey,* map series, 1953.

————, "Geology of the Valyermo quadrangle," *ibid.,* 1954.

————, "The San Andreas fault zone from Soledad Pass to Cajon Pass, California," *Calif. Dept. Nat. Resources, Div. Mines, Bull.* 170 (1954), Chapter IV, "Structural features," pp. 37–48.

Wallace, R. E., "Structure of a portion of the San Andreas Rift in southern California," *Bull. Geol. Soc. of Amer.,* vol. 60 (1949), pp. 781–806.

Allen, C. R., "Geology of the north side of San Gorgonio Pass, Riverside County," map sheet 20, with text, *Calif. Dept. Nat. Resources, Div. of Mines, Bull.* 170 (1954).

————, "San Andreas fault zone in San Gorgonio Pass, southern California," *Bull. Geol. Soc. Amer.,* vol. 68 (1957), pp. 315–350.

San Gabriel fault

Crowell, J. C., "Strike-slip displacement of the San Gabriel fault," *Calif. Dept. Nat. Resources, Div. Mines, Bull.* 170 (1954), Chapter IV, "Structural features," pp. 49–52. (Describes a probably active fault associated with the San Andreas system. Geological evidence for large right-hand slip.)

————, "Probable large lateral displacement on San Gabriel fault, Southern California," *Bull. Am. Assoc. Petroleum Geologists,* vol. 36 (1952), pp. 2026–2035.

Haywards fault

Buwalda, J. P., "Nature of the late movements on the Haywards rift, central California," *B.S.S.A.,* vol. 19 (1929), pp. 187–199.

Garlock fault

Hill, M. L., and Dibblee, T. W., Jr., "San Andreas, Garlock and Big Pine faults, California," *Bull. Geol. Soc. Amer.,* vol. 64 (1953), pp. 443–458.

New Zealand

Alpine fault

Wellman, H. W., and Willett, R. W., "The geology of the west coast from Abut Head to Milford Sound, Part 1," *Trans. Royal Soc. N. Z.,* vol. 71 (1942), pp. 282–306.

Wellman, H. W., "The Alpine fault in detail; river terrace displacement at Maruia River," *N. Z. Journ.,* Section B, vol. 33 (1952), pp. 409–414.

Munden, F. W., "Notes on the Alpine fault, Haupiri Valley, North Westland," *ibid.,* pp. 404–408.

Bowen, F. E., "Late Pleistocene and Recent vertical movement at the Alpine fault," *ibid.,* vol. 35 (1954), pp. 390–397.

Cotton, C. A., "The Alpine fault of the South Island of New Zealand from the air," *Trans. Royal Soc. N. Z.,* vol. 76 (1947), pp. 369–372.

Hope fault

Cotton, C. A., "The Hanmer plain and the Hope fault," *N. Z. Journ.,* Section B, vol. 29 (1947), pp. 10–17.

Baluchistan

McMahon, A. H., "The southern borderlands of Afghanistan," *Geographical Journ.,* vol. 9 (1897), pp. 393–415.

Turkey

Pamir, H. N., "Les séismes en Asie Mineure entre 1939 et 1944. La Cicatrice nord-anatolienne," *Proc. 18th International Geological Congress, Great Britain, 1948,* London, 1950, Part XIII, pp. 214–218.

Ketin, I., "Über die tektonisch-mechanischen Folgerungen aus den grossen anatolischen Erdbeben des letzten Dezenniums," *Geol. Rundschau,* vol. 36 (1948), pp. 77–83.

Sumatra

Westerveld, J., "Quaternary volcanism on Sumatra," *Bull. Geol. Soc. Amer.,* vol. 63 (1952), pp. 561–594.

Philippines

Willis, B., "Philippine earthquakes and structure," *B.S.S.A.,* vol. 34 (1944), pp. 69–81. (For the direct evidence of right-hand strike-slip, which is minimal, see pp. 76–77.)

King, Philip B., and McKee, Edith M., "Terrain diagrams of the Philippine Islands," *Bull. Geol. Soc. Amer.,* vol. 60 (1949), pp. 1829–1836. (Beautiful maps; no new data.)

Southwest China

Heim, Arnold, "Earthquake region of Taofu," *Bull. Geol. Soc. Amer.,* vol. 45 (1934), pp. 1035–1050.

Lee, S. P., "Tectonic relation of seismic activity near Kangting, East Sikang," *Journ. Chinese Geophys. Soc.,* vol. 1 (1948), pp. 43–50. (Cf. also Gutenberg and Richter, *Seismicity of the Earth,* 2nd ed., 1954, p. 73.)

Alaska

Sainsbury, C. L., and Twenhofel, W. S., "Fault patterns in southeastern Alaska" (abstract), *Bull. Geol. Soc. Amer.,* vol. 65 (1954), p. 1300.

St. Amand, Pierre, "The tectonics of Alaska as deduced from seismic data" (abstract), *ibid.,* p. 1350.

———, "Geological and geophysical synthesis of the tectonics of portions of British Columbia, the Yukon Territory, and Alaska," *Bull. Geol. Soc. Amer.,* vol. 68 (1957), pp. 1343–1370.

Scotland

Kennedy, W. Q., "The Great Glen fault," *Quart. Journ. Geol. Soc. London,* vol. 102 (1946), pp. 41–76 and plate.

Bartlett trough

Taber, S., "Jamaica earthquakes and the Bartlett trough," *B.S.S.A.,* vol. 10 (1920), pp. 55–89.

———, "The seismic belt in the Greater Antilles," *ibid.,* vol. 12 (1922), pp. 199–219.

Venezuela

Rod, Emile, "Strike-slip faults of northern Venezuela," *Bull. Am. Assoc. Petroleum Geologists,* vol. 40 (1956), pp. 457–476.

Other material

Larson, E. R., "Minor features of the Fairview Fault," *B.S.S.A.*, vol. 47 (1957), pp. 377–386.

Noble, James A., "Geology of the Rosamond Hills, Kern County," map sheet No. 14, with text, *Calif. Dept. Nat. Resources, Div. Mines, Bull.* 170 (1954).

Koch, T. W., "Analysis and effects of current movement on an active fault in Buena Vista Hills oil field," *Bull. Am. Assoc. Petroleum Geologists,* vol. 17 (1933), pp. 694–712. (See also Gilluly, Waters, and Woodford, *Principles of Geology,* W. H. Freeman, San Francisco, 1951, pp. 176–177.)

Wilt, J. W., "Measured movement along surface trace of active thrust fault in Buena Vista Hills, Kern County, California," *B.S.S.A.* (In press.)

Benioff, H., and Gutenberg, B., "The Mammoth 'earthquake fault' and related features in Mono County, California," *B.S.S.A.*, vol. 29 (1939), pp. 333–340.

Petrushevsky, B. A., "O svyazi seymischeskikh yavleniy na Uralo-Sibirskoy platforme i v Tyan'-shane s geologicheskoy obstanovkoy etikh territoriy," *Byull. Moskov. Obshch. Ispit. Prirodi, otdel. geol.,* vol. 30, No. 6 (1955), pp. 31–53.

———, "Znachenie geologicheskikh yavleniy pri seysmicheskom rayonirovanii," *Trudy Geofiz. Inst. Akad. Nauk SSSR,* No. 28 (155), (1955), pp. 1–59.

See also other references in Chapter 33.

CHAPTER 14

Tectonic Earthquakes— General

EARTHQUAKES AND FAULTING

Students tend to take for granted that an invariable connection between faulting and earthquakes is as incontrovertibly established as the law of gravitation. It is not quite so cut-and-dried as that. As Table 14-1 shows, only about forty earthquakes are known for which the evidence of faulting is individually clear; of these, half are incompletely known or inaccurately described. On this evidence alone, it is an enormous extrapolation, however well justified, to say that faulting accounts for the hundreds of large earthquakes and the millions of small ones which are added to the record every few years. Supporting evidence, geological and seismological, will follow in this chapter. It amounts nearly to proof, but the too readily convinced student will do well to refer to the chapter on deep-focus earthquakes (Chapter 19).

HISTORICAL NOTE

After the middle of the nineteenth century, it was gradually recognized that volcanic causes could account for only a small part of earthquake activity, and that nonvolcanic earthquakes occur in regions of geologically young mountain-building, especially in the vicinity of active faults. Montessus de Ballore pointed out that high relief, whether on land, or at the coast, or under the ocean, is an almost sure sign of high seismicity. However, for many years the only earthquakes which were generally recognized as having been accompanied by visible faulting were those of Cutch (India) in 1819 and New Zealand in 1855. These two instances were made known especially by Lyell, who described them in the later editions of *Principles of Geology*. The first interpretation put on the association of earthquakes with faults was not the one which the student now takes for granted. It was more like

189

the account of certain comparatively small and shallow disturbances given in Chapter 12—displacements affecting oil wells which are believed to be triggered by ordinary earthquakes originating at greater depth. At a later date Oldham took refuge in this kind of explanation for the great earthquake of 1897. The earthquake proper was supposed to be a violent subterranean disturbance (Oldham named it the "bathyseism"), which might be remotely related to volcanism; from this disturbance elastic waves were radiated which produced all the ordinary macroseismic effects and in addition precipitated fault movements.

Perhaps the strongest personal influence in maintaining this point of view was that of Eduard Suess, whose great work, *Das Antlitz der Erde* (*The Face of the Earth*), is one of the principal milestones of geology. Suess attributed all geological processes directly or indirectly to the action of gravity in the crust of a contracting and presumably cooling earth. This conviction committed him to opposing any evidence of forces acting to uplift against gravity, unless he could balance them by subsidence elsewhere. It led him to write long passages of special pleading in which he denied the uplift on the South American coast after the earthquakes of 1822 and 1835 and bitterly assailed the powers of observation and even the integrity of the witnesses (see Chapter 31). For the 1819 earthquake of Cutch, Suess argued for the evidence of downthrow and against that of uplift. Although in a careful study of the small shocks of northern Austria he did much to bring about recognition of the connection of earthquakes with faulting, his obsession with the force of gravity kept him from arriving at a modern interpretation.

ELASTIC REBOUND

Alternatives were still seriously considered until the close of the nineteenth century. John Milne, in discussing the Japanese earthquake of 1891, recognized the fault displacement as possibly the actual source of elastic waves and hence of macroseismic shaking, but he did not then take this position unequivocally, although Kotō did so in the plainest terms (Chapter 30). General acceptance gradually came about after the California earthquake of 1906, following precise theoretical formulation by H. F. Reid in several publications, among which was the concluding volume of the report of the California Earthquake Commission. Following Reid, we call this the *elastic-rebound* theory of earthquakes. The energy radiated as elastic waves is liberated in the process of faulting. The faulting causes the earthquake; the earthquake does not cause the faulting.

Fracturing

The energy source for tectonic earthquakes is potential energy stored in the crustal rocks during a long growth of strain. When the accompanying elastic stresses accumulate beyond the competence of the rocks, there is fracture; the distorted blocks snap back toward equilibrium, and this produces the earthquake. Energy is drawn from a wide zone on both sides of the actual fracture. Naturally minor and local shaking may be associated with irregular fault surfaces grinding against each other as fracture progresses and the original process normally continues through a series of aftershocks (Chapter 6). Nevertheless, the main features, both macroseismic and microseismic, are best accounted for in terms of a single principal event. Fracture takes place chiefly along already established weaknesses; the great active faults are wounds in the earth which have opened again and again.

Reid's Formulation

Many details of Reid's theory depended on work done by the U. S. Coast and Geodetic Survey after the 1906 earthquake. By repeating previous triangulation, and assuming that monuments over 30 miles from the San Andreas fault had not moved during the earthquake, it was found that displacements were maximal at the fault and decreased with distance, so that a previously straight line became curved. This is commonly illustrated by a set of three diagrams (Fig. 14-1A, B, C). Here A represents conditions

FIGURE 14-1

Illustrating Reid's elastic rebound theory. A, unstrained condition; B, strained condition; C, relief of strain by faulting and elastic rebound.

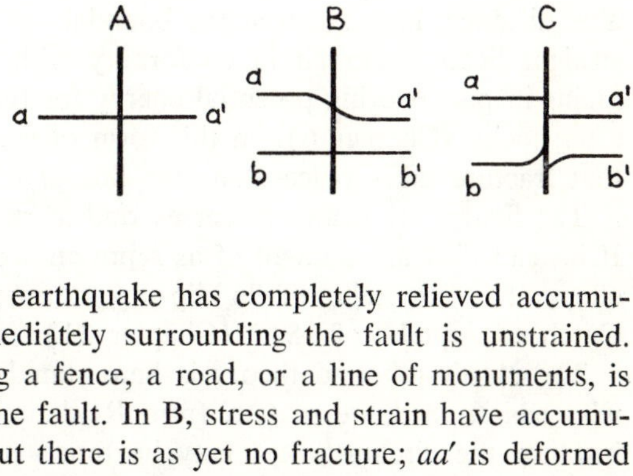

supposed to exist after a great earthquake has completely relieved accumulated stresses; the region immediately surrounding the fault is unstrained. A straight line aa', representing a fence, a road, or a line of monuments, is constructed at right angles to the fault. In B, stress and strain have accumulated; the region is distorted, but there is as yet no fracture; aa' is deformed into a curve. At this time a new straight line bb', fence or other marker, is constructed at right angles to the fault. A major earthquake now takes place, and fracture along the fault relieves the strain. The two parts of aa' now become straight but are offset, while bb' shows both offset and curvature. This last curvature fits the results of the resurvey after 1906, indicating that most of the regional distortion relieved in the earthquake had occurred before the survey monuments were set. This type of curvature must not be confounded with the very different effect of drag in the fault zone (Chapters 13 and 28).

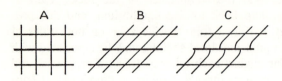

FIGURE 14-2

Elastic rebound due to regional deformation. A, unstrained condition; B, strained condition; C, relief of strain in vicinity of a fault.

An Alternative

In a sense the curvature at stage B suggests that nature is somehow aware in advance of where the fracture is to take place. This is odd when one considers the existence of other faults in the region. Thus the Haywards fault, also of strike-slip character, branches from the San Andreas fault and runs nearly parallel to it on the east side of San Francisco Bay. Another set of three diagrams (Fig. 14-2A, B, C) attempts to represent the main facts otherwise.

In Figure 14-2, A again represents the unstrained condition; for clearness the area is divided into rectangular blocks with one boundary along the fault. In B, the whole has been deformed by uniform shear with no local concentration, so that straight lines remain straight. In C, fracture has occurred along the central line, but the crustal blocks have not returned to perfectly rectangular form. Recovery is complete only near the fault; plasticity, or at least the lack of perfect elasticity, reduces the change to zero at an appreciable distance, here taken at the boundary of the figure. The lines previously straight become curved in conformity with observation. The distortion remains in part, storing potential energy for further earthquakes, especially for aftershocks (Chapter 6). In this form of statement one need only postulate that fracture takes place along the line of least resistance at the time.

The figures are drawn to correspond to strike-slip in the 1906 earthquake. If instead they are thought of as representing vertical faulting, they show the effect of elastic rebound in dip-slip, corresponding to observations of the Cutch earthquake of 1819 (Chapter 31).

The elastic-rebound theory is here stated in very simple form. Everyone who has studied the matter, from Reid on, has realized that in any given instance the circumstances may be much more complex. The 1906 earthquake probably represents a particularly uncomplicated case.

Preliminary Distortion

The figures as drawn call for progressive distortion at the surface preceding a great earthquake. This ought to be detectable by frequent retriangulation; indeed, it was thought that existing survey data indicated such motion before 1906. Accordingly, much triangulation has been carried out in the California region, but, in general, displacements have been found only after earthquakes, not before. A false alarm was created in 1925, when it appeared from preliminary reduction of data that Gaviota Peak, west of Santa Barbara, had shifted about 25 feet. This led to dire predictions of the imminence of a great earthquake, but the supposed displacement disappeared in the final revision of the triangulation work.

The same thing happened several times on a small scale when the men doing the field survey were convinced that minor systematic displacements had been found, but on reduction at Washington it was concluded that these were within the margin of error. Suspicion arose that these findings might be due to the method of reducing the observations, which distributes observed discrepancies evenly, so that a comparatively large shift at one or two points would be in effect explained away as if due to an accumulation of minor variations. Otherwise, acceptance of the results as given would require one to suppose that any gradual preliminary distortions took place at depth and that observable displacements occurred only with actual fracture, breaking its way up to the surface. This is not a satisfactory explanation; in detail it gives rise to many difficulties. The awkwardness was removed by a study made by Whitten in 1948 at the Washington office of the U. S. Coast and Geodetic Survey. He began this investigation by accepting the premise that gradual distortion of the California region is going on, and he raised two significant questions to be answered by computation. First, are the available triangulation data consistent with such a continuous movement? It was concluded that they definitely are. Secondly, if the movement is uniform, what is the best value for its rate? It was found that on this basis "the outer coastal area is moving northwestward at a rate of about 5 centimeters per year and has a total displacement of about 3 meters since 1880, the date of the first precise surveys." Displacements in the earthquakes of 1906 and 1940 were of the order of 15 feet or, say, 450 centimeters. This would call for accumulation during 90 years. One might then expect any given part of the region to be affected by a great earthquake roughly once a century. On this basis, how often such earthquakes might be expected in the California area depends on how far the effects of strain relief are thought to extend geographically. Since earthquakes with notable faulting occurred in California in 1857, 1868, 1872, 1906, 1940, and 1952, in Nevada in 1915, 1932, and 1954, and in adjacent Mexico in 1956, either the geographical extent in question is not great or the simple theory must be modified.

Whitten has also reported on later surveys. In 1951 two arcs of triangulation crossing the San Andreas and other faults were resurveyed. One, established in 1930, passes through Hollister; the other, set up in 1932, through Cholame.† In both arcs systematic displacement vectors were found consistent with general right-hand shearing of the region, which has changed the angles of parallelograms by about 2 seconds of arc in 20 years. Continuity

---

† Cholame is in northeastern San Luis Obispo County on the San Andreas fault zone, near the epicenter of the shock of March 10, 1922 (about 35° 43′ N).

in displacements near the San Andreas fault indicates that no significant slip occurred there in the interval.

Retriangulation in Imperial Valley, California (Figs. 14-3, 14-3A), was carried out in 1954, repeating surveys of 1941 (which followed the 1940 earthquake. See Chapter 28). In the 14 years points on the west have moved almost 4 feet relative to points on the east; compared with surveys of 1935–1939, the shift is almost 6 feet. Angular distortion is at the same rate as farther north, about 1 second of arc in 10 years. There is a discontinuity near Brawley which is interpreted as fault slip in that vicinity, probably attributable to the earthquakes of 1950 (Chapter 6).

In the summer of 1954 a first-order triangulation arc was extended eastward from Fallon, Nevada; on December 16 the region was affected by a major earthquake with large and extensive faulting (Chapter 28). The arc was reobserved in the summer of 1955. The main fault trends roughly north-south; points on the west shifted northward; points on the east, southward. The maximum on both sides was about 4 feet, making total relative right-hand slip about 8 feet. Displacements decrease away from the fault, as required by elastic rebound; points 15 to 20 miles east and west showed prac-

FIGURE 14-3   *Retriangulation results, U. S. Coast and Geodetic Survey, Imperial Valley, 1939–1941. [Redrafted after Whitten by C. R. Allen.]*

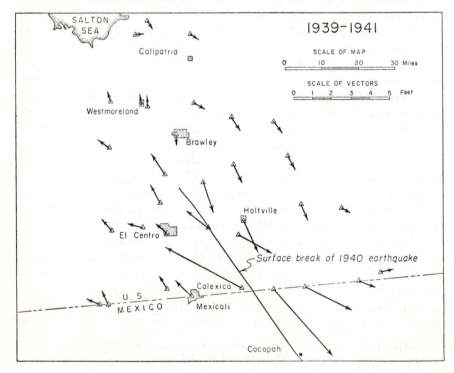

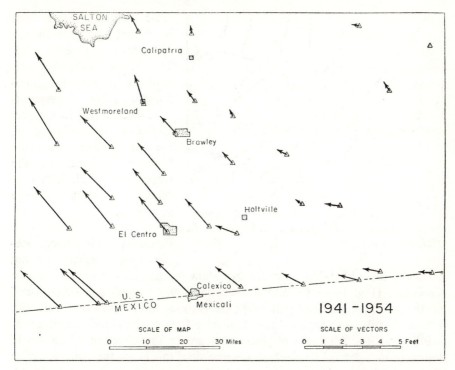

FIGURE 14-3A   *Retriangulation results, U. S. Coast and Geodetic Survey, Imperial Valley, 1941–1954. [Redrafted after Whitten by C. R. Allen.]*

tically no movement. Releveling established vertical displacements of as much as 7 feet.

For Whitten's report on resurveying after the Kern County earthquakes of 1952, see Chapter 28. Chapter 30 includes brief notice of the results of resurveying after the Japanese earthquakes of 1923, 1927, 1930, and 1948.

ADDITIONAL EVIDENCE

Thus, observations made after the 1906 earthquake have been confirmed on other occasions in California, Nevada, Japan, and elsewhere. The elastic-rebound theory is readily acceptable for earthquakes accompanied by observed faulting. Nevertheless it is a large step to apply this theory to tectonic earthquakes in general.

Macroseismic Data

Continuity of Displacement. Usually fault displacements accompanying an earthquake are of the same kind as those required to produce the large tec-

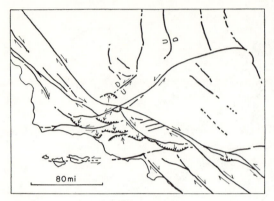

FIGURE 14-4 *Principal Southern California faults, with sense of displacement.* [*After M. L. Hill, modified.*]

tonic features. Thus fault scarps are usually upthrown in the same sense as the adjacent mountain blocks; they tend to increase the topographic relief (exceptions have been noted in Chapter 13). Observed strike-slip on the San Andreas and other faults in California is in the same sense (right-hand) as offsets of streams and large structures.

Regional Coherence. Regional displacements are not all of the same kind, but normally they show mechanical coherence. Thus in California (Fig. 14-4) the San Andreas fault and a number of others trend northwest with right-hand strike-slip; the Garlock fault trends northeast with left-hand strike-slip; the east-west Tranverse Ranges show evidence of north-south thrusting. All these displacements are in the direction which would follow from a general north-south compression; this is not the same as saying that all the features involved were originally produced by such a compression. The complicated tectonics of Japan shows similar coherence with east-west compression.

Microseismic Data

Continuity. An especially consistent group of microseismic data bears on this point. Many years ago Gherzi and Somville pointed out that the first motions of the *P* waves recorded at their stations (in China and Belgium respectively) were regionally consistent as to epicenter; that is, earthquakes of a given area normally record either with initial compressions or initial dilations at a given distant station. Similar regional distribution was later worked out for stations at Rome, Berkeley, Pasadena, and elsewhere. Vesanen at Helsinki established regional patterns for the whole wave form of the first part of the seismograms at that station. Such consistency indicates that tectonic processes in a given area do, in fact, continue in the same sense. There are sporadic exceptions, and sometimes a small region appears to behave differently from its surroundings; such effects may be compared to local eddies in the flow of a stream. There is ample geological evidence to support their existence on a large scale in both space and time; one instance is the small but very active tectonic arc of New Britain which stands athwart the general trend of the Pacific belt, another is the great swirl of structures in the Moluccas and about the Banda Sea.

Coherence in Small Shocks—California.   Application of the theory to small shocks involves further inference. Independent evidence shows that after-shocks, at least, are part of the same process which includes large earth-quakes (Chapter 6).† The study of initial recorded first motions can also be applied on the small scale. This was done very effectively by Gutenberg for southern California. The area southeast of Pasadena is favorable for a test of this kind, since it is traversed by several of the faults already men-tioned, running roughly northwest-southeast, and all characterized by right-hand strike-slip as far as the geological evidence goes. The method used will appear from the diagrams in Figure 14-5. In A, right-hand strike-slip occurs on the indicated fault, with epicenter at the center of the diagram. The first motion recorded at a given station will be compression or dilatation (up or down in the vertical component) according to whether the nearer crustal block is displaced toward or away from the station. Compression is indicated by a solid circle, dilatation by an open circle, at the locations of various sta-tions. It will be seen that compressions and dilatations are separated into alternate quadrants.

In Figure 14-5B, there is one recording station, located at the center of the diagram. Strike-slip of the same kind as in A now is supposed to take place on a number of faults, with epicenters indicated by circles. These cir-cles are solid when the station in the center should record a compression, open when it should record a dilatation. The resulting dots are separated by quadrants in the same way as in A, in spite of the inverted significance of the plot.

Figure 14-6 is a plot of the second type from Gutenberg's paper; it shows compressions (triangles) and dilatations (circles) as recorded for the indi-cated epicenters at the Riverside and La Jolla stations. The complete study involved determining more than 4200 first motions. The pattern shown shifted geographically with the recording station, as it should. Shocks north of the Transverse Ranges were also studied; while the type of motion recorded

FIGURE 14-5

Azimuthal distribution of compression (solid circles) and dilatation (open circles) due to right-hand strike-slip. A, as observed at points distributed round the epicenter; B, as observed at a station in the center for the indicated epicenters.

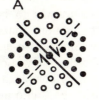
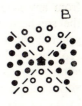

A   B

† Båth finds unexpected mechanical complexity among the aftershocks of the Kern County series in 1952 (Chapter 28).

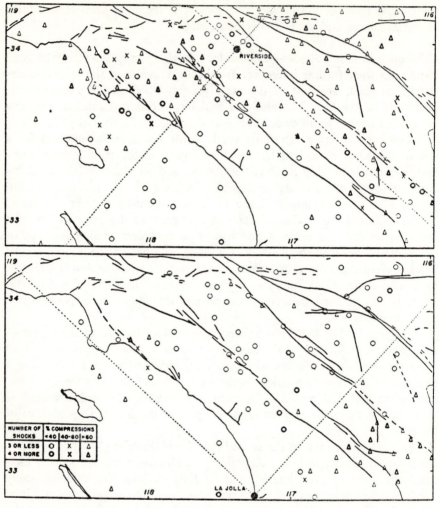

FIGURE 14-6   *Epicenters of shocks recording with initial compression (triangles)
and dilatation (circles) at Riverside and at La Jolla.* [*Gutenberg.*]

from any small area was consistent, the pattern was more complex than in
the southern region, as was to be expected from the known geology. In
general, the type of motions found corresponded with those inferred from
geological evidence. Simple two-dimensional patterns like those here figured
apply only to strike-slip on vertical faults. Where dip and dip-slip are involved, there are greater complexities. However, similar consistency is found
in such cases; outside of California the most extensive observations of this
kind come from Japan, where there are many stations and the method has
been applied to deep earthquakes as well as shallow ones (Chapters 19 and
30). Eiby has obtained corresponding results for New Zealand. No evidence

appears anywhere to conflict with our belief that the small non-volcanic earthquakes are mechanically like the large ones and are due to the same geological processes.

The alternative procedure—investigation of the distribution of compressions and dilatations in recording a given shock—is an active current field of research (Chapter 32).

ULTIMATE CAUSES

Our discussion thus far leaves open the question of the ultimate cause of earthquakes. We need only to assume that forces exist producing continuously increasing strains which are always of the same character, and that these strains, when relieved by fracture, immediately begin to accumulate anew. Nothing is implied about the nature of the forces; strictly speaking, this is a problem not in seismology, but in general geology and geophysics. The forces which bring about faulting and earthquakes are the forces which form mountains; and the nature and cause of mountain-building is the central outstanding problem of historical geology.

The days are gone when all was cheerfully explained in terms of a cooling and contracting earth. The implied continuous motion of great masses relative to one another fits well with the later theories of continental drift, especially in regard to motion of the continents relative to the Pacific Basin. There is a still more recent vogue for currents of plastic flow, probably thermal convection currents, in the mantle of the earth. Such currents, fixed like the established currents of circulation in the oceans, would produce continuous and coherent stresses in a given region. For this and other contemporary speculation the student should consult Eardley's excellent summary.

Seismology presents the geologist with a sort of snapshot, a comparatively detailed picture of processes now going on. In spite of the geological principle of uniformity, these present processes may not be altogether representative of what has gone on in the past; but they are nevertheless important data which must be accounted for by any inclusive theory. In the language of mathematics, they are boundary conditions which must be satisfied.

OBSERVATION OF FAULTING

Table 14-1 presents, in summary form, information on earthquakes associated with faulting, including some questionable instances. Its preparation has depended in part on books by Montessus de Ballore (1924) and Davison (1936), of which the former is the more reliable. Details of the individual events will be found arranged by regions in Part Two of this book. A bibli-

Table 14-1   Earthquakes with Known Faulting*

I. California and Nevada

    1857   Southern San Andreas fault   SAB

    1868   Haywards fault   SAB

    1872   Owens Valle   DSAB

    1875?   Mohawk Valley, Plumas County   DCB

    1903   Wonder, Nevada   B

    1906   Central San Andreas fault   SA

    1915   Pleasant Valley, Nevada   DA

    1932   Cedar Mountain, Nevada   SC

    1934   Excelsior Mountains, Nevada   DC

    1940   Imperial fault, California-Mexico   SA

    1947   Manix, Mojave Desert   SC

    1950   Fort Sage Mts   DCA

    1952   (July 21) White Wolf fault, Kern County, California   DA

    1954   (July 6) Rainbow Mountains, east of Fallon, Nevada   DC

            (August 23) same area as preceding; disturbance extended   DC

            (Dec. 16) Stillwater and Clan Alpine ranges, Nevada; east of the preceding   DSTA

II. New Zealand

    1848   South Island (Awatere fault?)   BC

    1855   The North Island   DA

    1888   The South Island (Hope fault, Amuri region)   SA

    1929   The South Island (Buller earthquake)   DSA

    1931   The North Island (Napier earthquake)   DRA

    1932   The North Island (Wairoa)   D(S)

    1942   The North Island   DC

III. Japan and vicinity

    Chronicled older events, involving changes in level on the south coast of Honshu   DB

    1847   Zenkōji (Nagano)   DB

    1891   Mino-Owari   SDA

    1894   Sakata[1]   DB

    1896   Riku-Ugo[1]   DTA

    1906   Formosa   SD⁻A

    1923   Kwanto   (S)RA

    1925   North Tajima[1]   S?D?C

    1927   Tango[1]   SDTA

    1930   Izɹ   SRA

    1935   Formosa   SDTA

    1943   Tottori[1]   SA

    1945   Mikawa   DA

    1946   Nankai   RA

    1948   Fukui[1]   RC(S)

    1951   (Oct. 22) Hwalien, Formosa   SA

IV. America, exclusive of I

    1811–1812 Mississippi Valley (New Madrid) **RB**

    1887 Sonora (Mexico) DA(TB)

    1899 Alaska (Yakutat Bay) DAB

    1934 Utah C

    1944 Argentina DAC

    1946 Peru DA

    1956 Baja California SDA

    1958 Alaska SA

    1959 Montana DA

V. Other regions

    1819 Cutch, India DA

    1887 Tian Shan DSB

    1892 Baluchistan SAB

    1892 Sumatra SAB

    1894 Greece DSC

    1897 Assam, India DRA

    1905 Mongolia (Tannu-ola) DSB

    1911 Tian Shan (Kebin) DA

    1928 East Africa DBC

    1928 Bulgaria DC

    1939 Turkey SA

    1942 Turkey SDA

    1943 Turkey D(S?)B

    1944 Turkey SDA

    1953 Turkey SA

    1957 Turkey S

    1957 Mongolia SDA

Notes.

D = dip-slip. S = strike-slip. T = two principal traces.

R = regional distortion with complex faulting.

A = clear and well authenticated. B = imperfectly known.

C = doubtful, or a borderline instance with small reported displacement, or otherwise complicated. Designations B and C do not necessarily reflect on the quality of the field investigation. One or two instances in which the report of faulting appears to be due to poor observation are omitted.

Earthquakes for which the nature and extent of faulting has been inferred exclusively from seismometric registration (such as the Long Beach, California earthquake of 1933) are not listed.

Strike-slip is nearly always observed to be accompanied by dip-slip, which characteristically reverses throw along the line of faulting.

* General references: Montessus de Ballore, La Géologie Sismologique (1924); Charles Davison, Great Earthquakes (1936).

[1] Localities near the north coast (Japan Sea) of Honshu.

ography in chronological order appears in Appendix XVI.

Limitations

The geographical distribution of the earthquakes listed in the table is peculiarly limited, owing to geological and human conditions. Observed faulting is largely confined to regions of block tectonics, and the information is comparatively fragmentary for most of the active arc structures (see Chapter 26). Partly for this reason, the types of faulting seen in the field are less inclusive than those which may be inferred from the evidence preserved in the rocks. Thus stratigraphy provides classical examples of large-scale low-angle thrusting, which is questionably represented, if at all, among observed earthquakes.

Effect of Hypocentral Depth.   Obviously, even a great earthquake originating at depth need not break the surface. Apart from the special classes of deep-focus earthquakes discussed in Chapter 19, shocks originate at depths somewhat greater than normal, in the range of 40 to 60 kilometers. In certain areas, such as the Marianas Islands and Chile, such shocks are the rule, and major faulting breaking the surface is correspondingly less likely. Apparently thrust faulting predominates in these regions. Such faulting, especially low-angle thrusting, is unlikely to produce an observable surface trace unless the thrust plate is broken up by fractures which appear at the surface, as in Oldham's original interpretation of the 1897 earthquake. However, thrusting may produce observable regional warping, particularly if it results in changes of level at the coast.

Large faulting of any kind may take place under a great thickness of alluvium, manifesting itself at the surface only in indirect and complicated fashion, as in the Mississippi Valley in 1811–1812, or in Bihar in 1934.

Inglewood Fault Zone.   Even where there is no alluviation, large faulting may go on in basement rock without directly breaking through the sedimentary cover. As usually explained, the Inglewood fault zone in southern California, the seat of the Long Beach earthquake of 1933, is determined by a major break in the basement. The displacement is held to be chiefly strike-slip, although at depth the basement level differs as much as 4000 feet on the two sides of the fault. This displacement decreases rapidly upward. No break comes directly to the surface; instead the overlying Tertiary rocks are folded. These folds account for the series of oil-producing anticlines which accompany the tectonic zone. Internal to the folds are many small faults, often arranged in echelon, and mechanically coherent with the main strike-slip distortion. This whole feature has been characterized as a gigantic "mole track" mechanically similar to the feature developed by strike-slip faulting in alluvium (Chapter 13).

Accessibility.   In general, to be observed, the fault break must be on land; this removes the possibility of observation for many of the most seismic areas of the world, notably about the Pacific. In the nature of things, most of the known instances occur in populated or readily accessible areas. Expensive expeditions to remote regions are not often undertaken for seismological purposes. Fresh scarps may attract attention in uninhabited areas, but strike-slip faulting is particularly difficult to establish unless artificial features like roads and fences are offset.

Dates and Data.   The human element is evident in the distribution of dates in Table 14-1. They reflect the rapid increase in scientific observation since 1800, and show that, as remarked in the introductory chapter, seismology is indeed a young science. The extent and reliability of the information vary widely; at one extreme are earthquakes like those of 1906 in California and of 1923 and 1927 in Japan, which have been described in detail and by numerous authors, at the other are the Sumatra earthquake of 1892 and the New Zealand earthquake of 1848, which were not formally investigated and for which our information is fragmentary. Indication of doubt is not necessarily a reflection on the quality of investigation. Even when field work has been thorough, uncertainty remains in many instances. Most of these are borderline occurrences; faulting which may have been large at depth has barely extended to the surface, or displacement has been so small as to be barely observable.†

Dip-Slip and Strike-Slip

The table indicates the nature of principal observed displacements, whether dip-slip or strike-slip. Where both were observed, that which appeared dominant is named first. Strike-slip occurs together with dip-slip in a large proportion of the earthquakes studied. There is often suggestion of strike-slip from indirect evidence such as echelon traces and tension cracks when more direct proof is lacking. When field observation establishes large strike-slip it is the most unequivocal proof of the true tectonic character of the displacement—more convincing than scarps and other dip-slip evidence, which may originate in a variety of secondary ways (Chapter 13).

The relative frequency of strike-slip in observed faulting gives us information only for shallow earthquakes in regions of block tectonics. Studies of recorded compressions and dilations, however, seem at present to extend this result generally, even to deep-focus earthquakes, as Scheidegger has emphasized lately (Chapter 32).

Complex Faulting

Tectonic events are rarely simple. Even when only one principal trace has been observed, in large earthquakes there is nearly always suggestion of further complexity. Often two principal traces are found, and the character of displacement on these generally differs—downthrow may be opposite, or one may be principally strike-slip while the other is chiefly dip-slip. There

---

† Examples: Tajima, Japan, 1925 (Chapter 30); Manix, California, 1947 (Chapter 28).

are several such instances in Japan and Formosa. With greater complexity, conditions approach those of regional distortion, perhaps associated with thrusting, as noted in Chapter 13.

Indirect Tectonic Effects

In assembling the data it became increasingly necessary to distinguish (as in Chapter 7) between: (1) primary faulting; (2) secondary effects, such as scarps, due to shaking; and (3) indirect effects mechanically connected with faulting in basement rock, but not constituting direct extension of breaking to the surface. Well-attested occurrences of group (3) are few. Earthquakes in Argentina in 1944 and in Japan in 1945 were described as showing low-angle reverse faulting incidental to (probably vertical) relative displacement of large blocks. This thrusting should be regarded as an effect rather than a cause of the earthquake. The surface expression of the Nevada earthquake of 1932, interpreted as the result of strike-slip faulting in the basement rock, was almost as complex as the features of the Inglewood fault zone. The dropped block in the Garlock fault zone in 1952 is probably an indirect effect of different type (Chapter 7). Such an event might be considerably delayed after the main shock, and then would not differ mechanically from a large slump. There is a suggestion of such an explanation for one of the traces of the Nevada earthquake of December, 1954, which probably originated in a large aftershock 4 minutes after the main shock (Chapter 28).

UNIDIRECTIONAL FAULTING

There is a common impression that faulting in large earthquakes commences near the center of the active fault segment and extends toward the two ends. This appears to have arisen partly from the shift in meaning of the term epicenter, noted in Chapter 2. In the pre-instrumental period of seismology it was usual to locate the epicenter at the center of figure of the isoseismals. Epicenters placed in this way later became identified with epicenters determined by instrumental means, which are believed to represent the points of initial rupture, although, when both macroseismic and microseismic data are available, it nearly always develops that the instrumental epicenter is eccentrically placed with reference to the isoseismals. As noticed in Chapter 4, Mallet arrived at such an eccentric epicenter by relatively crude means of physically interpreting macroseismic data.

When foreshocks and aftershocks can be assigned microseismic epicenters, foreshocks usually appear to originate close to the epicenter of the main shock, while aftershock epicenters extend along the whole active segment.

This fits in well with the idea of rupture at one end and unidirectional progression of faulting.

It is not intended to imply that unidirectional progression is universal. Where there is great complexity and a series of traces, as in the Nevada earthquakes of 1954, no such simple description can apply, although there is considerable significance in such results as that of Tandon for the Assam-Tibet earthquakes of 1950 (Chapter 5), in which the aftershock epicenters were distributed over a vast area with the epicenter of the main shock near one edge.

As instrumental seismology progresses, there are more and more instances in which actual faulting can be studied with the aid of the epicenters of the minor shocks, even when there is no surface break. This is true of the Long Beach earthquake of 1933, which followed this pattern: foreshock epicenter near that of the main shock, aftershocks scattered along the whole active segment, concentration of aftershocks near the two ends. Similar results were obtained for the smaller Parkfield earthquake of 1934, and for the larger Desert Hot Springs shock of 1948 (Chapter 28).

It is noteworthy that the California earthquake of 1906, exceptional in so many ways, appears to have been abnormal in this respect also. The seismograms, inferior though most of them are by present standards, make it difficult to place the point of initial rupture far north of San Francisco—certainly not as far as the northern extent of faulting; while the data would have to be badly forced to place an epicenter near the southern end of observed faulting. The dubious evidence of faulting progressing in two directions in the Imperial Valley earthquakes of 1940 should also be noted.

On the other hand, the remarkable series of earthquakes in Turkey beginning in 1939 show faulting progressing from east to west in successive occurrences along faults of the same strike-slip system.

EXCEPTIONAL OBSERVATIONS

In this and the preceding chapters a number of earthquakes have been mentioned for which either the faulting presented exceptional features or other unusual phenomena were observed. Almost every well-studied earthquake has some feature worthy of special attention; a brief list, excluding those in which the unique feature is historical interest, might run as follows:

1755	Lisbon: seiches
1897	Assam: regional distortion
1899	Alaska: great displacement; effects on glaciers
1906	California: great linear extent
1931	New Zealand: evidence of thrusting

1932	Nevada: complex surface expression
1934	Bihar, India: slump belt effects
1935	Formosa: fault scarp seen to form after earthquake
1939–1944	Turkey: faulting successively extending in repeated events
1944	Argentina, and
1945	Japan: thrusting incidental to motion of large blocks
1947	Manix, California: apparent fault trace nearly at right angles to line of epicenters
1952	Kern County, California: thrusting
1954	Nevada: repeated earthquakes and complex faulting

Details for most of these will be found in Part Two or in preceding chapters (Chapters 5, 9, and others). A few call for brief notice here.

The Manix earthquake of 1947 presented a trace along an old fault with left-hand strike-slip of only a few inches. Epicenter alignment suggested a buried fault on which right-hand strike-slip should be expected, since its strike was nearly at right angles to the exposed fault. It is probable that the observed displacement on the old Manix fault, though mechanically coherent, was of the indirect tectonic type noted earlier. There is a similar possibility in connection with some of the aftershocks of the 1952 Kern County series.

The at first astonishing report for the 1935 Formosa earthquake will be taken up again in its proper place in Part Two (Chapter 30). Here we need only note that there is nothing incredible about a scarp in alluvium being seen to form after felt shaking, since the break should extend upward through unconsolidated material with speed less than that of seismic waves. There apparently is a similar observation for the Tango earthquake of 1927.

THE LEGENDARY PAST

Earthquake observation of old date is so inadequate scientifically that faulting before 1800 assumes a legendary character. Table 14-1 would be greatly extended if it included earthquakes accompanied by local or regional subsidences and upheavals, particularly on seacoasts. Such events can be traced back through recorded history into the distant past; some writers would even include the Biblical deluge. Montessus de Ballore discusses the submergence of Helice during an earthquake in 373 B.C. Helice, one of the major cities of Greece, was located on the south side of the Gulf of Corinth; the catastrophe took place almost at the height of Greek civilization. It may have been subsidence by faulting and not sliding, but we lack the necessary details.

Imamura has a list of 26 Japanese earthquakes accompanied by subsidence and uplift, with dates extending centuries into the past. Most notable among these is one in 1703 in the same region as the great earthquake of 1923, but apparently with larger displacements; historical records and elevated strand lines with shells indicate uplifts locally reaching 5 meters. Still higher strand lines are correlated by Imamura with an earthquake in A.D. 818. Such evidence must be examined with care to distinguish it from the raised strand lines which many investigators have identified in different parts of the world and have attributed to general post-Pleistocene changes of sea level. One such level is about 10 feet above the present; a frequently observed strand at 25 feet is believed to be Pleistocene.

FAULTS IN REPORTING

One obstacle in assembling this material has been the occasional great disproportion between seismological data and geological speculation in published material. Use of earthquake observations to illustrate or support pet ideas is almost unavoidable in a textbook like this one—let the student take warning!—but it is out of place in research publications dealing with individual events. Geological background is needed, and geological interpretation desirable; but this should not result in devoting the whole space to geology, so that actual earthquake information is overcondensed or seriously incomplete. Reference to the work of other authors is not sufficient if, as sometimes happens, the detailed seismological study is published only in the proceedings of a local society, while the geological lucubrations appear in an internationally circulated periodical.† A seismologist has to be a passable linguist if he is to read the literature at all well, but sometimes the limited publication of details lays the curse of Babel on him with undue bitterness.

Especially in abbreviated and summarized descriptions, defects in illustration become serious. One would not expect experienced geologists to omit coordinates and important place names from their maps, or to obscure the mapping of fault lines with heavy contours and formation symbols—but they sometimes do!

The speculative geologist often forgets that, even in very seismic areas, minor earthquakes may have only a local significance; instrumental seismology shows that their epicenters often have little relation to the major faults and tectonic units. A great earthquake, on the other hand, is necessarily of great geological importance, but a great geologist may be needed to interpret it correctly.

References

Earthquake mechanism

Reid, H. F., *The Mechanics of the Earthquake, The California Earthquake of April 18, 1906, Report of the State Investigation Commission*, Vol. 2, Carnegie Institution of Washington, Washington, D. C., 1910. (See especially pp. 16–28.)

---

† Here the fault may lie not with the investigator but with an editor. General journals are sometimes inhospitable to local details.

———, "The Mechanics of Earthquakes. The Elastic Rebound Theory. Regional Strain," *Bull. Nat. Research Council,* No. 90, 1933, pp. 87–103.

Oldham, R. D., The Cutch (Kacch) earthquake of 16th of June, 1819, with a revision of the great earthquake of 12th June, 1897," *Mem. Geol. Survey India,* vol. 46 (1928), pp. 71–147. (Contains the hypothesis of the "bathyseism.")

Louderback, G. D., "Faults and earthquakes," *B.S.S.A.,* vol. 32 (1942), pp. 305–330. (See also papers by Benioff cited in Chapters 6, 22, 26.)

Results of resurvey

Whitten, C. A., "Horizontal earth movement, vicinity of San Francisco, California," *Trans. Am. Geophys. Union,* vol. 29 (1948), pp. 318–323.

———, "Crustal movement in California and Nevada," *ibid.,* vol. 37 (1956), pp. 393–398.

———, "Geodetic measurements," *B.S.S.A.,* vol. 47 (1957), pp. 321–325.

For corresponding results in Japan consult Appendix XVI for the years 1923, 1927, 1930, 1948; also: Tsuboi, C., "Deformations of the earth's crust as disclosed by geodetic measurements," *Ergeb. kosmischen Physik,* vol. 4 (1939), pp. 106–168.

Analysis of initial motion of seismograms (general papers)

Byerly, P., "Nature of faulting as deduced from seismograms," *Geol. Soc. Amer., Spec. Paper,* No. 62, 1955, pp. 75–86.

Scheidegger, A. E., "The physics of orogenesis in the light of new seismological evidence," *Trans. Royal Soc. Canada,* vol. 49 (Sec. IV), pp. 65–93.

These general references are among the latest discussions. See Chapter 32 for others, and for references depending on combining the initial motions of particular earthquakes at many stations.

Mapping and study of first motions at a given station

Gherzi, E., "Notes de séismologie," *Observatoire de Zi-ka-wei,* Nos. 4 (1923), 6 (1925), 10 (1929).

Somville, O., "Sur la nature de l'onde initiale des téléséismes enregistrés à Uccle de 1910 à 1924," *Publ. Bureau Central Séismologique International,* Ser. A, vol. 2 (1925), pp. 65–76.

di Filippo, D., and Marcelli, L., "Sul movimento iniziale delle onde sismiche registrate a Roma durante il periodo 1938–1943," *Ann. Geofisica,* vol. 2 (1949), pp. 589–606.

Byerly, P., and Evernden, J. F., "First motion in earthquakes recorded at Berkeley," *B.S.S.A.,* vol. 40 (1950), pp. 291–298.

Båth, Markus, "Initial motion of the first longitudinal earthquake wave recorded at Pasadena and Huancayo," *ibid.,* vol. 42 (1952), pp. 175–195.

Mühlhäuser, S., "Die Richtung der ersten Bodenbewegung (Kompression oder Dilatation) in Stuttgart für die Hauptbebengebiete der Erde, als Grundlage für grosstektonische Betrachtungen," *Zeitschr. Geophysik,* Sonderband, 1953, pp. 76–91.

Vesanen, E., "Über die typenanalytische Auswertung der Seismogramme," *Ann. Acad. Scientiarum Fenniae,* Ser. A, III, no. 5 (1942).

————, "On seismogram types and focal depth of earthquakes in the North Japan and Manchuria region," *ibid.,* No. 11 (1946).

————, "On Alaska earthquakes," *ibid.,* No. 14 (1947).

Jones, J. W., "Seismogram types and initial motion at Seattle," *Publ. bureau central séismologique international,* Ser. A, vol. 18 (1952), pp. 221–236.

Gutenberg, B., "Mechanism of faulting in Southern California indicated by seismograms," *B.S.S.A.,* vol. 31 (1941), pp. 263–302.

Eiby, G. A., "The direction of fault movement in New Zealand," *N. Z. Journ.,* Ser. B, vol. 36 (1955), pp. 552–556.

Earthquakes with known faulting

de Montessus de Ballore, F., *"La Géologie sismologique,* Librairie Armand Colin, Paris, 1924.

Davison, C., *Great Earthquakes,* Murby, London, 1936.

For references on individual earthquakes listed in Table 14 and elsewhere see the chronological bibliography (Appendix XVI).

Other references

Woodford, A. O., Schoellhamer, J. E., Vedder, J. G., and Yerkes, R. F., "Geology of the Los Angeles Basin," *Calif. Dept. Nat. Resources, Div. Mines, Bull.* 170, Chapter 2, pp. 65–81. (Includes description of the Inglewood fault zone and associated structures.)

Imamura, A., *Theoretical and Applied Seismology,* Maruzen, Tokyo, 1937. (Effects of earthquakes of 818, 1703, and 1923 compared on pp. 178–184.)

Fairbridge, R. W., *Multiple Stands of the Sea in Post-Glacial Time, Proc. 7th Pacific Science Congress,* vol. 3 (1952), pp. 345–347.

Eardley, A. J., "The cause of mountain building—an enigma," American Scientist, vol. 45 (1957), pp. 189–217. (An up-to-date report.)

Båth, M., and Richter, C. F., "Mechanisms of the aftershocks of the Kern County, California, earthquake of 1952," *B.S.S.A.* (in press).

Note added in proof: For faulting during 1957 see notes on pp. 616 and 625.

CHAPTER 15

Seismograph Theory and Practice

DEFINITIONS AND BASIC PRINCIPLES

In earlier chapters we have looked at earthquakes mainly from the macroseismic point of view. Here and there we have used instrumental results to clear up particular points, and we have referred to special microseismic features of certain earthquakes. We must now introduce more microseismic material, starting with an account of the instruments. Fundamental definitions are:

Seismoscope, a device which indicates the occurrence of an earthquake but does not write a record.

Seismograph, an instrument which writes a permanent continuous record of earth motion, a seismogram.

Seismometer, a seismograph whose physical constants are known sufficiently for calibration, so that actual ground motion may be calculated from the seismogram. (See also footnote to "Electromagnetic seismographs" on a later page.)

Seismoscopes have been invented again and again. The oldest on record was set up in China in A.D. 132 by Chang Heng.† It was quite ingenious and it had directional characteristics. Ornamental dragon mouths arranged in a ring each held a ball, which earthquake motion in that direction would drop into the waiting mouth of a carved toad. Present-day amateurs often set up "earthquake alarms" in basements or attics; and the principle has been applied commercially to automatic valves designed to shut off gas, electricity, or water in case of a strong earthquake.

An obvious improvement is to have the instrument make a mark on a rotating chronograph drum, thus recording the time of the earthquake. This is a first step toward the true seismograph. Such instruments were operating

---

† Chang Heng is better known for literary than for scientific attainments; apparently he was a genius of broad gifts like Leonardo da Vinci and Omar Khayyám. Textbooks commonly give his name as Chōkō, a Japanese form used in John Milne's writings, also written Tyōkō.

in the mid-nineteenth century, particularly in Italy. At that time they were often called seismographs. Instruments suiting our present definition were not developed until about 1880, in Japan first, then in Italy and elsewhere in Europe.

The seismometer raises some uncommon mechanical difficulties. In the physical laboratory we are accustomed to making precise measurements of the motions of bodies. Such measurements are made with reference to some fixed base, usually a pier set solidly in the ground. But an earthquake affects a large area in a very short time, and we then have no fixed base. We should like to measure earthquake motion relative to the whole mass of the earth, but there is no simple way to do it.

The first obvious suggestion is that of a "steady mass," a heavy body free to move relative to the ground. In the ideally perfect case, the ground moves relative to the steady mass, which because of its inertia does not move at all; the recording apparatus would then show this relative motion. This ideal is unrealizable; the practical substitute is a mass retained by a spring, or other restoring force, which tends to return the mass to its normal position relative to the ground but is not strong enough to interfere with it seriously. Such an apparatus is a pendulum, in the general sense; almost all seismometers now in use depend on some kind of pendulum for the sensitive element. Many types have been used, and their behavior has been further diversified by varying the manner in which the recording system responds to the motion of the pendulum.

A typical seismograph assembly consists of three connected units: a *pendulum,* a *chronograph,* and a *recording device* which writes a line representing the motion of the pendulum on the chronograph drum.

THE CHRONOGRAPH AND TIMING

Seismologists use a chronograph like those in service in other branches of physical science. It is ordinarily a drum rotating at carefully controlled constant speed and advancing with each revolution along a screw threaded on the drive shaft. The recording device writes a spiral line on a sheet mounted around the drum. This sheet when cut and opened shows a series of lines, straight and parallel when the instrument is undisturbed, deflected when there is an earthquake. A clock marks on the record, usually once or twice each minute. The drum carries the paper under the writing point at a constant speed, which in ordinary installations may be fixed as low as 7 millimeters per minute or as high as 60 millimeters per minute. At first the recording sheet was usually smoked paper, on which the record was scribed by a sharp stylus—the seismograph "needle" of newspaper headlines. Most seismographs now record on photographic paper by means of a small sharply focused spot of light. This calls for recording in a dark room or under a light-tight housing and handling the sheets in red light until they are developed and fixed. Seismograms have also been written with pen and ink, or with a hot stylus on chemically prepared paper such as is used for cardiograms.

Recorded earthquakes are timed with reference to the clock marks on the seismogram. Standard time is found by comparing the marking clock with a carefully regulated master clock or directly with radio time signals from the Naval Observatory or other equally reliable source. Accurate timing thus demands good performance from three clocks: the clockwork rotating the drum, the marking clock, and the master clock—the last, by means of radio transmission, may be at the Naval Observatory.

MAGNIFICATION AND RECORDING

Since actual ground motions are small, except in earthquakes which are locally strong, a mechanical or optical magnification, with or without further electromagnetic or electronic amplification, is used to get readable records. Magnifications generally in use range from 100 to 100,000; less for strong-motion recorders, somewhat more for the most sensitive systems at favorable locations.

Purely mechanical magnification involves a lever system. The pendulum must be heavy, so that when accelerated by small earth motion it will overcome friction in the lever system or at the writing point. A smaller mass can be used by recording photographically. In that case the pendulum, or some part rotated by its motion, carries a mirror. Light from a fixed lamp is reflected from this mirror through a lens system which focuses into the recording spot on the chronograph drum. When the record is developed it is technically a negative, appearing as a dark line on a white background.

In many seismograph assemblies motion of the pendulum generates electromotive force in a circuit through a galvanometer. The galvanometer mirror is then used for photographic recording. Photoelectric and various electronic amplification systems have been used; at Pasadena such installations have been used chiefly to get enough power to drive a pen recording in ink. The immediate reason for this is to have a visibly recording instrument which can be seen in the act of registering an earthquake. Photographic recording has the disadvantage that the seismograms are not available until they have been taken off the drums and developed. This is a hindrance in a local earthquake emergency, since it delays exchange of information with other stations, with the press, and with interested officials or private persons; such exchange is of great value in planning a field expedition. An advantage in having one visibly recording instrument is that during the working day any trouble with the drive, clocks, or pendulum can be detected promptly and corrected, instead of continuing until the next daily change of sheets. Dr. Benioff has recently set a pendulum to control a discriminator circuit by varying a capacity; power in the discriminator circuit is taken from the local supply current, so that it is sufficient to operate the pen recorder.

A cheaper way to set up an auxiliary visibly recording unit is to use one of the old mechanical instruments writing on smoked paper. However, the lever systems are almost as delicate and complicated mechanically as the more modern systems are electronically, and in addition one has to face the unpleasant messiness associated with smoked-paper recording.

In all simple seismographs, deflection of the recording index—light spot or writing point—on the seismogram has a simple relation to the displacement of the pendulum. Upward deflection on the seismogram as read may indicate motion of the ground upward, northward, or eastward relative to the pendulum mass, while downward deflection indicates ground motion down, south, or west. Some stations arrange their instruments opposite to this, and a small minority of horizontal-component instruments show motion otherwise than north-south or east-west.

When earthquake motion is large and lasts a long time, successive lines of recording on the seismogram are tangled. This is sad but it cannot be prevented without expensive waste of paper. Recording on magnetic tape is a natural suggestion; this method is still in the development stage, and thus far it has been successful only for special purposes.

Two serious risks of losing important data when recording local earthquakes are often neglected. Pendulum clocks are often stopped by strong earthquakes, or their rates seriously affected. Hence it is desirable in very seismic regions to record at least partly with spring clocks, chronometers of the balance-wheel type, or even electric clocks. The last, however, are exposed to the second chief risk, failure of the local electric power supply, particularly during or immediately after a strong earthquake. Such failure may put out the lights for photographic recording and may stop the chronograph drives if they use motors running off the power line. At Pasadena and its auxiliary stations power for drives and lights is drawn from storage batteries kept charged from the line, so that operation continues through temporary power failures. An alternative is to operate regularly from the line supply and switch over automatically to batteries in case of failure.

ELEMENTARY SEISMOMETER THEORY

Theory of the Undamped Seismometer

To register ground motion completely we need at least three instruments to measure displacement vertically and in two horizontal directions.† To record

---

† Perfectly general motion would also involve rotations about three perpendicular axes, and three more instruments for these. Theory indicates, and observation confirms, that such rotations are negligible.

one such component we set up a pendulum capable of moving only north and south with respect to its frame. We let

x = position of the center of gravity of the pendulum, referred to a fixed point of origin

X = position of a specified point on the frame, referred to the same fixed point of origin, so that $x = X$ in the position of equilibrium

K = a spring constant, so that the restoring force on the pendulum is $-K(x - X)$

m = mass of the pendulum

$u = x - X$ is the displacement of the pendulum relative to the frame; the deflection on the seismogram will be a measure of u

The equation of motion is

$$m\ddot{x} = -K(x - X) \tag{1}$$

If we bring in u in place of x and write $K/m = \omega^2$, we find

$$\ddot{u} + \omega^2 u = -\ddot{X} \tag{2}$$

This relates the seismogram deflection, represented by u, to the actual ground displacement X. If the ground is at rest, $X = 0$ and (2) represents a simple harmonic motion; that is, if $X = 0$,

$$u = B \sin (\omega t + b) \tag{3}$$

Here B and b are constants specifying amplitude and phase. If $\omega = 2\pi/\tau$ we find that τ is the period for one complete oscillation.

Now suppose that the ground is in motion, and begin by assuming that this motion is simple harmonic, with its own period T. Put $p = 2\pi/T$, and write

$$X = A \sin pt \tag{4}$$

It is possible (but not necessary) for the pendulum to respond with forced oscillations of period T, in phase or out of phase. To show this substitute into (2) the expression for X from (4), and also

$$u = B \sin pt \tag{5}$$

Carrying out the differentiations, and dividing out common terms, we find as a necessary condition for the form (5)

$$\frac{B}{A} = \frac{p^2}{p^2 - \omega^2} \tag{6}$$

The complete solution of (2) with $X = A \sin pt$ is

$$u = B \sin pt + C \sin (\omega t + c) \tag{7}$$

where B is given by (6) and C, c are arbitrary constants specifying amplitude and phase of a free oscillation superposed on the forced oscillation. This unwanted free oscillation corresponds to what we shall shortly be calling the transient; it distorts the wanted forced oscillation, can be started by any accident, and theoretically does not die out at all, although in practice it must eventually be extinguished by friction.

A second undesirable feature of the instrumental performance represented by (7) is the fact that B becomes theoretically infinite when $p = \omega$. This is the well-known "resonance catastrophe." In practice it is also limited by friction; but the actual effect may be bad enough to justify the remark of a distinguished seismologist on picking up for study a seismogram borrowed from a distant station: "Well, from this record we can find out the period of their pendulum, and nothing else!"

The quantity B/A is termed the dynamic magnification. Equation (6) may be written

$$\frac{B}{A} = \frac{\tau^2}{\tau^2 - T^2} \tag{8}$$

If T is much smaller than τ, this approaches $B/A = 1$; that is, if the pendulum period is much greater than that of the ground oscillation, the instrument becomes a displacement meter. If T is much larger than τ, we have approximately $B/A = -\tau^2/T^2$; but, since the acceleration is $4\pi^2 A/T^2$, B is proportional to the acceleration, and the instrument becomes an accelerometer. To make satisfactory use of either of these properties it is necessary to diminish the effects of resonance and of the transient free oscillation. This is done by introducing *damping,* by which we mean any force which opposes the pendulum motion and increases with its velocity. The friction naturally present in a mechanical system provides some damping, but usually it is necessary to add a specific device such as an air vane or a plunger immersed in oil. These methods do not completely meet the ideal requirement that the damping force shall be directly proportional to the pendulum velocity; this is usually assumed as a first approximation, since it simplifies mathematical treatment. It is not difficult to set up electromagnetic damping by eddy currents induced in copper; this is convenient in practice and closely approaches the mathematical requirement.

Theory of the Damped Seismometer

We introduce damping into (1) by adding a new term, so that it reads

$$m\ddot{x} = -K(x - X) - Q(\dot{x} - \dot{X}) \tag{9}$$

We put $u = x - X$, $K = m\omega^2$, $Q = 2hm\omega$, where h is a new instrumental constant:

$$\ddot{u} + 2h\omega\dot{u} + \omega^2 u = -\ddot{X} \tag{10}$$

If now we put $X = 0$ in order to investigate the free motion of the pendulum, we find that it is no longer a simple harmonic motion. A complete solution of (10) with $X = 0$ is

$$u = Be^{-h\omega t} \sin (j\omega t + b) \tag{11}$$

in which j (an abbreviation) is defined so that

$$h^2 + j^2 = 1 \tag{12}$$

while B and b are arbitrary constants. The solution can be verified by substituting (11) into (10) and carrying out the differentiations.

If $h < 1$, j is a real number and (11) represents a damped harmonic oscillation. If $h > 1$, j is imaginary; (11) is still formally a solution, but a more useful expression, which can also be checked by substitution, is

$$u = ae^{(-h + j)\omega t} + be^{(-h - j)\omega t} \tag{11a}$$

where a and b are arbitrary constants and

$$j^2 = h^2 - 1 \tag{12a}$$

There is then no oscillation; if the pendulum is drawn to one side, stopped, and then released it returns to zero without going farther. It is said to be overdamped.

The limiting case $h = 1$ is termed that of critical damping. Here either (12) or (12a) will give $j = 0$; the complete solution of (10) for $X = 0$ and $h = 1$ has the special form

$$u = Be^{-\omega t}(1 + bt) \tag{11b}$$

In practice overdamping is rarely used, and damping remains below the critical value, so that (11) applies directly.

Now, if we put $X = A \sin pt$, it is no use to try simply for $u = B \sin pt$, as we did when $h = 0$. We have to include a phase difference between the ground and the pendulum, so that

$$u = B \sin (pt + b) = B(\sin pt \cos b + \cos pt \sin b) \tag{13}$$

If we substitute (13) into (10) along with $X = A \sin pt$ and carry out the differentiations, we find that we have terms multiplied by either $\sin pt$ or $\cos pt$. If the equation is to hold for all values of t, the coefficients of $\sin pt$ on the left side must add up to A, and those of $\cos pt$ must add up to zero. We thus get two new equations; dividing out common factors, these are

$$(\omega^2 - p^2) \cos b - 2h\omega p \sin b = Ap^2/B \tag{14}$$

$$(\omega^2 - p^2) \sin b + 2h\omega p \cos b = 0 \tag{15}$$

The simplest way to find the value of A/B is to square (14) and (15) separately, add the squares, and then apply the fact that $\sin^2 b + \cos^2 b = 1$. The formidable-looking result is

$$\left(\frac{A}{B}\right)^2 = \left(1 - \frac{\omega^2}{p^2}\right)^2 + \frac{4h^2\omega^2}{p^2} \tag{16}$$

or

$$\left(\frac{A}{B}\right)^2 = \left(1 - \frac{T^2}{\tau^2}\right)^2 + \frac{4h^2T^2}{\tau^2} \tag{17}$$

As for the phase angle b, (15) gives directly on dividing by $\cos b$

$$\tan b = \frac{2h\omega p}{p^2 - \omega^2} = \frac{2h_\tau T}{\tau^2 - T^2} \tag{18}$$

The complete solution of (10) with $X = A \sin pt$ is

$$u = B \sin(pt + b) + Ce^{-h\omega t} \sin(j\omega t + c) \tag{19}$$

where B and b are calculated from (16) and (18), C and c are arbitrary constants, and j is given by (12).

We now see that damping takes most of the curse off the behavior of our pendulum. The unwanted oscillation with coefficient C no longer continues indefinitely; it falls off exponentially—rapidly if h approaches 1 and ω is not small—so that it is really a transient effect. The "resonance catastrophe" also disappears; for, if in (17) we put $T = \tau$, we find $B/A = 1/2h$, so that if h is not small B/A is not seriously exaggerated.

B/A is called the dynamic magnification. Its behavior in terms of T/τ is shown in Figure 15-1; the curves are drawn for several fixed values of the damping, from $h = 0$,

FIGURE 15-1 *Dynamic magnification* B/A *and phase displacement* b *as functions of the period ratio* T/τ *for given damping constant* h.

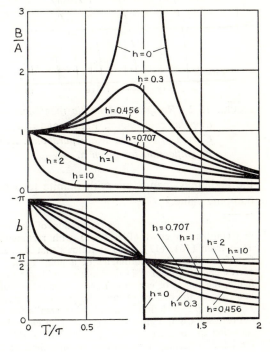

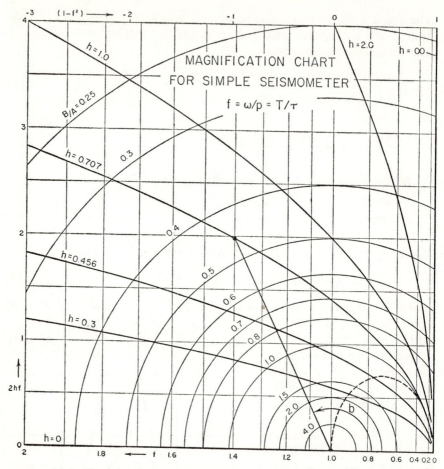

FIGURE 15-2 *Reciprocal dynamic magnification* A/B *and phase displacement* b *as polar coordinates. Values of* f = T/τ *indicated along axis of abscissas. Curves of constant* h *are parabolas. Dashed ellipse is locus of maximum magnification for given* h.

the undamped pendulum with infinite resonance peak, to $h = 1$, the critical damping value.†

† An elegant alternative representation of the result follows. Define $f = \omega/p = T/\tau$, and construct a chart (Fig. 15-2) with rectangular coordinates $x = 1 - f^2$, $y = 2hf$. Then from (16), (17), (18) it follows that $A/B = (x^2 + y^2)^{\frac{1}{2}}$ and $\tan b = y/x$; therefore from the polar coordinates of a point on the chart one can read off directly the corresponding (reciprocal) dynamic magnification and the phase angle. As shown in the figure, the curves of constant h are parabolas; by following one of them across the chart one can see the change of magnification and phase with varying f. Since $x = y = 0$ for $f = 1$ and $h = 0$, infinite magnification results for resonance with zero damping.

Professor Chuji Tsuboi has published an equivalent construction ("A graphical representation of forced oscillation amplitudes," *Zishin,* vol. 14, 1942, p. 76; in Japanese). He takes as rectangular coordinates our values of x and y divided by the sum of their

Because damping to some extent decreases the motion of the pendulum and the recorded amplitudes, the instrument appears less sensitive. Until about 1900 many seismographs were deliberately constructed with as little damping as possible. The seismograms were greatly distorted and were valuable only to indicate the mere occurrence of a shock and perhaps to time the arrival of the first large waves. Great progress in interpretation of seismograms followed the general use of damping adequate to furnish more trustworthy details of earth motion.

Instrumental Constants

The properties of a simple seismometer are commonly expressed by three numerical constants:

(1) The free period, which we have called τ; usually denoted by T_0 or simply T. When there is damping, a glance at equation (11) will show that the time interval between successive swings of the free oscillation is not τ, but τ/j. In practice, either the damping device is removed and the free period determined without it, or the damped oscillations are observed and the period corrected for damping.

(2) The damping constant, which may be given as h but is experimentally determined, and commonly reported, as the damping ratio ϵ. This is the ratio of two successive swings, in opposite directions, of the motion represented by (11); ϵ and h are connected by the relation

$$\epsilon = e^{\pi h/j} \tag{20}$$

(3) The static magnification, denoted by V. This is the contribution of the recording device. If the pendulum is deflected by an amount u and is held at rest in the deflected position, the writing point or light spot is deflected on the seismogram by an amount Vu. The magnification of the whole assembly is the product of B/A and V; that is, if the ground performs a simple harmonic motion of period T and amplitude A, the pendulum oscillates with period T and amplitude B, while the seismogram shows deflection with period T and amplitude VB.

Superposition of Solutions—Harmonic Analysis

So far we have supposed that the earth displacement is a simple harmonic motion. In an actual earthquake it is nothing of the kind; we have to extend the theory to cover more general motions.

---

squares; the result is to perform an inversion, in the mathematical sense, on our Figure 15-2. The phase angle is unchanged, but the radius vector becomes B/A instead of A/B. The curves of constant h are ovals of the fourth degree, and those of constant f are circles.

Charts drawn on similar principles are in use in electrical engineering.

Suppose that we have solved (10) for three different forms of X (not necessarily simple harmonic motions), $X = X_1$, $X = X_2$, $X = X_3$. Call the corresponding solutions $u = u_1$, $u = u_2$, $u = u_3$. Then from the rules for differentiating a sum of terms it follows that $u = u_1 + u_2 + u_3$ is a solution for $X = X_1 + X_2 + X_3$. This can be applied to the sum of any number of terms; it is a simple case of the general principle of superposition.†

Here we can apply one of the most valuable mathematical tools in the workbox of the theoretical physicist, Fourier's theorem, which states in effect that any arbitrary motion which is physically reasonable (that is, it does not have any fancy types of mathematical discontinuities) can be represented as a sum of superposed simple harmonic motions. Definite rules for finding the amplitudes, phases, and periods of these harmonic motions are laid down. The process of applying these rules is called harmonic analysis. Combining this with the principle of superposition, we theoretically have a means to solve the general problem of equation (10); that is, if the earth motion X is given as any arbitrary function of time, we can find the corresponding deflection of the pendulum, u, as a function of time.

The seismometer performs an automatic harmonic analysis of the motion X. To each harmonic component of period T it applies a magnification and a phase shift depending on T, according to equations (17) and (18). The resulting magnified and shifted harmonic motions are then synthesized into the motion u.

The seismologist's problem is to reverse this. Dividing the recorded deflections on the seismogram by the constant V, he finds the pendulum motion u. He carries out harmonic analysis of u, applies dynamic magnification and phase shift to each harmonic element in reverse, and synthesizes the individual results to find X. It is not usual to carry through this complex process completely, except for engineering purposes in studying records of locally strong earthquake motion. Such work often is done with automatic aids such as electronic computers.

The seismologist usually wants only the calculation for certain parts of the seismogram which can be isolated, such as the first few waves at the beginning or a series of regular waves with periods near 20 seconds. For many purposes exact timing is more important than exact amplitudes. Individual wave groups often arrive very abruptly; this has the effect of a large short-period element in the Fourier analysis, and such sharp impulses are best timed by instruments with high dynamic magnification for short periods. However, this often is not the best condition for exact determination of amplitudes.

In general, measurement of earth motion does not reach the level of precision expected in many other branches of physical investigation. High precision requires: (1) that T, h, V be determined accurately and main-

---

† Mathematically this is possible because (10) contains no products or powers of derivatives.

tained at known values; (2) that damping be closely proportional to $\dot{u}$; in general, that the behavior of the system conform very exactly to equation (10). This is technically difficult. Adjusting a seismometer, unlike most laboratory procedures, is not a matter of an individual test, but of continuous 24-hour operation year after year.

PRINCIPAL TYPES OF SIMPLE SEISMOGRAPHS

Until about 1920 seismograph design was dominated by an effort to lengthen the pendulum period to make the dynamic magnification nearly constant and get an accurate displacement meter.

The period of an ordinary gravity pendulum is $2\pi(L/g)^{\frac{1}{2}}$, where L is the effective length of the pendulum. The "seconds pendulum" of a grandfather's clock, for which the period is 2 seconds, has a length of about 1 meter. Hence to get a period of 10 seconds one needs the awkward length of 25 meters. This difficulty was solved by a group of British workers in Japan about 1880—Gray, Ewing, and Milne—who used only a small component of g to control a horizontal bracket pendulum (Fig. 15-3). The heavy pendulum mass is at the end of a horizontal boom, suspended from a point not quite above its pivot so that the whole is free to rotate about a slightly tilted axis. Rotation raises the boom and mass against gravity, producing a small restoring force. With proper design and mechanical skill a period of 12 seconds is readily obtained; this is the pendulum period of the photographically recording Milne-Shaw seismograph, a reliable instrument still operating at more than twenty stations in all parts of the world. The bracket pendulum for many years was the typical seismograph, appearing in dictionary illustrations and general handbooks.

The inverted pendulum (Fig. 15-3), developed especially by Wiechert about 1900, balances a heavy mass on a knife-edge and uses springs for the restoring force. These instruments nearly all register on smoked paper, and to overcome friction a very heavy mass was used, so that they were the giants of their day. One with a mass of 17 tons is still in operation at Tacubaya, near Mexico City; it has a spring system to take both horizontal components of motion from a single inertial mass. The logical final step

FIGURE 15-3

Diagrams illustrating the principle of seismographs using the horizontal bracket pendulum (Milne), the inverted pendulum (Wiechert), and the small torsion pendulum (Anderson-Wood, Nikiforov).

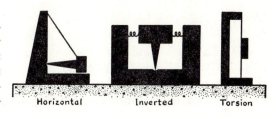

Horizontal   Inverted   Torsion

was taken in Switzerland by de Quervain and Piccard; they floated the heavy mass on springs so as to record the vertical motion as well, and their masterpiece, the "Universalseismograph," built in 1922 at Zurich, has a mass of 21 tons. In the vertical component this monster has a period of 1 second and a static magnification of 1500; the horizontal components have periods of 3 seconds and static magnifications of 1400.

Such great instruments are actually less effective than the relatively tiny torsion seismometer developed by Anderson and Wood in 1922. This instrument uses photographic recording; a static magnification of 2800 with a pendulum period of 0.8 second is obtained with an assembly which, case and all, stands only a foot high. A metal fiber suspension held taut by a weight is attached to the side of a copper cylinder (mass less than a gram!) which can rotate about the fiber against the restoring force of torsion. Damping, somewhat less than critical, is applied by eddy currents induced in the copper as it moves in the field of a permanent magnet. The copper cylinder carries a small mirror for recording. This instrument was very satisfactory in the investigation of local earthquakes, for which it was designed. A similar torsion instrument was independently designed and put to the same purpose by Nikiforov in the Soviet Union.

VERTICAL-COMPONENT SEISMOGRAPHS

The preceding theory was set up for a horizontal-component instrument. There is no difficulty in applying it to vertical motion. If a heavy mass is suspended by a coil spring which is allowed to extend until the elastic force balances gravity, a first-year physics calculation shows that the mass moves as if subject to an elastic restoring force acting symmetrically up and down toward the equilibrium position. This force and the mass fix the period of the pendulum, and the simple theory proceeds from that point. Practical difficulties are serious. If a long period is wanted, the spring must be weak. Even with a short period the position of equilibrium will be affected appreciably by temperature, so that the instrument must be thermally compensated or set up in a constant-temperature vault. Special alloys such as elinvar are used to reduce this effect. Another trouble is elastic fatigue; the spring, especially if it is a weak spring, yields slowly under the weight and settles to a definite equilibrium position only after a long time (if at all). Vertical-component instruments are the "problem children" in many stations, constantly getting out of adjustment. The higher the magnification sought, the worse these problems become. Almost all successful high-magnification vertical-component seismographs use electromagnetic registration, recording photographically from a galvanometer. Exception must be made for an extraordinary instrument with magnification 2000 set up by Straubel and Eppenstein at Jena about 1906. Its performance was far ahead of its time, and its recordings were not properly appreciated until years later.

SEISMOGRAPHS IN SERVICE IN 1920

In 1920 the majority of seismographs in regular operation were still recording on smoked paper. Many designers had applied their ingenuity to obtain reliable registration and high magnification while minimizing the effect of friction at the writing point and in the magnifying system. The Italian group was perhaps earliest in this field, and until about 1900 their instruments were considered the most satisfactory. Vicentini's seismograph was installed at many stations throughout the world; other well-known Italian instruments were those of Stiattesi, Agamennone, Alfani, and Cancani.

In Germany, Wiechert, Mainka, and Bosch developed seismographs which were widely used. In Japan, Omori devised several types; outside Japan these were chiefly represented by the Bosch-Omori.

Distribution of the various instruments was determined partly by commercial availability and cost. Thus, although by 1920 the Wiechert type was adopted by about 80 stations (more than any other), the majority of these instruments were small (pendulum mass about 80 kilograms, static magnification 80–100).

John Milne succeeded in distributing his seismographs to many distant parts of the world; by 1920 these were in operation at about 45 stations. They were then relatively obsolete, with low damping and low magnification for short periods. Between 1920 and 1930 more than half were replaced by the new Milne-Shaw instruments. Milne died in 1913, but in 1915 J. J. Shaw developed a modification of the Milne instrument with adequate damping and photographic registration. Because of their wide geographic distribution and reliability in operation, the Milne-Shaw instruments have contributed greatly to seismology. They were constructed with a pendulum mass of 1 pound (454 grams), a damping ratio of 20:1, a period of 12 seconds, and a static magnification of usually 150 or 250.

In 1920 only a few well-equipped stations were recording electromagnetically.

ELECTROMAGNETIC SEISMOGRAPHS†

It is not technically difficult to make a moving pendulum generate a small varying electric current. If this current is passed through a sensitive galva-

---

† Here let us note an awkwardness in terminology. Electromagnetic seismograph assemblies include the three normal elements: pendulum, recording system, and chronograph. However, the recording system divides naturally into a galvanometer and a gen-

nometer, it can be recorded photographically from the galvanometer mirror. The first effective seismographs using this principle were developed by Galitzin† about 1906. A complete theory would include the back effect of the galvanometer on the pendulum; this effect can be minimized by making the pendulum mass much greater than that of the moving parts of the galvanometer.

A variety of devices can be used for converting mechanical energy of the pendulum into electrical form; devices for this conversion, or the reverse, are now commonly called transducers. The electromotive force generated is generally proportional to the velocity of pendulum motion, which determines the rate at which lines of force are cut.

Without back effect the motions of the system are described by two equations:

$$\ddot{u} + 2h\omega\dot{u} + \omega^2 = -\ddot{X} \tag{21}$$

$$\ddot{x} + 2h'\omega'\dot{x} + \omega'^2 = k\dot{u} \tag{22}$$

Equation (21), which represents the motion of the pendulum, is identical in form and notation with (10). Equation (22) is essentially the equation of motion of the galvanometer; but the instrumental constant k may be taken to include optical lever magnification, and x then is the deflection of the recording spot on the seismogram. h' and ω' are galvanometer constants corresponding to h and ω for the pendulum. Usually both pendulum and galvanometer are critically damped, so that $h = h' = 1$. The complete assembly has no property corresponding to static magnification, for, if the pendulum is displaced and held in place, the galvanometer returns to zero.

To generate current Galitzin employed the inverse of the galvanometer principle; the pendulum was made to move a coil in a fixed magnetic field. To further simplify theory and computation Galitzin designed his pendulum and galvanometer to have the same free period ($\omega = \omega'$). These periods, τ and τ', were commonly near 12 seconds.

Galitzin established a chain of stations extending from Pulkovo across Asia to Vladivostok. These are still maintained by the Soviet government, together with many other stations having instruments of later design.

There is little difficulty in adapting this type of registration to the vertical component of motion, and Galitzin did so about 1911. The instrument differs only in construction detail from those used for horizontal motion; it

---

erator of electric current (a transducer) attached to the pendulum. The combination of pendulum and transducer is a unit in design, construction, and handling; but there is no generally acceptable term to refer to it. Manufacturers call it a seismometer, but their label conflicts with the established usage followed here. The corresponding unit used in seismic prospecting is called a geophone (or more familiarly a "jug"); but that name is inappropriate for the seismological unit designed to detect earthquakes.

† B. B. Galitzin was a prince of the Russian Empire. The transliteration is his own, but the Russian letters of his name would be better represented by Golitzin (or Golicyn, as it is sometimes printed).

is nearly as stable and can be given the same periods and response properties.

A new type of instrument for the vertical component was constructed at Pasadena in 1930 by Dr. Hugo Benioff after repeated efforts had failed to produce a stable vertical-component mechanical instrument with characteristics and magnification similar to those of the torsion seismometer. The Benioff transducer has extremely high efficiency in converting pendulum motion into electric current. The principle is that of the telephone transmitter; motion alters an air gap between armature and pole pieces of a magnet, inducing an electromotive force in coils wound round the pole pieces. In other phrasing, motion alters the reluctance of the magnetic circuit, so that for distinction from other types designed by Benioff (some of which use the moving-coil system) this is called the variable-reluctance seismometer. It was readily adapted to the horizontal component, and routine registration in all three components goes on satisfactorily at many stations using this instrument.

The initial successes of the Benioff system were with short-period units (those now in routine use at Pasadena and auxiliary stations have $\tau = 1$ second, $\tau' = 0.2$ second). These were originally designed to give high magnifications for local earthquakes. However, they proved unexpectedly effective for recording teleseisms. They provided recordings of types of seismic waves which were expected theoretically but had not been identified, such as $P'P'$ and $PKKP$ (see Chapter 17).

The Benioff seismometer adapts well to long-period recording; a short-period pendulum is associated with a long-period ballistic galvanometer ($\tau = 1$ second, $\tau' = 100$ seconds). This assembly is more stable than with a larger τ, and still gives high magnification for large ground period T. Instrumental development and seismogram study are now going on in the range of extremely large T, measurable in minutes.

In 1928 Frank Wenner, then at the U. S. Bureau of Standards, constructed an electromagnetic seismometer in which the back effect of the galvanometer on the pendulum is not negligible; the response is further modified by making h' large. Wenner instruments are in service at several stations operated by or associated with the U. S. Coast and Geodetic Survey. Wenner developed the complete theory of the response of such a system. In place of equations (21) and (22) there is a single fourth-order differential equation; the solution is not especially difficult, and it reduces readily to forms usable in practice. In 1935 this theory was generalized by Coulomb and Grenet in a paper which leads to a clear systematization of the responses of all possible combinations of a pendulum and a galvanometer. On this basis Grenet designed and constructed seismometers of various characteristics. Some of these, particularly those with high magnification for short periods, have lately gone into regular service at important stations in Europe, Algeria, and New Caledonia.

For new instruments in the USSR see Chapter 33. Seismometers designed by Willmore, with pendulum periods of 1 second and short-period galvanometers, are now in service in the West Indies and elsewhere.

Unusually stable long-period instruments were developed at Columbia University by Press and Ewing, beginning about 1952; a three-component set is now operating dependably at Pasadena, with $T_0 = 30$ seconds, $T_1 = 90$ seconds. (See Fig. 17-9.)

HIGH MAGNIFICATION AND NOISE LEVEL

Especially with smoked-paper recording, great ingenuity was applied to get high magnification. With modern electromagnetic recording, magnifications of 100,000 are readily obtained, and there would be no serious technical difficulty in going much higher if it were of any use. The practical limit is set, not by instrumental technique, but by noise in the general sense—minor and usually irregular background disturbance which is recorded along with the ground motion due to earthquakes. Most of this is also actual ground motion, classifiable under the general heading of microseisms (Chapter 23) but due to both natural and artificial causes. Raising the magnification increases the recorded background, which finally becomes so large that it interferes with the reading and interpretation of the seismogram.

Recording conditions can be improved by locating the station at a distance from persistent sources of disturbance. The chief artificial causes are traffic and machinery. Principal fixed natural causes are surf, especially on rough coasts; strong waves on lakes and ponds; waterfalls and rapid streams. Because strong winds are often very disturbing, sheltered situations are preferred. Wind may act by shaking towers, large doors, or other structures, and probably also trees. A comparatively small shift of the seismometer away from a particular spot has been known to result in a great decrease in background.

Noise level shows good correlation with the ground effect on macroseismic intensity. The worst noise is found on loose or unconsolidated ground. Tertiary sediments are often not much better; the best foundations for a station are granitic rock, solid metamorphics, or very solid Paleozoic sediments.

STRAIN SEISMOMETERS

In 1932 Dr. Benioff set up a new type of high-magnification seismometer.†
It is essentially a glorified strain gauge (Fig. 15-4). Two piers are set 60

---

† The principle had been used by John Milne in 1888 and by E. Oddone in 1900, but their magnifications were low, and Oddone's instrument did not even write a record.

FIGURE 15-4

Diagram illustrating the principle of the Benioff strain seismometer.

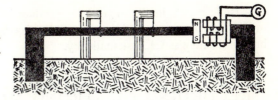

feet apart; a horizontal rod anchored to one pier extends not quite to the second pier. A variable-reluctance transducer unit is set to respond to variations in the gap between the second pier and the free end of the rod, and a record is written by a galvanometer. For seismic waves, with periods rarely less than 0.1 second, the rod has almost the effect of a rigid body. More precisely, the rod may be treated as a pendulum for oscillation along its length, like an open-end organ pipe; its fundamental free period is about $\frac{1}{70}$ second. This is so much shorter than that of most seismic waves that resonance may be neglected; damping, at first introduced into the system, was later abandoned.

Because of the manner in which the record is taken, the behavior is different from that of any pendulum seismograph. The response measures the linear strain tending to increase or decrease the distance between piers. Longitudinal waves approaching from any direction deflect the light spot up or down on the seismogram according as the volume alteration in the ground is compressional or dilatational. The maximum response to longitudinal waves is obtained, of course, when the waves are traveling parallel to the rod; while for transverse (shear) waves the response is at maximum when the rod is at 45° to the wave front. Waves of any type emerging nearly vertically, such as those reflected from the core of the earth, do not affect the strain seismometer at all.

When connected with a long-period galvanometer, the strain seismometer assembly is exceptionally stable, even more so than the long-period type of variable-reluctance seismometer. Its response is almost exactly that of a pendulum with the galvanometer period, so that it is an excellent displacement meter for short periods. Such a unit is very useful in the routine recording of the longer-period waves from teleseisms; it is also applied in studying waves of very long period, measured in minutes or even hours. The special directional properties are made use of by combining such units together or in association with pendulum seismometers for investigating the details of ground motion and the directions of the wave fronts. In newer installations rods of fused quartz are used instead of metal. One of these, in the granitic rock of the San Gabriel Mountains, is writing a continuous record of the local strain condition for eventual correlation with the general redistribution of strain in the region preceding and following large earthquakes. There are distant hopes that such data may eventually lead to some form of prediction, or at least to anticipation of the general course of seismic events in the area. A by-product is a continuous record of tides in the solid earth; such tides have been observed by various experimenters, but until lately never systematically.

RECORDING OF STRONG MOTION

The original purpose of the seismograph was neither the location of epi-centers, nor the investigation of the interior of the earth, but the detailed study of earthquake motion as such. This calls for a well-calibrated seis-mometer. Engineers are interested particularly in motion strong enough to damage structures. Such motion is usefully recorded only by "strong mo-tion" instruments with static magnifications in the range from 1 to 20. Since useful records will be written only now and then, recording usually is not continuous, but the apparatus is triggered by some auxiliary seismoscope when there is strong local shaking. Installations are made by preference in areas with a known history of strong earthquakes, and in communities with modern structures for comparison. An extensive program of this sort is op-erated by the U. S. Coast and Geodetic Survey; their list for 1956 shows 49 locations in California and 22 elsewhere. The program began in 1932–1933, in time to obtain highly valuable records of the destructive Long Beach earthquake, and has now accumulated a large file of important seis-mograms. Aside from direct engineering application to earthquake-resistant design of structures, these records are important in investigating the release of energy in earthquakes, and in examining the relation between intensity and acceleration. The increase of strong motion on unconsolidated ground is well confirmed by comparing the seismograms at different locations. The greater extent of motion on the upper floors of tall buildings has been thoroughly documented by comparing records taken on upper floors with others written in the basements. Some practical results of experience are:

(1) Because of the triggered operation, the beginning of earthquake motion is not recorded. This is usually of no great engineering importance, unless identification is doubtful (see under 2). In the Pasadena group of stations a number of instruments are continuously recording with low mag-nification on film, which is relatively cheap and can be discarded if no earthquake is registered.

(2) Because of practical difficulties the exact time is usually not re-corded. This sometimes leaves doubt as to what part of the earthquake motion is shown, and, if several strong earthquakes occur within a few hours, there may be questions of identification.

(3) If the trigger is too sensitive it will be set off accidentally, or by a series of earthquakes too small to give useful records. In this way the re-cording sheet or tape may run out just in time to miss a large shock.

(4) It is unwise to record two or more seismograms on one drum. When the motion is large, the overlapping traces are almost impossible to disentangle.

(5) Tilting, shifting, or internal mechanical readjustment may shift the zero position during a strong earthquake. If not detected or compensated for, this leads to large errors in the computed earth motion. Good design provides means for detecting and following such wandering of the zero.

(6) The components may not be cleanly separated. An instrument which responds only to north-south motion for small deflections may also respond partly to east-west or to vertical motion when accelerations approach or exceed $0.1g$.

(7) The effects of building vibration should be separated in study from those of ground motion. This can be done to some extent by comparing the upper-floor and basement records, but there still remains vibration fed back into the ground from the disturbed structure.

(8) A large number of recordings are needed for any safe generalization. Differences among the individual seismograms are great. This is no surprise to the seismologist, who is accustomed to working with records of thousands of earthquakes and is acquainted with their variety and the frequent appearance of exceptional characteristics.

LIMITATIONS OF THIS CHAPTER

The literature on seismometers is extensive and in at least three directions goes beyond the range of an elementary text:

(1) Constructional detail, some of it highly ingenious. Methods used to increase magnification without optical or electromagnetic means are now to some extent technically obsolete. However, the student interested in instrument design will find many ideas applicable in other ways. Probably the best compendium of such information is in *Handbuch der Geophysik* (Vol. 4).

(2) Techniques used to test, calibrate, and adjust seismometers. These sometimes differ widely between similar instruments of different manufacture.

(3) Adequate theory for electromagnetic instruments. This has been omitted from this book with regret. Sohon's book has a good account of the Galitzin instrument; Byerly's treatment is better and more thorough. The student planning research on instruments, or new developments, should by all means work through the paper by Coulomb and Grenet; the later paper by Grenet is a valuable supplement. Coulomb's section in *Handbuch der Physik* is excellent.

The development and discussion of equation (10) omits many technical details, such as proof of the completeness and generality of the solutions found. For these, refer to any of the standard mathematical treatises, where any equation equivalent to (10) is classed as the general linear ordinary differential equation of the second order with constant coefficients. Equation (10) as we have written it is often called the seismometer equation; in engineering contexts it usually appears as the indicator equation or the equation of forced oscillations.

References

General and theoretical

Berlage, H. P., Jr., "Seismometer, Auswertung der Diagramme," *Handbuch der Geophysik,* Borntraeger, Berlin, 1930, Vol. IV, No. 2, pp. 299–526.

Benioff, H., *Earthquake Seismographs and Associated Instruments, Advances in Geophysics,* Academic Press, New York, 1955, Vol. 2, pp. 219–275.

Byerly, P., *Seismology,* New York, 1942, Chapter VIII, pp. 104–151.

Coulomb, J., Séismométrie" in: *Handbuch der Physik,* Springer, Berlin, 1956, Vol. 47, *Geophysik I,* pp. 24–74.

Sohon, F. W., *Introduction to Theoretical Seismology,* Part II, Seismometry, Wiley, New York, 1932.

Electromagnetic instruments (theory)

Refer to Byerly and Coulomb as above, and:

Wenner, F., *Journ. Research Natl. Bur. Standards,* vol. 2 (1929), pp. 963–969.

Benioff, H., "A new vertical seismograph," *B.S.S.A.,* vol. 22 (1932), pp. 155–169.

Rybner, Jörgen, *G. Beitr.,* vol. 51 (1937), pp. 375–401; vol. 55 (1939), pp. 303–313.

Coulomb, J., and Grenet, G., "Nouveaux principes de construction des séismographes électromagnetiques," *Ann. physique,* Ser. 11, vol. 3 (1935), pp. 321–369.

Grenet, G., "Les caracteristiques de séismographes électromagnetiques," *Ann. géophys.,* vol. 8 (1952), pp. 328–332.

Other instruments

Refer to Wenner and Benioff as above, and:

Imamura, A., "Tyōkō and his seismoscope," *Japanese Journ. Astronomy Geophysics,* vol. 16 (1939), pp. 37–41.

Eppenstein, O., "Das Vertikalseismometer der seismischen Station zu Jena," *G. Beitr.,* vol. 9 (1908), pp. 593–604.

Anderson, J. A., and Wood, H. O., "Description and theory of the torsion seismometer," *B.S.S.A.,* vol. 15 (1925), pp. 1–72.

Benioff, H., "A linear strain seismograph," *ibid.,* vol. 25 (1935), pp. 283–309.

Willmore, P. L., "The theory and design of two types of portable seismograph," *M.N.R.A.S. Geophys. Suppl.,* vol. 6, pp. 129–137, 1950.

Press, F., Ewing, M., and Lehner, F. E., "A long-period seismograph system," *Trans. Am. Geophys. Union,* vol. 39 (1958) (in press).

For instruments now in use in the USSR see Chapter 33.

Installation

Carder, D. S., *The Seismograph and the Seismograph Station,* U. S. Dept. of Commerce, Washington, D. C., 1956. (Discusses station location and selection of instruments; describes instruments in use in the United States and Canada; has a section on amateur seismology.)

Strong-motion recording

"Earthquake investigations in California, 1934–1935," *U. S. Coast Geodetic Survey, Spec. Publ.* No. 201, Govt. Printing Office, Washington, D. C., 1936.

Ulrich, F. P., "The California strong-motion program of the U. S. Coast and Geodetic Survey, *B.S.S.A.,* vol. 25 (1935), pp. 81–95.

See also the serial, *United States Earthquakes,* for 1933 and following years, and progress reports for the U. S. Coast and Geodetic Survey published in *B.S.S.A.*

Classical treatises

These, although obsolete in some of their objectives, are sound in treatment and will repay study.

Wiechert, E., "Theorie der automatischen Seismographen," K. Gesellschaft der Wissenschaften zu Göttingen, Abhandlungen, math.-phys. Klasse, Neue Folge II, 1. Göttingen, 1903, 128 pp.

Galitzin (Golicyn), B., *Lektie po seismometrii,* St. Petersburg, 1912; German translation: Hecker, O., ed. *Vorlesungen über Seismometrie,* Teubner, Leipzig, 1914.

CHAPTER 16

Elasticity and Elastic Waves

STRESS AND STRAIN

The *stress* at a point in the interior of a body is determined by the system of forces acting in the vicinity of that point. The deformation of the body in the vicinity of a given point is termed *strain*. The concepts of stress and strain are fundamental to the theory of seismic waves, and also to the interpretation of the mechanism of earthquakes with reference to either macroseismic or microseismic data. Moreover, the structural geologist needs them in any discussion of the deformation of rocks. In this chapter the primary emphasis will be on essential physical and mathematical ideas. The student who wishes to go into detail will find material in other textbooks and in handbooks, many of which follow the general treatment and notation of Love's treatise on elasticity, either simplifying by leaving out detail and side issues, or expanding by filling in steps.

BRANCHES OF MECHANICS

In very elementary classes the student is sometimes left with the belief that when he has learned Newton's three laws he has assimilated all that is essential in mechanics and that what remains is mere detail. This is like imagining that one knows how to read when one has learned the alphabet.

Mechanics proceeds stepwise from simple systems to complicated ones. At every level there is a new group of ideas. They are introduced as naturally and gradually as possible, but the student should understand the generalization, and not just take it for granted.

The course of generalization in mechanics is not uniquely determined beforehand; it must be studied.† The customary order is mechanics of (1) a

---

† In 1927 Erwin Schrödinger was lecturing at Pasadena on quantum wave mechanics, which was then new. Having developed the theory in its simplest form, he went on to generalize to cover special relativity and electromagnetic forces. Wishing not to be dogmatic, he remarked: "Now, the generalization is not unique." Then a thought struck him; he looked at his audience with a mildly startled expression, and continued: "Of course not; otherwise it wouldn't be a generalization!"

232

particle, (2) a system of particles, (3) rigid bodies, (4) deformable bodies, including fluids and elastic solids. In the first two divisions we operate with forces as ordinary three-component vectors. In dealing with rigid bodies, as in many engineering problems, we think of forces as acting along particular lines, with specified points of application; and we work not with forces only, but with moments and couples as well.

MECHANICS IN GEOLOGY

Problems in structural geology are often handled by a combination of the mechanics of rigid and deformable bodies; but many students have come to grief by mixing the two methods indiscriminately. In dealing with deformable bodies we think best in terms of stresses and strains rather than of forces and displacements.

In reading the older geological literature, remember that the founders of geology, who were mostly well-trained men, had a clearer understanding of physical fact than some of their successors. At a time when the modern vocabulary of stress and strain had not been generally adopted, they wrote of "forces" when they were thinking, correctly, about stresses. With the next generation, geology entered an observational and fact-gathering stage. In order to go into the field and make maps showing the rocks exposed at the surface, the geologist does not have to think much about forces and stresses. But, when he begins to reason about what lies below the surface, his physics must be sound, and, if he tackles the serious stratigraphic and structural questions of how things got that way, he needs exceptional physical understanding and intuition. It is sad that many geologists who rightly stood high in their field have shown by their publications that their foundation in fundamental physics was not adequate. "Force," as used by the founding geologists, has often been reread in terms of the freshman physics definition. Dr. M. King Hubbert, among others, has lately been working hard to bring sound physics back into geology in order to meet the problems of interpretation that the great accumulation of observations has now brought back into the foreground of the science.

Textbook writers and others sometimes try to handle stress and strain problems by setting up special cases that can be dealt with by elementary means. Such simplifications are often unnecessarily hard to read. The physics of stress and strain is not more difficult than that of force and displacement, unless one insists on making it so. The ideas are more complicated in the sense that they have more working parts, as an automobile is more complicated than a pair of roller skates. It does not help much to understand the automobile if one concentrates exclusively on what it has in common with roller skates.

This chapter is a sketch of fundamentals only. Bulky proofs and formulas are either outlined here or relegated to Appendix IV.

ANALYSIS OF STRESS

Stress as usually defined has the physical dimensions of force per unit area. We consider at first only ideally elastic solids, which may be deformed by force, but will return perfectly to the initial condition when the deforming forces are removed. Suppose that such a solid is squeezed, twisted, and so on, then clamped in its deformed condition. It will adjust to a state of equilibrium under the existing forces, external and internal. We now wish to describe the physical conditions at a point within the body.

The simple idea of force is no longer sufficient. At any internal point there is equilibrium and the resultant force is zero, even for completely different kinds of deformation. We set up rectangular axes of x, y, z, with origin at the given point, and consider the body cut in two on the yz plane (perpendicular to the x axis). At the origin the two halves of the body act on each other with equal and opposite forces. These forces act in a line which may have any direction in space and may form any angle with the yz plane. The three components of such force, expressed in units of force per unit area of the plane, are called components of stress and are designated Xx, Yx, Zx [(outward directed forces, which are tensional, being taken as positive (Fig. 16-1)].

Similarly we may cut on the xz plane, perpendicularly to the y axis, and define stress components Xy, Yy, Zy; and, cutting at right angles to the z axis, we define Xz, Yz, Zz. The total of nine components completely specifies stress at the given point. We have assumed a state of equilibrium; there is a textbook proof that, if you consider a small cube with faces at right angles to the axes, the cube will be out of equilibrium and will be rotated unless $Xy = Yx$, $Xz = Zx$, $Yz = Zy$. Thus there are really only six different components of stress.

At the given point in the solid we might have oriented coordinate axes in three directions different from those chosen (still perpendicular, of course). If we denote these by x', y', z' and add corresponding dashes for the stress components,

FIGURE 16-1 *Definition of stress components. The Vector* Fx *is the force per unit area across a plane normal to the X-axis;* Xx, Yx, Zx *are its components.*

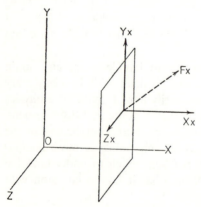

$$X'x' = a^2Xx + 2abXy + 2acXz + b^2Yy + 2bcYz + c^2Zz \qquad (1a)$$

$$X'y' = adXx + (ae + bd)Xy + (af + cd)Xz + beYy \\ + (bf + ce)Yz + cfZz \qquad (1b)$$

where a, b, c are the cosines of the angles between the new x' axis and the old axes of x, y, z, while d, e, f are the corresponding cosines for the y' axis. There are similar equations for transforming the other components of stress, so that, if the components for one set of axes are known, those for any other axes can be computed. This is what we mean by saying that the components for any one set of axes completely specify the stress at the given point.

There is an important theorem that at any given point one can always choose the axes so that $Xy = Xz = Yz = 0$. The axes in this position are called the principal axes of stress, and the three remaining components Xx, Yy, Zz are the principal stresses, which are either tensions or compressions.

If we can always reduce the stress to three components, why trouble with six? Answer: In general, the directions of the principal axes are different for different points, so that to describe stress in the whole body one still needs six components. At each point, in addition to the three principal stress components, three more numbers are necessary to specify the directions of the principal axes. The student will save much trouble by taking the six components of stress as circumstances to be accepted, worked with, and used— not as a complication to be dodged.

EQUATIONS OF MOTION

If forces are no longer in equilibrium, the elastic body will begin to deform, and we need the equations of motion for all of its points. Motion depends on the space rate of change, or gradient, of the stress, for, if the stress components are constant over a given distance interval, the portion of the body between the ends of the interval is acted on by equal and opposite forces. Let a small volume of mean density ρ be displaced from its equilibrium position, so that u, v, w are the components of the displacement vector. The components of its acceleration are the second derivatives of u, v, w with respect to time. The force acting parallel to the axis of x is determined by the derivatives of the three components of stress Xx, Xy, Xz, which represent forces acting in that direction, in such a way that the Newtonian equation of motion takes the form

$$\rho \frac{\partial^2 u}{\partial t^2} = \frac{\partial Xx}{\partial x} + \frac{\partial Xy}{\partial y} + \frac{\partial Xz}{\partial z} \qquad (2)$$

with two analogous equations for the forces parallel to y and z.

In Chapter 15 we began the theory of the seismometer with the equation of motion of a simple pendulum. After setting force equal to mass times acceleration, we had to introduce a law relating the force to the pendulum displacement. We took the force as proportional to the displacement, according to Hooke's law.

Now we need, similarly, to relate the stresses to the displacements u, v, w. Just as we could not operate simply with forces, but had to introduce stresses, so we cannot work directly with displacements, but must bring in their derivatives in the form of strains.

ANALYSIS OF STRAIN

Consider two neighboring particles of the solid at positions x, y, z and $x + dx$, $y + dy$, $z + dz$. As the body deforms, they will be displaced by amounts u, v, w, and $u + du$, $v + dv$, $w + dw$. If du, dv, dw are zero, so that the two particles move together, there is no resulting strain, and consequently no restoring stress. Hence the strain will depend on the (partial) derivatives of u, v, w, with respect to x, y, z; there are nine such derivatives. However, mathematical analysis shows that actual deforming strain depends on only six different combinations of derivatives. This is a matter of pure geometry and does not depend on the physical properties of the solid.

The six components of strain are ordinarily defined as

$$
\begin{aligned}
e_{xx} &= \frac{\partial u}{\partial x} & e_{yy} &= \frac{\partial v}{\partial y} & e_{zz} &= \frac{\partial w}{\partial z} \\
e_{xy} &= \frac{\partial u}{\partial y} + \frac{\partial v}{\partial x} & e_{xz} &= \frac{\partial u}{\partial z} + \frac{\partial w}{\partial x} & e_{yz} &= \frac{\partial v}{\partial z} + \frac{\partial w}{\partial y}
\end{aligned}
\tag{3}
$$

If we wish, we can set up nine components as we did with stress, and then eliminate three by symmetry: $e_{xy} = e_{yx}$, $e_{xz} = e_{zx}$, $e_{yz} = e_{zy}$. If one considers a small element of volume which in the unstrained condition has the form of a cube with sides parallel to the coordinate axes, e_{xx}, e_{yy}, e_{zz} measure the extension of the cube parallel to the axes. Change in volume of the cube, or of any small volume, is measured by the quantity $\theta = e_{xx} + e_{yy} + e_{zz}$. (In the language of vector analysis θ is the divergence of the displacement vector u, v, w). Change of form of the cube, without change of volume, is shearing strain and is governed by e_{xy}, e_{xz}, e_{yz}.

PRINCIPAL STRAINS; STRAIN ELLIPSOID

If we rotate the axes of x, y, z, the components of strain are transformed in a way very similar to the transformation of the components of stress. Given any condition of strain, it is always possible to choose the axes so that at any one point $e_{xy} = e_{xz} = e_{yz} = 0$. These axes are the principal axes of strain at the point, and the remaining components e_{xx}, e_{yy}, e_{zz} are the principal strains.

A small sphere in the unstrained state is deformed into an ellipsoid, the strain ellipsoid. This gives a convenient visualization of strain, and the various components can be derived from it quantitatively. There is a reciprocal strain ellipsoid, which is an ellipsoid in the unstrained state and is converted to a sphere in the strained state. The principal axes of the reciprocal strain ellipsoid (not those of the strain ellipsoid) are the same as the principal axes of strain.

There are other ellipsoids and more general quadric surfaces associated with the strain system, and a corresponding set associated with the stress system. These can be useful in analysis, but the student must always be sure which is being used.

TENSORS

To some students it is a horrible and mentally paralyzing shock to learn that stress and strain are tensors. This is the result of thinking of tensors only in connection with relativity, and with the myth of extreme difficulty which the popular press has built up about Einstein's work. Einstein did not invent tensors. He did not even name them; that was done by W. Voigt of Göttingen, in 1887. Voigt was investigating the physical properties of crystals, experimentally and theoretically. He had to deal with stresses, strains, and many physical quantities with similar mathematical properties. The name tensor, derived from the same root as tension, was proposed to suggest the relationship, particularly with stress.

Thus stress and strain not only are tensors—they are the original ones, the grandparents of all the rest. In present-day standard terminology, the stress components form a symmetrical tensor of the second order;[†] this means that there are six quantities associated with each point which transform to new coordinate axes according to definite rules for which we have already given examples (equations 1a, 1b). To make the components of strain fit the same definition, we would have to introduce a factor ½ into our expressions for e_{xy}, e_{xz}, e_{yz}. The existence of principal axes and of various associated ellipsoids are properties common to all tensors of this kind.

In the general theory of relativity there are special forms of notation and procedures in operation which constitute the tensor calculus, originally invented for purely mathematical purposes. These methods are applicable to the theory of elastic solids but are not needed for the purposes of this book.

---

† Vectors are tensors of the first order.

RELATION OF STRESS TO STRAIN; GENERALIZATION OF HOOKE'S LAW

Our equations of motion contain the stress components; to replace them by the strain components we have to relate stress and strain by some generalization of Hooke's law. With six components of each, the generalization is not unique. It would be nice to set each component of stress proportional to one component of strain, but experiment demonstrates that the elasticity of real solids is not that simple. Failing this, we start with the assumption mathematicians always try out first, that the relations between stress and strain are linear equations, at least to the first order of small quantities. Since any component of stress may depend on all six components of strain, we would have

$$Xx = Ae_{xx} + Be_{yy} + Ce_{zz} + De_{xy} + Ee_{xz} + Fe_{yz}$$

and five similar equations for the other components of stress. The coefficients A, B, C, D, E, F, etc., would be numbers characteristic of the particular solid, specifying its elastic properties. The six equations would thus give 36 elastic coefficients. Fortunately there are a great many mathematical symmetries which can be deduced from those existing in the known crystal systems. For the least simply symmetrical crystals, the triclinic system, the number of independent coefficients reduces from 36 to 21; for other systems it is smaller, and for cubic crystals it reduces to 3. Finally, if we have not a crystal, but an isotropic solid (one whose properties are independent of direction), 2 coefficients are sufficient (Appendix IV).

The choice of the two principal coefficients may be made in a variety of ways, giving different forms to the equations of motion. For mathematical purposes the most convenient choice is that made by Lamé. Denoting his two elastic coefficients by λ and μ, the stress-strain equations for an isotropic solid can be written

$$\begin{aligned} Xx &= \lambda\theta + 2\mu e_{xx} & Yy &= \lambda\theta + 2\mu e_{yy} & Zz &= \lambda\theta + 2\mu e_{zz} \\ Xy &= \mu e_{xy} & Xz &= \mu e_{xz} & Yz &= \mu e_{yz} \end{aligned} \tag{4}$$

This shows that the principal axes of stress are the same as the principal axes of strain; for, if $Xy = Xz = Yz = 0$, then $e_{xy} = e_{xz} = e_{yz} = 0$, and vice versa.

Note that, if we had set up the strain tensor by introducing the additional factor ½ into the definitions of e_{xy}, e_{xz}, e_{yz}, we could say that the stress consists of the sum of two parts, one obtained by multiplying the corresponding strain components by 2μ, the other a symmetrical (hydrostatic type) compression or dilatation of amount $\lambda\theta$.

COEFFICIENTS OF ELASTICITY

Of the two Lamé constants introduced in the preceding section, μ has a relatively simple physical significance. It measures the resistance of the elastic solid to shearing deformation and is termed the modulus of rigidity. Its value can be determined directly by experiment. The constant λ, on the other hand, is not simply related to experimentally observed quantities, and its value is usually calculated from those of μ and of one of the other experimentally determined coefficients to which both are related. The simplest such relation is

$$k = \lambda + \frac{2\mu}{3} \tag{5a}$$

in which k is the bulk modulus, or modulus of incompressibility, which can be determined by pressure experiments.

Another pair of important coefficients are Young's modulus E and Poisson's ratio σ. If a rod of unit cross section is subjected to tension, E measures the resulting elongation, while σ measures the accompanying decrease in cross section relative to the extension. In terms of λ and μ,

$$\sigma = \frac{\lambda}{2(\lambda + \mu)} \tag{5b}$$

and

$$E = 2\mu + \frac{\lambda\mu}{\lambda + \mu} = 2\mu(1 + \sigma) \tag{5c}$$

Any two of the elastic coefficients named in this section are sufficient to define the properties of an ideal isotropic elastic solid.

REDUCED EQUATIONS OF MOTION; ELASTIC WAVES

Usually the next step is to take the expressions for the stress components in terms of the strain components and elastic coefficients, and substitute these into the equations of motion. Expressing the strain components, as defined, in terms of the derivatives of the displacements, we arrive at three equations of motion which involve various second partial derivatives of u, v, w, with respect to x, y, z. In general, λ and μ are not constant; therefore the equations of motion will have terms involving the derivatives of λ and μ. These terms disappear if the elastic solid is homogeneous, so that the elastic coefficients are constant throughout; and the form which then results is the one in which the equations are most commonly presented (Appendix IV). This sets a theoretical trap; it is not safe to start with the equations which require constant λ and μ and then apply the results to non-homogenous solids. The literature contains some errors of this kind.

The most general mathematical statement of the properties of wave propagation is included in the standard partial differential equation

$$\frac{\partial^2 A}{\partial x^2} + \frac{\partial^2 A}{\partial y^2} + \frac{\partial^2 A}{\partial z^2} = \frac{1}{C^2} \frac{\partial^2 A}{\partial t^2} \tag{6}$$

This may be a single equation, in which case A is a scalar quantity; or there may be a set of three equations in each of which A is one component of a vector. C is the phase velocity of propagation of waves.

In Appendix IV it is shown that there are two simple methods of combining derivatives of the three equations of motion into an equation or a set of equations of the form of the wave equation. In one of these A is the scalar quantity θ, and $C^2 = (\lambda + 2\mu)/\rho$. The equation then represents waves which are physically identical with sound waves and ordinarily constitute the initial earthquake motion on recorded seismograms, including those designated by P. The vibration is longitudinal to the direction of propagation, along the ray; the corresponding deformation is compressional or dilatational, and in mathematical terminology the wave is irrotational.

The second type of wave derived from the equations of motion is a vector wave, so that there is a set of three equations, in each of which A is represented by one component of a vector formed from the derivatives of u, v, w, and in all of which $C^2 = \mu/\rho$. Waves of this type, which can exist only in solids, are sound waves or not, according to one's taste in philosophy. The vibration is transverse, at right angles to the ray; the deformation is a shear, or mathematically equivoluminal. Such waves include the S waves of seismology.

The existence of both types of waves is demonstrated in the experimental laboratory and by microseismic observation. Theory shows that only these two types can exist in an ideally elastic isotropic and homogeneous solid of infinite extent. However, theory also shows that, as soon as boundaries or discontinuities exist, waves of other types may arise; the observation and interpretation of such waves is an important branch of seismology.

With a little algebra the student can show that Poisson's ratio can be calculated from the ratio of the velocity of longitudinal waves to that of transverse waves, and vice versa. A simple assumption, fairly close to physical fact and convenient in theory, is that $\sigma = \frac{1}{4}$; this corresponds to taking $\lambda = \mu$, and the ratio of the two velocities then becomes $\sqrt{3}$, or 1.732. . . .

SURFACE WAVES

It was Lord Rayleigh who first proved the possibility of a special type of elastic waves propagated along the surfaces of a bounded elastic solid. The most important practical application is to waves along the surface of the earth.

The original theory made many simplifying assumptions. Curvature was neglected; the waves were propagated along a flat surface. There was an infinite train of sine waves of uniform amplitude, without beginning or end in either time or space. The wave front was plane and vertical. The amplitude of displacement decreased exponentially from the surface downward.

Rayleigh found that such waves could theoretically exist. He did not approach the problem of how these waves could be set up by the action of a subterranean source such as an earthquake. This difficult question has been partially dealt with by later investigators.

The velocity of propagation of Rayleigh waves depends on Poisson's ratio; for $\sigma = \frac{1}{4}$ it is 0.9194 . . . times that of transverse waves (this numerical factor is the square root of the quantity $2 - 2/\sqrt{3}$). There is no transverse vibration; displacements are exclusively vertical and in the direction of propagation. Each individual particle describes an elliptical orbit which is retrograde.†

Waves of this general type are actually recorded; but the large waves of teleseisms, which represent the surface wave phases, frequently show motion transverse to the direction of propagation, so that the Rayleigh theory does not cover all the facts. A. E. H. Love, the authority on elasticity, investigated a further hypothesis. Starting from the seismological evidence (Chapter 18) that at least in continental regions there is a horizontal discontinuity at a depth of the order of 60 kilometers, he investigated the possibility of wave propagation. The resulting theory shows the possibility of what are now called Love waves (to the delight of punning students). In these the displacement is altogether transverse, without either vertical or longitudinal components. The velocity is equal to that of transverse waves in the medium below the discontinuity. Waves of this general type are also recorded, and their superposition on Rayleigh waves accounts for most of the observations. These were the first of the general group of guided waves to be identified; others have lately assumed importance in seismology (see the following sections and Chapter 17).

---

† This means that when the particle is at the top of its ellipse it is moving opposite to the direction of propagation of the wave.

THEORETICAL LIMITATIONS

The ideal elastic solid is a very special and rather artificial construction, and it is a little surprising that theory based on its properties suits the observations. Deviations from the theoretical results are to be expected, and some are definitely observed. Various natural extensions of the theory have been worked out. Thus the properties of the earth might be expected to deviate from ideal isotropy, since under the action of gravity vertical and horizontal elastic properties might differ, but no observations have been clearly correlated with this possibility. There appears to be a tendency for the period of a given type of seismic waves to increase with distance, but this does not fit the ordinary theory. If the effect is not due to differential absorption or to the normal spreading of a pulse, it seems to call for some kind of viscosity, and much theoretical work has been done in that direction. Even for the ideal elastic solid, the usual theory is incomplete since it considers only small quantities of the first order and assumes linear relation between stress and strain; this is an unlikely assumption near the source of an earthquake, and the gradual progress in setting up theories which take account of higher-order terms, such as Murnaghan's "theory of finite strain," seems promising when compared with observations.

In the next chapter we shall have to consider the propagation of seismic waves in the interior of the earth, when the elasticity and density, and consequently the velocities of the longitudinal and transverse waves, vary with depth. This leads to the collision with theory already mentioned, since the existence of such waves is usually demonstrated only for the homogeneous case, when the equations of motion contain no terms including the derivatives of the coefficients of elasticity. More exact work shows that this is not a serious incompleteness; the two principal types of waves still exist, and their behavior does not diverge much from that given by the simple theory so long as the wave lengths are long in comparison to local irregularities and to significant changes in elastic properties with depth. Fortunately, most of the results needed can be derived rather simply from the general geometry of wave propagation.

Similar considerations based on variation of velocity with depth apply to the theory of surface waves, whether of Rayleigh's type or of Love's. The recorded surface waves have particularly long wave lengths, so that their velocity of propagation, for example, results from a sort of natural integration of the varying elastic properties down to a depth comparable with the wave length. Study of their velocities, accordingly, provides a means for distinguishing large differences in regional structure, as between continents and oceans.

GUIDED WAVES; CHANNEL WAVES

One of the most fruitful new fields of seismological investigation is the general theory and systematic observation of guided waves, which include channel waves and many types of surface waves. The reader will find details in the contribution by Ewing and Press, and the book by Ewing, Jardetzky, and Press, cited in the references at the end of this chapter. Here only a general statement is possible.

Guided waves exist in a layered medium in the general sense, such that its properties change only vertically and not horizontally. In addition to body waves of the general types of P and S, energy may be propagated horizontally, concentrated near certain levels determined by the functional form of the relation between the elastic properties and the vertical coordinate. The vicinity of such a critical level constitutes a channel. Displacements in such horizontal waves diminish rapidly (usually exponentially) with distance outside the channel.

The cases most significant in seismology are those in which there is a channel between two discontinuities, of which the upper is often a free surface (such as the surface of the earth), and the lower usually represents an abrupt change in elastic properties, like the Mohorovičić discontinuity at the base of the continental crust.

The velocity of guided waves depends on the elastic constants within and adjacent to the channel (not merely on their values at a given point, as for P and S). Usually the velocity depends on wave length; this constitutes dispersion,† and we must consider both phase velocity and group velocity.

GROUP VELOCITY

The increase in velocity of observed surface waves with wave length constitutes *dispersion*. It significantly affects the interpretation of seismograms of these waves. Waves of P and S type show such small dispersion that special investigations for the purpose have failed to determine it.

In studying wave propagation where there is dispersion it is necessary to distinguish between the *phase velocity* and the *group velocity*. The former is that which is derived from the equations of motion, like the value 0.9194 $(\mu/\rho)^{\frac{1}{2}}$ for the velocity of Rayleigh waves when Poisson's ratio is ¼. The latter is essentially that with which the radiated energy travels; it controls the

---

† The theoretically ideal Rayleigh wave on the surface of a homogeneous medium should not show dispersion; but the corresponding waves observed in seismology are dispersed because of the change in elastic properties with depth.

response of recording instruments and is more likely to be found correctly from seismograms than the phase velocity, where they differ significantly.

In general, if C is the phase velocity, C' the group velocity, and L the wave length,

$$C' = C - L \frac{dC}{dL} \tag{7}$$

C' thus may be either greater or less than C. For seismic surface waves, since dC/dL is positive, C' is less than C. Roughly stated, if one considers a limited wave train, a group of waves, the train as a whole is propagated with velocity C'; but individual crests and troughs travel through the group, appearing at the tail of the train and disappearing at its head, so that their speed of travel in space is the higher velocity C. For further details see Appendix IV.

COMPUTERS AND MODEL SEISMOLOGY

The equations of motion for elastic solid bodies are so complicated that it is often difficult to get exact solutions, even for the homogeneous case. Accordingly, modern computing-machine techniques are being applied.

Mathematical processes, with sufficient analysis, can be reduced to combinations of the elementary operations of arithmetic. Digital computers, which perform those operations at high speed, can be used to solve differential equations, and are applicable to such problems as tracing the propagation of elastic waves. Analog computers can be applied more directly.

Most analog computers involve the mathematical correspondence between mechanical and electrical oscillation systems. The equation governing current I in a circuit with inductance L, resistance R, and capacitance C, subject to electromotive force E, is

$$L\ddot{I} + R\dot{I} + \frac{I}{C} = \dot{E}$$

Except for notation, this is of the same form as our equation (10) of Chapter 15, which represents the free and forced oscillations of a mechanical system. By proper choice of electrical quantities, the circuit may be made to imitate the behavior of a given mechanical oscillator. This can be used, for example, to integrate a seismogram and obtain the ground displacement as a function of time. In engineering application one may consider a tall building or other structure as a pendulum of given period (generally the fundamental or longest period; see Appendix II), and study its behavior under applied oscillation by setting up the electrical analog.

A recent development of the analog computer, in which mechanical elements play the principal part, is model seismology. An apparently small technical advance which has facilitated much new experimentation is the availability of barium titanate crystals, which have the piezoelectric property of oscillating mechanically in phase with oscillating electromotive forces applied to attached electrodes. This substance is more effective than Rochelle salt, formerly used for such purposes. The crystals are used both as emitters and as receivers.

Pioneering experiments in model seismology were carried out as early as 1927 by Terada and Tsuboi, who produced Rayleigh waves in the laboratory. A more generally applicable technique was reported in 1953 by Northwood and Anderson, who used it to study propagation of a pulse along a free plane surface and also to investigate the type of reflection of the waves pP and PP (Chapters 17 and 19), which represent minimum and maximum times respectively.

In 1954 the research group at Lamont Observatory (Columbia University) reported results in two-dimensional seismology; they used a plastic disk to study Rayleigh waves as well as waves reflected from the rim (both longitudinal and transverse waves were observed, with satisfactory analogy to those observed in the earth). By fabricating disks from sheets of different materials, refracted waves in layered media were modeled.

Since 1954, papers in this field have appeared in steadily increasing number. No summary discussion has been published. Special mention may be made of the work of Tuve and Tatel (Geophysical Laboratory, Carnegie Institution of Washington) and of Knopoff (University of California at Los Angeles).

Recent results by Press at Pasadena bear on the use of seismograms to determine the mechanism of distant earthquakes (Chapter 32).

References

General

Love, A. E. H., *A treatise on the Mathematical Theory of Elasticity,* 4th ed., Cambridge University Press, 1927; reprinted, Dover Publications, New York, 1944. (This is the standard manual, including everything except experimental material. Some parts are difficult for the student, and it is not always easy to find what is wanted.)

Macelwane, J. B., *Introduction to Theoretical Seismology, Part I, Geodynamics,* Wiley, New York, 1936; reprinted, St. Louis University, 1949. (The treatment of elasticity follows that of Love, filling in much elementary detail.)

Bullen, K. E., *An Introduction to the Theory of Seismology,* 2nd ed., Cambridge University Press, 1953. (The treatment of elasticity and elastic waves is very clear but condensed.)

Gutenberg, B., "Theorie der Erdbebenwellen," *Handbuch der Geophysik*, Borntraeger, Berlin, 1929, vol. 4, Sec. 1, pp. 1–150. (Very thorough; differs in order of treatment and in notation from the preceding; contains many references to the older literature.)

Original contributions and special material

Lord Rayleigh (J. W. Strutt), "On waves propagated along the plane surface of an elastic solid," *Proc. London Math. Soc.*, vol. 17 (1885), pp. 4–11. (The original paper on Rayleigh waves.)

Love, A. E. H., *Some Problems of Geodynamics*, Cambridge University Press, 1911. (Contains the theory of Love waves.)

Ewing, M., and Press, F., "Surface waves and guided waves," in: *Handbuch der Physik*, Springer, Berlin, 1956, vol. 47, *Geophysik I*, pp. 119–139. (An excellent modern summary.)

Ewing, M., Jardetzky, W. S., and Press, F., *Propagation of Elastic Waves in Layered Media*, McGraw-Hill, New York, 1957. (Thorough and inclusive discussion.)

Murnaghan, F. D., "Finite deformations of an elastic solid," *Am. Jour. Math.*, vol. 59 (1937), pp. 235–260. (The first of many papers on this subject by Murnaghan and others.)

————, *Finite Deformation of an Elastic Solid*, Wiley, New York, 1951. (A textbook.)

Model seismology

Terada, T., and Tsuboi, C., "Experimental studies on elastic waves, Part I," *Bull. E.R.I.*, vol. 3 (1927), pp. 55–65; Part II (Tsuboi), *ibid.*, vol. 4 (1928), pp. 9–20.

Northwood, T. D., and Anderson, D.V., "Model seismology," *B.S.S.A.*, vol. 43 (1953), pp. 239–245.

Oliver, J., Press, F., and Ewing, M., "Two-dimensional model seismology," *Geophysics*, vol. 19 (1954), pp. 202–219.

Press, F., Oliver, J., and Ewing, M., "Seismic model study of refractions from a layer of finite thickness," *ibid.*, pp. 388–401.

Tatel, H. E., "Note on the nature of a seismogram: II," *Journ. Geophys. Research*, vol. 59 (1954), pp. 289–294.

Knopoff, L., "Seismic wave velocities in Westerly granite," *Trans. Am. Geophys. Union*, vol. 35 (1954), pp. 969–973.

CHAPTER 17

Seismic Waves In and Over the Earth

ANALYSIS OF SEISMOGRAMS

Exact analysis of seismograms written at distances up to a few hundred kilometers from the epicenter calls for very accurate timing and for study of finer detail than the corresponding analysis for teleseisms. In consequence, instrumental seismology is in many ways on a more settled basis for study on the world scale than for local and regional investigation.

Observed Complexity

The theory of seismic waves has progressed together with the interpretation of seismograms as the improvement of instruments made it possible to observe more accurately. Any seismogram represents a complicated disturbance, which for a teleseism may be of long duration. At first this complication was ascribed to the earthquake source—a natural procedure in view of the field evidence of tectonic complexity in large earthquakes.

An objection came to light as soon as files of seismograms began to accumulate. At any one station the records of earthquakes originating in the same distant region are often surprisingly alike. Seismograms written months or years apart can often be matched almost exactly, wave for wave. This implies a high degree of mechanical parallelism between two tectonic events, unless one adopts the alternative hypothesis that the originating disturbances were both relatively simple and of short duration; this means that the recorded complication arose between hypocenter and station, with waves of different character traveling along different paths and at different speeds.

Early Analysis and Notation

This second alternative stands for the main course of revision in interpretation from 1900 to 1940. The first stage began when seismograms from instruments with adequate damping and magnification revealed details previously blurred by resonance or lost below the threshold of response. It was seen that the large waves of the conspicuous part of the seismogram, which continued to be called the "principal earthquake," were ordinarily preceded by as much as half an hour of smaller and shorter-period motion, then named the "preliminary tremor." Soon it was observed that there was usually a division into a first and second preliminary tremor.

An international terminology, based on Latin designations, was adopted for reporting the normal type of seismogram. First and second preliminary tremors were denoted by the letters P and S, which stood simply for *undae primae* and *undae secundae* (*unda:* wave). The large, long-period waves of the principal phase were lettered L (*undae longae*); the maximum of the seismogram was M. The decreasing later waves are termed the *cauda, C* (or *coda,* which is Italian in place of Latin); F (for *finis*) was used to report the approximate time of the end of the recorded disturbance.

At first P and S were taken to originate from minor preliminary fracturing of the rocks in anticipation of the principal events, represented by L and M. Some years of research were needed to establish that the P and S groups consist, at least at their beginnings, of longitudinal and transverse waves through the body of the earth, while L, M, C consist primarily of surface waves, including some of approximately the types predicted by Rayleigh and Love. All these wave groups result from a single disturbance beginning at the hypocenter, with a duration much shorter than that of the seismogram at teleseismic distance.

Reflections and Multiple Shocks

P and S waves are repeated by reflection at the surface of the earth or of its core. Milne once wrote approvingly of an assistant who could find evidence on the seismograms of two or more earthquakes where only one was evident at first examination; he must have identified reflected waves such as PP and SS as the P and S of separate earthquakes. The same humorous history followed with the sharp reflected waves characteristic of deep-focus earthquakes (Chapter 19), so that the remark "two shocks" in published station bulletins became a signal for later workers to look for evidence of depth.

Reflections and refractions at the surface of the earth's core give rise to still other types of recorded waves, especially at great distances; this will be taken up later.

Because of the many causes of complexity in seismograms of a single earthquake, it is a common saying at seismological stations that "two shocks are the last refuge of a seismologist." Indeed, when the records of only one station are available, two shocks from different epicenters will explain almost any series of recorded phases; accordingly every effort is made to be sure that one shock cannot account for all the readings.

Of course, multiple shocks do occur. As noted in Chapter 6, some regions, like New Britain, habitually produce complex earthquakes at a relatively high level of magnitude. Aftershocks are often large enough to complicate the seismogram of the preceding main shock. Occasionally, by coincidence, two good-sized earthquakes originate at widely different epicenters within a few minutes; the resulting confusion sometimes gets cleared up only with the aid of the stations nearest one or the other source.

Seismology is now far enough advanced so that, when all the data are in hand, the effect of repeated action at the earthquake source can usually be separated from the multiplicity caused by wave propagation; but such decisions are still far from infallible. Unsuspected peculiarities in the propagation of seismic waves, and new types of waves as yet unidentified, await detection by alert observers. As this book is being prepared, such discoveries are often being reported, particularly in the field of surface waves and guided waves.

SEISMIC WAVES INSIDE THE EARTH

Standard Tables and Deviations

A remarkable finding of teleseismology is the high degree of spherical symmetry in the distribution of the velocities of seismic waves, as shown by their transit times from hypocenter to station. For earthquakes at normal shallow depth the transit time for the first recorded P waves is the same to all stations at the same angular distance Δ from the epicenter, independently of direction, with deviations ordinarily not over 2 seconds. Tables have been constructed which give standard transit times to the second (Appendix VIII). Deviating observations, when not due simply to error at the stations or in locating the epicenter, can usually be explained in one of three principal ways:

(1) The hypocenter may be deeper than normal. If it is, the first waves at all distant stations arrive systematically early relative to the near ones. The deviation, which increases with distance, can be calculated and corrected for. However, it is often impossible to distinguish it from the effect of a slightly earlier origin time for the earthquake. Decision is reached, if at all, only by closely investigating times of recorded arrival for all the chief types of seismic waves.

(2) The earth is not an exact sphere. The diameter through the poles is about 1/297 shorter than that in the plane of the equator; the difference is

21 kilometers (13 miles). The effect on the time of seismic waves is small but observable, amounting to as much as 2 or 3 seconds in the transit time of *P*. To reduce this systematic error, distances are stated in terms of the true angle at the center of the earth between the radii to hypocenter and station, computed with proper allowance for the ellipticity. This demonstrably increases the general concordance of data when the readings of many stations are used. It has been said that, if we had no other evidence of the ellipticity of the earth, we could deduce it and roughly estimate it quantitatively from seismological data alone.

(3) At some stations there appears to be systematic mean deviation from the standard travel times of *P*. Where the tendency is for the arrival to be late, various explanations can be suggested. Where it is 2 or 3 seconds early this may be due to especially good equipment, the first small motion being read and timed when less sensitive instruments would furnish the time of a later larger wave. However, it is nearly certain that in some areas, like the central Mississippi Valley, there is a real geophysical cause.

Velocity and Depth

The fact that the transit times in general do not depend on path, but only on distance, shows that velocities within the earth depend primarily on depth alone. With one chief exception (the surface of the core, discussed later) and some minor ones still under investigation, velocity is found to increase regularly with depth. If we now look at theoretical expressions for the velocities of elastic waves:

$$V = \left(\frac{\lambda + 2\mu}{\rho}\right)^{\frac{1}{2}} \quad \text{for longitudinal waves} \quad (P) \tag{1a}$$

and

$$v = \left(\frac{\mu}{\rho}\right)^{\frac{1}{2}} \quad \text{for transverse waves} \quad (S) \tag{1b}$$

it is clear that the increase in velocity with depth is not due to the increase in density ρ, since that alone would have the opposite effect. To account for the facts we conclude that elasticity increases faster than density as we go deeper into the earth.

Paths of Waves in the Earth

In elementary optics we consider light rays passing through lenses and prisms, or being reflected from mirrors, without referring to the wave properties of light. Similarly in seismology we get a good enough theoretical approximation to the paths and times associated with a given velocity-depth relation by considering rays only. Our first problem is: What are these rays and times inside a sphere when the velocity V depends only on r, the distance from the center?

If velocity varies continuously with depth, the rays will generally be curved. At any point on such a ray we can define the angle of incidence i, which is the angle between the radius through the point and the tangent to the ray. Moreover, it follows from symmetry that a ray passing through any two points of the sphere will lie wholly in the plane passing through those two points and the center. It remains to find a relation between r, i, and V.

Suppose that, instead of a continuous variation of V with r, we have a sphere made up of thin concentric shells in each of which V is constant. If the velocities in two adjacent shells are V_1 and V_2, a ray passing through the boundary will be refracted, having angles of incidence i_1 and i_2. Applying Snell's law of refraction from elementary optics, we have

$$\frac{\sin i_1}{\sin i_2} = \frac{V_1}{V_2} \tag{2}$$

which shows that the quantity $\sin i / V$ is the same on both sides of the boundary. Now within each shell the velocity is constant, and the ray consequently is a straight line. Along a straight line $r \sin i$ is constant (this is proved by drawing radii to two ends of a straight line segment and applying the law of sines from trigonometry); hence in each shell $r \sin i / V$ is constant. But, at the point of refraction from one shell to the other, $\sin i / V$ is unchanged, as we have seen. Hence the quantity $r \sin i / V$ is the same in both shells and, by extension of the argument, in all the shells through which the ray passes. Now let the shells become thinner and thinner, with less and less velocity difference between successive shells; in the limit we pass over to the continuous variation of V as a function of r, and the general relation

$$\frac{r \sin i}{V} = \text{constant} \tag{3}$$

along a given ray is established.†

When the velocity V increases with depth, the rays will be convex downward; each ray will have a deepest point. At this point i is $90°$ and $\sin i = 1$; if we denote the corresponding values of r and V as r_m and V_m we have

$$\frac{r \sin i}{V} = \frac{r_m}{V_m} \tag{4}$$

for any one ray.

As rays emerge at the surface and are recorded on seismographs, we obtain data for calculating the velocity as a function of r. At the surface, r is the known radius of the earth; V can be determined from observations on ex-

---

† Note that there is no dropping of small quantities in passing to the limit. The equation applies equally well to the continuous and the discontinuous case.

plosions and local earthquakes; and i can be calculated from tables such as those of Appendix VIII which give the time of arrival at the surface in terms of distance from the epicenter. Denote the apparent surface velocity from these tables as V'; then at the surface sin $i = V/V'$. Calculation of V as a function of r involves a complicated and rather artificially constructed numerical integration. (Appendix VI.) Once the velocity function is known, transit times can be calculated for all distances and all hypocentral depths, for refracted and reflected waves as well as for direct waves. An independent determination can be made for velocity distribution for transverse waves, and this forms a basis for calculating still further transit times.

THE INTERIOR OF THE EARTH

Crust, Mantle, Core

The paths of seismic waves are affected by two chief discontinuities within the earth, at depths of about 30 to 60 kilometers and of 2900 kilometers, respectively. The former, associated with the name of Mohorovičić, is of especial importance in interpreting seismograms at short epicentral distances up to a few hundred kilometers. Evidence for the latter was discovered by Wiechert and by Oldham; but the correct identification and determination of the depth of this deep discontinuity was carried out by Gutenberg.

These two discontinuities divide the earth internally into an outer shell, the *crust*,† an intermediate shell, the *mantle,* and a central core.

Within the crust, discontinuities and irregularities exist, some of only local or regional extent. Within the mantle there are several levels near which there is departure from the regular increase of velocity with depth; these levels act as minor discontinuities. Within the core there is an interior division; the inner core may differ considerably from its surroundings in physical characteristics.

The major divisions are shown in Table 17-1 and Figure 17-1. The Mohorovičić discontinuity at the base of the crust will be discussed in the next chapter in connection with the recording of local earthquakes. Its level varies widely; in some continental areas it is as deep as 60 kilometers, in others as shallow as 30 kilometers, while in oceanic areas it rises to within 10 or even 5 kilometers of the sea surface. This variation partly explains the minor deviations from mean expected transit times of seismic waves from teleseisms as recorded in certain localities. The irregularity of the base of the crust is per-

† The term *crust* or *continental crust* has been used with various meanings. In this book "crust" means only that part of the earth which is above the Mohorovičić discontinuity. Some authors identify the crust in this sense with the "lithosphere" of classical geology; this concept involves further hypotheses.

Table 17-1   The Earth's Interior

Depth (km)	Depth (mi)		Velocities (km/sec) V	Velocities (km/sec) v	Radius (km)	Radius (mi)
		SURFACE				
0	0		5	3	6370	4000
		Continental crust	6	3.5		
30+	20+	MOHOROVIČIĆ	7?	4?	6340—	3980—
		DISCONTINUITY	8.2	4.5		
		Mantle				
2900	1800		13.5	8	3470	2200
			8	—		
		Core				
5000	3100		10	—	1400	900
		Inner core				
			11.5	?		
6370	4000	CENTER			0	0

V = velocity of longitudinal waves (P in crust and mantle, K in the core).
v = velocity of transverse (S) waves.
Depth of the Mohorovičić discontinuity differs widely from region to region and may be double that shown in the table.

haps the principal departure from spherical symmetry in the physical properties of the earth.

The level of the boundary between mantle and core appears to be nearly constant at 2900 kilometers. However, there exists as yet no published investigation which systematically compares the depth of the core under different regions of the earth. The level of the transition from the outer to the inner core is known to be near 5000 kilometers, but at present it cannot be fixed with exactness.

Velocities of longitudinal waves at various levels are indicated in Table 17-1. Those for the continental crust are mean values only, and subject to revision; for details refer to Chapter 18. Velocities within the mantle are known with comparative precision. Velocities of transverse waves within the core are not given, since at least in the outer core such waves do not exist.

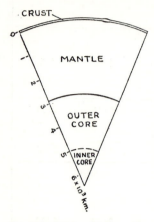

FIGURE 17-1 *Major discontinuities in the earth.*

Nature of the Core

The tides of the oceans are familiar; but the solid earth also yields, in a much smaller but measurable degree, to the deforming tidal forces produced by the attraction of the sun and moon. This measurement leads to a calculation of the mean rigidity μ for the whole earth. The same elastic constant can be calculated from the slight displacements of the axis of rotation of the earth which produce the effect of variation of latitude; the numerical results of the two methods are consistent. It appears that the earth has a mean rigidity of the same order as that of steel. This is small; for it is possible to calculate the mean rigidity of a mass of rock the size of the earth and composed of material similar to the rocks of the crust, and the result is much larger than the observed mean. Wiechert concluded that there must be a large interior region of the earth where the rigidity is very low; its properties should approach those of a liquid, or more generally a fluid, and it might be in the form of a central core. This hypothesis is very different from the notion of a "thin" crust overlying a molten interior which was current and popular two generations ago. The surface of the core, when finally found by Gutenberg, proved to be nearly halfway down to the center. Nevertheless, the core is large; it is larger than the planet Mars, and its radius is a little greater than the diameter of the Moon.

There is direct evidence that the core is at least partly fluid, as expected by Wiechert; it does not transmit transverse waves. S waves arriving through the mantle at the boundary of the core do not enter it, but they may be transformed by refraction into longitudinal waves (K) in the core. Seismologists have sometimes claimed to identify, on seismograms written at large distances, motion indicating transverse waves through the core; but the several identifications are not mutually consistent, and most of the observations can be explained in other ways. If transverse waves do pass into the core, they must do so with low energy and low velocity, corresponding to a small value of the rigidity; whereas S waves which have passed through the mantle just outside the core are usually large and readily identified. Furthermore, the wave ScS, which is a transverse wave reflected back into the mantle from the core boundary, is very strong at short distances; in some Japanese deep shocks it has been observed almost at the epicenter. This strong reflection is consistent with low rigidity in the core.

Bullen and others have suggested that below the 5000-kilometer level which bounds the inner core the material may again be solid and transmit transverse waves. This interesting possibility is difficult to check by observation, and the point thus far remains undecided.

BODY WAVES OF TELESEISMS

Notation

In general, the various individual wave groups on seismograms represent as many possible paths for a single initial disturbance. For body waves, these possibilities represent various combinations of reflections and refractions; each one is denoted by a particular combination of symbols, chiefly letters, which stand for successive segments of the ray in order from source to station. The type of wave is shown by P for longitudinal and S for transverse waves in the mantle, by K for longitudinal waves in the core (for longitudinal waves in the inner core, the letter I has lately been approved officially). P' is an abbreviation for PKP; P'' and P_2' are specializations of P', explained under the next heading.

Reflection at the surface of the earth is indicated simply by the succession of the chief symbols, as PP, SS, but reflection at the outer surface of the core is shown by interposing c, as PcP, ScS.

The smaller letters p and s, as in pP, sP, are used to distinguish a special type of reflection at the surface of the earth, observed chiefly for deep-focus earthquakes and described in Chapter 19.

The use of the symbol K makes it unnecessary to indicate refraction at the core boundary. It replaces an earlier additional use of the symbol c according to which SKS was written as $ScPcS$.

Observed Types

The possible types of observed waves may be classified most conveniently by the number of reflections along the path. Reflected waves are sometimes large, but those waves which are not reflected at all are generally the largest and most important recorded phases; they are P, S, PKP (P'), PKS, SKP, SKS.

Waves reflected only once may be classed according to the point of reflection. Those for which it occurs at the obvious point on the earth's surface between source and station are PP, PS, SP, SS, $SKSP$. Reflections at the outer surface of the core are PcP, PcS, ScP, ScS; those at its inner surface are $PKKP$, $PKKS$, $SKKP$, $SKKS$. Reflections also take place at points on the surface of the earth very distant from both epicenter and recording station: those most often observed are $P'P'$ and $SKPP'$. In practice it is usually not possible to distinguish $SKPP'$ from $PKSP'$, $P'SKP$, and $P'PKS$, which should arrive about the same time unless the hypocenter is very deep. A similar nearly simultaneous group sometimes observed consists of *PKSPKS, PKSSKP, SKPPKS, SKPSKP*. Finally, there are some observations of *SKSSKS*.

Reflection comparatively near the epicenter is also possible. Such waves are observed chiefly for deep-focus earthquakes; their notation begins with *p* or *s* prefixed to the symbol of any other type of body wave (Chapter 19; see Fig. 19-3).

Repeated reflections are not uncommon; the observed motion decreases in amplitude and in clearness with each repetition. *PPP* and *SSS* are often observed. *PPS* (arriving nearly together with *PSP* and *SPP*) is sometimes large. *P'P'P'*, although usually small and seen only on seismograms of deep shocks, is of special interest. *PcPP'* is a conspicuous phase on seismograms written a few degrees from the antipodes of the epicenter.

Paths of Waves without Reflection

P and S. The paths of *P* and *S* through the mantle from the hypocenter to the recording station differ only a little, since the ratio of their velocities changes slowly with depth. Because of the increase in both velocities with depth, the rays are curved and convex downward. As the angle of incidence at the hypocenter decreases, the rays descend more steeply and to greater depth and emerge at the surface at increasing distance. When this distance corresponds to a central angle of about 103° between hypocenter and station (for ordinary shallow earthquakes) the *P* ray grazes the core (Figs. 17-2, 17-3). Beyond this distance the amplitude of recorded *P* decreases rapidly; but there is not a sharp shutting off, and *P* waves, especially of long period, continue to be recorded up to angular distances of at least 130°. This is a wave phenomenon not explained in terms of geometrical optics; it corresponds to diffraction of the *P* waves round the core boundary, analogous to the diffraction of light into a shadow. The effect for *S* is even stronger, but not so easy to observe.

PKP. At the angle of incidence for which *P* grazes the core, the first refracted wave enters the core and the resulting ray is the first for *PKP* (*P'*). It emerges at a central angle about 10° in excess of 180°. Because of symmetry, a wave of this type appears at all points distant 10° from the antipodal point, or at $\Delta = 170°$.

For the moment we shall neglect the effect of the inner core, and describe what follows as the initial angle of incidence decreases. The central angle described by the ray then decreases to and below 180°, while the effective angle Δ increases to 180° and then decreases again; the transit time of *PKP* shortens, so that the representative curve on the transit-time chart (Fig. 17-4) descends, at first toward the right and then toward the left. This continues until Δ reaches 142°, which is an angle of minimum deviation similar

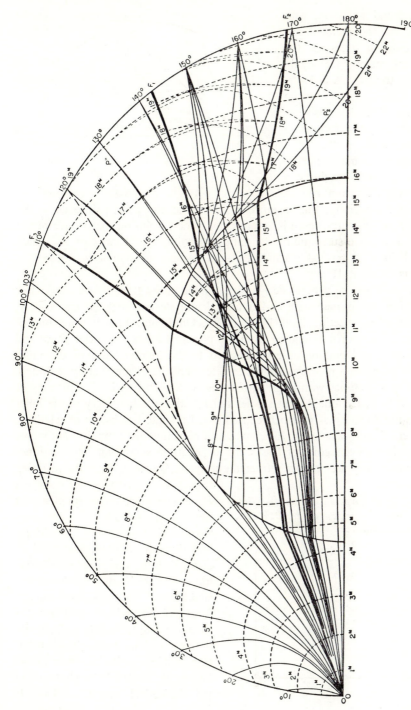

FIGURE 17-2 *Longitudinal waves in the earth. Rays and wave fronts.* [*Gutenberg and Richter, 1939.*]

to that found in working with prisms. As the incident ray becomes still steeper, the point of emergence again travels toward $\Delta = 180°$, which it reaches finally when the angle of incidence is zero, so that the ray follows a diameter of the earth.

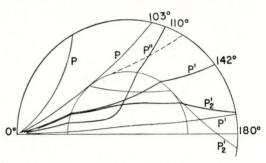

FIGURE 17-3   *Simplified diagram, rays for* P, P′, P″.

Because the minimum deviation at 142° produces a strong concentration of energy in that vicinity, this distance is known as the principal focus of *P′*. Because of symmetry, there is actually not a focal point on the surface of the earth but a circle surrounding the antipodal point at a distance of 38°. Such a focal curve is termed a caustic. Again because of symmetry, there is a focusing effect at 180°.

PKP or *P′* is the general designation for the entire transit-time curve. For distinction, the branch with earlier arrival and steeper angle of incidence is $P_1′$; the other, meeting it at 142°, is $P_2′$. On Figure 17-4 a third branch marked *P″* is shown extending from the principal focus to shorter times and distances; this group of observations led to the identification of the inner core. Rays passing near the 5000-kilometer depth level are sharply deflected, emerging, sometimes with large amplitudes, at Δ as low as 110°. This may or may not be reflection. Recent studies indicate that the boundary of the inner core is not so sharp as that between core and mantle. For waves of relatively long period and wave length it behaves as a reflecting discontinuity; but waves with periods of a second or less seem to penetrate somewhat into the inner core, where they are sharply refracted out again. On either basis the resulting transit-time curve for the complete *P′* phase is much more complex than the three branches in our figure, consisting of a series of flat loops. Notations have been suggested to distinguish the various parts of these loops, but this is of only theoretical significance. The expected curves, focal points, and points where the loops cross differ widely with slightly different assumed velocity distribution; and in practice it is often almost impossible to decide in what part of the theoretical tangle a given *P′*

FIGURE 17-4

Longitudinal waves in the earth. Simplified time-distance curves for P, P′, P″.

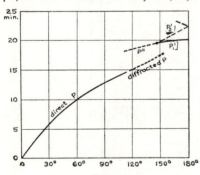

observation belongs. Consequently P'' is used for any P' reading for Δ less than 142°.

SKP. The phase *SKP* is in many ways analogous to *PKP*. The chief difference is that, while longitudinal waves change direction sharply on encountering the core and being greatly slowed down, there is relatively small difference between the velocity of S just outside the core and that of K just within. This means that there is little change of direction at refraction. The transit-time curves are shifted on the chart relative to those of P'; the principal focus of *SKP* is near $\Delta = 133°$. At larger distances there are two principal branches like those of P', and there is an extension to shorter distances due to the effect of the inner core which corresponds to P''. Recording of *SKP* is complicated by that of *PKS*, which would have the same transit time for a surface source and should arrive a few seconds later for ordinary shallow shocks.

SKS. For *SKS* there is no great deflection of the ray either on entering or on leaving the core. The recorded phase is often large, and for distances over 84° it precedes S, adding difficulty to the interpretation of seismograms.

Reflected Waves

The paths of waves which undergo one or more reflections can be understood by combining segments of those for the unreflected waves. It must be remembered that Snell's law of sines applies equally well to reflection as to refraction. Where the wave type is unchanged at reflection, as for *PcP* and *SS*, the familiar law of reflection results; the angles of incidence and reflection are equal. Where the wave type, and consequently the velocity, change, this is no longer so. Thus at the reflection which constitutes *PS*, the S wave, having the lower velocity, has the smaller angle and leaves the surface more steeply than the P wave approached it. This explains why some genuine-looking combinations of symbols do not represent real waves, so that, while *SKSP* is an observed wave, *PKPS* cannot exist.

Surface Reflections. The groups of waves resulting from repeated reflections at the surface are often characterized by large amplitudes with long periods, especially in earthquakes of large magnitude recorded at great distances. Near $\Delta = 120°$, *PP* and *PS* are so prominent that they are sometimes misidentified as P and S, which would imply a much shorter distance. *SS* and *SSS* are often seen as large long-period waves between the S group and the arrival of the first surface waves. Sometimes one of these waves is taken for a surface wave; this error opens the way to a wide variety of seismological and geophysical mistakes.

Core Reflections.   The core reflections PcP, PcS, ScP, ScS are quite strong when Δ ranges from 30° to 40°, and the seismogram has a very complicated appearance. ScP is sometimes taken for S, sometimes (because of its prevailing short periods) as the beginning of a separate earthquake. It has already been mentioned that ScS is large at short epicentral distance; but usually this can be observed only for deep-focus shocks. As P approaches grazing incidence, PcP approaches P; sometimes it can be identified as a large impulse only a few seconds after the first motion. It is often difficult at first to distinguish such a PcP from the phase pP characteristic of deep shocks.

P′P′ and Related Waves.   The waves $P'P'$, $SKPP'$, and $P'P'P'$ were discovered at Pasadena in 1933. They are characteristically small, short-period phases resembling the P phases of poorly recorded teleseisms. Since they follow large shocks, they were at first taken for aftershocks. However, in routine measurement and tabulation it was observed that there were too many of these apparent aftershocks coming at intervals of about half an hour ($P'P'$ or $SKPP'$) and 45 minutes ($P'P'P'$) after the P of the corresponding large shock. It was recognized that these must be late phases, not separate shocks. After thorough investigation and testing, the new phases were identified and their travel times established. They are most readily observed on the records of deep shocks, since the surface waves of shallow shocks disturb them. They are largest when the reflection occurs near the focal distance of P' (142°); consequently for a surface source $P'P'$ should be expected particularly for Δ near $360° - 284° = 76°$, and $P'P'P'$ near $426° - 360° = 66°$. It was found that observations of $P'P'$ had been made and reported by many seismological stations, usually as P of a separate shock. In some cases these supposed separate shocks had been located as individual earthquakes in the International Seismological Summary. On the time-distance chart $P'P'$ reproduces the details of P' on double the scale; its transit time generally decreases with increasing distance, while that of $P'P'P'$ increases.

PKKP and SKKP.   The discovery of the $P'P'$ group depended largely on the high magnification for short-period motion provided by the Benioff vertical-component seismometer. This also led to the first acceptable observations of $PKKP$ and $SKKP$, which have the same short-period characteristics as $P'P'$. They were found on seismograms written at large distances; $PKKP$ has a focus near 120°. Many readings published for $PKKP$ relate to the large long-period reflected phases, such as PS and PPS, which arrive near the same time and can often be seen on the Benioff records together with $PKKP$.

PcPP′. $PcPP'$ was also a Pasadena discovery; this was due not so much to instrumentation as to the fortunate occurrence of several large shocks in the Indian Ocean not far from the antipodes of the station. There is a focus near

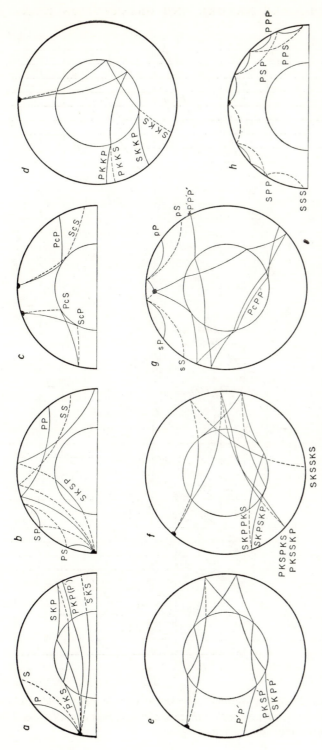

FIGURE 17-5 *Paths of body waves of teleseisms, with letter symbols. Longitudinal wave ray segments shown as full lines; transverse wave ray segments shown dashed.*

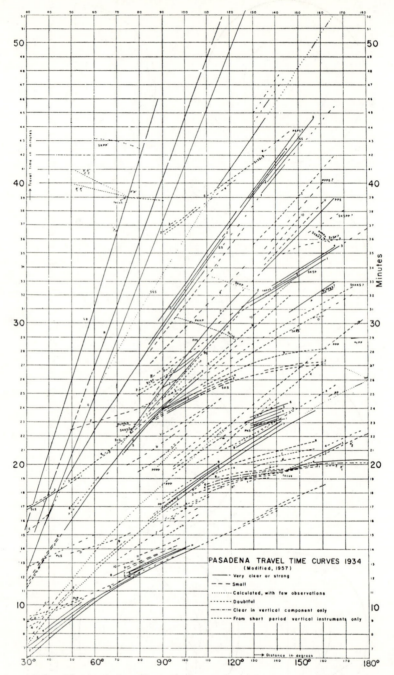

PASADENA TRAVEL TIME CURVES 1934
(Modified, 1957)

——— Very clear or strong
— — — Small
·········· Calculated, with few observations
-·-·-·- Doubtful
—·—·— Clear in vertical component only
- - - - - From short period vertical instruments only

FIGURE 17-6 *Observed time-distance curves for seismic waves.* [*After Gutenberg and Richter, 1934.*] *Note that these show what was actually found and are not theoretically constructed.*

175°, and the readings would be valuable for many purposes were it not for difficulty in estimating the effect of the boundary of the inner core.

Paths of the principal body waves, and the corresponding time-distance curves, are illustrated in Figures 17-5 and 17-6.

INTENSITIES OF REFLECTION AND REFRACTION

When an elastic wave, longitudinal or transverse, strikes the boundary between two regions in which the wave velocities are different, four derived waves (in general) come into existence: two reflected waves, longitudinal and transverse, and two refracted waves, also longitudinal and transverse. The energy of the incident wave is partitioned among the four derived waves in proportions which depend on the elastic constants and densities in the two regions, as well as on the angle of incidence. (As remarked on a preceding page, the angles of both refraction and reflection are determined by Snell's law when that of incidence is given.) Assuming elastic solids obeying the equations discussed in Chapter 16, it is theoretically possible to compute the energy partition and to compare it with observation of the amplitudes recorded for reflected and refracted seismic waves.

Computations of this sort were originally carried out by Knott, working with potential functions. Zoeppritz later used a method which operates throughout with the amplitudes (displacements u, v, w) in the wave motion; these amplitudes are most directly related to instrumental registration. The result of the theory is expressed in a system of four equations (Appendix VII); these are simple linear equations in the ratios of the amplitudes of the four derived waves to that of the incident waves, so that solution is a matter of elementary algebra. However, the coefficients are complicated, involving trigonometric functions of the angles of incidence, reflection, and refraction, as well as the elastic constants and densities. A great deal of numerical labor is needed to get the amplitude ratios for any one assumed angle of incidence. There are very few short cuts, and generalization is difficult because slight changes in the assumptions often lead to large changes in the results.

This is an excellent instance of a problem which is "solved" in theory but not in practice. Several investigators have made large groups of calculations covering the more important conditions; but the choice of parameters thus far has not altogether covered the range of seismological interest. This is a game any number can play; and anybody can get into it who is willing to do a large amount of time-consuming and carefully checked calculation.

One general result is associated with further difficulty in computation. As in optics, when the velocity in the second medium is greater than in the first, there is a critical angle of incidence beyond which total reflection occurs. Computation in such a case involves using complex quantities and separating the real and imaginary parts; the work to be done for each assumed value of the angle of incidence is approximately doubled. The total reflection phenomenon appears in connection with both waves in the second medium, if both velocities are larger than that of the incident wave in the first medium.

Under conditions approximating those in the solid earth, and excluding grazing incidence or angles for which there is total reflection, the transmitted wave of the same type as the incident wave (longitudinal or transverse) carries more energy than any of the other three derived waves (Gutenberg).

Calculation of amplitudes inevitably involves algebraic sign, which shows whether or not there is change of phase (from compression to dilatation, for example) at reflection; this is often of importance in comparing with observations.

SURFACE WAVES AND RELATED WAVES

The surface wave section of the seismogram of a large shallow teleseism is often very complicated. Occasionally there is an appearance of simplicity in the maximum phase, with a long train of almost uniformly harmonic waves of period near 20 seconds. Almost surely this is not related to any regularity in the generating mechanism of the earthquake. Rather it is a phenomenon of resonance, in the most general meaning of that term; it may be compared to producing a musical tone by blowing across the mouth of a bottle.

For rough purposes such as issuing preliminary station bulletins, the old notations L and M are still in use for the beginning and maximum of the long-period surface wave phase. Closer study and the recognition of detail have led to other symbols.

The G Wave

Seismograms of large teleseisms often show the L group beginning with waves of large actual amplitude and very long period (ranging from 1 minute up to at least 4 minutes). These early waves are usually horizontal and transverse; the vertical component and the horizontal motion in the direction of propagation are nearly zero. These are the characteristics predicted for Love waves; Gutenberg, who first described the very long waves systematically, so identified them. In order to distinguish these in notation from the ordinary L group of shorter period, Byerly proposed the notation G (for Gutenberg). With decreasing magnitude these G waves decrease much more rapidly than the 20-second surface waves of the M group. An instructive published instance refers to shocks in the Solomon Islands on October 3, 1931. The records of an aftershock 3 hours later than the main earthquake can be compared with it wave for wave, with amplitudes about one-sixth as large including the L and M groups; but the G wave is an exception. Seismograms on which the two shocks appear, showing the G wave for the main shock with large recorded amplitudes corresponding to ground motion of half a millimeter or more, give no indication of a G wave in the smaller shock. This naturally must be explained in terms of the originating mechanism of the main earthquake; one thinks of release of strain and relative motion of large blocks, favoring the radiation of waves with long periods and wave lengths. (See Fig. 17-9A.)

Long Rayleigh Waves

When a station records G, a good vertical-component seismograph frequently records the surface waves as beginning, several minutes after the time of G, with a similar long-period wave. Horizontal-component instruments then show that motion in this wave is in the direction of propagation and not transverse to it. This corresponds to the theoretically expected Rayleigh wave and is designated by R. (British authors commonly use LQ and LR for Love and Rayleigh waves; other notations are also in use.) Identification is sometimes difficult because the wave fronts may approach from a direction appreciably different from that of the epicenter. This is interpreted as due to horizontal refraction where the waves cross an important structural boundary; it is very noticeable on the North American west coast, where the large surface waves of many teleseisms arrive after crossing the boundary between the Pacific basin and the continental area. On rare occasions, when the waves are traveling nearly along such a boundary, two successive G waves and two successive R waves are recorded because of passage along different paths with different velocities.

The velocity for the fastest G waves is about 4.5 kilometers per second, over paths in all parts of the world; the corresponding R velocity is about 4.2 kilometers per second. These fastest and longest waves are followed by others of the same type with lower velocity; this dispersion is predicted by theory, as indicated in the preceding chapter. Since the speed of the long waves is determined by a sort of mechanical averaging of the transverse wave velocity in a range of about 1 wave length down from the surface (for a velocity of 4.5 kilometers per second and a period of 1 minute the wave length is 270 kilometers), as the wave length decreases the speed decreases, corresponding to the lower transverse wave velocities nearer the surface. Now, if the distribution of S velocity with depth differs in two regions, one should get different dispersion curves by plotting wave velocity against period. Gutenberg and others have obtained this result. The dispersion curves for trans-Pacific paths generally differ from those for continental paths, such as those for shocks in eastern Asia recorded in Europe. The two curves approach closely for the longest waves, indicating that the difference in structure between continental and oceanic areas disappears at great depths.

The Lg Wave

A new tool for finding the boundary of continental structure is a wave called *Lg* by Ewing and his co-workers at Columbia, who have identified it as a guided wave in the continental crust. It is a relatively small short-period disturbance, easily seen superposed on the longer surface waves, which travels over long continental paths with relatively little loss of energy but is cut off abruptly when the path has even a small oceanic segment.

The T Wave

In 1940 Father Linehan described a remarkable short-period wave group found on seismograms at Weston, Massachusetts, of earthquakes in the Caribbean area, which he lettered *T* (for *undae tertiae*). Other workers on the Atlantic coast had found similar phases on their seismograms before this explicit recognition; at Berkeley, California, large *T* waves were found to have been recorded for Hawaiian earthquakes, notably one on January 23, 1938, which was so large and short-period that it was supposed to be of local origin. Ewing, Press, and Tolstoy showed that the *T* wave is in effect a sound wave in the sea which is propagated from the shore into the continental interior with the velocity of ordinary seismic waves. The lower velocity in water accounts for the late arrival at distant stations. *T* waves have now been reported at many stations, in Japan, Sweden (Kiruna and Uppsala), and Australia (Riverview, near Sydney); they are noticeable on many seismograms at Huancayo, Peru.

Repetitions

Large surface waves of *G*, *R*, and *L* type often show repetition due to traveling repeatedly around the earth. The *L* group from a teleseism distant 90° may be followed about an hour later by the similar wave which has traveled the long way round by way of the antipodes, covering an arc of 270°; this is called W_2. Next comes W_3, which has covered a complete circuit of 360° plus the 90° from source to station; it is followed by W_4, which has gone round in the other direction. Repetitions of this kind have been observed up to W_8, and in rare cases to W_{15}, with alternate (odd- or even-numbered) arrivals separated by 2 to 3 hours. The large, long-period waves at the beginnings of the groups are often recognizable and are designated G_2, G_3, . . . or R_2, R_3, . . .

THE FAMILY OF GUIDED WAVES

Love waves, Lg, and T are three examples of the general group of guided waves. Love waves and Rayleigh waves in their shorter-period range are channel waves in the crust (continental or oceanic). Ewing and Press suggest that the G wave, with its long period and great wave length, is essentially a channel wave in the mantle; they class the corresponding long-period R as a mantle Rayleigh wave. Lg is a channel wave in the crust, which Gutenberg attributes to his "lithosphere channel" needed to explain the phase $\bar{P}$ in local earthquakes (Chapter 18). Phases discovered and named Pa, Sa by Caloi are believed by him to be guided by a low-velocity channel in the mantle, which Gutenberg identifies as his "asthenosphere channel" (*cf*. Fig. 18-4), postulated to explain the behavior of seismic waves emerging at epicentral distances near 20°; Ewing and Press, who discovered the same phases independently, have suggested an alternative explanation in terms of repeated reflection from the base of the crust.

Lg, which is primarily a transverse wave similar to L, often is observed to be doubled—$Lg1$ and $Lg2$. Båth has worked extensively on Rg which is the corresponding wave analogous to R, with longitudinal and vertical motion. Recently he has observed another continental-restricted guided wave, with velocity near 3.8 kilometers per second, which he attributes to an intermediate level immediately above the base of the continental crust and designates Li.

Oliver and Ewing find that it is possible to explain both Lg and $Lg2$ as surface waves corresponding to higher modes of oscillation in a crust with velocity increasing downward, not requiring a low-velocity channel within the crust. Gutenberg and Press have lately identified a small but consistently observable phase which bears the same relation to Lg that Pa does to Sa; they have named it Πg.

A low-velocity channel in the ocean, the Sofar channel, transmits energy horizontally with high efficiency and has been used for long-distance signaling. The T wave is essentially a guided wave in the sea; impinging on the continent, it gives rise to P, S, and surface waves which are recorded by seismographs as the complex and irregular T phase.

Extension of the theory of guided waves to the scale of the globe makes it applicable to the fundamental oscillations of the earth as a whole; it is believed that such oscillations may be excited by the greatest earthquakes.

INTERPRETATION OF SEISMOGRAMS

The Problem

To fix ideas, let us at first restrict this discussion to seismograms of tele-seisms originating at normal shallow depth. General principles will apply also to deep-focus earthquakes and to shocks in the local region; more detail will be found in Chapters 18 and 19.

Problems of interpretation arise at two stages: that of preliminary report and that of revised publication. Most good stations send out first accounts of their readings monthly, weekly, or even daily; these are exchanged with other stations either directly or through one of the national or international offices (Chapter 21). The preliminary report is often based on information found only on the seismograms themselves. Exceptions occur when the press or official sources promptly furnish information about earthquake percep-tibility and effects or readings at other stations. The press associations often act as valuable go-betweens, relaying the readings of one station to others; such information is usually incomplete and subject to many kinds of error, but nevertheless it is often available in time to send out a better report than could be made from the data of the single station.

First Measurements

If no outside data are at hand, the first step is to go over the seismograms and determine the times of arrival of all clearly distinguishable wave groups (seismic phases). For a large and complex record, judgment is needed not to make too many readings; usually only past experience shows which elements of motion are likely to prove significant. As little reference as possible should be made to any presumed distance of the recorded shock. Every large and clear phase should be noted, whether the worker can explain it or not. (Inex-perienced workers often mistakenly measure times when the motion decreases instead of those where it increases.) Times normally should be read to the second. Where the beginning of a large wave group is very gradual or is dis-turbed, it is read to the tenth of a minute; this applies especially to the time of first arrival of surface waves, which is valuable for preliminary purposes. If times of beginning of G and R are read, further timing within the surface waves is not of immediate interest. Short-period phases like $P'P'$, ScS, Lg may arrive after the surface waves begin and should be timed if possible.

Fitting the Chart

The measured times, corrected for clock error, should be written down in order, with notes as to which waves are large or small and which components

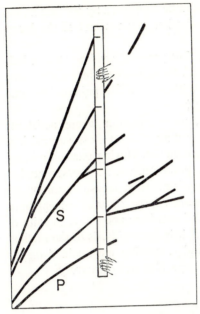

FIGURE 17-7 *Fitting readings to the transit-time chart.*

they are measured for. The next step is to apply a time-distance chart for shallow earthquakes, like that in Figure 17-6, but preferably on a larger scale. On the straight edge of a long strip of paper the recorded times are plotted from some arbitrary zero on the same time scale as that of the reference chart. The strip is held vertical on the chart, parallel to its time scale, and it is moved about until plotted marks on its edge fit transit time curves on the chart (Fig. 17-7). Distance Δ is then read directly from the horizontal chart scale; the origin time of the earthquake can be found by subtracting transit times on the chart from the corresponding times plotted on the strip.

Precautions and Errors

This procedure is simple in principle, but many errors are possible. It will not do to assume that the first motion identified on the seismograms is the earliest phase shown on the chart at a given distance, whether P or P'. Often the true first motion is small and may be masked by disturbance or microseisms—or it may simply be overlooked, while some more conspicuous later wave is taken as the beginning of the seismogram. It is often hard to decide whether the first motion is P, P', or PP, and all three possibilities must be considered. If the station lacks a sensitive vertical-component instrument, the first measurable motion, even for a large teleseism, may be S, SKS, or SKP.

For well-recorded shallow teleseisms the time of onset of surface waves is a good safeguard. As the chart shows, the time interval between the first motion and either G or R increases rapidly with distance; if these are plotted on the test strip, gross errors are less likely. Sometimes long-period phases like SS or SSS are mistaken for the beginning of surface waves.

An established station readily accumulates a file of seismograms for teleseisms at known distances. Studying these, the seismologist learns the appearance of the principal seismic waves as recorded by the instruments used, and even comes to recognize the probable epicentral region from the character of the seismogram. Errors are more common at new stations; a seismologist with experience elsewhere may be misled by some unfamiliar local circumstance or by the characteristics of unfamiliar instruments. Often the seismologist is as new as his station, and he must feel his way through a succession of mistakes. His apprenticeship can be shortened by correlating his results as quickly as possible with those of other stations and of the international centers.

These remarks ought to be superfluous; but years of reading station bulletins will convince anyone that they are not. Some workers repeat the same obvious misinterpretations year after year. With some it is a point of personal prestige not to admit they were ever wrong.[†]

Still worse is the unscientific practice, most common where assistants carry on the work of measuring and interpreting without close supervision, of forcing the seismogram by selective measurement. Having reached a hasty conclusion as to distance and general interpretation, such a worker applies tabulated or charted times directly to the seismogram, measures minor arrivals near the expected moments, and ignores large or even conspicuous waves which do not fit. Such procedure continues here and there, in spite of public condemnation in the strongest terms;[‡] it is the antithesis of the method of the research worker, who rejoices to find something unexplained on his records and reports it meticulously in the hope of a new discovery.

Revision

These remarks are still more pertinent when preliminary reports from a station are succeeded by a final and presumably revised bulletin. For purposes of revision the reports of the international office and of other stations are available, and in considerable detail if the interval between provisional and final reports is about a year. To correct the first interpretations, epicenters and origin times from these sources can be combined with transit-time charts, using the known distance from epicenter to station. Here, again, the worker should not look for what he ought to find but should seek to interpret correctly what he has already measured. On the other hand, since no one is infallible, he should occasionally expect to find errors in the information he receives. The more scientifically the revision is done, the more valuable the material will be when it is included in the *International Seismological Summary*.

Direction

The methods of interpretation described thus far yield distance, but not direction. When an earthquake has just recorded, the seismologist is nat-

---

[†] On September 1, 1923, the director of a certain station gave out a grossly wrong distance for the great Japanese earthquake; apparently he had misidentified *PP* as *P*. When the discrepancy was brought to his attention, he is said to have responded, "I am a man of science. I will not change my findings!"

[‡] "Unsitte vieler Stationen, die Diagramme nach den Laufzeitkurven zu bearbeiten"— Gutenberg. The German has an untranslatable pungency.

urally curious as to its direction and probable epicenter, particularly if he is exchanging information with the press. Whether direction can be given, even roughly, depends on circumstances beyond control. If recording begins with a large sharp P wave, so that the direction of the first impulse can be seen in all three components, the seismologist is in luck. The ray lies in the vertical plane through epicenter and station. If the recorded initial impulse is a compression, the first motion will be upward and horizontally away from the epicenter; if a dilatation, the first motion will be downward and toward the epicenter. Thus, motion up, north, and east recorded on a set of three seismographs indicates that the epicenter is southwest of the station.†

If the initial impulse is not clear, it may be possible to compare directions of displacement in the three components at some later instant during the group of P waves (or PP, or even sometimes P'). If the timing is very careful, and the three instruments are alike, the result is the same as for an initial P. Direction cannot be derived so simply from S because it may arrive polarized in any plane.

Love (or G) waves and Rayleigh waves can be used to some extent to determine direction; the motion in Love waves is transverse to the direction of propagation, so that such a wave arriving from the west should record only in the north-south component. The motion in Rayleigh waves being longitudinal and vertical, under the same circumstances R should be absent from the north-south component and recorded only on the east-west and vertical-component seismograms. G has a decidedly higher velocity than the corresponding long-period R, so that the two usually arrive well separated in time. Careful analysis of the phase relations between vertical and horizontal components of R (corresponding to the retrograde elliptical orbit of the particle) will discriminate between the two otherwise possible directions 180° apart. A general difficulty with this method is horizontal refraction, since the wave fronts swing through a wide angle in crossing major boundaries like that between the Pacific and the continents.

When direction is uncertain and there are only a few active seismic areas at the given distance, the temptation is to guess or to offer alternatives. This is risky in dealing with the press; it often leads to misunderstandings, giving popular commentators and columnists the opportunity to compare conflicting guesses from several stations, and imply that "the experts are all wrong." It is safer to hold off until times or distances from other stations are available and then combine data, using the globe before naming the epicenter.

If the initial motion is a clear compression or dilatation, this should be included in preliminary published bulletins; if magnitude is estimated, data used for the purpose should be given.

---

† Along a great circle, of course. The seismologist works with a globe to avoid the deceptive appearance of ordinary world maps.

APPEARANCE OF SEISMOGRAMS AT VARIOUS DISTANCES

The fact that a single time-distance chart applies to seismic waves for earth-quakes all over the world is paralleled by uniformity in the appearance of such waves as recorded at a given distance—their gradual or sharp commencement, the predominance of long or short periods, and their general amplitude, large or small. There are peculiarities in some regions and along some paths. The instrumental factor is much more important; divergent descriptions of seismograms from the same distance often turn out to be based on seismographs of different characteristics. The following applies to earthquakes originating at normal shallow depths, presumed to be near 25 kilometers; for the special problems of deep focus see Chapter 19. Distances Δ will be given in degrees of a great circle. See Figures 17-8, 17-9, and 17-9A.

0–1°: Sharp P and S groups; interval from 2 to 15 seconds.

1–5°: Seismograms retaining local earthquake appearance, but with P and S groups complicated, increasing in amplitude from phase to phase.

5–12°: Amplitudes and definiteness of the P and S groups generally decrease with distance, especially toward the end of the range. The aspect of the seismograms often differs extremely along different paths. Sometimes near 9° there are very sharp short periods in the P and S groups. There is often a large long-period surface wave with relatively abrupt onset (iG); readings given as for S may refer to this, or to the short-period Lg waves which appear superposed on it along continental paths.

FIGURE 17-8   *Typical record of Milne undamped seismograph.* [*As reproduced by International Seismological Association.*]

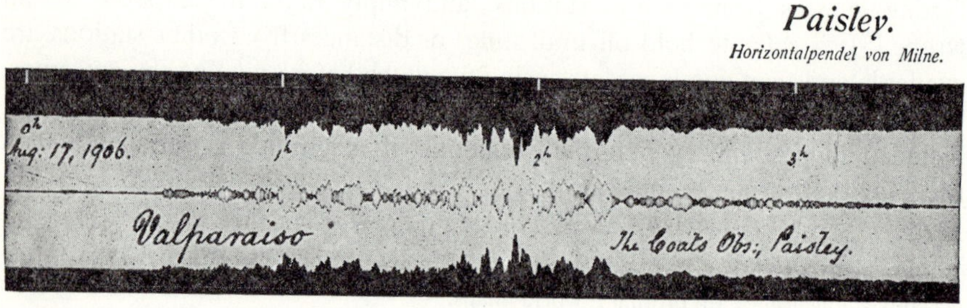

Paisley.

Horizontalpendel von Milne.

13–26°: *P* and *S*, which have almost vanished at 12°, reappear—*P* near 13° and *S* near 18°—with relatively large amplitudes and long periods.†

26–37°: *PcP, ScP, ScS*, are often large, the first two especially in the vertical component. *ScS* can occasionally be caught as a sharp impulse of relatively short period riding the long waves of the *G* and *R* groups. *S* shows rather long periods, especially beyond 33°; it is usually larger horizontally than vertically (this helps to distinguish it from *ScP*). The two transit-time curves intersect near 38°; misidentification of *ScP* as *S* is a common error which leads to great confusion when the two phases are not close together.

38–84°: Here at first *P* is small but increases rapidly with larger distance. The beginnings of *P* and *S* are both usually clear in this range, so that the corresponding time interval *S-P*, which ranges from about 6^m 10^s to 10^m 15^s, is a reliable measure of distance. Long periods prevail in *S* from about 40° to 55°; they may result in very small recorded amplitudes when short-period seismographs are used.

In all but the last few degrees of this range *PP* is large, particularly when the point of reflection is in a continental area.

P'P' has a focus near 75°; on rare occasions it records with amplitudes as large as *P* for the same shock, but generally with distinctly longer prevailing periods. *SKPP'* follows *P'P'* by several minutes; it is often larger and may be mistaken for *P'P'*. The focus of *P'P'P'* is near 65°, but this phase is usually seen only on seismograms of deep-focus earthquakes.

84–103°: It is usually easy to identify a seismogram as belonging to this range of distance; but it is difficult to place it more closely. *SKS* here precedes *S*, and follows *P* after an interval of about 10^m 20^s to 10^m 30^s; any reported *S* following *P* by such an interval is to be regarded with suspicion, although it may be correctly identified. *SKS* is merely the earliest of a complicated succession of phases, as difficult to separate as to identify, including *SKKS, S*, and *SP* or *PS*. Because *SKS* passes through the core as a longitudinal wave, it is always polarized in the vertical plane, so that it has no horizontal component transverse to the great circle from epicenter to station. This fact can be used to prove that a given arrival with a transverse component is not *SKS*, but *SP* and sometimes *S* show the same longitudinal property as *SKS*. *SKS* is relatively small up to 90°, but then increases and is large to well beyond 103°. *S* is large, particularly near 95°.

PP in this range is often larger than *P*, which it follows after about 4 minutes, and for which it is sometimes mistaken. From about 95° onward, short-period vertical instruments begin to record *PKKP*; if this phase can be found, it helps to fix the distance.

---

† British seismologists, led by Jeffreys, attributed this effect to a minor discontinuity in the mantle, "the 20° discontinuity," an interpretation based almost wholly on the recorded times. Gutenberg and others, considering the relative amplitudes, have preferred a low-velocity channel within the mantle, which Gutenberg has lately termed the asthenosphere channel. There appear to be guided waves associated with this channel, as with the lithosphere channel discussed in the next chapter.

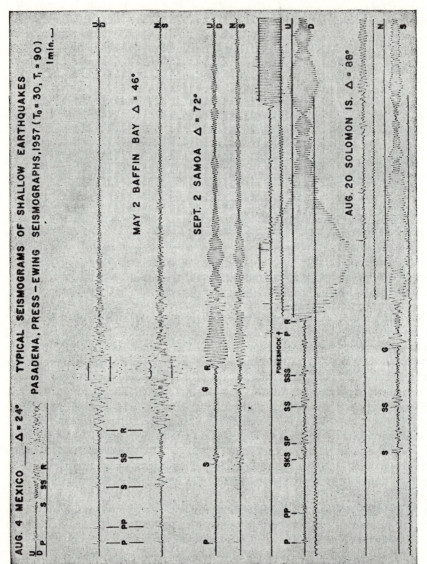

FIGURE 17-9   *Typical seismograms of shallow teleseisms.*

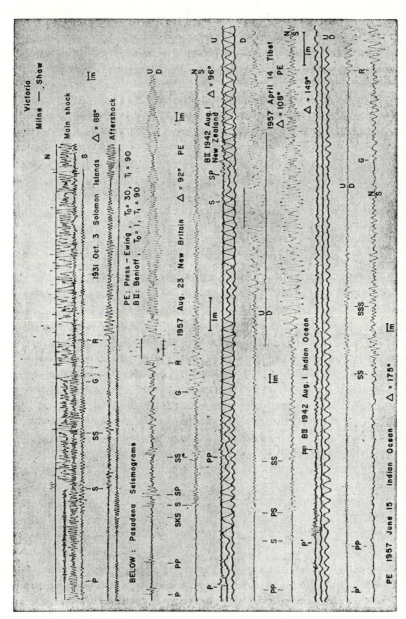

FIGURE 17-9A *Typical seismograms of shallow teleseisms.*

105–128°: The first large phase is usually *PP*. It is preceded by *P″*, which may appear as a large short-period phase in the vertical component. The diffracted *P*, about 4 minutes earlier than *PP*, is normally a long-period motion, sometimes consisting of only one or two waves, and recorded for large earthquakes only. *PKKP* has a focus near 120° and is usually found on short-period records (especially vertical). *SKS* is large, decreasing slowly with distance. Short-period instruments occasionally show a small clear phase which corresponds to *SKP* as *P″* does to *P′*.

PP and *PS* are large, especially in the range 115° to 125°. A common elementary error is to identify these waves as *P* and *S*, with a false distance of about 80°. This results generally from ignoring the evidence of surface waves, for at 80° these begin about 10 minutes after *S* and reach a maximum in a few more minutes, while at 120° the first surface waves are about 20 minutes after *PS* and the maximum may not be reached for another half hour.

129–141°: This range includes the focus of *SKP*, at about 132°. Near that distance the group of *SKP* and *PKS* waves is usually the largest phase in the early part of the seismogram, and the combined motion is almost always large in the horizontal component.

P″ shows evidence of the complexity which theory requires; short-period instruments usually record 5 to 10 seconds of small motion before the beginning of the chief phase. *PP*, which is usually clear but not large, follows after about 2 minutes, with *SKP* about a minute later. Beyond 135° there are usually no definite phases between *SKP* and the rather large *SS*, except for *PKKP* and *SKKP*; these are found only on short-period records, and in this range *SKKP* is the larger.

142–160°: As its focal distance is reached, *P′* suddenly appears with spectacularly large amplitudes compared with the rest of the seismogram. It may be clearly recorded without any trace of later motion, even surface waves. *SKP* is not seen; near 142° it is covered by *PP*, which continues as a large phase. *SKKS*, *SKSP*, and *PPS* are clear at first but fade out toward 160°; near that distance it is difficult to find any clear phases other than P_1', P_2', *PP*, and the long-period motions beginning with *SS* and continuing into the surface waves. The interval between the two *P′* phases, if they are rightly identified, fixes the distance rather closely.

160–180°: The interval of relative quiet following the clear phases near the beginning of the seismogram and preceding the surface waves, which has been increasing steadily through the shorter distance ranges, now approaches an hour, so that the seismogram looks like a recording of two separate disturbances. Usually P_1', P_2', *PP*, and *PPP* are all conspicuous; and there is a large wave, often erroneously reported as *S*, which may be *SKKS* but in that case is unaccountably late. *PP* for central angles over 180° is often clearly recorded. *PcPP′* is large for a short range of distance near its focus at 175°.

CONCLUDING NOTES ON MISINTERPRETATIONS

Most of the common errors of interpretation are those which result from using antiquated handbooks and tables and from trying to force all seismograms into the simple scheme represented by the letters *P*, *S*, *L*. Frequently the chief error is in misidentifying as *S* some other phase, such as *PS*, *SKS*, *ScP*, or the phase at large distances which is apparently late for *SKKS*; *SKP* is also sometimes reported as *S*. These errors are still more frequent in dealing with deep-focus teleseisms, or with shallow teleseisms of magnitude less than about 6; surface waves are then recorded poorly or not at all, so that there is no simple guide to the general range of distance to be considered.

Treating deep-focus shocks as if they were shallow leads to a special set of errors, some of which will be discussed in Chapter 19. Lately a converse type of misinterpretation has been seen occasionally: obvious shallow shocks with large surface waves are treated as of deep focus because some large and sharp phase following the first motion, such as *PcP* or P_2', is taken for the phase *pP* characteristic of deep shocks. It is equally unsafe to suppose that a shock is deep merely because surface waves are relatively small or are not found; this can be an effect of peculiar path, or of small magnitude, or of very large amplitude for a particular wave like *P'* near the principal focus.

Proper interpretation, in short, depends on considering all the evidence on the seismogram, not on a hasty selection of data made in routine manner.

References

Theory of seismic rays in the interior of the earth

Adequate discussion may be found in any of the general handbooks; instructors may wish to review the Göttingen papers, "Über Erdbebenwellen" (see references to Chapter 1). Special features are well discussed by Bullen:

Bullen, K. E., *Theoretical Seismology*, 2nd ed., Cambridge University Press, 1953, Chapter VII, pp. 108–122.

———, "Features of the travel-time curves of seismic rays," *M.N.R.A.S. Geophys. Suppl.*, vol. 5, pp. 91–98, 1945.

Inglada, V., *Estudio sobre la propagación de las ondas sísmicas*, Instituto geográfico y catastral, Servicio de sismología, Madrid, 1945, 347 pp. (Recommended to those who wish to study a treatment in Spanish.)

For calculation of velocities see Appendix VI; for coefficients of reflection and refraction see Appendix VII.

Structure of the earth (detail; for general references see Chapter 1)

Mohorovičić, A., "Das Beben vom 8 X 1909," *Jahrbuch des meteorologischen Observatoriums in Zagreb (Agram) für das Jahr 1909,* vol. 9 (1910), Part 4, pp. 1–63. (The original paper establishing the discontinuity at the base of the crust.)

Poldervaart, A., ed., "The crust of the earth," *Geol. Soc. Amer., Spec. Paper* 62 (1955).

Other references bearing on the Moho and the crust follow Chapter 18.

Oldham, R. D., "The constitution of the earth, as revealed by earthquakes," *Quart. Journ. Geol. Soc.,* London, vol. 62 (1906), pp. 456–475. (Recognizes the effect of the core on seismograms, but misinterprets the observed waves in detail, obtaining a wrong radius for the core.)

Gutenberg, B., "Über die Konstitution des Erdinnern, erschlossen aus Erdbeben-beobachtungen," *Phys. Zeitschr.,* vol. 14 (1913), pp. 1217–1218. (Preliminary publication; correct radius for the core.)

————, "Über Erdbebenwellen, VIIA. Beobachtungen an Registrierungen von Fernbeben in Göttingen und Folgerungen über die Konstitution des Erd-körpers," *Göttinger Nachrichten* 1914, pp. 1–52, with chart.

————, "The structure of the earth as viewed in 1957," *Scientia* (in press).

Lehmann, Inge, *Publ. bureau central séismologique international,* Ser. A, vol. 14 (1936), pp. 87–115. (Proposes an inner core.)

Gutenberg, B., and Richter, C. F., "*P'* and the earth's core," *M.N.R.A.S. Geophys. Suppl.,* vol. 4, pp. 363–372, 1938.

Jeffreys, H., "The times of the core waves," *ibid.,* vol. 4, pp. 548–561, 594–615, 1939.

Bullen, K. E., "Compressibility-pressure hypothesis and the earth's interior," *ibid.,* vol. 5, pp. 355–368, 1949. (Suggests that the inner core is solid.)

————, "The rigidity of the earth's inner core," *Ann. geofisica,* vol. 6 (1953), pp. 1–10.

Press, F., "Rigidity of the earth's core," *Science,* vol. 124 (1956), p. 1204.

Gutenberg, B., "The 'boundary' of the earth's inner core," *Trans. Am. Geophys. Union,* vol. 38 (1957), pp. 750–753.

See also other references on core waves, below. Observed Seismic waves (primarily for shallow teleseisms)

General

Jeffreys, H., and Bullen, K. E., *Seismological Tables,* British Association for the Advancement of Science, London, 1948.

Jeffreys, H., "The times of *P, S* and *SKS,* and the velocities of *P* and *S,"* *M.N.R.A.S. Geophys. Suppl.,* vol. 4, pp. 498–534, 1939.

Neumann, F., "Principles underlying the interpretation of seismograms," *U. S. Coast and Geodetic Survey, Spec. Publ.* No. 254, Govt. Printing Office, Washington, D. C., 1951.

Gutenberg, B., and Richter, C. F., "On seismic waves," *G. Beitr.,* vol. 43 (1934), pp. 56–133; vol. 45 (1935), pp. 280–360; vol. 47 (1936), pp. 73–131; vol. 54 (1939), pp. 94–136.

Wadati, K., *et al.,* "On the travel time of earthquake waves," *Geophys. Mag.,* vol. 7 (1933); Part I, pp. 87–99 (with K. Sagisaka and K. Masuda); Part II, pp. 101–111; Parts III, IV, pp. 113–137, 139–153 (with S. Oki); Part V (with K. Masuda), pp. 269–290.

Byerly, P., "The first preliminary waves of the Nevada earthquake of December 20, 1932," *B.S.S.A.,* vol. 25 (1935), pp. 62–80.

Core waves

References under "Structure of the earth" as above, and:

Lehmann, Inge, *"ScPcS,"* *G. Beitr.,* vol. 23 (1929), pp. 369–378.

Gutenberg, B., and Richter, C. F., "On *P'P'* and related waves," *ibid.,* vol. 41 (1932), pp. 149–169.

Gutenberg, B., *"PKKP, P'P',* and the earth's core," *Trans. Am. Geophys. Union,* vol. 32 (1951), pp. 373–390.

Denson, M. E., Jr., "Longitudinal waves through the earth's core," *B.S.S.A.,* vol. 42 (1952), pp. 119–134.

Forester, R. D., *"SKP* and related phases," *ibid.,* vol. 46 (1956), pp. 185–201.

Båth, M., "The density ratio at the boundary of the earth's core," *Tellus,* vol. 6 (1954), pp. 408–414.

Unexplained phases

Gutenberg, B., "Unexplained phases in seismograms," *B.S.S.A.,* vol. 39 (1949), pp. 79–92. (Documents a few well-established transit-time curves, some of which may be due to reflection at minor discontinuities in the mantle.)

Surface waves and guided waves; general discussion

Consult references to Chapters 1 and 16, and:

Gutenberg, B., "Channel waves in the earth's crust," *Geophysics,* vol. 20 (1955), pp. 283–294.

———, "Wave velocities in the earth's crust," *Geol. Soc. Amer., Spec. Paper* 62, 1955, pp. 19–34.

T waves

Linehan, D., "Earthquakes in the West Indian region," *Trans. Am. Geophys. Ur ion,* 1940, pp. 229–232.

Tolstoy, I., and Ewing, M., "The T phase of shallow-focus earthquakes," *B.S.S.A.*, vol. 40 (1950), pp. 25–51.

Ewing, M., Press, F., and Worzel, J. L., "Further study of the T phase," *ibid.*, vol. 42 (1952), pp. 37–51.

Ewing, M., and Press, F., "Mechanism of T wave propagation," *Ann. géophysique*, vol. 9 (1953), pp. 248–249.

Byerly, P., and Herrick, C., "T phases from Hawaiian earthquakes," *B.S.S.A.*, vol. 44 (1954), pp. 113–121.

Coulomb, J., and Molard, P., "Propagation des ondes séismiques T dans la mer des Antilles," *Ann. géophysique,* vol. 8 (1952), pp. 264–266.

Wadati, K., and Inouye, W., "On the T phase of seismic waves observed in Japan," *Proc. Japan Acad.*, vol. 29 (1953), pp. 47–54. Second paper, *Geophys. Mag.*, vol. 25 (1954), pp. 159–165; reprinted, *Proc. 8th Pacific Science Congress,* vol. 2A, pp. 783–792.

Båth, M., "A study of T phases recorded at the Kiruna seismograph station," *Tellus,* vol. 6 (1954), pp. 63–72.

Burke-Gaffney, T. N., "The T-phase from the New Zealand region," *Journ. and Proc. Royal Soc. New South Wales,* vol. 88 (1954), pp. 50–54.

Lg and Rg; Pa and Sa; Li, IIg, etc.

Press, F., and Ewing, M., "Two slow surface waves across North America," *B.S.S.A.*, vol. 42 (1952), pp. 219–228.

Båth, M., "The elastic waves Lg and Rg along Euroasiatic paths," *Arkiv Geophysik,* vol. 2 (1954), pp. 295–342.

———, "Some consequences of the existence of low-velocity layers," *Ann. geofisica,* vol. 9 (1956), pp. 411–450.

———, "A continental channel wave guided by the intermediate layer in the crust," *Geofisica pura e applicata,* vol. 38 (1957), pp. 19–31.

Oliver, J., Ewing, M., and Press, F., "Crustal structure of the Arctic regions from the Lg phase," *Bull. Geol. Soc. Amer.,* vol. 66 (1955), pp. 1063–1074.

Oliver, J., and Ewing, M., "Higher modes of continental Rayleigh waves," *B.S.S.A.*, vol. 47 (1957), pp. 187–204; vol. 48 (1958), pp. 33–49.

Caloi, P., "L'astenosfera come canale-guida dell' energia sismica," *Ann. geofisica,* vol. 7 (1954), pp. 491–501.

Press, F., and Gutenberg, B., "Channel waves IIg in the earth's crust," *Trans. Am. Geophys. Union,* vol. 37 (1956), pp. 754–756.

Variation of latitude

Lambert, W. D., "The Variation of Latitude," Chapter 16, pp. 245–275, in *Natl. Research Council, Bull. 78; Physics of the Earth, II, Figure of the Earth,* Washington, D. C., 1931.

Wahl, E., "Über die periodischen Eigenschaften der Polbahn," *Zentr. Geophysik, Meteorologie Geodäsie,* vol. 4 (1939), pp. 1–10.

Monographs

These few publications, not otherwise discussed in detail in this book, are examples of research involving collection of seismograms from all available stations for a single earthquake.

Rudolph, E., and Szirtes, S., "Das kolumbianische Erdbeben vom 31. Januar 1906," *G. Beitr.,* vol. 11 (1911), pp. 132–199, 208–275. (Selected as an early instance.)

Hodgson, E. A., "The *P* and *S* curve resulting from a study of the Tango earthquake, Japan, March 7, 1927," *B.S.S.A.,* vol. 22 (1932), pp. 38–49 (see also pp. 270–287).

Byerly, P., "The Montana earthquake of June 28, 1925," *ibid.,* vol. 16 (1926), pp. 209–265.

Lehmann, Inge, "The earthquake of 22 III 1928," *G. Beitr.,* vol. 28 (1930), pp. 151–164.

Macelwane, J. B., "The South Pacific earthquake of June 26, 1924," *ibid.,* vol. 28 (1930), pp. 165–227.

See also references to Chapters 18, 19, and the chronological bibliography in Appendix XVI.

CHAPTER 18

Seismic Waves at Short Distances

Pn AND $\bar{P}$

Observation

In 1909 A. Mohorovičić made a fundamental discovery on seismograms of an earthquake with epicenter in Croatia not far from his station at Agram (now Zagreb). Similar observations have since been made in other areas, including California, New Zealand, Japan, and Central Asia. There is a critical epicentral distance, generally in the range from 100 to 150 km. Stations at shorter distances register the P and S groups as initial sharp phases followed by smaller motion. Beyond the critical distance, P and S begin with relatively small and long-period motion, designated Pn or Sn, followed soon by at least one larger and sharper impulse of shorter period. Mohorovičić found that the time-distance curve of Pn is continuous with that of the first P wave recorded at teleseismic distances. He identified the later sharp phase, which he called $\bar{P}$, with the sharp first arrival at the shortest distances. The apparent velocities of $\bar{P}$ and Pn, as read from their time-distance curves, are about 5.5 and 8.2 kilometers per second, respectively. The latter velocity is higher than that found in 1909; there has been a steady slight upward revision due to the use of more sensitive instruments, which show an earlier arrival for the relatively small first motion.

The simplest form of the explanation adopted by Mohorovičić is shown in Figure 18-1. Assume that the velocities 5.5 and 8.2 kilometers per second are constant above and below a horizontal surface. The diagram and discussion neglect the curvature of the earth and of this surface; correcting for this would reduce the calculated velocity in the lower medium slightly below the apparent observed velocity of 8.2. $\bar{P}$ is interpreted as the direct ray to the recording station; Pn is refracted horizontally below the discontinuity, where with increasing distance higher velocity eventually compensates for longer path, so that there is a critical distance beyond which Pn arrives

282

before $\bar{P}$. Since the angle of refraction is 90°, the angle of incidence above the boundary is fixed by sin $i =$ 5.5/8.2 = 0.67, whence $i = 42°$ approximately. Theoretical time-distance curves are shown in Figure 18-1. That for Pn is straight, but that for $\bar{P}$ is a hyperbola asymptotic to a line through the origin with slope corresponding to 5.5 kilometers per second. Simple formulas for times and distances of Pn are given in Appendix IX. Putting $\Delta = 0$ does not give $t = 0$ for Pn; the straight line does not pass through the origin. The extrapolated time has occasionally been called "epicentral time"; this is very confusing, since

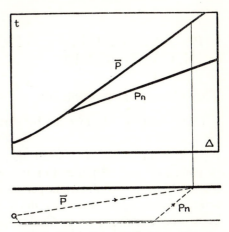

FIGURE 18-1 Pn *and* $\bar{P}$. *Time-distance curves and interpretation according to Mohorovičić.*

the wave which arrives at the epicenter is $\bar{P}$, and its time does not coincide with the extrapolated Pn.

Theory

Refraction like that postulated for Pn has serious theoretical complications. If one assumes a train of plane waves refracted at the given angles, calculation shows that energy returning to the surface should be very small. While Pn is not a strong wave relative to the rest of the seismogram, its recorded amplitudes are too large to suit such a simplified theory. The conflict is partly removed by considering a point source and an originally spherical wave front. Theory then yields more acceptable amplitudes, but the mathematics is of the most difficult character. Refraction of this type, in which an originally spherical wave front is refracted into a conical one which rises to the surface, has been demonstrated experimentally by von Schmidt and others.

The Moho

The surface of separation between the regions of lower and higher velocity is termed the Mohorovičić discontinuity. Lately the convenient but inelegant abbreviation "the Moho" has become current; it is used here with apologies to the reader and to the memory of the distinguished discoverer. As used in this book, the region between the Moho and the earth's surface is the crust.

*P** AND CRUSTAL STRUCTURE IN CALIFORNIA

The Conrad Discontinuity

Later study of seismograms by many investigators revealed further complexity, usually explained by subdividing the crust into more layers. In Europe, Conrad observed a small sharp impulse between *Pn* and $\bar{P}$ which he named *P** and attributed to refraction through an intermediate layer with a velocity near 6.5 kilometers per second. This has been called the Conrad layer, and its upper boundary the Conrad discontinuity; these terms have been used elsewhere, as in California, with doubtful propriety, since it is nearly certain that the detailed structure of the crust differs radically from region to region.

Jeffreys and others accepted the Conrad discontinuity as separating a predominantly granitic layer above it from a basaltic layer below, and proposed such notations as *Pg* and *Pb*. This usage was influenced by earlier geological speculation, but it was probably a misdirection. If the two layers are real, it is doubtful that "granitic" and "basaltic" describe them in a way which would satisfy a petrologist. For granitic rock under pressure, laboratory data generally lead to velocities of longitudinal waves nearer 6 than 5.5 kilometers per second. The velocity 8 kilometers per second, approaching that of *Pn*, is found in dunite and other olivine-rich rocks which have often been supposed to originate at great depth.

Layers in California

In California, Gutenberg found four groups of waves which could be attributed to refraction; this led to assuming two intermediate layers. The last revised form of this interpretation, dated 1944, is represented in Figure 18-2. Although it represents the observational data fairly well, and epicenters were located satisfactorily with its aid, this scheme is now obsolete. It is given here for the benefit of the student who may encounter references to it. Hypocenters were taken to be at the level of the first discontinuity. The

Thickness	V	v
17½ km.	5.58 km./sec.	3.26
14½ km.	6.03	3.64
4 Km.	6.91	4.08
	8.0	4.4

FIGURE 18-2

Structure and seismic wave velocities, Southern California, as given by Gutenberg, 1944 (now obsolete).

thickness of the layers is that given for the coastal region of southern California; the upper two were not varied, but more to the east the third layer was taken as nearer 10 than 4 kilometers thick, and in the vicinity of the Sierra Nevada it was given a thickness of about 30 kilometers; this placed the Moho there at a depth of over 60 kilometers instead of 37 kilometers.

Root of the Sierra Nevada

Seismological evidence for this "root" of the Sierra Nevada was first given by Byerly. Investigation of seismograms for earthquakes in the California Coast Range showed that epicenters clearly too far west would result unless the arrivals of first motion at the stations in Owens Valley (Tinemaha and Haiwee) were being delayed by as much as 2 or 3 seconds. Such a delay would result if the continental rocks of the crust were to descend to relatively great depth under the Sierra Nevada, interposing rock through which the velocity would be appreciably lower than 8 kilometers per second. Pn would thus be delayed, or else it would be cut off, and the first arrival would be one of the later waves studied by Gutenberg. This existence of a thick "root" of continental rock under a great mountain system is consistent with those observations on gravity which lead to the doctrine of isostasy. The root of the Sierra Nevada has been confirmed by H. W. Oliver, using gravity readings at 1500 locations.

REFLECTED WAVES

The 1944 synthesis represented in Figure 18-2 did not depend simply on identification of refracted waves; there are strong reflections from the Moho and apparently from the next discontinuity above it. It has been mentioned in Chapter 4 that reflections from the Moho were observed for the Whittier earthquakes of 1929 (Fig. 4-6). Similar reflections, of both longitudinal and transverse waves, have been observed regularly at the southern California stations; a good example recorded in 1953 is shown in Figure 4-7. The facts were published repeatedly, so that workers at Pasadena were surprised to learn that certain distinguished seismologists were either unaware of the observations or unwilling to accept them; even in lately published discussions doubt is being raised as to the sharpness or even the reality of the Moho in California. Such remarks are sometimes documented by referring to the failure of Tuve and Tatel, working with artificial explosions in southern California in 1949, to find waves reflected from the Moho. This was literally a failure, and no more; not only are the earthquake records convincing, but in 1953 Dr. G. G. Shor also recorded reflections (though less sharply defined) of longitudinal waves originating in quarry blasts. The principal reflection arrived at recorders near the quarry about 10 seconds after detonation; a smaller reflection, about 1 second earlier, probably corresponds to a shallower discontinuity. Compare the discussion in Chapter 4.

REINTERPRETATION

Observed High Velocities

Since 1944 the hypothesis of several internal discontinuities in the crust has largely been abandoned. This revision was compelled by accumulating observations of the velocity of the first recorded P wave from artificial explosions, such as quarry blasts, timed accurately at the firing points. This velocity was persistently found to be near 6 kilometers per second rather than 5.5 kilometers per second. Lower velocities were, of course, found at very short distances, where the paths were usually through sedimentary or even unconsolidated material, but, with blasting strong enough to send waves to greater distances, the higher velocity almost invariably appeared. The first such observations were made in northern Germany, in a Mesozoic area; accordingly, they occasioned no concern, since laboratory data indicated that higher velocities might be expected in some sedimentary rocks than in granitic rock. But the first wave recorded at Pasadena from a blast in 1931 near Victorville, along a path largely through the crystalline rocks of the San Gabriel Mountains, arrived with a mean velocity near 6 kilometers per second. Hodgson derived nearly the same velocity for rockbursts in the area of the Canadian Shield. In 1949 a large quarry blast near Corona, southern California, recorded at numerous permanent and temporary stations, showed the 6 km./sec. velocity along many paths. Such results made it hard to appeal to any special local condition for explanation.

It is not easy to reconcile these data with the well-observed velocity of approximately 5.5 kilometers per second for $\bar{P}$. The naïve proposal that velocity decreases with depth, so that the higher value is observed for surface explosions and the lower for earthquakes, leads instantly to an absurdity. With such a velocity distribution, rays emerging toward the surface from earthquake hypocenters should curve downward, and the velocity of 5.5 kilometers per second could not be observed.

$\bar{P}$ is not p

Gutenberg cut the knot by deciding that $\bar{P}$ is not the direct wave from the hypocenter. The velocity of $\bar{P}$ is derived mainly from records at relatively large distances (over 150 kilometers), where it follows Pn. Few good data are available for determining directly the velocity of the P wave at distances less than 100 kilometers. However, in 1949 a closely spaced net of stations was set up in southern California surrounding an area where small shocks are frequent. In a few months readings for each of seven minor earthquakes were available at 6 to 8 stations within 120 kilometers of the epicenters. These data could be used to find the velocity of the first wave with the single assumption of constant velocity; the whole group of readings was found to fit best for a velocity near 6.34 kilometers per second. This was taken to be a mean velocity over the depth range between surface and hypocenter; the corresponding mean hypocentral depth was found to be 16 km. The direct wave at short distances is now denoted by p, and $\bar{P}$ is retained for the 5.5 kilometers per second wave.

Work with later records, notably those of the Kern County earthquakes of 1952, shows that this result fits well for most shocks recorded in southern California. For a few there is indication of depth nearer 10 than 16 kilometers, and at least in limited areas there is indication of a relatively low

FIGURE 18-3 *Effect of low velocity channels on wave propagation.* [*Gutenberg.*]

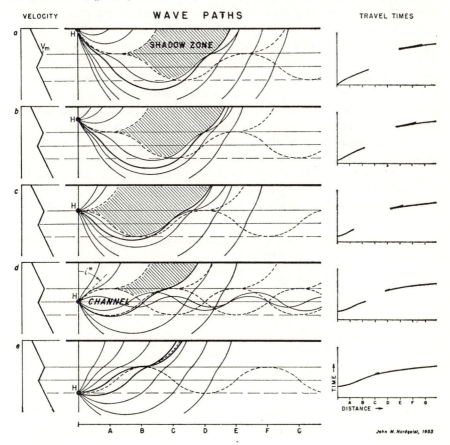

VELOCITY   WAVE PATHS   TRAVEL TIMES

John M. Nordquist, 1953

velocity near the surface, approaching 5 kilometers per second. (For short paths near the surface in the almost unweathered granite rock of the Yosemite region, a velocity of 5.25 kilometers per second was determined by using small explosions.) This low velocity is of no use in explaining $\bar{P}$.

$\bar{P}$ as a Guided Wave

Gutenberg regards $\bar{P}$ as a channel wave guided by a horizon of relatively low velocities within the crust (see Fig. 18-3) which he calls the lithosphere channel. Such a channel wave may travel with a velocity lower than the minimum velocity for body waves at the channel level, just as the velocity of Rayleigh waves is less than that of the corresponding body waves. The energy of a channel wave falls off with distance less rapidly than that of a body wave; this is in accordance with the extensive recording of $\bar{P}$, which should represent leakage of energy from the channel up to the surface. A low-velocity channel suggests a level of weakness within the crust, and this correlates well with the placing of earthquake hypocenters at that depth.

On this new basis little can be asserted about velocities or physical conditions between the lithosphere channel and the Moho. The wave originally called *Pm*, with a velocity near 7 kilometers per second, may really be refracted through a layer a few kilometers thick above the Moho, whose upper surface accounts for the earlier of two reflected waves; but its complexity on the records suggests that it, too, may be a guided wave.

Little reinterpretation of the Moho itself is required. The assumption of higher velocities above it necessitates bringing it nearer the surface, to a depth in southern California of about 29 rather than 37 kilometers. The interpretation of reflected waves is otherwise unaffected.

A tentative velocity-depth chart by Gutenberg is given as Fig. 18-4.

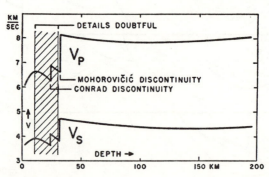

FIGURE 18-4   *Suggested velocity distribution in the upper 200 kilometers of the earth.* [*Gutenberg, 1955.*]

THE MOHO IN CONTINENTAL AREAS

Enough local samples of the Moho are now available to permit tentative generalization. Under what may be considered normal continental areas, the Moho is found by various methods at depths ranging from about 30 to 40 kilometers. (Tuve and Tatel find approximately the same level under the high Colorado Plateau.) At least under the higher and younger mountains it dips to something like 60 kilometers, as under the Sierra Nevada and perhaps the Alps.† Depths of 50–60 kilometers are also found under the Caucasus and in Central Asia (including the Tian Shan).

Under the Appalachians there is at present evidence of descent of the Moho only to about 45 kilometers. In most of eastern North America the discontinuity seems to be near the normal level. In the western Transvaal (South Africa) a depth of 35 kilometers has been found. In Alaska Tatel and Tuve report depths ranging from 31 to 34 kilometers.

A shallow level for the Moho in the continental borderland off the coast of southern California (Chapter 27) follows from a recent study by Frank Press of the phase velocity of Rayleigh waves reaching Pasadena and other California stations on paths crossing the Pacific; the crust offshore may be only half as thick as its "normal" value in the area immediately surrounding Pasadena. A shallow level for the Moho in New Zealand, near 20 kilometers, has been derived by Bullen from earthquake data and by Officer from blast records. A more complete publication by Eiby reports four layers with total thickness slightly over 18 kilometers. From recordings of atomic explosions the depth of the Moho in Central Australia is found to be 35 kilometers or slightly greater.

OCEANIC AREAS

Because the few good seismological stations in oceanic regions are isolated, it is difficult to use their records for analyzing local structure. Great progress has been made in the last few years by recording artificial explosions; most of these are small charges set off near the ocean surface and recorded on board distant vessels. Waves refracted through the underlying rock can be recognized, and their velocities determined to a good approximation.

In most oceanic areas thus far investigated, the Moho is found almost

---

† Present evidence, far from complete, suggests a distinction; apparently under the Alps the whole crust is thickened and downwarped, while under the Sierra Nevada the thickening affects only that part of the crust below the channels carrying $\bar{P}$ and Lg. If verified, this can be attributed to difference in later tectonic history in the two regions. Under the Tian Shan Gamburtsev found a "granitic" layer only 15–20 kilometers thick; the deeper layer extending to 60 kilometers he considered basaltic.

uniformly at about 10 to 11 kilometers below the ocean surface. Perhaps the best observations on the transition from continental to oceanic level were made off the eastern coast of the United States, where the Moho rises from about 30 to 15 kilometers in a horizontal distance of 200 kilometers, and then more gradually to the 10 kilometer level about 500 kilometers off-shore. Apparently the great oceanic trenches show an opposite anomaly; under the axis of the Tonga Trench the Moho descends to at least 20 kilometers below sea level (about 10 kilometers below the bottom of the trench), although on the east flank it is found at 12 kilometers. Raitt reports the Moho at depths near 6 kilometers below sea level at about 20 points in the Pacific basin, and near 12 kilometers at six others.

The mid-Atlantic Ridge (Chapter 25) is a submerged mountain range, and as such it partakes of continental character. There are not as yet adequate data to settle whether or not the Moho there descends to continental levels; less accessible ridges in the Indian and other oceans are probably of the same character.

Eaton reports the Moho as 30 kilometers deep under Hawaii.

SEISMOGRAMS AT SHORT DISTANCES

Instrumental Requirements

Comparatively little can be accomplished with seismograms belonging to the local earthquake range of distance (up to 1000 kilometers or 9°), unless recording is on an open time scale so that times can be read to the nearest tenth of a second. This is much more exacting for instruments and personnel than work with teleseisms, where timing to 1 second is generally sufficient.

The first motion of the P group, which is the most important single event to be timed, is often small, particularly in the horizontal components. Exact work thus calls for instruments of high magnification, especially in the vertical component, for periods in the range from 0.05 to 2 seconds prevailing in local earthquakes. When such instruments record the larger local shocks, motion after the first impulse is usually photographically underexposed or off scale. Additional seismographs operating at lower magnification are necessary elements in a complete program, especially if any work is to be done with the S group of waves.

If standard transit times for the principal recorded waves can be established in a given area, epicenters can often be located by routine methods with sufficient accuracy for many uses, although often not closely enough for important geological purposes. Setting up such standards, bitter experience has shown, calls for a large group of stations with accurate timing, constituting a network with average spacing not much over 20 kilometers, continuously maintained and further supplemented by additional emergency installations to record aftershocks and large artificial explosions. Such an extended effort is only practicable in a region at least as active as California, where earthquakes are frequent enough to yield results in a limited number of years. The following discussion is based principally on work in California; comparable experience elsewhere may yet lead to different methods and different results.

Interpretation

As with teleseisms, problems of interpretation arise in both preliminary and final form. The work for a net of stations may be done chiefly at the headquarters station. For urgent immediate purposes after a large earthquake, preliminary determinations must often be made without waiting for other data. This is done not only for the benefit of the press and the public; prompt decision must be reached in sending out field investigators and emergency recording equipment to avoid loss of valuable data.

For preliminary purposes, distance is estimated from the time interval between P and S groups, with as much attention to detail as is possible in haste. For large earthquakes, sensitive instruments record only the time and direction of first motion, and then go off scale. Missing details, such as the time of S, must be supplied from seismograms of low-magnification (strong-motion) instruments or from the recording of aftershocks.

Direction can be estimated from the azimuth of first motion, as for teleseisms. Here considerable error may result, since horizontal refraction occurs in regions of complicated structure. Thus at Pasadena an apparent direction near northeast has often been found for epicenters which were actually due east or even a little south of east, in the area including Riverside. This suggests a relatively high velocity under the mountains north of the direct path. Sassa found horizontal deflections of seismic rays of as much as 40° in the vicinity of the volcano Aso and correlated them with great fractures.

Even for small shocks which constitute no emergency, preliminary interpretation is usually carried out at the chief station without waiting for the records from others. For obvious scientific reasons it is best to measure the seismograms at several stations independently, and only then correlate them to detect errors of interpretation and measurement. The final correlation usually occurs during the process of using all the readings to locate the epicenter, when any serious discrepancies should lead to re-examination of the original seismograms.

Routine Procedure

Where hundreds or thousands of seismograms are being measured annually it is necessary to set some reasonable limit to the amount of detail considered. Epicenter location and research work are usually hampered by the entry of too many unidentified readings on the card file or ledger. The information most regularly needed includes times for:

(1) The first motion recorded
(2) Any large, clear phase in the P motion
(3) Phases between P and S likely to be identifiable as reflected waves
(4) The earliest identifiable S wave and any clear large phases shortly after it
(5) Anything outstandingly large or unusual

The record should show clearly which phases are large or small, and which are sharp or gradually emergent. Amplitudes should be read sufficiently for assigning magnitudes. In general P phases are most clearly recorded in the vertical component and S phases in the horizontal component, provided that the vertical-component instrument is comparable in response and magnification with the others (which is far from the case at many stations).

Appearance of Seismograms

Even for the California region, no generally applicable description of appearance of seismograms at various distances can be offered, as was done on

FIGURE 18-5   *Seismograms at short distance, showing a large false S phase.*

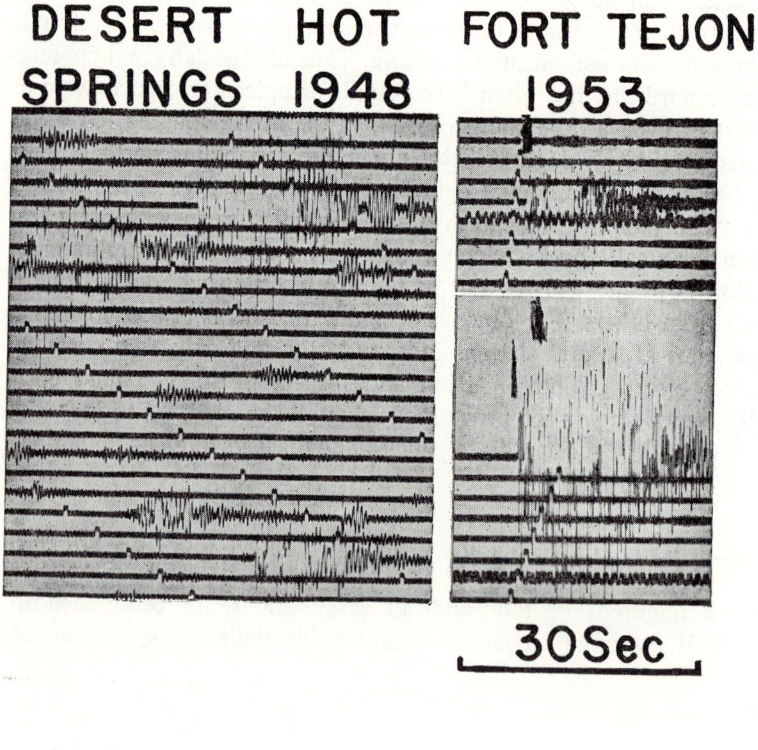

earlier pages for teleseisms. There are outstanding local differences; one is the effect of the Sierra Nevada. Another is that earthquakes originating in the Coast Range structures give seismograms at Pasadena which are less clear, involving a great many more small arrivals, than those from shocks in the eastern area including the Mojave Desert. Imperial Valley shocks are sometimes hard to work with because of their tendency to come in swarms. What here follows should be taken only as a rough guide and example.

At distances less than 100 kilometers the beginnings of the P and S groups are often quite sharp, although the sharp beginning of S on the horizontal components may be measurable earlier than what looks like the same wave in the vertical component. At very short distances there is a pitfall which has been mentioned in other chapters: an apparent S following P after 1 to 1.5 seconds may not be the true S phase (Fig. 18-5). Sometimes the genuine S can be seen following a few seconds later. Occasionally the short

FIGURE 18-6   *Long Beach aftershock seismogram at Pasadena.*

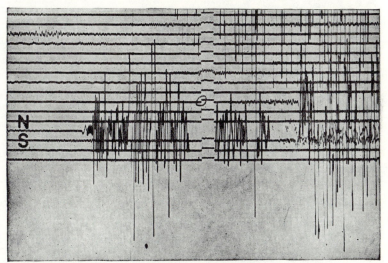

Pasadena 1933 March 11

Torsion seismograph($T_0 = 0.8$)

showing aftershocks of
Long Beach earthquake

1m

S-P interval is consistent with other data for the same shock, which is then to be considered unusually shallow. The ambiguous interpretation makes it unwise to locate epicenters on the basis of small S-P intervals alone.

Up to about 100 kilometers the worker should look for sharp phases between the P and S groups, especially in the vertical component. These may be reflected waves, and if they fit standard transit times they help fix the distance. The S phase may be followed by a corresponding reflection, sometimes very sharp. Some seismograms at short distances show peculiarities due to local structures; thus, at Pasadena in 1933 and thereafter, shocks originating near the epicenter of the principal Long Beach earthquake could be recognized by four sharp phases arriving 3, 6, 9, and 12 seconds after P (Fig. 18-6).

FIGURE 18-7   *Typical seismograms of local shocks.*

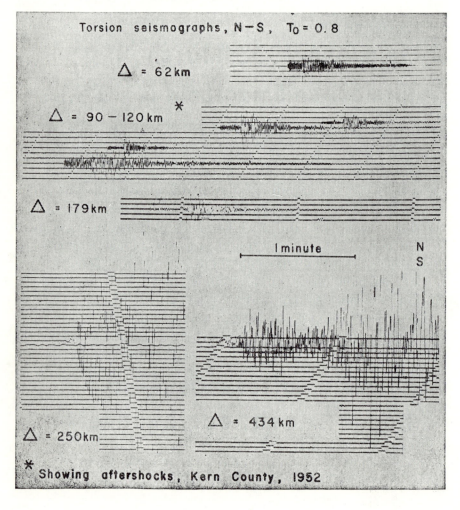

Beyond 100 kilometers in southern California the amplitude of the first motion decreases, and the small initial P is followed by a larger wave. This looks like Pn followed by $\bar{P}$, but analysis of recorded times indicates that the small first motion is still the direct wave from the hypocenter, while the later large motion, whether by coincidence or not, often fits the calculated time for Pn. Near 140 kilometers Pn becomes the first motion, and it is followed by one or more large phases of the P group, which often are not the same wave at stations recording the same shock at different distances. The corresponding complications in the S group can be traced to some extent but are harder to follow because the initial complication of S is obscured by earlier motion, and because the various possible polarizations often make the three components very different in detail.

$\bar{P}$, with a velocity between 5.5 and 5.6 kilometers per second, is best observed at distances from 200 to 300 kilometers, along with the corresponding $\bar{S}$.

At distances of several hundred kilometers seismograms are difficult to interpret precisely; the beginnings of both P and S are often lost or hard to identify (see Fig. 18-7).

SEISMIC PROSPECTING

Prospecting for oil and minerals by geophysical methods is a large and rapidly changing subject. Commercial secrecy often delays publication of new techniques until they are beginning to be obsolete. However, the two fundamental procedures for prospecting by the seismic method can be described in a few words.

Seismic waves are generated by the detonation at or near the surface of the earth of a small amount of explosive, sometimes only a few sticks of dynamite, and registered by seismographs of special design, called *geophones,* with high magnification for short-period waves and recording speeds great enough so that times can be read to the thousandth of a second.

In *refraction shooting* the geophones are spaced, if possible, along a straight line from the shot point to establish a time-distance plot for the first recorded wave. Ideally this plot can be represented by a series of straight-line segments with slopes successively correponding to the velocities in individual layers.

Given the distribution of velocity among the various layers, as determined by refraction shooting, the levels of these layers can be found by *reflection shooting.* Geophones over a wide range of distances from the shot point are made to record in parallel on a single tape or sheet. Reflections from deep levels arrive nearly simultaneously at all the geophones and can be picked out by the alignment of similar motion across the record. From the given velocities the depths of reflection can be computed.

These simple procedures are greatly complicated by the effects of dip and of irregularities in subsurface structure.

References

FOR REFERENCES on artificial explosions and on rockbursts see Chapter 12; for waves reflected from the Moho see Chapter 4.

Theory, etc.

Jeffreys, H., "On compressional waves in two superposed layers," *Proc. Cambridge Phil. Soc.,* vol. 23 (1926), pp. 472–481.

Muskat, M., "Theory of refraction shooting," *Physics,* vol. 4 (1933), pp. 14–28.

von Schmidt, O., "Über Knallwellenausbreitung in Flüssigkeiten und festen Körpern," *Zeitschr. tech. Physik,* vol. 19 (1938), pp. 554–561. (Contributes important experimental material.)

Joos, G., and Teltow, T., "Zur Deutung der Knallwellenausbreitung an der Trennschicht zweier Medien," *Phys. Zeitschr.,* vol. 40 (1939), pp. 289–293.

Cagniard, L., *Réflection et Réfraction des Ondes séismiques progressives,* Gauthier-Villars, Paris, 1939. (Also bears on the theory of surface waves. Mathematically original and extremely hard to read.)

Dix, C. H., "The method of Cagniard in seismic pulse problems," *Geophysics,* vol. 19 (1954), pp. 722–738. (An introduction to the preceding reference.)

Sandner, A., *Das Problem der seismischen Grenzschichtwelle bei der Behandlung der Wellengleichungen,* Freiberger Forschungshefte, C 26. Akademie-Verlag, Berlin, 1956. (Reviews Cagniard's method and correlates it with Sommerfeld's.)

The Moho and crustal structure; general

Mohorovičić, A., "Das Beben vom 8 X 1909," *Jahrbuch des meteorologischen Observatoriums in Zagreb (Agram) für das Jahr 1909,* vol. 9 (1910), Part 4, pp. 1–63. (The fundamental paper. Publications in this field are numerous; the following include some of the earlier and some of the most recent.)

Gutenberg, B., "Die mitteleuropäischen Beben vom 16. November 1911 und vom 20. Juli 1913; 1. Bearbeitung der instrumentellen Aufzeichnungen," *Veröffentlichungen des Zentralbureaus der internationalen seismologischen Assoziation,* Strassburg, 1915.

———, "Der Aufbau der Erdkruste," *Zeitschr. Geophysik,* vol. 3 (1927), pp. 371–377.

———, "Der Aufbau der Erdkruste in Europa," *Geol. Rundschau,* vol. 19 (1928), pp. 433–439.

Jeffreys, H., "On near earthquakes," *M.N.R.A.S. Geophys. Suppl.,* vol. 1, pp. 385–402, 1926.

———, "A rediscussion of some near earthquakes," *ibid.,* vol. 3, pp. 131–156, 1933.

———, "A further study of near earthquakes," *ibid.,* vol. 4, pp. 196–255, 1937.

———, "The times of *P* up to 30°," *ibid.,* vol. 6, pp. 348–364, 1952.

———, "The times of *P* in Japanese and European earthquakes," *ibid.,* vol. 6, pp. 557–565, 1954.

Ewing, M., and Press, F., "Geophysical contrasts between continents and ocean basins," *Geol. Soc. Amer., Spec. Paper* 62, 1955, pp. 1–6.

———, "Structure of the earth's crust," in *Handbuch der Physik,* Vol. 47, *Geophysik I,* 1956, pp. 246–257.

Raitt, R. W., "Seismic refraction studies of the Pacific ocean basin. Part I: Crustal thickness of the central equatorial Pacific," *Bull. Geol. Soc. Amer.,* vol. 67 (1956), pp. 1623–1640.

Gutenberg, B., "Crustal layers of the continents and oceans," *Bull. Geol. Soc. Amer.,* vol. 62 (1951), pp. 427–440.

———, "Channel waves in the earth's crust," *Geophysics,* vol. 20 (1955), pp. 283–294.

Press, F., Ewing, M., and Oliver, J., "Crustal structure and surface-wave dispersion in Africa," *B.S.S.A.,* vol. 46 (1956), pp. 97–104.

Byerly, P., "Subcontinental structure in the light of seismological evidence," in *Advances in Geophysics,* vol. 3, Academic Press, New York, 1956, pp. 105–152. (A very good survey of the literature.)

Gutenberg, B., "Wave velocities in the earth's crust," *Geol. Soc. Amer., Spec. Paper* 62 (1955), pp. 19–34.

Roots of mountains

Byerly, P., "Comment on 'The Sierra Nevada in the light of isostasy,' by Andrew C. Lawson," *Bull. Geol. Soc. Amer.,* vol. 48 (1938), pp. 2025–2031.

Gutenberg, B., "Seismological evidence for roots of mountains," *Bull. Geol. Soc. Amer.,* vol. 54 (1943), pp. 473–498, 3 pls.

———, "Zur Frage der Gebirgswurzeln," *Geol. Rundschau,* vol. 46 (1957), pp. 30–38.

Oliver, H. W., "Isostatic compensations for the Sierra Nevada, California," abstract. *Bull. Geol. Soc. Amer.,* vol. 67 (1956), p. 1724.

Southern California; structure and wave velocities

Gutenberg, B., "Earthquakes and structure in southern California," *Bull. Geol. Soc. Amer.,* vol. 54 (1943), pp. 499–526, 1 pl.

———, "Travel times of principal *P* and *S* waves over small distances in southern California," *B.S.S.A.,* vol. 34 (1944), pp. 13–32.

————, "Reflected and minor phases in records of near-by earthquakes in southern California," *ibid.,* pp. 137–159. (The three preceding papers present the former synthesis, revised in the following three.)

————, "Travel times from blasts in southern California," *ibid.,* vol. 41 (1951), pp. 5–12.

————, "Revised travel times in southern California," *ibid.,* pp. 143–163.

Richter, C. F., "Velocities of *P* at short distances," *ibid.,* vol. 40 (1950), pp. 281–289.

Press, F., "Determination of crustal structure from phase velocity of Rayleigh waves. Part I: Southern California," *Bull. Geol. Soc. Amer.,* vol. 67 (1956), pp. 1647–1658.

————, "Velocity of *Lg* waves in California," *Trans. Am. Geophys. Union,* vol. 37 (1956), pp. 615–618.

New Zealand

Hayes, R. C., "Seismic waves and crustal structure in the New Zealand region," *Dominion Observatory (Wellington) Bull.* No. 101 (1936).

Bullen, K. E., "On near earthquakes in the vicinity of New Zealand," *N. Z. Journ.,* vol. 18 (1936), pp. 493–507.

————, "Note on New Zealand crustal structure," *Trans. Royal Soc. N. Z.,* vol. 82 (1955), Part 5, pp. 995–999.

Officer, C. B., "Southwest Pacific crustal structure," *Trans. Am. Geophys. Union,* vol. 36 (1955), pp. 449–459.

Eiby, G. A., and Dibble, R. R., "Crustal structure project," N. Z. Dept. of Sci. and Industrial Research, Geophysical Memoir No. 5, 1957.

Africa

Willmore, P. L., Hales, A. L., and Gane, P. G., "A seismic investigation of crustal structure in the Western Transvaal," *B.S.S.A.,* vol. 42 (1952), pp. 53–80.

Gane, P. G., Atkins, A. R., Sellschop, J. P. F., and Seligman, P., "Crustal structure in the Transvaal," *ibid.,* vol. 46 (1956), pp. 293–316.

Press, F., Ewing, M., and Oliver, J., "Crustal structure and surface-wave dispersion in Africa," *ibid.,* pp. 97–104.

Australia and Hawaii

Doyle, H. A., "Seismic recordings of atomic explosions in Australia," *Nature,* vol. 180 (1957), pp. 132–134.

Eaton, J. P., "Seismometric results from recent Hawaiian earthquakes" (abstract), *Bull. Geol. Soc. Amer.,* vol. 68 (1957), p. 1853.

Eastern North America, and Alaska

Hodgson, J. H., "A seismic survey in the Canadian Shield," *Publ. Dominion Observatory*, Ottawa, vol. 16 (1953), pp. 113–163, 169–181.

Katz, S., "Seismic study of crustal structure in Pennsylvania and New York," *B.S.S.A.*, vol. 45 (1955), pp. 303–325.

Tatel, H. E., Adams, L. H., and Tuve, M. A., "Studies of the earth's crust using waves from explosions," *Proc. Amer. Phil. Soc.*, vol. 97 (1953), pp. 658–669.

Tatel, H. E., and Tuve, M. A., "Seismic crustal measurements in Alaska," *Trans. Am. Geophys. Union*, vol. 37 (1956), p. 360 (abstract).

Caucasus and Central Asia

Rozova, E., "Contribution to the question of the structure of the earth's crust in Central Asia" (in Russian), Trudy Seys. Inst. Akad. Nauk SSSR, No. 94 (1939), pp. 1–15.

———, "Contribution to the question of the deep-seated structure of the Caucasus" (in Russian), *ibid.* pp. 16–34.

Gamburtsev, G. A., and Veytsman, P. S., "Sopostostavlenie dannikh glubinnogo seysmicheskogo zondirovaniya o stroenii zemloy kori v rayone severnogo Tyan-Shanya s dannimi seysmologii i gravimetrii," Izvestiya Akad. Nauk SSSR, ser. geofiz. (1956), pp. 1035–1043.

Italian region

Caloi, P., "Struttura geologica-sismica dell' Europa centro-meridionale . . . ," *Ann. geofisica*, vol. 5 (1952), pp. 507–518.

di Filippo, D., and Marcelli, L., "Struttura della crosta terrestre in corrispondenza dell' Italia centrale," *ibid.*, pp. 569–579.

Horizontal refraction

Sassa, K., "Anomalous deflection of seismic rays in volcanic districts. Geophysical studies on the volcano Aso, Part 3," *Mem. Kyoto Imp. Univ. Coll. Science*, Ser. A, vol. 19 (1936), pp. 65–78.

Seismic prospecting

Ewing, M., and Press, F., "Seismic prospecting," *Handbuch der Physik*, vol. 47, *Geophysik I*, pp. 153–168.

Dix, C. H., *Seismic Prospecting for Oil*, Harper, New York, 1952.

CHAPTER 19

Deep-Focus Earthquakes

RECOGNITION OF EARTHQUAKES originating at great depths is a late development in seismology; consequently the discussion can be presented in one chapter.

EARLY OBSERVATIONS AND THEORY

Mallet and many of his contemporaries were convinced that ordinary earthquakes originate at moderate depths. They reasoned from the observation that the meizoseismal area is relatively small and apparent intensity falls off rapidly outside it. This can be accounted for on extremely simple assumptions—constant velocity, straight rays, disturbance falling off according to an inverse square law—if the hypocentral depth is of the same order as the radius of the meizoseismal area, say 15 to 30 kilometers or 10 to 20 miles.

Extreme instrumental demands must be met to determine depths accurately; but many microseismic methods give depths to order of magnitude, and these tend to confirm the rough conclusions from macroseismic data. Moreover, the Mohorovičić discontinuity is placed by all evidence at depths from about 30 to 60 kilometers, and most ordinary earthquakes appear to originate above it.

Geological and geophysical theories, in harmony with the observations bearing on isostasy, indicate that at depths near 100 kilometers the rocks yield plastically; therefore large, slowly accumulating strains should be impossible. It was long held that earthquakes at great depth, if they occur at all, must be fundamentally explosive in nature, involving sudden changes of volume, rather than the result of fracture involving shear. Such earthquakes were called "crypto-volcanic" earthquakes. (A later name suggested for deep shocks, "plutonic" earthquakes, which is otherwise appropriate and attractive, remains little used because of the implied association with igneous processes.) Even as late as 1949 a paper based on experimental laboratory work suggested that deep earthquakes "more probably" result from abrupt volume changes (see the section on mechanism).

300

DISCOVERY

Turner's Work and Early P'

Deep-focus earthquakes were discovered by H. H. Turner in 1922 in the course of editing the *International Seismological Summary*. By that time it was generally accepted that ordinary earthquakes were of shallow origin; for such earthquakes standard transit-time tables had been constructed, and were regularly used by Turner. He noted that for a comparatively few earthquakes, some of which were large and well observed, the arrival of P' ($= PKP$) at distant stations was abnormally early relative to that of P at short distances; if origin times were calculated from S-P, the recorded P' was sometimes a minute or more earlier than expected. Deep focus readily explains this, since in that case the transit time of P to short epicentral distances should increase, while that of P' should decrease.

In 1922 the standards of timekeeping and of instrumental magnification were much lower than in, say, 1932, but repeated discrepancies of the order of a minute, especially when the observed times were early, were too large for ordinary error of timing and measurement. However, Turner was working with reported readings, not with original seismograms; those not accepting his results emphasized the possibility of outright mistakes—like a wrong count of minutes on a seismogram—or they exaggerated the imperfections of the standard transit time tables.

"High Focus"

Unfortunately, Turner also reported instances of "high focus" in which the arrival of P' was relatively late; he placed certain earthquakes at a level 200 kilometers above that of the standard shock represented by the tables. This would make ordinary earthquakes originate at depths of at least 200 kilometers, a conclusion almost unacceptable. Later special investigations by several seismologists showed that at least some of the "high" shocks were multiple events; a small earlier shock was registered at the nearer stations, while the recorded P' referred to a following larger earthquake. It is now generally agreed that the ordinary transit-time tables represent earthquakes originating at depths of the order of 25 kilometers, and that "high focus" is largely illusory.

On the other hand, deep focus could not be explained away. Three papers by Wadati, Scrase, and Stechschulte presented evidence that removed all reasonable doubt.

Wadati

The paper by K. Wadati (1928) referred to shallow and deep earthquakes in and near Japan. Taking advantage of the large number of recording stations there, he drew charts representing the increase of the time interval *S-P* and of the time of arrival of *P* with increasing epicentral distance. Figure 19-1 shows the appearance of these charts for a shallow shock (the North Tajima earthquake of 1925; Chapter 30) and a deep shock in 1927 with nearly the same epicenter. For the shallow shock *S-P* decreases to a

FIGURE 19-1   *Curves of equal* S-P *and of equal time of arrival of* P; *shallow and deep shocks in Japan.* [*Wadati.*]

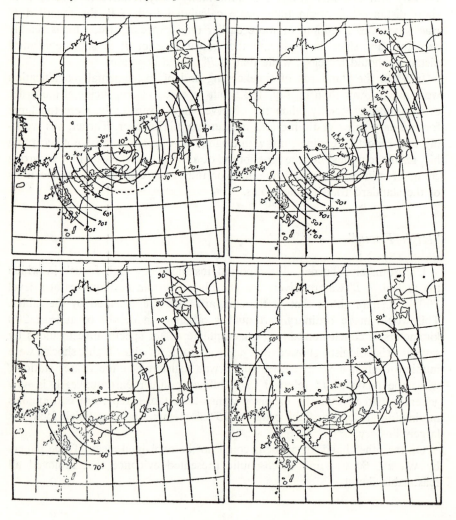

few seconds near the epicenter, and increases rapidly with distance. For the deep shock it is nowhere less than 40 seconds, and it increases more slowly with distance, as the wider spacing of successive curves shows. This difference in spacing also appears in the curves for equal times of *P*; these are *co-seismal lines*.† As one moves away from the epicenter, direct distance increases more slowly for a deep source than for a shallow one; the curvature of the rays due to increasing velocity with depth enhances this effect. To reject Wadati's conclusions one would have to reject all the timing of *P* on the one hand, and all the identifications of *S* on the other. This is awkward, since, as Wadati showed by publishing seismograms, large deep shocks are characterized by particularly sharp and clear *P* and *S* phases.

Macroseismic perceptibility of deep earthquakes shows decided but easily understood peculiarities. Wadati remarks that, although some of these earthquakes, as indicated by their seismograms, were of larger magnitude than many destructive shallow shocks, they were nowhere of high intensity, although often reported felt at points quite far from each other and from the epicenter. In such cases isoseismals are nearly impossible to draw; the areas of perceptibility are isolated patches, which are often nearly the same for shocks with quite different epicenters and show a distribution more or less related to the local differences in crustal structure. Wadati comments especially on the earthquake of January 21, 1906, which had an epicenter just off the Pacific coast of Honshu, a depth of 340 kilometers, and one of the largest magnitudes (8.4) ever observed for a deep shock. It was felt strongly in many parts of Japan. Both Japanese and European seismologists tried unsuccessfully to reconcile the microseismic data with each other or with the macroseismic facts. Irregularity in surface distribution of intensity is evident for shocks at depths of the order of 60 kilometers; it accounts in part for the difficulty of relating intensity distributions to instrumental epicenters, in Chile for example.

Scrase and Stechschulte

F. J. Scrase and V. C. Stechschulte took the course of collecting seismograms for some of Turner's instances of deep focus; their results were published at about the same time in 1931. Scrase worked with a limited number of seismograms for each of several earthquakes, Stechschulte with a collection of seismograms from 69 stations, and reported times from many others, for a single earthquake. Not long afterwards Scrase also published a study from all available data for a single deep shock. The results of the two investigators were in excellent agreement. Deep-focus earthquakes were con-

---

† The drawing of supposed coseismal lines, connecting points at which the earthquake disturbance arrived simultaneously, was a favorite device of the elder seismologists. The observations were usually quite inaccurate, and many odd conclusions as to velocities and epicenters were derived in this way.

firmed, the characteristics of their seismograms described, and corrections to the standard transit times to allow for deep focus were worked through on lines already pioneered by Turner.

Once the main facts were determined, investigation of deep earthquakes swiftly reached a solid footing. All the necessary procedures were ready for application. Transit times could be calculated theoretically from the velocity-depth distributions already worked out from the data of shallow earthquakes. Indeed, because of the general sharpness of recording of deep shocks, their recorded times could be used to refine the standard data. Within a few years after 1931 this branch of investigation had accounted for the observations almost disappointingly well; no new major problems were raised for study.

MECHANISM

The cause of deep earthquakes, on the other hand, has been vigorously debated since about 1932. It soon became clear that it would be necessary to revise our ideas of the physical properties of material at depths of a few hundred kilometers. The hypothesis of an explosive type of source, associated with an abrupt change of volume, was soon ruled out by three lines of evidence.

FIGURE 19-2 *Earthquake of September 21, 1931 in Japan. Distribution of initial compressions* (u) *and dilatations* (d). [*Honda.*]

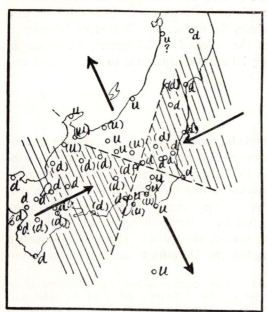

(1) Deep-focus teleseisms regularly record with large *S* waves having amplitudes at least as large, relative to those of the *P* group, as on seismograms of shallow shocks. This is quite contrary to expectation for an explosive source, as is verified by seismograms of large artificial explosions (on which *S* often cannot even be identified).

(2) Compression and dilatation in the *P* phase show a quadrantal distribution in azimuth irreconcilable with an explosive source, but easily explained in terms of shearing fracture. In Japan

(Figs. 19-2, 19-2A), where stations are often well distributed about the epicenter, some of the plots showing this effect also exhibit a circle of reversal from compression or dilatation inside to their opposites outside, as theoretically should be expected for a deep shock at the distance where the ray leaving the hypocenter horizontally reaches the surface.

(3) Either compression or dilatation tends to be recorded regularly at a given station for teleseisms originating in a given region. For deep and shallow shocks with epicenters in the same area, the first motion of both is ordinarily of the same type; thus at Pasadena earthquakes at all depths in and near the Andes in Chile and Argentina usually record with initial dilatation. Such observations point toward persistent mechanical coherence between the processes producing deep and shallow shocks in the same region.

All this evidence indicates that deep-focus earthquakes originate in a process involving shear and elastic rebound and of the same general nature as that causing shallow shocks. The problem of plastic flow at great depth then must be faced; the apparent contradiction is resolved by appealing to a time parameter. Slowly accumulating strains will be relieved by flow before they can arrive at fracture, but rapidly accumulating strains may progress until fracture is reached. The behavior may be compared with that of a block of wax, which flows gradually under pressure or even under its own weight but fractures sharply if struck with a hammer.

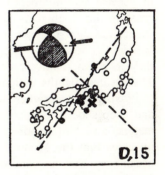

FIGURE 19-2A *Distribution of compressions (solid circles) and dilatations (open circles) in Japanese deep-focus earthquakes.* [*Honda and Masatuka.*]

CLASSIFICATION

Turner referred to earthquakes which fit the standard time tables as "normal," and others have followed this usage. Wadati in 1928 divided Japanese earthquakes into three groups: shallow, intermediate, and deep. The boundary between the intermediate and deep classes he set at 300 kilometers, since epicenters of earthquakes originating below this level showed a geographical distribution distinct from the rest. Gutenberg and Richter have generally followed Wadati's classification, because in some regions, such as the west coast of South America and the Marianas Islands, ordinary shallow shocks are the exception, and the "normal" depth of earthquakes there is about 50 to 60 kilometers. Thus the general term "deep-focus earthquake" includes the two classes, intermediate and deep (in the restricted sense).

The level of separation near 300 kilometers is probably not the same everywhere. In some areas foci slightly deeper than 300 kilometers clearly belong to the belts of intermediate shocks, while elsewhere shocks somewhat shallower than 300 kilometers are geographically associated with the deep class.

In several regions hypocenters extend to 650 kilometers or deeper. While in general the number and magnitude of deep shocks decrease with increasing depth, there is no gradual tapering off such as might lead one to expect occasional deeper hypocenters. In practically every active zone there is a lowest level for which several large shocks are on record; apparently they locate a relatively definite "floor" for the active part of the mantle. The greatest depths assigned with any confidence are near 720 kilometers.

Provisionally a rather arbitrary boundary has been set between shallow and deep shocks. Gutenberg and Richter apply "shallow" down to 60 kilometers, and "intermediate" from 70 to 300 kilometers. Present evidence indicates that the Moho is a level of discontinuity in all parts of the earth, oceanic as well as continental, although its depth varies widely. In some future time, when we have reasonably good information for all parts of the earth, it should be possible to divide all earthquakes into shallow, originating above the Moho, and deep-focus, originating below it.

The discovery that deep-focus earthquakes originate in well-defined zones and belts launched a general restudy of earthquake geography; some of the results are set forth in Part Two.

CHARACTERISTICS OF SEISMOGRAMS

Small Surface Waves

The most obvious mark of the seismogram of a large deep teleseism is the small amplitude, or even the absence, of surface waves. P and S may record with amplitudes which, for a shallow earthquake, would be followed by a long train of large L and M waves; yet no such waves are found.

Theoretically this is to be expected. From such theories as those of Rayleigh and Love it follows that, if the hypocenter of a shock of given energy is displaced to greater depth H, the amplitude of surface waves should fall off proportionally to $\exp(-cH/\lambda)$, where λ is the wave length and c is constant. This was once used as an argument against Turner's discovery of 1922, on the plea that all large earthquakes show surface waves. Byerly then pointed out that occasionally there were well-recorded large shocks with small surface waves; some such seismograms had even been reproduced in publication. In fact, the circumstance was noted by Angenheister as early as 1905, and Zoeppritz soon after suggested that these earthquakes might have an exceptional cause, such as unusually rapid tectonic displacement (he did not suggest deep focus). The matter would probably have come to general attention sooner if workers at many stations had not formed the habit of first searching their seismograms for surface waves, and only then picking up P and S; earthquakes represented only by P without surface waves might be overlooked, or taken for accidental disturbances, or dismissed as small and unworthy of attention.

Persistence of G and R

As focal depth increases, G and R waves of very long period decrease much more slowly than the ordinary M waves with periods of 20 seconds (because H/λ is smaller, the exponential decrease with depth is less rapid). Intermediate shocks frequently show one clear G or R wave standing out on the seismogram with no evidence of following surface waves. In view of the absence of G waves in many shallow shocks of moderate magnitude, this indicates that at least in these intermediate earthquakes the generating process is extensive, involving the displacement of large blocks. Occasionally intermediate shocks, and rarely even deep shocks, show the shorter-period surface waves with measurable amplitudes when they "ought" to be absent; this may be explained in terms of faulting extending from the deep focus up toward the surface, or perhaps in terms of the setting off of another shock at shallow depth.

The waves SS, SSS, and sSS often show very long periods on these seismograms, and they must be distinguished from surface waves with care.

Reduplicated Phases

Interpreting seismograms of deep teleseisms depends largely on a characteristic reduplication of the principal phases, caused by additional reflections from the surface of the earth which are not readily observed for shallow shocks. This effect could occur even if velocity were constant within the earth (Fig. 19-3). From a given source below the surface there would then be three reflected rays reaching a given point of observation at the surface; as indicated, these correspond

FIGURE 19-3 *Paths of pP, PP, and P'P' in a sphere of constant velocity.*

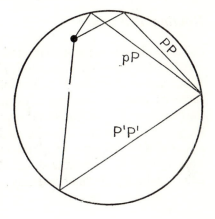

for Δ near 60° to *PP*, *P′P′*, and the deep-focus characteristic phase *pP*, which is reflected from a point relatively near the epicenter. The increase of velocity with depth in the actual earth only modifies this in detail. Students familiar with Fermat's principle "of least time" will find it interesting to prove that in the sphere with constant velocity the "ordinary" reflection *PP* represents a maximum time, while the actual minimum is given by *pP*. Jeffreys has correlated this with the fact that *pP* is usually sharp and clear, while *PP* is often less sharp and may be complex; although in actual recording there are numerous exceptions to both rules, the distinction applies in general. The variation of velocity with depth complicates the theory without changing its main result.

This wave *pP* arrives at a distant station following *P* by an interval which changes slowly with distance but is a good measure of depth; for the deepest known shocks this interval is over 2 minutes. The related phase *sP* corresponds to *pP* as *SP* to *PP*; the upward ray to the point of reflection represents a transverse wave and is therefore even steeper than that for *pP*.

Since the reflections of *pP* and *sP* take place near the epicenter, the coefficient of reflection may be expected to be influenced by local structure in the vicinity. Shocks at depths of 60 to 80 kilometers under Japan often record at distant stations with *pP* or *sP* so much larger than *P* that the first wave is overlooked in measurement. In other instances the succession *P*, *pP*, *sP* is often so sharp as to give the impression of three independent shocks. Since a similar repetition is possible for all the chief phases of the seismogram, so that there may be *S*, *pS*, *sS* or *PP*, *pPP*, *sPP*, deep-focus shocks have often been misinterpreted as multiple events (double if only *pP* or *sP* was noted, triple if both were considered). In general, both these reflections and the principal phases tend to be more sharply recorded than the phases on seismograms of shallow shocks of equal magnitude.

INTERPRETATION OF SEISMOGRAMS

Routine Procedure

The problems of preliminary and revised interpretation for recorded deep-focus earthquakes arise in about the same way as for shallow teleseisms (discussed in Chapter 17). Complication and difficulty are increased because an additional unknown quantity, the depth of focus, must be fixed. Furthermore, the lack of surface waves deprives the seismologist of a valuable guide to epicentral distance and exposes him to the risk of spectacular mistakes such as only carelessness and inattention could bring about in working with large shallow shocks. The writer has several such misinterpretations painfully on his conscience.

A file of seismograms for hypocenters at known distance and depth is of great assistance, especially in preliminary work. Because of the rarity of deep shocks compared to shallow ones, and of the great effect of different depth, such a file accumulates rather slowly.

Even the most preliminary interpretation demands some means for determining the depth, at least approximately. If pP can be identified, this is more than adequate; but misinterpretations are hard to avoid. The characteristic pattern of successive P, pP, sP, in which the second interval is somewhat less than the first, can often be recognized with fair probability (it helps to notice that sP is more likely than pP to be clear in the horizontal components). More commonly there is only one sharp phase in the minute or two following P; this may be pP, sP, PcP, or PP (the identification of PcP as pP is the cause of many errors). Or the distance may be unexpectedly large, and the two sharp phases at the beginning of the seismogram may be P_1' and P_2' or one of many other combinations. Tables for determining depth from pP-P and sP-P (or pP'-P' and sP'-P') are given in Appendix VIII.

When a depth has been selected, the determination of distance (and interpretation of the various phases) can be carried out by the same method of transit-time chart and movable strip used for shallow teleseisms. It is necessary to have separate transit-time charts for depths at intervals not wider than 100 kilometers down to at least 700 kilometers. Charts for 150 kilometers, 250 kilometers and so on are also useful. At Pasadena tables and a chart have been constructed for 60 kilometers instead of 50 kilometers. Charts for a fixed distance with variable depth can be constructed, but they are more confusing than helpful to most workers. Special charts allowing for both distance and depth have been made up in various ways; but they nearly always assume positive identification of pP, or S, or both, and often this is the information that is lacking in first-day work. Further notes on this point will be found in Appendix VIII.

For routine work in preparing preliminary bulletins, it is helpful if the seismograms or even the readings of a closely spaced group of stations can be compared. Astonishing variations in the clearness of pP are often found, so that identification and measurement may be uncertain at one station and perfectly clear at another. Knowledge of the times of several stations at progressively increasing epicentral distances helps to separate pP and sP from other phases—especially from PcP, since with increasing distance the interval PcP-P decreases rapidly, while pP-P increases slowly.

Appearance at Various Distances

No such orderly description of the appearance of seismograms at various distances can be given as was done for shallow shocks in Chapter 17. At any one station, Pasadena for example, the distances and depths from which

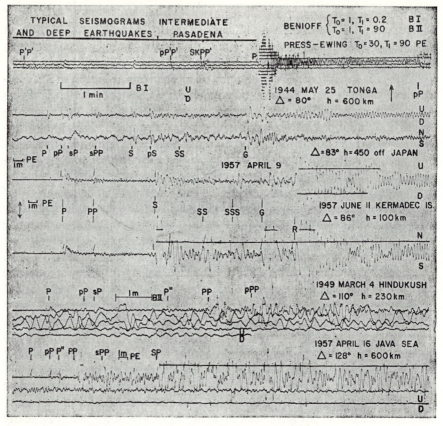

FIGURE 19-4   *Typical seismograms; deep-focus teleseisms.*

deep shocks are recorded are limited by their restricted geographical distribution. The following few notes are based on recordings at Pasadena and Huancayo (Peru), with a little information from other sources. (Fig. 19-4.)

At short distances there is no conspicuous difference in appearance between the seismograms of deep and shallow shocks, except that the P and S groups for deep shocks are generally sharper and shorter in period. Often depth can be established by the appearance of the strong phase ScS, recorded only on horizontal-component seismograms, following a few minutes after S and looking like S of an aftershock. The phase sScS also has often been identified.

At distances ranging from 60° to 80° the phases P'P', SKPP', P'P'P' can be found, especially on short-period vertical-component seismograms for the larger shocks. The interval P'P'-P (or SKPP'-P) varies slowly with distance; in combination with S-P it will fix both distance and depth if there is no error in phase identification. Care should be taken not to mistake SKPP' for P'P', which arrives earlier and is often less conspicuous. At larger depths it is frequently hard to separate SKPP' from pP'P'.

The complication caused by the arrival of *SKS*, *S*, and *PS* close together exists for deep-focus shocks as well as for shallow ones, although the generally clearer recording of deep shocks makes it less difficult to separate the phases (with inter-mediate shocks matters are worse because of *pS*, *sS*, etc.). The transit-time curves for *SKS* and *S* continue to intersect between 80° and 85° as the depth increases; but the time interval *S-P* (or *SKS-P*) at the intersection decreases from $10^m\ 20^s$ for shallow shocks to $9^m\ 30^s$ for a depth of 600 kilometers, so that *S-P* intervals for deep shocks demand more suspicious attention.

In the writer's experience the most confusing seismograms have been those of teleseisms at intermediate depth in the distance ranges near 35° and 115°. These are unexpectedly similar, and, although they show distinct and sharp phases, proper adjustment of depth and distance will fit a given series of meas-urements to either. Near 35° the phases *PcP*, *ScP*, *PcS*, *ScS*, which are often very sharp in shallow teleseisms, are augmented by *pPcP*, *sPcP*, *sScP*, *sScS*. Near 115° the seismogram begins with *P″* followed by *PP*, *SKP*, *PPP*, *SKS*, etc., with the additional phases due to deep focus. It has happened too often that such a record has been further confused by the arrival of a *P* phase from some other earthquake. In such circumstances all that can be done is to make careful note of all the clear phases with a view to interpretation when data from other sta-tions or from the international centers become available.

The phase *sPP* is often much better registered than *pPP*, and *sS* is more often found than *pS*. Long-period phases such as *SS*, *sSS*, *SSS* are sometimes taken for surface waves, and thus deep focus is overlooked.

Depths are often seriously overestimated by taking *sP* for *pP*. Sometimes a majority of stations recording a given earthquake report a large *sP* and overlook or fail to find *pP*. This is common with intermediate shocks under the Hindu Kush.

When reports of many stations can be combined, confusion is decreased; what appeared as a hopeless complication falls into definite order. The student can convince himself of this by plotting distance against time for all the readings of available stations as reported in the *International Summary* for a large deep earthquake. Identifications of the various phases, which are sometimes repeated from the station bulletins by the compilers of the *Summary* without correction, may be wide of the mark, but the actual readings usually fit definite and readily interpreted time-distance curves. Sometimes *P′P′* and similar phases are reported as separate earthquakes which may be taken as aftershocks or given entirely different epicenters.

References

Discovery and discussion

Turner, H. H., "On the arrival of earthquake waves at the antipodes, and the measurement of the focal depth of an earthquake," *M.N.R.A.S. Geophys. Suppl.*, vol. 1, pp. 1–13, 1922.

Banerji, S. K., "On the depth of earthquake focus," *Phil. Mag.*, vol. 49 (1925), pp. 65–80. (Objecting that deep shocks should not show surface waves.)

Byerly, P., *B.S.S.A.*, vol. 15 (1925), pp. 148–152. (Critical review of the preceding.)

Wadati, K., "Shallow and deep earthquakes," *Geophys. Mag.* (Tokyo), vol. 1 (1928), pp. 162–202; vol. 2 (1929), pp. 1–36; vol. 4 (1931), pp. 231–285.

Scrase, F. J., "The reflected waves from deep-focus earthquakes," *Proc. Royal Soc.* (*London*), Ser. A, vol. 132 (1931), pp. 213–235.

Stechschulte, V. C., "The Japanese earthquake of March 29, 1928, and the problem of depth of focus," *B.S.S.A.*, vol. 22 (1932), pp. 81–137.

Analysis and transit-time studies

Jeffreys, H., "Some deep-focus earthquakes," *M.N.R.A.S. Geophys. Suppl.*, vol. 3, pp. 310–343, 1935.

Gutenberg, B., and Richter, C. F., "Materials for the study of deep-focus earthquakes," *B.S.S.A.*, vol. 26 (1936), pp. 341–390; vol. 27 (1937), pp. 157–183.

————, "Données relatives à l'étude des tremblements de terre à foyer profond," *Publ. bureau central séismologique international*, Ser. A, vol. 15 (1937), pp. 1–70.

Miyamoto, M., "On the *ScS* waves of deep-focus earthquakes observed near the epicenter and their applications, "*Geophys. Mag.*, vol. 8 (1934), pp. 77–101.

Honda, H., "On the *ScS* waves and the rigidity of the earth's core," *ibid.*, pp. 165–177.

Jeffreys, H., "On pulses whose travel times are not true minima," *Proc. Cambridge Phil. Soc.*, vol. 39 (1943), pp. 48–51. (Bears on the sharpness of *pP* and *PP*.)

Honda, H., and Itō, H., "On the reflected waves from deep focus earthquakes," *Science Repts. Tôhoku Univ.*, Ser. 5, Geophysics, vol. 3 (1951), pp. 144–155.

Hayes, R. C., "A new phase in deep-focus earthquakes," *N. Z. Journ.*, vol. 17 (1935), pp. 553–562. (Reports *sScS*.)

Monographs

Szirtes, S., "Seismogramme des japanischen Erdbebens am 21. Januar 1906," *Zentralbureau der internationalen seismologischen Assoziation*, Ser. A, 1909, Strassburg.

Scrase, F. J., "The characteristics of a deep-focus earthquake," *Trans. Royal Soc.* (*London*), Ser. A, vol. 231 (1931), pp. 207–234. (Study of the Manchurian earthquake of Feb. 20, 1931.)

Wadati, K., and Isikawa, T., "On deep-focus earthquakes in the northern part of the Japan Sea," *Geophys. Mag.*, vol. 7 (1933), pp. 291–305. (Deals largely with the earthquake of February 20, 1931.)

Kawasumi, H., and Yosiyama, R., "On the mechanism of a deep-seated earthquake as revealed by the distribution of initial motion at stations throughout the world," *Proc. Imp. Acad. (Tokyo)*, vol. 10 (1934), pp. 345–348. (Reports on the earthquake of February 20, 1931.)

Brunner, G. J., "The deep earthquake of May 26, 1932, near the Kermadec Islands," *G. Beitr.*, vol. 53 (1938), pp. 1–64.

Caloi, P., and Giorgi, M., "Studio del terremoto delle isole Lipari del 13 aprile 1938," *Ann. geofisica*, vol. 4 (1951), pp. 9–26.

Lynch, W. A., and Dillon, V., "The deep-focus earthquake of May 19, 1940, in the Sea of Okhotsk," *B.S.S.A.*, vol. 33 (1943), pp. 251–267.

"High focus"

Tillotson, E., "The 'high focus' earthquakes of the *International Seismological Summary*," *G. Beitr.*, vol. 52 (1938), pp. 377–407.

————, "The African Rift Valley earthquake of 1928 January 6," *M.N.R.A.S. Geophys. Suppl.*, vol. 4, 1937, pp. 72–93 and note, *ibid.*, p. 315.

CHAPTER 20

Locating Earthquakes

ACCURACY OF EPICENTERS

Like other specialists, seismologists occasionally find their results being misapplied. This happens most often in public debates on earthquake risk (Chapter 24), but scientific discussion is sometimes distorted by misunderstanding of the methods used to locate epicenters and determine depths, or of the accuracy reasonably to be expected.

The reader of the preceding chapters should have an idea of the gross and even ludicrous mistakes that can occur in preliminary first-day estimates of earthquake locations, when only a small part of the necessary data is available. The popular press usually presents only these first guesses; by the time accurate locations can be made the earthquake is no longer news. Technically interested persons should beware of forming conclusions about an earthquake from what they read in newspapers; information for scientific or other serious purposes should be taken from technical journals or station bulletins or obtained directly by correspondence with seismologists. Reasonable time should be allowed for collecting and reducing the data.

Precision in placing an epicenter depends on its situation with respect to good seismological stations. It should be well surrounded by stations, local or distant, in different directions. If the stations are all more or less to one side, accuracy is much lower; but, since it is then usually possible to fit an epicenter to the data with high apparent consistency, the reliability of the result is often overestimated even by experienced seismologists. Magnitude is an important factor in setting the attainable accuracy of location, since small earthquakes are not clearly recorded at distant stations.

An earthquake not near any station, but large enough to be well recorded as a teleseism in different directions, can eventually be placed with an error not over 25 kilometers or 15 miles. A few seismic regions, such as that in the far southern Pacific, are so remote that only rare shocks are large enough to record adequately.

Preliminary locations for teleseisms, when readings for only a few stations are at hand, are never so accurate. The distance of a clearly recorded teleseism from a single station may possibly be determined with an error not over about a degree (say 100 kilometers or 60 miles); if such determinations are available for a few good stations, the epicenter may be fixed with no greater uncertainty.

The best epicenter locations depend on recorded times for at least a few stations within 300 kilometers. When an epicenter is within an active network of stations, as in California, Japan, the USSR, or central Europe, greater precision is attainable. Preliminary estimates of distance then can often be made with an error not over 10 kilometers, and final locations for well-recorded shocks are often trustworthy within about 5 kilometers (3 miles). Closer placing than this is exceptional, and results from special investigation or the use of special equipment, such as portable instruments taken into an area where aftershocks are occurring.

Without a group of several good stations, velocities and crustal structures are uncertain and location of epicenters is unreliable. This obvious fact is often ignored, to the great confusion of non-seismologists. Perusal of the literature sometimes suggests that, the fewer the available stations and the worse their equipment, the more far-reaching are the conclusions drawn by certain seismologists, presumably because their imaginations are unhampered by having too much data to reconcile. A pardonable error, committed in the early stages of many local-earthquake programs, is to assume that most of the small earthquakes originate along the principal and geologically evident faults. Rough station distances are then applied to locate epicenters along these faults. With better data it is usually discovered that epicenters of small shocks occur most often on minor faults, and published maps of located epicenters pass from a neat alignment of spots along the main faults to a general peppering over the whole active region. (See Fig. 28-3.)

DEPTHS

The depth of hypocenters is of great geological and geophysical interest, but accuracy in determination is rare. Normal, or shallow, earthquakes originate above the Mohorovičić discontinuity; a common average hypocentral depth assumed for standard transit-time tables is 25 kilometers. Deep-focus earthquakes are known to occur at depths near 700 kilometers (450 miles); these depths can often be assigned within 10 kilometers even when the epicenter is uncertain by 100 kilometers or more. The methods then used are usually not applicable to shallow earthquakes.

High accuracy for depths should eventually be obtainable where earthquakes are recorded by a network of local stations with precise timekeeping. At present the unsettled interpretation of crustal structures and of seismic waves at short distances introduces uncertainty which can only be removed by recording more earthquakes and investigating local structures in detail. In the present circumstances, the geologist should be particularly wary of using local-earthquake epicenters and depths to determine such details as the dip of important faults, unless special seismological investigation appears to justify the procedure. An example of what can be done in the present state of knowledge is provided by the Kern County earthquakes of 1952 (Chapter 28).

LOCATING EPICENTERS

General Principles

The simplest method of location, useful for rough purposes, depends on finding the *S-P* time interval at several stations, deriving the corresponding Δ for each, and combining these distances on a globe or map. *S-P* ranges from a few seconds in local earthquakes up to more than 11 minutes in teleseisms. Since the standard transit times for both *P* and *S* are tabulated to the nearest second for teleseisms, the method would be capable of considerable accuracy if *P* and *S* were always correctly identified and closely timed.

As indicated in Chapter 17, misidentification of *S* is one of the commonest errors. Even when the true *S* is recognized, there may be an uncertainty of many seconds as to the precise moment when it emerges from the previous motion. Finally, the simple *S-P* method of finding distance cannot be applied much beyond $\Delta = 100°$. For good results it is necessary to make use of the actual times of recording at the various stations.

Any fixing of distance Δ, even that from the *S-P* time interval, implies a determination of the origin time, *O*, at which the shock occurred.† It is easy to construct tables for passing from *S-P* to *P-O* (Appendix VIII); subtracting *P-O* from the recorded time of *P* gives *O*. As noted in Chapter 17, when distances are determined by means of a movable strip laid on the transit-time chart, a determination of origin time follows directly and has the advantage of considering all the recorded phases.

An origin time is derived from the data of each station. The times should be consistent. If there is a gross discrepancy at one station, it may be due either to time error or to misidentified phases; sometimes a little study will clear it up. If there is no such difficulty, an origin time is selected which is the mean of the several determinations, weighted if desirable according to the character and reliability of the data. From this chosen origin time the distances of the several stations are determined, using the tables for *P-O* in terms of Δ when possible. These distances are struck off as arcs on a

---

† This book uses for origin time the symbol *O*, generally accepted until a few years ago when some groups introduced new symbols on the plea that *O* was being misused. Since this misuse was due to errors of interpretation which could occur no matter what the symbol, there appears no reason to discard a good notation.

globe; if the arcs pass nearly through one point, that point is determined as the epicenter. It sometimes happens that all the arcs fall short of intersection; this can be due to having chosen too late an origin time, but (for teleseisms) it also results from deep focus. The reverse result, in which all the arcs appear to overshoot, is generally due to too early an origin time, unless deep focus has been assumed; however, it can happen for a badly recorded or complex shock when many stations have missed the first motion.

Teleseisms—Shallow and Deep

Most of what has been said will apply with minor changes to work with local earthquakes or with deep-focus teleseisms. Let us at first confine further discussion to ordinary shallow teleseisms. Refining epicenter determinations then depends on accurately determining distances from epicenter to stations. If extreme precautions are taken, this can be done by measurement on a globe, but most commercial globes are not accurate spheres, and the application of the map to the surface is usually inexact. The safe procedure is to calculate Δ from the coordinates of epicenter and station, using the formulas of spherical trigonometry or an equivalent. A vast amount of such computation is carried out for the *International Summary*; the results printed there can be used by an alert seismologist to lighten his labors.

Geocentric Latitudes. In Chapter 17 there was brief reference to the effect of the ellipticity of the earth. The quantity Δ, which we call the distance from epicenter to station, is or should be the angle at the center of the earth between radii to those points. It can be computed from the longitudes and from the *geocentric latitudes,* which are angles between the radii and the plane of the equator; it is an error to use the ordinary geographic or *geodetic latitude,* which appears on maps and is usually the one employed by a station in reporting its position.† The geodetic latitude is the angle between the normal to the

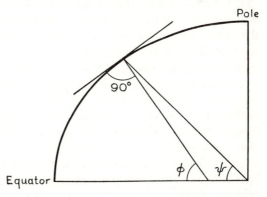

FIGURE 20-1   *Geocentric and geodetic latitudes.*

---

† Still a third is the *astronomical latitude,* which is the angle between the plumb line and the plane of the equator. In mountainous areas especially, the irregularities of gravity deflect the plumb line. At Mt. Wilson Observatory the astronomical latitude is 25.15 seconds of arc less than the geodetic, corresponding to 775 meters on the ground. This is one of the largest differences known.

earth's surface (more precisely, to a spheroid) and the equatorial plane. Figure 20-1, which exaggerates the ellipticity for the purpose, shows the difference between the geocentric latitude ψ and the geodetic latitude ϕ. This is given quantitatively by

$$\tan \psi = (1 - f)^2 \tan \phi = 0.993277 \tan \phi$$

where f is the flattening of the earth, taken as $1/297$. Approximately,

$$\phi - \psi = 11.7 \sin 2\phi$$

in minutes of arc. The maximum difference is at latitude $45°$; if both station and source are at such latitudes, the two corrections may add up to 23.4 minutes of arc, which amounts to 43 kilometers on the globe.

Direction Cosines.   In the *International Seismological Summary* and in many other publications, Δ is calculated from the standard direction cosines of the epicenter and station. These are the cosines of the angles which the radius through the given point makes with three rectangular coordinate axes fixed in the earth, with origin at its center, extending in order through the points on the equator at the Greenwich meridian and the meridian of $90°$ E longitude and through the north pole. The geocentric direction cosines for most of the active seismological stations are published through the international Seismological Summary office; the three cosines are lettered a, b, c. The corresponding cosines for epicenters used in the *Summary* are printed there at the head of the data for each shock, with letters A, B, C. With this notation

$$\cos \Delta = Aa + Bb + Cc$$

(Note that $A^2 + B^2 + C^2 = a^2 + b^2 + c^2 = 1$.)

If the worker finds that a shock in which he is interested has an epicenter near one used in the *International Summary,* he may take distances Δ from the *Summary* for stations shown there, and may calculate others from the given A, B, C and those of additional stations. This will shorten his work toward a revised epicenter. Whether this is done or not, the general procedure is fixed by a number of not very easy steps.

Routine Procedure.   (1) Find a first approximation to epicenter and origin time by working with globe and transit time charts or by any suitable form of lucky guessing.

(2) Calculate (or take from the *Summary*) distances Δ to the stations to be used, and note the corresponding azimuths of the stations about the epicenter (north is 0, east $90°$, south $180°$, west $270°$). (Azimuths are given in the *Summary,* but they are easily read from a globe with a stretched string and a protractor—with an error not over a few degrees, which is accurate enough for the purpose.)

(3) Using the chosen origin time, find *P-O* for each station, and read the corresponding distances Δ in degrees and tenths from the tables, interpolating where necessary.

(4) From the distances found in 3 subtract the corresponding calculated distances from the assumed epicenter. Plot the resulting residuals, with their proper signs, as ordinates against station azimuth as abscissa.

(5) Fit a sine curve to the plotted residuals; from this read corrections in latitude and longitude, giving the corrected epicenter and, if necessary, correction to the assumed origin time. (The curve will give both north-south and east-west corrections in degrees of a great circle; the latter must be divided by the cosine of the latitude to obtain the correction in degrees of longitude.)

(6) With the corrected epicenter and origin time, compare the transit times of *P'* or of *SKS* to distant stations in order to verify the assumed shallow depth or to detect overlooked deep focus (depths of the order of 60 kilometers are particularly hard to fix).

A detailed example of this procedure is given in Appendix X. It can be applied with proper modification to deep-focus earthquakes. Preliminary origin times can be determined from *S-P* intervals; use of the tables constructed for shallow shocks will introduce only a small systematic error, or better tables may be set up for various depths. Usually enough readings of *pP*, or determinations of depth from other phases, are available for at least a preliminary assignment of depth, which should be fixed to the nearest 10 kilometers. Distances are calculated as for shallow shocks, using the co-ordinates of a preliminary epicenter; the distances calculated from transit times and used to determine residuals are interpolated from tables, assuming the preliminary depth. If the sine curve is not symmetrical with respect to the line of zero residuals, correction is required to origin time, depth, or both.

For a well-recorded shock the next step is to apply the times of *P'*. Distances from stations reporting *P'* are determined, using the corrected epicenter, and transit times *P'-O* are used to find the corresponding depths. If these depths are not in harmony with those from *pP-P*, change is necessary. It may turn out that *pP* has been misidentified; but usually an adjustment in origin time will bring the data into harmony.

Instead of *P'*, *SKS* may be used for this check on depth. The phase is often sharply recorded; at distances beyond 90° it is well ahead of *S* and associated phases, so that misidentification at several stations is not likely. Moreover, the transit time *SKS-O* varies almost twice as fast with depth as *P'-O*, so that the former is a more sensitive measure of depth; however, the period of the *SKS* wave is correspondingly longer, and, although easily picked out on the seismogram, it may not be timed as exactly as *P'*. The net result is that *SKS* and *P'* are about equally useful for fixing depth.

Misinterpretations and Difficulties.　The procedures described are more or less routine; they presuppose a well-recorded shock, with P and P' correctly measured and identified at a large number of stations. They are not so readily applicable to small earthquakes, or even to major earthquakes recorded earlier than 1920. Unrecognized instances of deep focus contribute markedly to the confusion for earlier dates. Details of some of the complications encountered will be found in publications on the geography of deep and shallow earthquakes.

All the possibilities of misinterpretation set forth in Chapters 17 and 19 need to be weighed. As the number of available observations decreases, even single clerical errors become important; if one key station has reported the time of P one minute wrong, it may be impossible to identify the mistake and properly determine the epicenter.

Local Earthquakes

General Principles.　Epicenters for local shocks also ultimately depend on using times of P, accepting an origin time arrived at by various approximations. The method of fitting a strip to a chart giving standard transit times for the principal phases is useful for routine purposes when minor shocks are being located for inclusion in current reports. The first motion may be small and not clearly recorded, but usually interpretation of seismograms at several near stations can be adjusted to give consistent origin times, resulting in distance arcs which indicate a single epicenter. The origin time from this process of trial and error can then be used for further approximations or to improve empirically the standard time-distance chart. With better-recorded shocks and more stations, more direct procedures are available and preferable.

Because of the effect of uncertain depth, epicenters can be located best from stations at distances which are appreciably larger than the probable depth, say 60 to 100 kilometers, if the shock is large enough to be sharply recorded. If it were not for one doubtful assumption, epicenters could be located independently of hypocentral depth with considerable accuracy from times of Pn, if these are dependably recorded at stations in widely different directions. The time difference between arrivals at any two stations is multiplied by the apparent velocity of Pn (8.2 kilometers per second for southern California). This gives the difference in epicentral distance, which determines a hyperbola on the map. The epicenter may be determined graphically as the intersection of such hyperbolas; or analytical methods may be used, operating numerically with coordinates and times. The doubtful assumption mentioned is that the large structure of the crust is the same in all directions from the epicenter, and especially that the Moho is a level surface; but in many areas, as near the Sierra Nevada, the Moho almost certainly dips notably. The effect of a small dip would be to displace the calculated epicenter in the direction of dip. Note that this source of error would not seriously affect the relative position of adjacent epicenters; this was important in studying the Manix earthquakes of 1947 (Chapter 28).

Good epicentral locations are made by using stations in different azimuths at distances less than 100 kilometers, so that the seismograms begin with the direct wave *p*. Unless several stations are nearly at the same distance, so that their times of first motion are close together, the epicenter found will be appreciably affected by slight changes of assumed hypocentral depth and origin time, even on the simple assumption of a constant known velocity for *p*. There are various geometrical constructions with equivalent analytical forms. The following straightforward procedure assumes constant velocity.

Routine Procedure. (1) Times of the first arrivals of the direct longitudinal and transverse waves, *p* and *s*, at several stations are measured or guessed at. The time interval *s-p* is plotted as ordinate, with the actual time of arrival of *p* as abscissa. The points should lie on a straight line which crosses the zero line for *s-p* at the abscissa taken as the origin time. (This method is theoretically valid even if the velocities vary, providing Poisson's ratio remains constant.†)

(2) The time intervals *p-O* are multiplied by the best estimate for the mean velocity of *p* (for southern California, 6.34 kilometers per second) to give supposed straight-line distances from hypocenter to stations. Call any such distance *D*.

(3) Suppose that the stations are all at nearly the same elevation. (Differences in elevation can be corrected for geometrically.) About any two stations on the map strike circles whose radii are the corresponding values of *D*. These circles are, theoretically, surface traces of spheres on which the hypocenter must lie; hence the circles should intersect and their common chord should pass through the epicenter. If the corresponding circle is drawn for a third station, there result three common chords which geometrically must pass through a single point. This point is the epicenter. The corresponding hypocentral depth *h* is found (graphically if preferred) by combining the surface distance Δ from epicenter to station with the hypocentral distance *D*:

$$h^2 = D^2 - \Delta^2$$

(4) If data from four or more stations are available, there is a whole system of common chords which should pass through the epicenter. The adjustment of discrepancies can be carried out analytically by using the method of least squares.

(5) The fit can often be improved by a slight change of origin time; this

---

† A simpler way is to assume Poisson's ratio as ¼; then each interval *s-p* may be multiplied by 1.37 to determine *p-O*, and the mean of the resulting origin times adopted for use at step 2.

is legitimate, since the origin time derived from *s-p* intervals is not necessarily better than a good approximation. Such a shift moves the point of intersection of common chords for any three given stations along a straight line which passes through the center of the circle drawn through the three stations. If more than three stations are used, the intersections of such straight lines theoretically determine origin time as well as epicenter, the velocity always being assumed as given. Similar procedures apply if the origin time is held fixed and the velocity is varied.

(6) In routine work at Pasadena the hypocenter is assumed to be at a depth $h = 16$ kilometers. With this assumption, Δ can be tabulated as a function of $P\text{-}O$; circles struck with radius Δ should intersect at the epicenter. If the circles systematically fall short or overshoot, a change in assumed depth or origin time is indicated, and the data are examined further.

Depth. A study of the geometry involved will show that, while an epicenter is best fixed by stations well distributed in azimuth and not at very short distances, depth determination by these methods demands at least one station near the epicenter. This means that depth determinations can be made with reasonable accuracy only in rare cases;† almost all these are the result of setting up temporary stations in an area in which aftershocks of a large earthquake are occurring, as has been done in California and Japan.

The transit time of *Pn* also varies with hypocentral depth, but, while this can be used to compare shocks at different depths with nearly the same epicenter, exact interpretation of *Pn* depends on the depth of the Moho and the entire velocity distribution above it. When a few hypocenters have been placed by using direct *p*, others in the same area can then be studied by means of *Pn*. Times of earlier recorded shocks can then be reinterpreted, and the entire accumulation of years of recording becomes available for detailed investigation of the occurrence of earthquakes as well as the local crustal structures.

An example of detailed work in locating a small shock is given in Appendix XI.

Calculation of Small Distances. For many purposes it is desirable to calculate short distances from the coordinates rather than to measure on a map. Up to about 500 kilometers the effect of curvature of the earth is small, and distances can be calculated within about 0.1 kilometers by the formula

$$\Delta^2 = \Delta x^2 + \Delta y^2$$

with $\Delta x = A\Delta\lambda$ and $\Delta y = B\Delta\phi$. Here λ and ϕ are longitude and geodetic latitude. $\Delta\lambda$ and $\Delta\phi$ are differences between longitudes and latitudes of two given points; if these are expressed in minutes of arc, A and B are the lengths

---

† This does not apply to deep-focus earthquakes, for which the interval *pP-P* and other data are used.

in kilometers of 1 minute of longitude and latitude respectively. An accurate value of Δ is most easily obtained by using A and B corresponding to the mean latitude between the two points; this corresponds to navigation by "middle latitude sailing." The length B is near 1.85 kilometers, and varies slowly with latitude because of the ellipticity of the meridian. The length A is near $1.86 \cos \phi$, where the numerical factor also varies slowly with latitude; because of the cosine factor, for rapid work larger tables are needed for A than for B. Where distances are repeatedly calculated from a given point, as from an established seismological station, special tables can be constructed which will speed up the work. General tables for A and B are given in Appendix XII.

CHAPTER 21

International Seismology†

THE WORLD ORGANIZATION

Exchange of Data

Seismology for the whole earth depends on international cooperation. Many of the principal advances have resulted from collecting, at one research center, the seismograms (either original records or photographic copies) recorded at stations all over the world for a single important earthquake and studying them together. The possibility of doing this depends on the international courtesy of science, which has continued in spite of political disturbances and wars.

Exchange of data in the form of readings taken from seismograms proceeds on a large scale, and on the whole with increasing effectiveness. This exchange is sponsored and encouraged by the International Union of Geodesy and Geophysics. There are four principal stages of publication: preliminary station bulletins, preliminary international bulletins, revised station bulletins, and the *International Seismological Summary,* which has definitive status (as a repository of data; its epicenter determinations and other interpretations need repeated re-examination as the science progresses).

Until lately, preliminary station bulletins were issued monthly or quarterly, and circulated by ordinary mail. A more rapid procedure is now followed by many of the better-equipped stations. It is possible to send a single-sheet air letter anywhere in the world for 10 cents (or equivalent), and such letters are now being used to issue provisional readings at intervals of about 10 days. Many stations report by telegraph and teletype, sometimes daily, to such central offices as that of the U. S. Coast and Geodetic Survey at Washington, and the international central seismological office (Bureau Central Internationale Séismologique) at Strasbourg. For each earthquake sufficiently well-reported, the Washington office determines epicenters, origin times, and depths and circulates the results about 10 days after the date of occurrence. The Strasbourg office, after a few months, circulates a bulletin giving all the

---

† Conditions change rapidly. This chapter reports them as of July 1, 1956, in so far as information has been available.

principal readings reported by the stations for a full month, with determinations of epicenters, origin times, depths, and magnitudes from various sources (including many first worked out at Strasbourg). This publication and the *International Summary* are now supported by UNESCO.

Most stations, after receiving the international bulletins and those of other stations, issue a revised bulletin; phase identifications are then corrected in the light of better information, and new data are added. The southern California group of stations, for example, issues provisional data chiefly for the head station at Pasadena, but also includes readings for auxiliary stations in the revised bulletin.

By general agreement, revised bulletins are issued in time for the compilation of the *International Seismological Summary*. This can be done without haste, since the *Summary* (in part as a result of war) is operating well behind current date (now about 8 years; the issue covering earthquakes during the last quarter of 1948 was sent to press in August 1956). The *Summary* repeats readings given by the stations for earthquakes well enough recorded for location; it also gives carefully revised epicenters, origin times, depths, and many other useful details (now including macroseismic data). but omits amplitudes and periods.

Forms of Publication

General. Provisional reports take all manner of forms, determined largely by convenience at the issuing station. Revised bulletins (Figs. 21-1, 21-1A) conform more or less closely to internationally approved standards. Figure 21-1 gives a good typical example. Note that the language of the headings and remarks is French, a common choice for international circulation. The first column gives the calendar date† and also, in this bulletin, the component, N-S or E-W, to which the readings refer. The phase column gives designations of *P*, *S*, etc., usually qualified with *e* (for emergence) or *i* (for impulse). Times are given as Greenwich Civil Time‡ (abbreviated G.C.T., or sometimes G.M.T.; U.T., for Universal Time, denotes the same thing); readings are to the nearest second, or occasionally to the tenth of a minute. Wave periods are given in seconds; and ground amplitudes *A*, computed from trace amplitudes and the constants of the seismograms, are tabulated in three columns for the three components. Columns are provided for distance Δ and for remarks.

Many stations shorten this form by not giving all the readings successively for the separate instruments. Since amplitudes are calculated at most for the larger and more important phases, the amplitude columns are often omitted,

---

† This and some other bulletins have the annoying convention of entering only the day of the month in the date column, so that one often has to search through several pages to discover the name of the month. The year, 1951, is taken care of by the fact that this is an annual issue.

‡ The G.C.T. day begins at midnight of the zero meridian. International usage naturally ignores daylight-saving or summer time.

BUDAPEST

Date	Phase	Heure de Greenwich			Période	Amplitude			Δ	Remarques
						A_N	A_E	A_Z		
		h	m	s	s	μ	μ	μ	km	
31.	P	21	8	23					83,8°	Philippines
N—S	e		9	37					9300	
	iS		18	42	5	5				
	e		19	12						
	e			55						
	eSSS		26	42						
	eL		42							
	M		49		16	5				
	F	22	10							
E—W	iP	22	8	24						
	i			39						
	PP		11	39						
	iS		18	39						
	ePS		19	15						
	PPS			53						
	SS		24							
	eL		47							
	F	22	10							
Juin 5.	P	17	10	2					97,5°	Iles de
N—S	ePP		13	27					10800	Kiou-Siou
	SKKS		20	14						
	iS			39						
	ePS		22	5						
	ePPS			38						
	eSSP		27	30						
	eSSS		31	23						
	eL		38							
	M		43		24	78				
	M		48		17	47				
	M		50,5		14	27				
	M		52		16	43				
	M		53		16	40				
	F	18	30							
E—W	P	17	10	2						
	i			15						
	ePP		12	52						
	SKKS		20	16						
	iS			40						
	e		21	20						
	ePPS		22	49						
	eSSP		27	33						
	eSSS		30	5						
	eL		38							
	M		48		16		35			
	M		50		14		51			
	M		52		15		58			
	M		53		14		17			
	F	17	40							
6.	P	16	16	41					32,5°	
N—S	i		17	8					3610	
	e		21	38						

FIGURE 21-1 *Page from typical seismological station bulletin, Budapest, June 5, 1951.*

Грузинский Филиал
Академии Наук СССР

Тбилисский
Геофизический
Институт

სსრკ მეცნიერებათა აკადემიის
საქართველოს ფილიალი

თბილისის
გეოფიზიკური ინსტიტუტი

Filiale Géorgienne
de l'Académie
des Sciences de l'URSS

Institut Géophysique
de Tbilissi

№ 1—2

(Janvier—Juin 1936)

თბილისის გეოფიზიკური ინსტიტუტის
ცენტრალური სეისმური სადგურის

კვარტალური ბიულეტენი

КВАРТАЛЬНЫЙ БЮЛЛЕТЕНЬ
ЦЕНТРАЛЬНОЙ СЕЙСМИЧЕСКОЙ СТАНЦИИ
ТБИЛИССКОГО ГЕОФИЗИЧЕСКОГО ИНСТИТУТА

BULLETIN TRIMESTRIEL
DE LA STATION SÉISMIQUE CENTRALE
DE L'INSTITUT CÉOPHYSIQUE DE TBILISSI (Tiflis)

Tbilissi — 1937

FIGURE 21-1A *Title page, bulletin for Tbilissi (Tiflis, Georgian Republic). Note the use of three alphabets.*

Pasadena and auxiliary stations, 1951   Page 42

Date	Sta.	Phase	h	m	s
		June (continued)			
5	T	iP	01	41	53
		iPcP		43	57
		USCGS: 9½ N 86 W,			
		0 = 01:34:20, h = 60 km.			
5	MW	eP	03	15	16
	R	eP			17
		e			22
	Pr	iP			18
		i			22
	CL	iP			31
	T	iP			26
		i			31
5	P	iP	17	10	32 d
		i			43
		iPP		14	05
	PX	e(S)NE		21	16
		iNE			35
		eNE		22	32
		eGNE			34.0

			A	T
	PZ		1	3
	PH		½	3
	PPZ		2½	3
	SH		7	8
	MH		20	20

Date	Sta.	Phase	h	m	s
	R	eP	17	10	35
	Pr	iPNZ			38
		iPPNEZ		14	11
		eSE		21	38
	CL	iP		10	29 d
		i			42
		i			56
	T	iP			25 d
		i			42
		Magnitude 7			
		USCGS: 30 N 132 E,			
		0 = 16:57:47, h = 100 km.			
		CMO: 29.8 N 131.7 E,			
		h = 90 km.			
6	CL	iP	05	54	34
		i			46
		i			58
	T	iP			30
		USCGS: 37½ N 142 E,			
		0 = 05:42:45			
		CMO: 37.8 N 142.0 E,			
		h = 60 km.			
6	P	iPEZ	16	21	21 c
		i			28
		e			33
		ePP		23	26
	PX	eSNE		30	04
		eSSNE			33.3
		iNE			36.6
		eGNE			37.2
		eR		40	

			A	T
	ePZ		0.1	1½
	iPZ		½	2
	PH		0.15	1
	SH		10	10
	MH		140	20
	MZ		120	17

(continued)

Date	Sta.	Phase	h	m	s
		June (continued)			
6	R	iP	16	21	21
		i			27
		i			34
	Pr	iP			25
		i			33
		i			40
		iPP		23	54
	CL	iP		21	10
		i			18
		i			24
		i			29
		iPP		23	35
		eP'P'		50	34
	T	eP		21	04
		Magnitude 6¾-7			
		USCGS: 71 N 8 W,			
		0 = 16:10:52, h = 60 km.			
		BCIS: 72½ N 8½ W,			
		0 = 16:10:49, h = 60 km.			
7	MW	eP	05	38	45
	R	eP			42
	Pr	eP			39
	CL	eP			58
7	P	i(P)	23	11	29
	PX	iSKSNE		21	52
		eG			37.4

			A	T
	PZ		½	4
	SKSH		2	7
	MH		15	20
	MZ		12	18

Date	Sta.	Phase	h	m	s
	R	eP	23	11	25
		e			30
	Pr	iPNEZ			30
	CL	eP			30
		i			37
		i			51
		i		12	09
	T	eP		11	32
		Magnitude 6¾ - 6¾			
		USCGS: 27½ S 176 W,			
		0 = 22:59:00			
		BCIS: 26½ S 176½ W,			
		0 = 22:59.0			
8	CL	e	00	16	54
		e		17	05
		e			12
		Aftershock			
8	CL	i(P")	19	13	39
		e(PP)		15	53
		South Atlantic?			
8	MW	eP	22	33	29
		e			45
	R	eP			29
	Pr	eP			33
		e			48
	CL	eP			35
		USCGS: 26 S 176 W,			
		0 = 22:21:19, h = 100 km.			
8	P	e(P)	22	47	09
	R	e		47	01
	CL	eP		46	29
9	MW	eP	04	03	13
		epP?		05	21

(continued)

FIGURE 21-2   *Page from bulletin for Pasadena and auxiliary stations, June 1951.*

and the corresponding information inserted under remarks. The long eclipse of interest in amplitudes in favor of better timing caused this part of the report to be increasingly neglected; now that more and more work is being done with magnitudes and energies, stations are encouraged to report amplitudes and periods. Many give amplitudes only for the maximum surface waves; this is incomplete information even for shallow shocks. For deep-focus earthquakes magnitudes can be assigned only if amplitudes and periods for body waves are known.

The annual report from Budapest includes similar data for auxiliary stations on separate pages. Other groups of several stations report them together. An example is the revised bulletin for Pasadena and its auxiliary stations (Fig. 21-2). This bulletin, which for economic reasons takes a somewhat condensed form, has a column headed "Sta." in which are entered abbreviations for the station names (P for Pasadena). Amplitudes and periods (microns and seconds) for various recorded phases (including the maximum of surface waves) at Pasadena appear in the columns following the reported times for that station. Magnitudes, epicenters, and origin times quoted (USCGS: U. S. Coast and Geodetic Survey; BCIS: Bureau Central Internationale Séismologique), or determined at Pasadena, and other information or remarks, follow the data for each shock.

Figure 21-3 shows the form of the *International Seismological Summary*. The report for each shock is headed with its calculated origin time and epicenter. $A, B, C, \ldots$ are geographical data for calculating distances and azimuths which appear for each station in the columns headed "Δ" and "Az." Here Δ is the geocentric central angle between epicenter and station. The azimuth is that of the station at the epicenter, zero when the station is to the north, and running to 360° clockwise, beginning toward the east. Data in columns headed "P," "S," "Supp." are minutes and seconds of elapsed time from the given origin time to the various station readings. The "L" column gives similar times for surface waves, usually to the tenth of a minute. Columns headed "O-C" are residuals for the observed times in the P and S columns against those taken from standard tables for the calculated Δ.

Reporting Local Earthquakes. The majority of stations, not having highly sensitive instruments and not being located in very seismic areas, include in their bulletins readings for local earthquakes along with those of teleseisms. Most of the Japanese stations devote their bulletins primarily to local earthquakes. Wherever local shocks are frequent, a problem of selection arises.

The bulletin issued for Pasadena and auxiliary stations (Fig. 21-2) shows readings chiefly for teleseisms, which number about 1500 annually; for these, times are given only for some of the better equipped of the 15 stations in operation. Readings for the larger local shocks are also included, generally those of magnitude 5 and over, which are likely to be recorded at distant stations.

1946   **155**

April 13d. 18h. 57m. 50s.   Epicentre 19°·0S. 175°·5W.   Depth of focus 0·030.

A = − ·9433, B = − ·0742, C = − ·3236;   δ = +4;   h = +5;
D = − ·078, E = + ·997;   G = + ·323, H = + ·025, K = − ·946.

	Δ	Az.	P.		O−C.	S.		O−C.	Supp.			L.
	°	°	m.	s.	s.	m.	s.	s.	m.	s.		m.
Apia	6·3	35	—		—	e 2	37	− 7	—		—	—
Auckland	19·7	204	i 3	16	−58	8	32	+53	i 4	15	P	—
Arapuni	20·5	201	1	22	?	7	58	+ 5	—		—	—
Wellington	23·7	199	4	52	− 1	8	45	− 4	15	32	S꜀S	—
Christchurch	26·4	200	6	32	PPP	9	14	−19	7	32	?	10·9
Riverview	33·1	237	i 6	16k	− 1	i 11	13	− 5	i 7	34	pP	—
Santa Barbara z.	75·1	46	c 11	17	− 2	—		—	—			—
Pasadena	75·9	47	i 11	25	+ 1	i 20	46	0	i 12	24	pP	—
Mount Wilson	76·1	47	i 11	27	+ 2	—		—	i 12	26	pP	—
Palomar ·Z.	76·4	48	i 11	28	+ 2	—		—	i 12	29	pP	—
Riverside z.	76·4	47	i 11	28	+ 2	—		—	i 12	29	pP	—
Shasta Dam	77·2	39	i 11	31	0	e 21	5	+ 5	i 12	32	pP	—
Tinemaha	77·6	45	i 11	44	+11	c 21	43	SP	i 12	45	pP	—
Boulder City	79·2	46	i 11	44	+ 2	e 21	26	+ 5	i 12	46	pP	—
Overton	79·8	46	c 11	39	− 6	—		—	—			—
Pierce Ferry	79·9	47	i 11	47	+ 2	c 21	34	+ 6	i 12	49	pP	—
Tucson	80·1	51	i 11	48	+ 2	—		—	i 12	49	pP	—
Grand Coulee	83·6	34	i 12	4	0	—		—	i 13	7	pP	—
St. Louis	98·1	52	c 14	14	+62	c 24	24	+ 8	e 26	10	PS	e 42·2
La Paz z.	100·4	112	i 17	38	PP	—		—	—			—
San Juan	113·6	77	—		—	i 24	36	[+ 1]	—		—	—
Tashkent	121·1	307	e 19	43	PP	c 25	0	[− 2]	—		—	—
Sverdlovsk	124·1	326	e 20	11	PP	—		—	—		—	—
Baku	135·8	309	c 22	7	PP	—		—	—		—	—
Leninakan	140·1	311	c 18	59	[− 3]	—		—	—		—	—
Yalta	144·5	322	19	9	[− 1]	—		—	—		—	—
De Bilt	147·0	358	i 19	16k	[+ 1]	c 41	10?	SS	i 20	23	pPKP	e 60·2
Collmberg N.	147·1	350	c 19	16	[+ 1]	—		—	—		—	—
Jena	147·6	350	c 19	14	[− 2]	—		—	—		—	—
Uccle	148·3	1	e 19	21k	[+ 4]	c. 41	28	SS	e 20	33	pPKP	—
Cheb	148·4	350	c 12	10?	?	—		—	—		—	—
Ksara	148·4	304	i 19	19	[+ 2]	—		—	i 20	32	pPKP	—
Strasbourg	150·4	355	c 19	26	[+ 7]	—		—	—		—	—
Basle	151·4	356	c 19	23	[+ 3]	25	44	[− 19]	—		—	e 85·2
Zürich	151·5	356	c 19	29k	[+ 8]	—		—	—		—	—
Zagreb	151·6	343	c 19	26	[+ 5]	—		—	e 20	36	pPKP	—
Chur	151·9	353	c 19	22k	[+ 1]	—		—	—		—	—
Clermont-Ferrand	153·3	2	c 19	33	[+10]	—		—	—		—	—
Helwan	153·3	297	19	31	[+ 8]	29	45	?	—		—	—
Rome	156·2	346	e 19	29	[+ 2]	—		—	—		—	—

Additional readings —
Wellington PcP = 6m.31s., ScP = 9m.26s., SS = 11m.4s.
Riverview iEZ = 7m.43s., isS?N = 13m.50s., iSSZ = 14m.2s., iSSN = 14m.5s., iScS?EN = 16m.21s.
Pasadena iPcPZ = 11m.50s., esSN = 22m.42s.
Mount Wilson iPcPZ = 11m.52s.
Palomar iZ = 11m.56s.
Riverside iPcPZ = 11m.53s.
Shasta Dam eS = 21m.20s.
Tinemaha iPcPZ = 12m.10s.
Boulder City e = 21m.40s., ePS = 23m.14s.
Tucson i = 12m.14s., esP = 13m.24s., epPP? = 15m.57s., ePPP = 16m.41s.
St. Louis eSKSE = 23m.22s., esSKS?E = 25m.24s., eN = 27m.22s., eSSN = 31m.12s.
Collmberg eE = 19m.19s. and 19m.35s.
Jena eEN = 19m.18s., eN = 19m.30s., eE = 19m.37s.
Uccle ePS? = 30m.59s.?
Ksara ipPP = 23m.54s.
Strasbourg e = 19m.30s.
Zagreb e = 20m.41s.
Chur i = 19m.30s.ₐ.
Helwan i = 19m.45s.
Rome e = 32m.34s.
Long waves were also recorded at Sofia.

FIGURE 21-3   *Page from the International Seismological Summary.*

Smaller local shocks in southern California, down approximately to magnitude 3, are reported in a separate bulletin which gives only determined epicenters, origin times, and magnitudes. This lists about 300 shocks in a normal year; the individual station readings are not published.

The bulletin issued by the Japan Meteorological Agency† at Tokyo reports in general only the chief shocks in the area of the Japanese home islands. For 1955, 195 were reported in detail. Most of these were at least of magnitude 5 and were recorded at stations in Europe and America. Readings for a small number of teleseisms were also included. For a large shock the J.M.A. bulletin often gives readings from 80 to 100 Japanese stations.

SEISMOLOGICAL STATIONS OF THE WORLD

It is impossible to give the precise number of active seismological stations; even conservative estimates are surprisingly large. A list issued in 1951 by the *International Seismological Summary* staff names 602 stations; some of these had already ceased operation, but others then in existence were not listed. A list published in 1953 by the Royal Observatory of Belgium (at Uccle), under the auspices of the international association, includes 518 stations, of which 26 were noted as out of operation and 20 as merely projected. In all probability the number of stations in regular operation is between 500 and 600, excluding amateur stations. (Fig. 21-4.)

Except for preliminary purposes, comparatively few stations report individually. The collecting, interpreting, and editing of preliminary material at Washington and Strasbourg has been noted; a similar correlated bulletin, after interruption, is again being issued from the headquarters of the Jesuit Seismological Association at St. Louis University. The Washington office issues a definitive bulletin for 20 stations maintained by or cooperating with the U. S. Coast and Geodetic Survey; outside the 48 states this includes Balboa Heights (Canal Zone). College (University of Alaska, near Fairbanks), Honolulu, San Juan (Puerto Rico), and Sitka. The Strasbourg office reports similarly for 7 French stations and 12 in distant dependencies. The Japan Meteorological Agency (Tokyo) reports for more than 100 stations; bulletins from the Soviet Union give data for 73 stations there in operation at the end of 1954. Other groups reporting together are listed in Table 21-1. Some of the stations in these groups also report individually. A selected list of stations, with their coordinates, appears in Appendix XIII.

Of the stations now active, about 90 are in the United States, excluding Alaska (5) and Hawaii (2). The total for North America is about 130, for South America 15. The latest available total for the Soviet Union is 76. For Europe, excluding the USSR, a rough count based on the 1953 list published

---

† Known as the Central Meteorological Observatory before January, 1956.

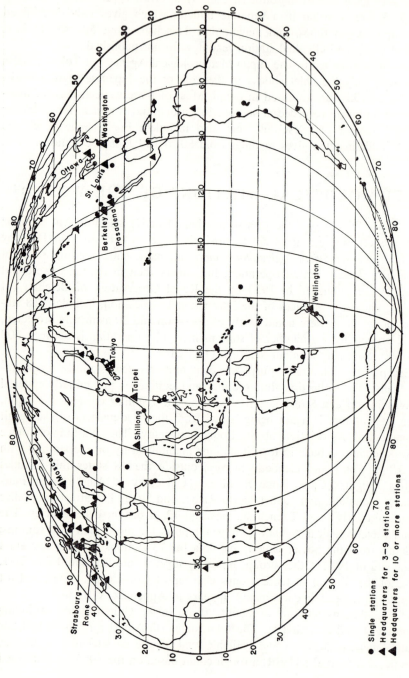

FIGURE 21-4 *World map of seismological stations.*

Table 21-1   Groups of Three or More Seismological Stations Reporting Together or From One Office (some also independently) as of July 1, 1956

Headquarters (Also a station unless otherwise noted)	Location of Headquarters	Number of Stations
Japan Meteorological Agency (formerly Central Meteorological Observatory)	Tokyo, Japan	108
Institut fiziki Zemli Akademii Nauk	Moscow, USSR	76
U. S. Coast and Geodetic Survey	Washington, D. C.	20
Jesuit Seismological Association	St. Louis, Missouri	19
Bureau Central Séismologique Française	Strasbourg, France	19
California Institute of Technology	Pasadena, California	16
Istituto Nazionale di Geofisica	Rome, Italy	14
India, Government Meteorological Department	Shillong, Assam	14[1]
Taiwan Weather Bureau	Taipei, Formosa	14
University of California	Berkeley, California	11
New Zealand, Department of Scientific and Industrial Research	Dominion Observatory, Wellington, New Zealand	11[1]
Seismological Service of Canada	Dominion Observatory, Ottawa, Canada	11[1]
Universidad Nacional de México	Mexico City[2]	9
Earthquake Research Institute	Tokyo University	9
Instituto Sismológico, Universidad de Chile	Santiago, Chile	7
Observatorul din Bucuresti	Bucharest, Romania	6
Imperial College of Tropical Agriculture (Volcanological Research Department)	Trinidad, B.W.I.	7
Geophysical Research Institute	Johannesburg,[2] South Africa	5
Institut National Séismologique de Hongrie	Budapest, Hungary	4
National Geophysical Institute	Praha, Czechoslovakia	4
Instituto Geográfico y Catastral	Madrid,[2] Spain	4
Seismological Observatory	Warsaw, Poland	4
Schweizerischer Erdbebendienst	Zurich, Switzerland	4
Institut de Météorologie et de Physique du Globe de l'Algérie	Algiers, Algeria	3
Institut pour la Recherche Scientifique en Afrique Centrale	Bukavu,[2] Belgian Congo	3
Instituto Geofísico de los Andes	Bogotá, Colombia	3
Geodetic Institute	Copenhagen, Denmark	3[1]
Meteorological and Geophysical Service	Djakarta, Indonesia	3
Kyoto University	Kyoto, Japan	3

Headquarters (Also a station unless otherwise noted)	Location of Headquarters	Number of Stations
Geophysical Research Laboratory	Quetta, Pakistan	3
Ve ∂urstofan	Reykjavik, Iceland	3
Württembergischer Erdbebendienst	Stuttgart, Germany	3
Seismological Laboratory	Uppsala, Sweden	3

Table 21-1   Continued

[1] Reports for India include readings for Colombo, Ceylon. New Zealand reports from Wellington include readings for Suva (Fiji). Canadian reports were given for an eastern division, with headquarters at Ottawa (including Resolute Bay, now Resolute, on Cornwallis Island) and a western division, with headquarters at Victoria. In 1956 the two reports were combined.

[2] No seismographs are operated at Madrid, Bukavu, or Mexico City (although the principal Mexican station, at Tacubaya, is in the Federal District). The station at Johannesburg is at the Union Observatory.

at Uccle gives 85. Japan has about 120; excluding these and the USSR, the count for Asia is about 40. There are 18 in Africa (including one on Madagascar), and 5 on the Australian continent. In the Pacific there are stations on Hawaii, Fiji, Guam, New Britain, New Caledonia, Palau, and Samoa. Atlantic stations include those on Greenland, Iceland, Bermuda, and the Azores. The most northerly station is Resolute Bay (lately renamed Resolute), on Cornwallis Island in the Canadian Arctic; others north of 66° N are Scoresby-Sund on the east coast of Greenland and Kiruna in northern Sweden.

As of July 1, 1956, no station was operating south of the Antarctic Circle, although installation of several was planned for the International Geophysical Year, 1957–1958. Four earlier expeditions have operated seismographs on the Antarctic continent: the Scott expedition in 1902–1903, the Byrd expedition (Rockefeller Mountain, near Little America) in 1940, the Ronne expedition (Marguerite Bay) in 1947–1948, and the French polar expedition (Adelie Land) in 1951–1952. In 1956 the southernmost operating station was at B. O'Higgins, which is the Chilean Antarctic base (63° 20′ S, 57° 54′ W); the next most southerly was on Macquarie Island, southwest of New Zealand.

THE INTERNATIONAL GEOPHYSICAL YEAR

This book will be published during the International Geophysical Year (abbreviated IGY), which began on July 1, 1957, and which will be extended to December 31, 1958. It is a vast international undertaking, an augmented successor of the International Polar Years of 1882–1883 and 1932–1933.

Although the program is world-wide, there is still special concentration on the polar regions, and on this occasion most of all on Antarctica.

As we go to press it is possible to speak only of plans. Results will be available only in fragments until long after the close of the program.

Popular attention has been caught by the plan to launch artificial satellites; but every branch of geophysics is represented.

In seismology, plans include adding new and special equipment to existing stations; but new stations are being installed, at least for the duration of the IGY, in remote areas. In Antarctica, American parties will operate seismographs at the South Pole, in Marie Byrd Land, and on the Knox Coast (design of equipment and training of personnel for the Knox Coast station was carried out at Pasadena). A British station will operate in Graham Land. A station set up by New Zealand workers, at Scott Base, was reporting some readings by April, 1957, and regularly beginning with July. Chile, Argentina, and the USSR were also planning to operate seismographs in Antarctica.

Some of the special equipment installed is intended for the study of strain accumulation on the Benioff system, some for recording *Lg* and other waves which may give data on the structure of Antarctica and other continents. Seismological surveys of ocean basins are in progress, a world-wide program of observing microseisms is organized, and there will be many new gravity surveys.

The USSR has an intensive program of geophysical investigation, including seismology, in the region of Kamchatka and the Kurile Islands, representing two of the most typical active Pacific arcs. Two new Soviet stations are in operation north of the Arctic Circle, and one, Mirny, in Antarctica.

CHANGES

Some formerly important stations now contribute comparatively little; this is true of Wiechert's station at Göttingen, where many of the foundations of present instrumental seismology were laid. Some were casualties of World War I; one of Milne's stations, on Cocos Island, was mistaken for a military installation and shelled out of existence by the German raider *Emden*. Many more stations in Europe and Asia were damaged or destroyed in World War II. Perhaps the worst loss was that at Manila in 1945; not only the instruments and equipment were destroyed, but also the irreplaceable file of seismograms and documents accumulated for 60 years. The staff of the Manila Observatory has made its new seismological headquarters at the former outpost station of Baguio; the present Manila station is operated by the Weather Bureau.

Changes in names of established stations, mostly for political reasons, are often confusing. Such are the changes from Batavia to Djakarta, and from

La Plata to Eva Perón and back; Galitzin's station, Ekaterinburg, became Sverdlovsk under the Soviet regime. A station initiated under Hungarian auspices as Ogyalla has operated under Czechoslovakian administration as Stará Ďala and Hurbanovo. Before World War I a station in the Sudeten area in the territory of Austria-Hungary was known as Eger; under Czechoslovakia it became Cheb; Hitler changed the name back to Eger, but it is now once more Cheb.

Different transliterations of Russian and Japanese station names have sometimes made one station appear as two. Japanese names beginning alternatively with H or F, as Hukuoka or Fukuoka, are especially bothersome in indexes and alphabetical lists. For other notes on Japanese spellings see Chapter 30.

AMATEUR SEISMOLOGY

Amateur seismology is not so rewarding as amateur astronomy. A good seismograph is as hard to construct as a good telescope; to buy one ready-made is expensive, and sometimes difficult. Moreover, an amateur seismograph is of little scientific value unless great pains, perhaps beyond the amateur's technical skill, are taken to ensure precise timekeeping.

It is not too difficult to set up an instrument with pendulum period of about 10 seconds and static magnification of 200 to 300, recording on smoked paper. Such a seismograph will write spectacular records of large teleseisms; but it will disappoint the amateur and his friends by not showing small local earthquakes which can be felt, and a strong local shock will put it out of action. The necessary equipment for recording local shocks involves short-period instruments with high magnification; so far no one has found a cheap and effective way of accomplishing this.

The amateur who has visions of contributing to science will find it not worth while to set up a seismograph in a large center of population, especially where there is a good station for teleseismic recording not far away. To be sure, in a local-earthquake investigation there is no limit to the number of stations that can be used, even when separated only by a few miles; but the amateur will probably find the necessity for time to the tenth of a second too exacting.

A reasonably good teleseismic station hundreds of miles from others can contribute appreciably; there is plenty of room for such stations in the wide open spaces of the central and western United States. In a seismic area like California, an amateur without an instrument can make a contribution by reporting earthquakes felt and not felt, if he happens to live in a community far away from the large cities where there is usually enough of such information.

From the seismologist's point of view, most amateur stations suffer from lack of continuity; they stop operating over week ends, or holidays, or vacations, or during illnesses, or whenever the amateur's interest flags. Moreover, few amateurs study the subject enough to interpret their seismograms competently; therefore, if the amateur publishes his own readings they are apt to be lacking in value.

There are honorable exceptions; some amateur stations with good equipment are operated continuously with high standards and contribute effectively to international seismology. Generally, they send their readings or seismograms to a station in the same area which incorporates the results in its bulletins.

References

REFERENCES for Chapters 17, 18, 19 provide numerous instances of international seismology in action; note especially the monographs on individual earthquakes. An excellent presentation of the state of seismology in 1954 will be found in the national and regional reports included in:

Rothé, J. P., ed., *Union géodésique et géophysique international, Association de séismologie et de physique de l'intérieur de la terre. Comptes rendus des séances de la dixième conférence réunie à Rome*, Strasbourg, 1955.
A similar publication may be expected for the 1957 meeting at Toronto.
Other specific descriptions include:

Neumann, F., "Earthquake investigations in the United States," U. S. Coast and Geodetic Survey, *Spec. Publ.* 282, Washington, 1952.

Macelwane, J. B., ed., *Jesuit Seismological Association, 1925–1950*, St. Louis University, 1950.

Istituto nazionale di geofisica, *Supplement to Annali di geofisica*, Rome, 1954.

Carder, D. S., *The Seismograph and the Seismograph Station*, U. S. Dept. of Commerce, Washington, D. C., 1956. (Has a short but excellent section on amateur seismology.)

For current information on the International Geophysical Year refer to the IGY Bulletin, prepared by the National Academy of Sciences. Bulletins Nos. 1 and 2 appear in *Trans. Amer. Geophys. Union*, vol. 38 (1957), pp. 611–626 and 627–641. The official central record of IGY activity will be included in *Annals of the International Geophysical Year*, in course of publication by Pergamon Press, London.

CHAPTER 22

Magnitude, Statistics, Energy

THE MAGNITUDE SCALE

Origin

The idea of an earthquake magnitude scale based purely on instrumental records arose naturally out of experience familiar to working seismologists. No one can spend many days at a seismological station without being impressed by the outrageous discrepancy that sometimes exists between the amount of popular excitement and alarm touched off by an earthquake and its actual character as indicated by seismograms. A small shock perceptible in the Los Angeles metropolitan center will set the telephone at the Pasadena laboratory ringing steadily for half a day; while a major earthquake under some remote ocean passes unnoticed except for seismograph readings, and rates a line or two at the bottom of a newspaper page. A few examples follow.

On and after August 5, 1949, an earthquake of magnitude 6¾ in Ecuador filled pages of the press with news items and pictures of appalling devastation —nearly all of which was due to a secondary effect, an enormous slide that overwhelmed a populated valley. In the excitement a shock of magnitude 7.5 a few hours later in the southwest Pacific was either ignored by the press or confused with the Ecuador shock.

It was hard to persuade some persons in southern California that the destructive Long Beach earthquake of 1933 was a minor event compared to the California earthquake of 1906. This misunderstanding became serious when it was publicly argued that southern California in 1933 had had "a great earthquake," that no more important shocks need be expected for many years, and that, consequently, safety precautions could be relaxed.

In the minds of the populace at Bakersfield, California, the major earthquake (magnitude 7.7) of July 21, 1952, was much less important than the comparatively minor aftershock (5.8) on August 22, which originated closer to that city and was accompanied by much more local damage, representing higher local intensity (VIII as against VII—although it is hard to separate the effects of the later earthquake individually from the cumulative effect of the major earthquake and numerous aftershocks). Figure 28-24 shows the

338

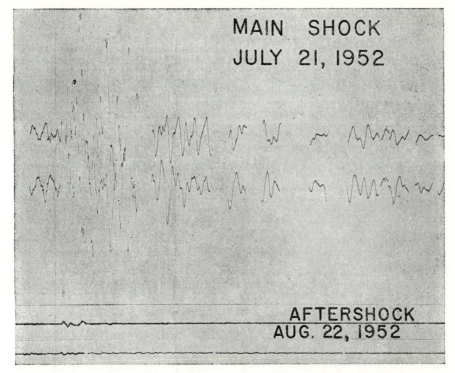

FIGURE 22-1 *Pasadena strong-motion records of the Kern County earthquake of July 21, 1952 and its aftershock on August 22 (two records of each).*

much wider extent of the isoseismals for the larger shock; Figure 22-1 gives a direct instrumental comparison. (See also Chapters 6, 8, and 28.)

Even seismologists have been misled. A French worker once published a paper on the peculiarities of earthquakes in western America as recorded in Europe, a principal point being that the Long Beach earthquake wrote much smaller seismograms than the Nevada earthquake of 1932, "which attained much lower intensity." His ideas of maximum intensity were based on press accounts, and he failed to consider the relative population density of Los Angeles County, California, and Nye County, Nevada (magnitudes: Long Beach, 6.3; Nevada, 7.3). On November 18, 1929, some seismologists in the eastern United States believed that they were recording one of the greatest earthquakes; actually, a rather ordinary major shock (magnitude 7.2) had occurred in the Atlantic unusually near their stations.

In 1931 the Pasadena center was about to issue its first listing of earthquakes located in southern California. To list two or three hundred such shocks in a year, without indicating their magnitude in some rational way, might lead to serious misinterpretation. Rating in terms of intensity was hardly possible; many of the shocks were not reported as felt, and some obviously large ones were reported only as barely perceptible at points distant from the epicenter, which might be in a thinly populated desert or mountain area or off the coast.

To discriminate between large and small shocks on a basis more objective than personal judgment, a plan was hit upon which succeeded beyond expectation. The very simple fundamental idea had already been used by Wadati to compare Japanese earthquakes.

Construction of a Scale—First Form

Of two earthquakes having the same hypocenter and recorded at the same stations, the larger should write larger seismograms at any one station. If the epicenters differ, the smaller shock may be so much closer to a given station that it writes larger seismograms there. Accordingly, the general procedure is first to determine the epicenter for each shock and then to plot the maximum ground motion at each station as ordinate with the corresponding epicentral distance as abscissa. Of two curves thus plotted for different earthquakes, one will probably be higher than the other, indicating that it represents the larger event.

Calculation of maximum ground motion calls for careful application of the principles given in Chapter 15. For the immediate practical purpose in southern California, considerable simplification was possible. All the stations used were equipped with torsion seismometers designed to have the same constants: $T_0 = 0.8$ second, $V = 2800$, $h = 0.8$. They should have had the same magnification for ground motion of the same character. For this reason, the plots were made using not computed ground motion but trace amplitudes in millimeters and tenths as measured directly on the seismogram; this method materially reduces working time when considering the numerous local earthquakes. Amplitudes were plotted on a logarithmic scale; since the measurements ranged from 0.1 millimeter to 10 or 12 centimeters, this gave a more manageable chart than a linear scale. Moreover, the results could now be stated numerically.

Figure 22-2 shows logarithmic plots for the data of observed earthquakes. The representative curves are roughly parallel to each other and to the dashed curve which is drawn at an arbitrary level. If this parallelism were exact, the difference between the logarithms of amplitudes of any two given shocks would be independent of distance; the amplitudes themselves would be in constant ratio. We could define a quantity M, to be called magnitude, given by

$$M = \log A - \log A_0 \tag{1}$$

where A is the recorded trace amplitude for a given earthquake at a given distance as written by the standard type of instrument, and A_0 is that for a particular earthquake selected as standard. The magnitude is thus a number

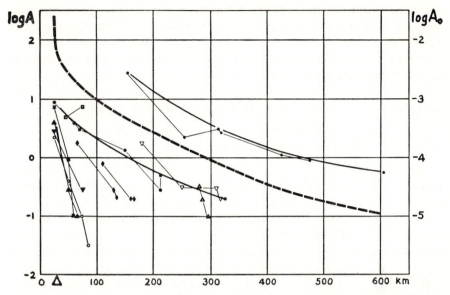

FIGURE 22-2 *Origin of the magnitude scale. Data for Southern California earth-quakes of January, 1932. [Redrafted from the original notes.]*

characteristic of the earthquake and independent of the location of the recording stations.

The term magnitude was selected by analogy with the corresponding usage in astronomy. The scale of star magnitudes is also logarithmic, though on a less simple basis; in a sense it is reversed, since the greater the magnitude the fainter the star.† The earthquake magnitude scale follows the more obvious course of assigning the larger number to the larger earthquake. Logarithmic scales are in use in other fields; examples are the decibel scale in acoustics and the pH scale for hydrogen-ion concentration.

Three arbitrary choices enter into the definition of M: (1) the use of a particular type of seismometer with specified constants; (2) the use of ordinary logarithms to the base 10; (3) selection of the standard shock whose amplitudes are represented by A_0. This standard shock has also been called the zero shock, since, if $A = A_0$, $M = 0$. This clearly does not mean "no earthquake"; a small earthquake might conceivably record with amplitudes smaller than those of the standard shock, which would give it a negative magnitude. The zero level was intentionally chosen low enough to make the magnitudes of the smallest recorded earthquakes positive.

The zero level A_0 can be fixed by naming its value at a particular distance. This was taken to be one thousandth of a millimeter at a distance of 100 kilometers from the epicenter; an equivalent statement is that an earthquake

---

† Earthquake magnitude corresponds logically to "absolute" stellar magnitude; apparent star magnitudes correspond to earthquake intensity.

recording with trace amplitude of 1 millimeter measured on a standard seismogram at 100 kilometers, is assigned magnitude 3. Naturally, shocks of magnitude near zero can be found recorded only on seismograms written at very short distances; their study requires additional instruments with magnification higher than the standard.

Assignment of Magnitude in Practice; Local Shocks

Applicability of the magnitude scale depends on establishing standard values of log A_0 as a function of distance Δ (see Table 22-1).

Table 22-1   Logarithms* of the Amplitudes A_0 (in millimeters) with which a Standard Torsion Seismometer ($T_0 = 0.8$, $V = 2800$, $h = 0.8$) Should Register an Earthquake of Magnitude Zero

Δ (km)	$-\log A_0$	Δ (km)	$-\log A_0$	Δ (km)	$-\log A_0$
0	1.4	150	3.3	390	4.4
5	1.4	160	3.3	400	4.5
10	1.5	170	3.4	410	4.5
15	1.6	180	3.4	420	4.5
20	1.7	190	3.5	430	4.6
25	1.9	200	3.5	440	4.6
30	2.1	210	3.6	450	4.6
35	2.3	220	3.65	460	4.6
40	2.4	230	3.7	470	4.7
45	2.5	240	3.7	480	4.7
50	2.6	250	3.8	490	4.7
55	2.7	260	3.8	500	4.7
60	2.8	270	3.9	510	4.8
65	2.8	280	3.9	520	4.8
70	2.8	290	4.0	530	4.8
80	2.9	300	4.0	540	4.8
85	2.9	310	4.1	550	4.8
90	3.0	320	4.1	560	4.9
95	3.0	330	4.2	570	4.9
100	3.0	340	4.2	580	4.9
110	3.1	350	4.3	590	4.9
120	3.1	360	4.3	600	4.9
130	3.2	370	4.3		
140	3.2	380	4.4		

* Since A_0 is less than 1, its logarithm is negative, and the table shows values for $-\log A_0$.

It is necessary to know the epicentral distance of the recording station, at least approximately. (Small errors in distance affect the magnitude only slightly.) The maximum trace amplitude on a standard seismogram is then measured in millimeters, and its logarithm taken. To this is added the quantity tabulated as $-\log A_0$ for the corresponding distance. The sum is a value for M.

This procedure is adequate for assigning magnitudes to earthquakes recorded at short distances. In using the data of a station with standard seismographs recording both horizontal components, it is correct to determine magnitude independently from each and to take the mean of the two determinations. This method is preferable to combining the components vectorially, for the maximum motion need not represent the same wave on the two seismograms, and it even may occur at different times. Rough rules like this are necessary for routine work in assigning magnitudes to hundreds of earthquakes.

A correction is applied for each station, or still better for each instrument. It is determined by examining statistically the magnitude determinations for a large number of shocks and finding the systematic deviation of the magnitude determined for any one instrument from that found from the mean result for all instruments. This procedure attaches to each instrument a correction similar to the "personal equation" of an individual observer. It is probably related chiefly to the local conditions of ground and installation. These corrections as now used for the southern California network are:

N-S component: Pasadena +0.2, Riverside +0.2, Santa Barbara −0.2, Tinemaha −0.2, Haiwee 0, Barrett −0.2.

E-W component: Pasadena +0.2, Riverside +0.2, Santa Barbara −0.2, Tinemaha −0.2, Haiwee 0, Woody −0.1.

An example of magnitude determination is given in Table 22-2. The mean of the several magnitude determinations is 4.26, and deviations from this are only a few tenths. The station corrections have been incorporated; if they

Table 22-2   Magnitude Determination—Earthquake of January 15, 1955, 01:03 G.C.T.

Station	Amplitude N-S (mm)	Amplitude E-W (mm)	Δ (km)	M from N-S	M from E-W
Pasadena	8.4	6.0	114	4.2	4.1
Riverside	7.9	8.5	179	4.5	4.5
Santa Barbara	24.5	30	90	4.2	4.3
Tinemaha	8.1	7.0	246	4.5	4.4
Barrett	0.8	—	328	3.9	—
Woody	—	16.8	84	—	4.0

had not been used, the dispersion would have been greater. The earthquake used for this example was a late aftershock of the Kern County series beginning in 1952. This is a fair example of the consistency of magnitude determination in routine work with local earthquakes. Magnitudes are easily assigned to the nearest half unit and can ordinarily be given to the tenth with an uncertainty not much exceeding one tenth. Shocks thus dealt with generally range in magnitude from 2 to 5, so that they are distributed among not less than 15 distinguishable levels of magnitude. The smallest identifiable earthquakes have magnitudes just above the zero of the scale; the largest are over magnitude 8, which means that at a given distance there is a ratio of not less than 10^8 between the disturbances produced by the largest and the smallest earthquakes. This shows the necessity, in a complete seismological program, of working with instruments of several different ranges of magnification.

Implied Assumptions

The magnitude scale as originally set up for local earthquakes worked out much better than expected. Instead of a rough separation of large, medium, and small shocks, it yielded a rather finely divided grading. That this is possible is due to the really enormous range between the largest and the smallest earthquakes, so that subdivision is possible even on the crudest assumptions.

On analysis, use of the scale implies that earthquake records at a given distance are all alike except for amplitude, so that the seismogram of a large shock can be obtained from that for a small one by multiplying all the motion with a constant factor. That this is not so is demonstrated by even moderate experience with seismograms, but the error does not begin to be troublesome until one works with larger shocks than are ordinarily handled in local-earthquake routine.

The relation between the magnitude of an earthquake and the amplitude of its seismograms should be like that between the power of a radio station and the signal strength on a particular receiving set. The signal depends on both power and distance of the transmitter and varies with conditions along the path of transmission and at the receiving station, including the amplification and selectivity of the receiver. Similarly the recorded amplitude of an earthquake should not only depend on magnitude and distance, but should also vary with physical conditions along the path and ground conditions at the recording station, as well as with the characteristics of the seismograph used. Moreover, in earthquakes as in radio transmission there are strong directional effects; much more energy may be radiated in some directions than in others, so that an acceptable assignment of magnitude calls for data from a number of stations surrounding the epicenter. The last requirement is often overlooked even by experienced seismologists, and magnitudes assigned from the readings of a single station, which in the nature of things can never be more than preliminary estimates, are handled as if they were final determinations.

Limitations

The magnitude scale was originally dependent on a particular instrument, the standard torsion seismometer. Theoretically it should be possible to use any other seismometer whose constants are known, for the motion of the ground may be calculated and from this the expected response of the torsion seismometer may be computed. The chief difficulty is that the wave which appears as maximum as written by the torsion instrument with its short free period of 0.8 second will not be the wave of maximum amplitude on the seismogram of a more usual type of instrument with comparatively long period. Instead, the corresponding motion will be written as apparently small, short-period oscillations superposed on larger, long-period swings. If the amplitudes and periods of these large swings are determined, and the corresponding response of the torsion instrument computed, application of Table 22-1 will give an incorrect magnitude (for most local earthquakes, too small a magnitude).

Another limitation of Table 22-1 is that without further evidence it cannot be assumed to apply outside the California area. The variation of recorded amplitude as a function of distance must change with altered hypocentral depth. Table 22-1 cannot be applied to deep-focus earthquakes; moreover, the mean depth for California earthquakes (about 16 kilometers) is less than the probable average depth for shallow shocks (taken as 25 kilometers in setting up transit-time tables). The relation between distance and amplitude may also be affected by differences in crustal structure or materials.

MAGNITUDE SCALE FOR TELESEISMS

Use of Surface Waves

It is difficult to extend Table 22-1 much beyond 600 kilometers for the California area; earthquakes favorably placed for the purpose are not frequent, and there is good evidence of differences along different paths. In general, assignment of magnitude at distances between 600 and 2000 kilometers is subject to numerous uncertainties, involves the effect of local crustal structure, and can be approached only by way of special investigation for each recording station.

For larger epicentral distances consistent assignment of magnitudes is possible. Dr. Gutenberg and the writer carried this out at first only for shallow earthquakes, using the computed horizontal ground amplitude, in microns, of the large surface waves with periods near 20 seconds which form the maximum of the normal seismogram (the two horizontal components are combined vectorially). The result as last revised appears in Table 22-3.

| *Table 22-3* | Logarithms of the Maximum Combined Horizontal Ground Amplitude A (in microns) for Surface Waves with Periods of 20 Seconds Produced at the Given Distance by a Standard Shock Taken as Magnitude Zero. (Correlation with Table 22-1 for short distances imperfect and now under investigation) |

Δ (degrees)	$-\log A$	Δ (degrees)	$-\log A$
20	4.0	90	5.05
25	4.1	100	5.1
30	4.3	110	5.2
40	4.5	120	5.3
45	4.6	140	5.3
50	4.6	160	5.35
60	4.8	170	5.3
70	4.9	180	5.0
80	5.0		

Magnitudes determined by using this table independently at different stations show good agreement. Station corrections can be derived as for the local magnitude scale; for Pasadena the correction is +0.1. These corrections include what was earlier termed the "ground factor" of the individual station.

Geographical Corrections. Systematic investigation also shows what appears to be an effect of path; earthquakes in certain areas record with abnormally large or small surface waves at a given station. In part, such effects may be due to the mechanism of the shock; Gutenberg's investigation of the major Kern County earthquake (Chapter 28) clearly showed systematic variation of energy radiated as surface waves with azimuth, which was easily correlated with the progression of faulting on that occasion. Such effects, when detected, must be allowed for in assigning final magnitudes.

Systematic geographical corrections for Pasadena were published by Gutenberg chiefly to show the loss of energy in surface waves traveling along the boundary of the Pacific basin. They were determined by finding residuals for magnitudes derived from the data for a given station against the magnitudes established from all available data. When these residuals are plotted on a map they show significant geographical distribution, which should result either (1) from effects along the path or (2) from azimuthal preference in the radiation of surface waves due to the tectonic characteristics of the epicentral region. Such plots have been constructed by Peterschmitt for Strasbourg, by di Filippo and Marcelli for Rome, by Båth for Uppsala, and by other workers for their own stations. This procedure should eventually improve the assignment of magnitudes from surface waves.

Relation to the Local Scale. The general precision of results derived from Table 22-3 (without geographical corrections) is not quite so high as that for the local-earthquake scale. While magnitudes are not uncommonly assigned by this method to the tenth of a unit, they are rarely dependable to better than ¼ unit. This form of scale has now come into general use; it is slightly disappointing that the latest revision indicates that it is not wholly consistent with the original scale for California local shocks. Earthquakes to which magnitudes can be assigned by both methods are chiefly in the range of magnitude from 6 to 7. Table 22-3 accordingly was adjusted to agree with Table 22-1 for such shocks. The latest investigation indicates that the table for teleseisms gives slightly lower values of magnitude, especially for the smaller shocks (too small to record legibly as teleseisms), where the discrepancy appears to reach half a magnitude. The exact correction to be applied between the two tables is not yet stabilized; as this is written, recording has commenced in southern California with instruments having the response curve of the standard torsion seismometer except for lower magnification. This should eventually remove much of the remaining uncertainty.

Use of Body Waves

Deep-focus earthquakes commonly do not register surface waves with periods near 20 seconds and measurable amplitudes; even when such motion is recorded in the expected part of the seismogram, much of it appears to be due to body waves, especially S waves, repeatedly reflected at the surface. Any magnitude scale for deep-focus earthquakes must be founded on the recorded body waves.

This at first looked very unpromising, for the amplitudes of P and S were known to vary rapidly and rather irregularly with increasing distance. Slight irregularities in the variation of velocity with depth lead to relatively rapid changes in the angle of incidence at the surface, resulting in concentration of energy at certain distances and dispersal at others.

The procedure adopted was to investigate the variation of amplitude of P, S, and PP with distance for shallow earthquakes whose magnitudes had been determined from surface waves. Once these were standardized, it was possible to proceed from shallow to intermediate and deep shocks. This laborious investigation was carried out by Dr. Gutenberg with complete success. It resulted in a set of charts from which a logarithmic constant for the earthquake of magnitude zero can be read as a function of depth and epicentral distance for P, S, or PP (Appendix VIII). However, this additive constant is not $-\log A$ but $-\log (A/T)$, where T is the period, in seconds, of the corresponding wave and A represents ground amplitude in microns computed from the trace measurement and the constants of the instrument used.

The passage from shallow to deep-focus earthquakes is partly theoretical. The velocity distribution in the earth being known with considerable accuracy, it is possible to compute the effect on the amplitude of seismic waves arriving at a given point assuming that the depth of the hypocenter varies without any other change.

The shift in basis from A to A/T was dictated partly by the considerable variation in prevailing periods of the principal wave groups for different earthquakes and partly by the necessity of using amplitudes and periods determined from instruments of widely different characteristics. The quantity A/T has the advantage of being simply related, in theory at least, to the kinetic energy of its wave train.

Relation to the Surface-Wave Scale.   Magnitudes determined in this way from body waves of shallow earthquakes were adjusted to agree with those found for surface waves when the magnitude is near 7. For larger magnitudes it soon developed that there is a systematic deviation between the results of the two procedures, and this was compensated for by adding approximately $\frac{1}{4} (M - 7)$ to the magnitude found from body waves. The amount of this correction was modified from time to time as more information accumulated; eventually it was regularly applied to deep-focus earthquakes as well as to shallow shocks. The latest form of this linear relation is

$$m = 2.5 + 0.63M \qquad \text{or} \qquad M = 1.59m - 3.97 \qquad (2)$$

where m and M are magnitudes derived from body waves and surface waves respectively. The two values agree at $m = M = 6.75$; above this $M > m$, below it $M < m$. The full implication of the latter result was not realized for some time, and a number of earthquakes of magnitude lower than 6.75 were suspected of having focal depths in excess of 25 kilometers simply because the surface waves appeared small relative to the body waves. More recent investigation strongly suggests that linear relationship (2) is valid over the whole range, so that small surface waves for small shocks are not evidence of deep focus. The physical conclusion is that the efficiency of generation of surface waves varies more rapidly with magnitude than that of body waves. Gutenberg suggests that this may be correlated with extension of faulting to the surface in large shocks, as well as with larger and slower displacements favoring generation of long surface waves.

Use of Two Scales.   The deviation between m and M raises the question as to which is, objectively, the better standard. If practicable, it would be preferable to reduce all magnitude determinations to the original scale for local earthquakes, represented by Table 22-1. At present there are not enough good data for the purpose, but there is the prospect that low-magnification instruments lately installed will supply it. Pending such an adjustment, magnitudes continue to be reported (as in this book) on the old basis (M for teleseisms, and the original scale for local earthquakes; the two agree closely enough in the magnitude range where both are applicable). However, for theoretical purposes, especially where calculations of energy are involved, Dr. Gutenberg and the writer have agreed upon the use of what he has termed a "unified magnitude scale" in which magnitude determinations from all sources are essentially reduced to the body wave basis by such methods as seem provisionally most reliable. Magnitudes on this basis are designated by m, those on the old (and imperfectly consistent) basis by M, from whatever source derived.

In practice, magnitude determination from body waves always yields m. From surface waves, Table 22-3 will yield M; but this can be reduced to m by applying equation (2), or a special table can be constructed to read m directly. Values of m from body waves and surface waves may then be combined to give a final adopted value of m. It would be equally practicable to reduce m derived from body waves to a corresponding value of M, as was usually done prior to Gutenberg's erection of m into the "unified magnitude scale." Gutenberg now regularly reports his results as m; in this book these have been replaced by M according to equation (2) with rounding off as shown in Table 22-4.

Table 22-4   Relation Between m and M for Large Magnitudes (as adopted in this book)

m	M	m	M
6.8	6.8	7.6	8.1
6.9	7.0	7.7	8.3
7.0	7.1	7.8	8.4
7.1	7.3	7.9	8.6
7.2	7.5	8.0	8.7
7.3	7.6	8.1	8.9
7.4	7.8	8.2	9.0
7.5	7.9		

Magnitudes below 6½ are in general repeated from *Seismicity of the Earth*; some of these may be altered if and when revision in the light of the latest investigations is undertaken.

Since the magnitude scale is no longer local in application, it is desirable to free it from dependence on local California conditions. The ultimate definition may be in terms of the amplitude and period of the maximum of the P group at a specified distance. At present this appears preferable to a definition in terms of surface waves; for, although data on surface wave maxima are more plentiful, being published regularly by many stations, experience indicates that they furnish no more reliable a measure of magnitude than do the body waves, while body wave data must be used in working with deep shocks. Earnest and repeated efforts have been made to encourage stations to determine and publish amplitudes and periods of at least P, S, and PP for the principal registered shocks. Of course, this adds appreciable labor to the preparation of bulletins, and too often such reporting is undertaken for a year or two and then dropped. (The same thing occurs with the reporting of times and other seismometric details.) It is hoped that the future will see more and better data available for magnitude determinations.

REVISED MAGNITUDES—1957

Magnitudes reported in *Seismicity of the Earth* were determined by Gutenberg (with assistance, chiefly clerical, from the writer and others) in a project carried on for about 15 years. In the course of this work the relation between body wave and surface wave amplitudes developed gradually, and it

Table 22-5   Largest Shallow Shocks—1904–1957
($m = 7.9$, $M = 8.6$ or over)

Date	G.C.T.	Epicenter		m from Body Waves	m from Surface Waves	Adopted m	Corresponding M (Table 22-4)
1905 Apr. 4	00:50.0	33° N,	76° E	8.0	7.8	7.9	8.6
July 23	02:46.2	49° N,	98° E	8.1	(7.8)	8.0	8.7
1906 Jan. 31	15:36.0	1° N,	81½° W	8.2	8.1	8.1	8.9
Aug. 17	00:40.0	33° S,	72° W	?	7.9	7.9	8.6
1911 Jan. 3	23:25:45	43½° N,	77½° E	8.1	7.9	8.0	8.7
1917 May 1[1]	18:26.5	29° S,	177° W	7.9±	?	7.9±	8.6±
June 26	05:49.7	15½° S,	173° W	8.0	8.0	8.0	8.7
1920 Dec. 16	12:05:48	36° N,	105° E	7.9	7.9	7.9	8.6
1929 Mar. 7[1]	01:34:39	51° N,	170° W	7.9	(7.8)	7.9	8.6
1933 Mar. 2	17:30:54	39¼° N,	144½° E	8.2	8.0±	8.1	8.9
1938 Feb. 1	19:04:18	5¼° S,	130½° E	8.0	7.7	7.9	8.6
Nov. 10	20:18:43	55½° N,	158° W	8.2	7.8	8.0	8.7
1941 June 26[1]	11:52:03	12½° N,	92½° E	8.0	7.9±	8.0	8.7
1942 Aug. 24[1]	22:50:27	15° S,	76° W	7.8	8.0	7.9	8.6
1950 Aug. 15	14:09:30	28½° N,	96½° E	8.0	8.0	8.0	8.7
1952 Mar. 4	01:22:43	42½° N,	143° E	8.0	7.8	7.9	8.6

[1] Shocks at depths of the order of 50 to 60 kilometers.

was not practicable to modify previous determinations constantly. In his most recent research, directed toward clarifying the magnitude-energy relation and setting up a unified magnitude scale, Gutenberg has revised determinations for all shocks previously assigned magnitude $M = 7$ or over, as well as for a number of others which revision elevates to that level. The work with body waves involves using the revised charts given in Appendix VIII, which at some points differ significantly from those previously published and used.

Special interest attaches to the magnitudes of the largest shallow shocks. Details of Gutenberg's revision are given in Table 22-5 for those assigned $m = 7.9$ (hence $M = 8.6$) or over from 1904 to date; this is the interval for which fairly adequate data are available. The great shocks of 1897–1903, listed in Appendix XIV, are assigned magnitudes on a less certain basis—the records of Milne's instruments. Six of these earthquakes, listed in Table 22-6,

Table 22-6   Largest Shallow Shocks—1897–1903

Date	G.C.T.	Epicenter	m	M
1897 June 12	11.1	26° N, 91° E	8	8.7 ±
Aug. 5	00.2	38° N, 143° E	8	8.7 ±
Sept. 20	19.1	6° N, 122° E	7.9 ±	8.6
Sept. 21	05.2	6° N, 122° E	8 ±	8.7 ±
1899 Sept. 10	21:41	60° N, 140° W	7.9	8.6
1902 Aug. 22	03:00	40° N, 77° E	7.9	8.6

are comparable with those of Table 22-5. The epicenters from both tables are mapped on Fig. 22-3.

In this book Gutenberg's m is generally not reported, being replaced by M; but to illustrate his procedure both m and M are given in Tables 22-5 and 22-6. In Table 22-5 Gutenberg's values of m derived from surface waves and from body waves are given separately, followed by his adopted m, and finally by the corresponding M from Table 22-4.

An incidental effect of the revision has been to increase the values of M for most of the largest shocks. Whereas in *Seismicity of the Earth* only two (January 31, 1906, Colombia-Ecuador; August 15, 1950, Tibet-Assam) were assigned $M = 8.6$, Table 22-5 has 16 entries. Two of these reach $M = 8.9$; one is the same shock of 1906, the other is the great Sanriku (Japan) earthquake of 1933, which recorded with outstandingly large body waves. The 1950 earthquake is now matched by five others.

Table 22-7, with only four entries, shows the comparative rarity of intermediate shocks of the largest magnitudes. No corresponding table is given for deep shocks, since there is only one known shock deeper than 300 kilo-

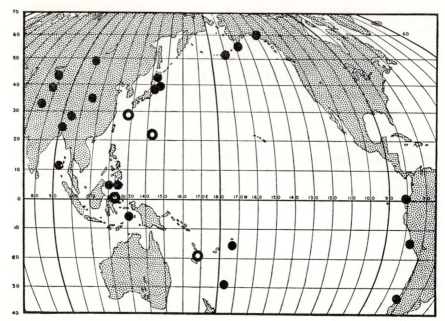

FIGURE 22-3   *Largest known earthquakes* (M = 8.6 *or over*), 1897–1956. *Solid circles, shallow; open circles, intermediate depth.*

meters as large as $m = 7.8$, $M = 8.4$—that of 1906, January 21, 13:49:35, at a depth of 340 kilometers under Japan (34° N, 138° E).

More extensive tables of large shocks, giving values of M only, appear in Appendix XVI.

Table 22-7   Largest Known Intermediate Shocks—1904–1957 ($m = 7.9$, $M = 8.6$, or over)

Date	G.C.T.	Epicenter	Depth (km)	m	M
1910 June 16	06:30.7	19° S, 169½° E	100	7.9	8.6
1911 June 15	14:26.0	29° N, 129° E	160±	8.0	8.7
1914 Nov. 24	11:53:30	22° N, 143° E	110±	8.0	8.7
1939 Dec. 21	21:00:40	0° N, 123° E	150±	7.9	8.6

OBSERVED INTENSITIES IN RELATION TO MAGNITUDE

After the magnitude scale was set up in California, it was seen that magnitude can be correlated roughly with the maximum intensity in the meizoseismal area when sufficient macroseismic data are available. The smallest shocks reported felt by persons are near magnitude 2; those causing damage to weak structures (intensity VII) are usually of magnitude 5 or over; destructive earthquakes generally exceed magnitude 6. For ordinary ground conditions in metropolitan centers in California, the following figures are representative:

Magnitude	2	3	4	5	6	7	8
Maximum Intensity	I–II	III	V	VI–VII	VII–VIII	IX–X	XI
Radius (km)	0	15	80	150	220	400	600

The radius given here is the mean epicentral distance of the limit of perceptibility to persons. This is extremely rough, since the isoseismals are usually elongated and show additional irregularities; it neglects the effects of increased intensity on unconsolidated ground and of lowered intensity on firm rock. Great caution should be used in applying even this rough tabulation elsewhere than in California; in other regions the prevailing hypocentral depth and differences in crustal structure may result in a different relation between magnitude and intensity distribution. In the central United States earthquakes which barely reached intensity VII near the epicenter have been perceptible over wide areas; for most of these no instrumental magnitudes can be given, but the Oklahoma shock of April 9, 1952, had a magnitude of 5½ or less and yet was felt in seven states and to distances of over 600 kilometers.

This disposes of the suggestion that, at least for public use, the magnitude might be replaced by an estimate of epicentral intensity. Not only would such an estimate be extremely unreliable, but also it would almost certainly be confused with actual intensities observed in the field. Appreciable risk of error is already taken in issuing preliminary magnitude estimates from the data of single stations; however, these are usually good enough for ordinary purposes, and they can be revised in time for use in seismological or engineering publications.

"CONFUSION"

The magnitude scale had been in use for over 15 years before it received much attention, except from laboratory seismologists. Magnitudes came into general notice about 1950; a number of stations used them in reporting earthquakes to the press. The scale thus gained wide circulation, for journalists soon learned by experience that magnitudes assigned by independent workers were usually in good agreement and bore some relation to the possible importance of earthquakes as news.†

---

† Misinterpretations of the scale are frequent. The hardest to combat is the assertion that magnitudes are "on a scale of ten," a fossil preservation of the Rossi-Forel intensity scale.

Lately there have been complaints that the use of the magnitude scale is confusing, or at least that the reporting of magnitudes in the newspapers "confuses the public." Sometimes it is said that the double use of magnitude and intensity is confusing, and seismologists should stick to one; this is like saying that a police description of a thief is confusing because it gives both his height and weight.

Much of the confusion complained of exists in the minds of the objectors. It is confusion of the physical concepts involved in magnitude and intensity, which existed before the magnitude scale was set up, and is merely exposed by its introduction. In the past, seismologists and engineers have on occasion used intensity both in its proper significance and with the meaning of magnitude. Earthquakes have been referred to as of "intensity X" with any one of the following meanings.

(1) Intensity X manifested at a specified location.

(2) Intensity X at the largest and best-known center of population (San Francisco, for example).

(3) Intensity X manifested as the observed maximum, usually on unconsolidated ground.

(4) Intensity X as the observed maximum, excepting areas of unstable ground.

(5) Intensity X as the probable maximum at the epicenter, estimated from effects at a distance.

Of these, only the first corresponds to the definition of intensity. The last is a crude equivalent for magnitude. The other three are commonly substituted for these in description and discussion; some writers shift from meaning to meaning between one sentence and the next.

Field workers have sometimes been misled by attempting to correlate observed intensities too simply with magnitude, without due allowance for the extreme effects of variation in ground and in construction, or for the usually eccentric position of the instrumental epicenter with respect to the isoseismals. Thus there was disappointment that the local effects of shaking in the 1952 Kern County earthquake were not much stronger than the highest intensity manifested in the much smaller Long Beach earthquake of 1933; yet the 1952 shock developed a visible fault trace, and the geographical extent of all the effects was far greater in 1952 than in 1933. On the other hand, faulting in the Nevada earthquake of December 16, 1954, was so spectacular that investigators at first were disinclined to accept an assignment of magnitude less than that for the Nevada earthquake of 1915; however, the isoseismals of 1954 were of slightly lesser geographical extent than those of 1915, indicating a slightly lower degree of radiation of energy in seismic waves.

A more serious type of confusion results from the publication of lists of supposedly larger earthquakes based on maximum reported intensity which ignore the instrumental evidence that earthquakes of all magnitudes are much more frequent than non-instrumental evidence, however carefully gathered, can hope to show. Such lists overemphasize the seismicity of populated districts and thus distort statistics. In regions like California, where population is now spreading into previously unoccupied and in some places highly seismic areas, the public is thus being not merely confused, but dangerously misled.

MAGNITUDE AND STATISTICS

Large Shocks

The magnitude scale has placed earthquake statistics on a new basis, much more uniform than any previously established. Careful investigators have

Table 22-8 Annual Numbers, $m = 6.9$–7.4 ($M = 7.0$–7.8) Inclusive

A. Shallow Shocks (Depths not over 60 km)

1918	8	1926	13	1934	13	1942	15	1950	20
1919	8	1927	13	1935	17	1943	27	1951	11
1920	4	1928	21	1936	23	1944	22	1952	11
1921	6	1929	16	1937	14	1945	21	1953	12
1922	8	1930	12	1938	20	1946	24	1954	10
1923	19	1931	19	1939	11	1947	24	1955	12
1924	10	1932	9	1940	10	1948	19	1956	8
1925	20	1933	11	1941	17	1949	20		

B. Intermediate Shocks (Depths 70–300 km)

1904	0	1915	5	1926	7	1937	10	1948	9
1905	2	1916	7	1927	6	1938	9	1949	12
1906	1	1917	2	1928	3	1939	8	1950	9
1907	1	1918	2	1929	4	1940	11	1951	7
1908	2	1919	2	1930	3	1941	8	1952	2
1909	6	1920	0	1931	4	1942	10	1953	6
1910	9	1921	3	1932	2	1943	14	1954	4
1911	6	1922	4	1933	3	1944	13	1955	3
1912	7	1923	0	1934	9	1945	4	1956	3
1913	6	1924	3	1935	5	1946	8		
1914	7	1925	1	1936	4	1947	2		

C. Deep Shocks (Depths over 300 km)

1904	0	1915	2	1926	1	1937	1	1948	2
1905	1	1916	2	1927	1	1938	0	1949	2
1906	0	1917	1	1928	1	1939	3	1950	4
1907	2	1918	3	1929	3	1940	3	1951	1
1908	0	1919	1	1930	1	1941	3	1952	1
1909	0	1920	1	1931	2	1942	1	1953	0
1910	4	1921	0	1932	4	1943	2	1954	2
1911	4	1922	1	1933	3	1944	1	1955	2
1912	2	1923	0	1934	3	1945	2	1956	5
1913	0	1924	4	1935	2	1946	4		
1914	0	1925	0	1936	1	1947	1		

always attempted to avoid confusing small and large earthquakes in their counts. Without instrumental records this is difficult, especially in an area where the density of population varies from sparse to concentrated, as in California. Minor earthquakes felt in centers of population are much more likely to be catalogued than similar shocks elsewhere; and a comparatively large shock may be reported only at a few isolated points and fail to impress the cataloguer. Even with good seismograms, some systematic elimination of the effect of distance on amplitude is needed; otherwise there is a false appearance of concentration of seismicity in the vicinity of well-equipped recording stations (see Chapter 25).

For statistical details the reader is referred to *Seismicity of the Earth* and to recent papers by Gutenberg. Certain broad outlines are clearly apparent from the revised lists of large shocks (Appendix XIV). Unpublished lists by Gutenberg show annual numbers of shocks assigned $m = 6.9 - 7.4$ ($M = 7.0 - 7.8$) inclusive; see Table 22-8. Tabular totals from this table and from Appendix XIV may be summarized as in Table 22-9.

Table 22-9   Numbers of Large Shocks

	1897–1903	1904–1917	1918–1955
Shallow Shocks			
M = 8.6 or over	6	7	9
7.9–8.5	32±	25	66
7.0–7.8	—	—	570
Intermediate Shocks			
M = 8.6 or over	—	3	1
7.9–8.5	—	12	8
7.0–7.8	—	61	214
Deep Shocks			
M = 7.9–8.5	—	4	4
7.0–7.8	—	18	66

For 1897–1903, valid statistics for the world are available only for the largest shallow shocks. Two or three large intermediate shocks are known in this interval, but the listing is made possible more or less by accident and certainly is not representative. For 1904–1917, no effort has been made to force statistics for shallow shocks below $M = 7.9$ out of the imperfect data. In the same interval, large, intermediate, and deep shocks have been searched for with care, but listing is incomplete. These years precede those covered by the *International Summary*; its first issues applied to the year 1918, since when statistics for all large shocks have improved (slowly for the first few years, and with deficiencies during World War II).

If the totals for 1904–1917 are multiplied by $^{38}\!/_{14}$, the products are generally somewhat below totals for 1918–1955, as one might expect from the imperfection of the data for the earlier period. There are conspicuous and significant exceptions for shallow shocks of magnitude 8.6 and over, and for intermediate shocks of magnitude 7.9 and over. There is no escape from the conclusion that the greatest shallow shocks were more frequent before 1918 than afterward; this obviously cannot be attributed to deficiencies in cataloguing, which should produce the opposite statistical effect. It is almost as certain that there was a long interval of lowered seismicity at the intermediate depth level, with no shocks there exceeding $M = 7.9$ between 1914 and 1939. Large fluctuations are evident in the annual totals for shocks of lower magnitude at all depths. A change in seismicity about 1922, evident in many detailed analyses, is here apparent only as an increase in the number of shallow shocks. There is decrease in general seismicity after 1950. (See also the concluding pages of this chapter.)

Geography

Geographical distribution of the larger earthquakes is shown on the world maps (Figs. 22-3, 25-3, 25-4). Numerical totals, derived from the tabulations, are as follows.

Great shallow earthquakes, $M = 8.6$ and over, 1897–1955: Circum-Pacific belt† 15, Alpide belt 3, Pamir-Baikal zone 3, Kansu (China) 1.

Shallow shocks, $M = 7.9$–8.5, 1904–1955: Circum-Pacific belt 76, Alpide belt 8, Pamir-Baikal zone 3, Kansu 1, Indian Ocean 3.

Shallow shocks, $M = 7.0$–7.8, 1918–1955: Circum-Pacific belt 480, Alpide belt 49, Pamir-Baikal zone 8, Atlantic Ocean 11, Indian Ocean 11, elsewhere 11.

Great intermediate earthquakes, $M = 8.6$ and over, 1904–1955: Circum-Pacific belt 4, elsewhere none.

Intermediate shocks, $M = 7.9$–8.5, 1904–1955: Circum-Pacific belt 17, Alpide belt 3 (2 in the Hindu Kush, 1 in the eastern Mediterranean), no others.

Intermediate shocks, $M = 7.0$–7.8, 1918–1955: Circum-Pacific belt 199, Alpide belt excluding Hindu Kush 8, Hindu Kush 7.

Deep shocks, $M = 7.9$–8.5, 1904–1955: Circum-Pacific belt 8, no others.

Deep shocks, $M = 7.0$–7.8, 1918–1955: Circum-Pacific belt 65, Alpide belt 1 (Spain, 1954).

Smaller Shocks

For the world at large, statistics for magnitudes less than 7 are difficult to set up. Counts for 1935, 1936, 1937 were carried out by Gutenberg, and for 1938 by Lomnitz (Table 22-10, with counts for larger shocks repeated from other tables for comparison). Gutenberg's listing is for $m = 6$ to 7.

---

† For geographical particulars of the various active belts see Part Two and *Seismicity of the Earth*.

Table 22-10   Earthquake Numbers—1935–1938

Date	Magnitude 6–7			Magnitude 7–7.8		
	Shallow	Intermediate	Deep	Shallow	Intermediate	Deep
1935	143	30	12	17	5	2
1936	144	31	12	23	4	1
1937	136	34	8	14	10	1
1938	122	41	9	20	9	0

Statistics involving smaller shocks can be set up for limited areas. The following tabulation, taken from *Seismicity of the Earth,* gives mean annual numbers of shallow shocks, within ±¼ unit of the indicated magnitudes, for southern California (including a small area in adjacent Mexico north of 32° N), and for a limited part of the seismic area of New Zealand:

	Magnitude M								
	8	7½	7	6½	6	5½	5	4½	4
Southern California	0	0	0.09	0.2	0.5	1.4	3.4	11.5	33
New Zealand	0.02	0.04	0.09	0.0	0.6	1.8	6.0	16.2	46

The southern California data are based on shocks from January, 1934, through May, 1943; the area covered is about 300,000 km². The New Zealand data for the smaller magnitudes are taken from bulletins issued from Wellington from October 1940 through January 1944; the area is about 225,000 km².

For January, 1953, through June, 1956, the following totals are available:

	Magnitude M			
	6.0 and over	5.0-5.9	4.0-4.9	3.0-3.9
Kern County aftershocks	0	3	35	390
Southern California otherwise	1	7	80	442

This covers a slightly smaller area than the 1934–1943 count and excludes Mexico.

Counts of still smaller earthquakes have been carried out for limited areas. Studies of swarms of earthquakes recorded near the stations at Riverside, Haiwee, and Woody in California showed that the smallest earthquakes recognizable on the best seismograms were of magnitude near the zero of the scale, and that the frequency of occurrence appeared to be increasing regularly with decreasing magnitude down to this limit of recognition. Analogous results have been obtained in Japan.

Earthquake Frequency and Magnitude

The general distribution of earthquakes over the observed range of magnitudes can be represented rather simply. For the world at large, and for most of the limited areas which have been studied, the frequency of shocks at any given magnitude level is roughly 8 to 10 times that about one magnitude higher. This may be represented by

$$\log N = A - bM \qquad \text{or} \qquad N = 10^A\, 10^{-bM} \qquad (3)$$

where N is the number of shocks of magnitude M or greater per unit time, A and b are constants, and logarithms are taken to the base 10. Writing N' for dN/dM and q for $\log_{10} e$, we find

$$\log(-N') = a - bM \qquad (4)$$

where $a = A + \log(b/q)$. Numerically $\log(1/q) = 0.3622$.

Using the previously tabulated data for shallow shocks of magnitude 7 and over during 1918–1955, and of magnitudes 6–7 for 1935–1938, and taking the time unit as 1 year, we have:

	M			
	6	7	7.9	8.6
N	153.22	16.97	1.97	0.24
$\log N$	2.19	1.23	0.29	−0.62

These figures are closely represented by $A = 8.2$, $b = 1.0$, equivalent to $a = 8.56$, $b = 1.0$.

In *Seismicity of the Earth,* Gutenberg derived from nearly the same data, considered in greater detail (individualizing steps of 0.1 magnitude), results equivalent to $a = 7.72$, $A = 7.41$, $b = 0.90$.[†] These coefficients require modification to take account of the general revision in the higher magnitudes; in later papers Gutenberg represents the data by two linear formulas, giving $\log(N'/10) = 7.04 - 0.92m$ below $m = 7.1$, and $\log(N'/10) = 12.6 -$

[†] If we denote the coefficients lettered a and b by Gutenberg as x and y, then in our present notation

$$\log(N'/10) = x + y(8 - M) \qquad (4a)$$

$1.7m$ for higher magnitudes. Applying $m = 2.5 + 0.63M$, we get

$$\log N = 7.81 - 0.58M \tag{5a}$$

and

$$\log N = 9.1 - 1.1M \tag{5b}$$

for smaller and larger M, respectively.

The data given above for southern California may be summarized as follows:

	M			
	3.75	4.75	5.75	6.75
N	38.59	5.59	0.79	0.09
log N	1.58	0.75	−0.10	−1.0

These values fit rather closely to $\log N = 4.77 - 0.85M$.

For New Zealand we have:

	M			
	3.75	4.75	5.75	6.75
N	68.75	8.55	0.75	0.15
log N	1.84	0.93	−0.12	−0.82

Here the rule fits less well; approximately, $\log N = 5.2 - 0.9M$. The logarithm of the ratio of the number of smaller shocks in the world to those in these two smaller regions may be approximated by taking the difference between the constant term $A = 8.2$ for the world and those for the two regions, which gives 3.4 and 3.0; this makes the ratios of numbers about 2500 and 1000 to 1. However, the result for larger shocks differs; if we extrapolate the above linear formulas to 7.9, we find for southern California $\log N = -1.94$ and for part of New Zealand $\log N = -1.9$; that is, shocks of magnitude 7.9 and over should occur in these regions on the average of once in 80 to 90 years. This result corresponds reasonably with historical data.

Reducing the data given above for 3½ years, 1953–1956 yields:

	Magnitude M			
	3	4	5	6
Kern County aftershocks, N =	122.28	10.86	0.86	0
Southern California otherwise, N =	151.43	25.14	2.29	0.29
Kern County aftershocks, log N =	2.08	1.03	−0.07	—
Southern California otherwise, log N =	2.17	1.40	+0.36	−0.54

Both sets of data fit individually to log $N = 4.8 - 0.9M$, or near enough with the older data for southern California, suggesting that the aftershock activity in the small area affected by the Kern County shocks of 1952 represented approximately a temporary doubling of the normal seismicity of the region.

Gutenberg's tabulations for the area of Japan (see Chapter 30) list 85 shallow, 20 intermediate, and 17 deep earthquakes of magnitude 7 and over for 1918–1955, and 27 of magnitude 8 and over for 1897–1955. Those for shallow earthquakes may be represented as follows:

	M	
	7	8
N	2.24	0.14
log N	0.35	−0.87

Taking $b = 1$, we find roughly log $N = 7.2 - M$; that is, the Japanese region accounts for roughly 10 per cent of the seismicity of the earth.

Tsuboi, correlating magnitudes M with amplitudes recorded at six principal stations in Japan, identified 38 earthquakes of magnitude $M = 7.0$ and over for 1931–1950 inclusive; this would give $N(7) = 1.90$ as compared with 2.24 above, but the area covered is slightly smaller. Working with 735 earthquakes in all, he set up linear relations for log $(N'/10)$ in the form used by Gutenberg, with coefficients equivalent to the following in our notation:

	a	A	b
Area A	7.88	7.50	1.06
Area B	5.19	4.97	0.72
Area C	4.67	4.49	0.66
A + B + C	7.75	7.39	1.01

Areas A and B are the Pacific coastal regions of Japan northward and southwestward from the Fossa Magna; area C is the Japan Sea coastal belt. The results for area A, and for the three areas combined, agree closely with the result given above ($A = 7.2$, $b = 1$).† However, in a more recent paper Tsuboi gives quite different figures, equivalent to $a = 5.68$, $A = 5.46$, $b = 0.72$.

Dr. Båth has fitted the aftershocks of the great Kamchatka earthquake of November 4, 1952 to a linear formula with $b = 1.5$.

Statistics of Deep-Focus Earthquakes

In regions where intermediate and deep shocks occur, the frequency of occurrence generally decreases with increasing depth; in most areas there is

---

† Note that coefficients a and A cannot simply be added together to combine adjacent regions, since they represent logarithms.

a fairly definite lowest horizon, usually in the range of 600 to 700 kilometers, below which no shocks are known. Seismicity reaches this level without any evident tapering off, since many of the largest deep shocks occur at such depths. The deepest known shocks (under the Flores Sea at depths near 720 kilometers) include one of magnitude 7.1. Especially noteworthy is the Spanish shock of March 29, 1954, one of the two known deep shocks outside of the Pacific area, which had a depth of 640 kilometers and a magnitude of 7.1.†

In some regions deep shocks are especially frequent at certain depths, as near 350 kilometers under Japan. Elsewhere, shocks in certain ranges of depths are not known, as in South America between 300 and 550 kilometers.

Gutenberg and others have set up linear relations between log N (or log N') and M for intermediate and deep shocks; because of the smaller numbers and the more restricted geographical distribution, the coefficients are less well determined.

Statistics of Other Forms

Counts of earthquakes for a given area and time interval are necessarily incomplete for the smallest magnitudes. On the other hand, statistics of the seismicity of a given area may be deficient for the largest magnitudes, owing to the accident of no large shock having occurred during the years of observation; yet, if a large shock occurs, it accounts for most of the energy release in its interval. Statistics for this type of problem, in which the data are likely to be deficient for extreme values, were developed by E. J. Gumbel for the study of such events as floods. Nordquist has applied Gumbel's method to California earthquakes and to the large shocks of the world. The resulting plots are orderly and consistent, agreeing satisfactorily with the requirements of Gumbel's theory.

THE LARGEST POSSIBLE EARTHQUAKES

It is evident that there must be some upper limit to earthquake magnitude; careful analysis of the statistics indicates falling off of earthquake frequency as the highest observed levels are approached. Thus, if log $N = 8.2 - M$, we should expect a shock of magnitude 10 or over at an average interval in years whose logarithm is $10 - 8.2 = 1.8$, or about every 90 years. Such an event would be so catastrophic that it could scarcely escape notice, even if centered in a remote region; but there is no historical seismic event to which we are inclined to assign magnitude above that (8.9) assigned to the largest shocks of the past 50 years.

---

† There is no falling off in facility of detection at large depths; a small shock near 600 kilometers is noticed as readily as one near 200 kilometers in the same region.

A physical upper limit must be set by the strength of the crustal rocks, in terms of the maximum strain which they are competent to support without yielding. Tsuboi, using reasonable numerical assumptions, has thus calculated energies for the largest possible earthquakes which are in accord with those for the largest observed as derived from the energy-magnitude relations set up in a later section. These limiting strengths and energies should decrease with increasing hypocentral depth; in this way an explanation is provided for the observed result that the frequency of intermediate and deep earthquakes falls off rapidly at magnitude levels lower than those of the largest shallow shocks.

OTHER MAGNITUDE SCALES

The concept of magnitude as such is in no way restricted to the scale originally set up by the author, or any of its later modifications. Just as several successive or contemporary intensity scales were set up by investigators who used different terms to define the various grades (cf. Chapter 11), so magnitude scales may be constructed on any basis logically consistent with the original scale. Owing to Gutenberg's adoption of the "unified" scale, represented here by m, we are at present operating with two explicitly different magnitude scales for teleseisms, while the relation of these to that still used for local earthquakes is imperfectly known.

The scale used for local earthquakes in New Zealand is intended to coincide with the original California scale, being based on the same tabulations and on torsion seismometers of the same design.† Workers elsewhere have used one form or other of the scale for teleseisms or have developed local scales of their own, based on other instruments.

Thus di Filippo and Marcelli devised for Italy a local-earthquake magnitude scale based on the seismograms of 200-kilogram Wiechert instruments ($T_0 = 4.3$ seconds, $V = 250$, $h = 0.33$). This was done very carefully, paralleling the procedure used to set up the California scale; the results appear satisfactorily coherent with determinations of magnitude elsewhere and by other means. Discrepancies only arise if it is assumed that the maximum wave recorded on the seismogram of the Wiechert instrument is the same as the wave which would be recorded as maximum by the standard torsion under the same circumstances; it is usually a wave of longer period with larger amplitude and smaller acceleration.

At Praha, magnitudes have been determined since 1950. Zátopek, Vaněk, and Kárník have each made valuable contributions. Teleseism magnitudes have been studied using surface and body waves. For local earthquakes, short-period seismographs have been used, and a new paper by Kárník, considering

---

† However, magnitudes reported from Wellington prior to 1949 should be increased by 0.30 to correct for a factor 2 in the static magnification.

periods as well as amplitudes, is an important forward step, and bears on the magnitude-energy problem.

Tsuboi's earlier assignments of magnitudes for Japan are obtained by using a pair of formulas which can be taken as independent definitions of magnitude:

$$M = 0.20\Delta + 0.67 \log A + 3.80 \text{ for } \Delta < 500 \text{ km}$$
$$M = 0.03\Delta + 0.60 \log A + 5.00 \text{ for } \Delta > 500 \text{ km}$$

where Δ is measured in units of 100 kilometers and A is the maximum ground amplitude in microns. Tsuboi has now replaced these formulas by

$$M = 1.73 \log \Delta + \log A - 0.83$$

Kawasumi set up a scale of different form which depends on macroseismic data but none the less represents magnitude. Kawasumi's magnitude is simply the number representing the intensity on the Japanese scale (which runs from 0 to 6) at an epicentral distance of 100 kilometers; observations at other distances may be reduced to 100 kilometers by appropriate curves or calibration, exactly as is done for instrumental data in other magnitude scales. Naturally the correlation between this scale and the others is highly irregular. Kawasumi's procedure is not far removed from the old practice of rating an earthquake in terms of its epicentral intensity, but it has the advantage of using data for all available distances and consequently eliminating the effect of ground and other local circumstances in the vicinity of the epicenter. It fails completely for deep-focus earthquakes.

A recurrent proposal is to redefine magnitude in terms of energy, giving an "absolute" magnitude scale. In the present state of knowledge this is premature, as will appear from the discussion which now follows.

MAGNITUDE AND ENERGY

The original intention of the magnitude scale was to take some of the nonsense out of earthquake statistics, and secondarily to combat errors caused by evaluating earthquakes in terms of their effects at particular localities without reference to epicentral distance, ground, or type of structures affected. It was realized that magnitude should eventually be related to energy, but that long investigation and accumulation of data would be required. Nevertheless, this possibility attracted a great deal of attention from the start; thus several preliminary and tentative correlations with energy were seized upon and applied far beyond their limits of validity. There is a tendency to blame the resulting contradictions on the magnitude scale itself.

Magnitude is evidently related to that energy which is radiated from the earthquake source in the form of elastic waves. Part of the original potential energy of strain stored in the rock must go into mechanical work, as in raising crustal blocks against gravity, or in crushing material in the fault zone; part must be dissipated as heat.

Reid estimated the work done in displacing crustal blocks during the California earthquake of 1906 as 1.75×10^{24} ergs. Energies of a number of earthquakes have been estimated from seismograms, for it is fairly well established that there is relatively little absorption of seismic waves after they leave the vicinity of the hypocenter. Consequently the energy in the expanding wave front, which can be estimated from the recorded amplitudes and periods, represents most of the energy radiated. In this way Jeffreys derived from the surface waves of the Pamir earthquake of 1911 (magnitude 7.6) and of the Montana earthquake of 1925 (magnitude 6¾) energies of about 10^{21} ergs. Dr. Båth at Uppsala, and Dr. C. Lomnitz working at Pasadena, lately applied the same theory to about 40 earthquakes. Sagisaka, working with the P and S waves of a shock in Japan on June 2, 1929, with a depth of 360 kilometers and magnitude 7.1, calculated an energy of 3.1×10^{20} ergs.

Energy in an elastic wave of given period is proportional to the square of the amplitude. If seismograms of different earthquakes at a fixed distance actually differed only in amplitude, the periods would be unchanged, and we should have

$$\log E = c + 2M \tag{6}$$

where c is constant. Preliminary work using the results of Jeffreys and others gave $c = 8$, but this value gives incredibly small energies for the smallest recorded shocks. More elaborate calculations by Gutenberg and Richter led to $\log E = 11.3 + 1.8M$; introducing overlooked factors, and a little further hypothesis, produced the formula

$$\log E = 12 + 1.8M \tag{7}$$

which was used in *Seismicity of the Earth*. Especially for the larger shocks, energies given by this formula are too high. In the interim di Filippo and Marcelli published a calculation which led to

$$\log E = 9.15 + 2.15M \tag{8}$$

All these formulas depend on theoretical study of the radiation of energy at short distances, near the epicenter. In a recent revision (1955) Gutenberg and the writer made extensive use of seismograms written by the strong-motion instruments operated by the U. S. Coast and Geodetic Survey, including those for the Kern County earthquakes of 1952. The remaining uncertainties of this method have been a principal factor in Gutenberg's preference for the "unified magnitude" m derived from body waves recorded at teleseismic distances. The relation of m to the radiated energy can be set up with less theoretical difficulty and a minimum of observational inaccuracy; it takes the form

$$\log E = 5.8 + 2.4m \tag{9}$$

Since $m = 2.5 + 0.63M$, this is equivalent to

$$\log E = 11.4 + 1.5M \tag{10}$$

In equation (10), M is at least an approximation to the magnitude determined from surface waves of shallow teleseisms. Dr. Båth's latest results, calculated from an entirely different set of data, confirm equations (9) and (10) very closely.

Gutenberg has used every available means to relate m to the magnitude M_L derived in the original manner from local-earthquake records in California. His preferred result is

$$m = 1.7 + 0.8M_L - 0.01M_L{}^2 \tag{11}$$

which leads to

$$\log E = 9.9 + 1.9M_L - 0.024M_L{}^2 \tag{12}$$

The terms in $M_L{}^2$ are highly empirical in nature and difficult to interpret satisfactorily in terms of physical dimensions. The relation between m and M_L, and consequently that between $\log E$ and M_L, will probably be modified soon by new data.

Putting $M = 8$ in equations (7), (8), and (10) gives $\log E = 26.4$, 26.35, and 23.4; thus revision leads to greatly reduced values for the energies of the largest shocks. However, the values of M have generally been increased, so that it would be better to put $M = 8.5$ in equation (10), giving $\log E = 24.15$. Since most of the energy of all earthquakes is in such shocks, revision materially reduces estimates of the annual total energy of seismic activity. On the earlier basis this energy was given in publications as 1.2×10^{27} ergs per year. Since the energy of the annual flow of heat from the interior through the surface of the earth is roughly 8×10^{27} ergs, the two numbers were close enough to suggest various geophysical speculations. Revision for the seismic energy now gives a figure near 9×10^{24} ergs per year, which is hardly more than a thousandth of the heat energy.

A further point of chiefly journalistic interest relates to comparison between large earthquakes and atomic bombs. The official figure for the energy released by a "nominal" atomic bomb of the Hiroshima type is 8×10^{20} ergs; a very large earthquake, on the old basis, might have an energy of 8×10^{26} ergs, hence comparable with a million atom bombs. On the new basis the largest earthquakes are found to have an energy not much over 10^{25} ergs, roughly equivalent to 12,000 of the nominal bombs.

The revisions in the magnitude-energy relation somewhat affect Dr. Benioff's work on the release of strain in earthquake sequences; but the actual modification is not large. Dr. Benioff takes the strain release as proportional to the square root of E; on the logarithmic scale this divides the coefficient of M in the formula for $\log E$ by 2; the quotient for the various formulas varies only from about 0.9 to 1.1.

In 1955 N. V. Shebalin, a young worker at the Geophysical Institute in Moscow, published a note on relations between energy, intensity and focal depth. From a study of 56 earthquakes he derived

$$0.9 \log E - I = 3.8 \log h - 3.3 \tag{13}$$

and

$$0.9 \log E - I = 3.1 \log h - 4.4 \tag{14}$$

where E is the energy in surface waves measured in megajoules (units of 10^{13} ergs), h is the hypocentral depth in kilometers, and I is the maximum intensity at the surface (M.M.). Equation (13) applies to hypocenters from the surface down to depths of 70 kilometers, equation (14) to depths of 80 kilometers or more.

V. I. Bune, publishing from Stalinabad, discusses methods of estimating energy from recorded amplitudes of P and S waves, and points out the crudity of the assumptions made by previous workers; eventually in order to derive actual results he has to introduce equally rough approximations. The second part of his paper deals with the Stalinabad earthquake of February 27, 1952. Approximate methods are used to integrate the seismograms of 7 stations; the energies calculated are of the order of 10^{19} ergs, but individual results diverge as much as 70 per cent.

Bune combines Shebalin's equations with a result given by Gutenberg and Richter in 1942:

$$\log E = 8.8 + 2 \log h + 1.8M \tag{15}$$

where E is in ergs and M is the magnitude (later work suggests significant revision in this formula). In this way Bune finds $M = 4.5$.

In a final section Bune attempts to compare his theoretical results with energies of large explosions calculated from seismograms; the ratios agree well with those of the potential energies of the charges.

SMALL SHOCKS AS A "SAFETY VALVE"?

Although the frequency of earthquakes increases rapidly with decreasing magnitude, the energy released in the individual shocks, computed by any of the formulas in the preceding subsection, decreases yet more rapidly. In consequence, if the earthquakes of any given area, or of the world, are examined for a limited time interval, it is generally found that the release of energy takes place principally in the relatively few shocks of largest magnitude. This has a practical bearing on the popular idea that small earthquakes may operate as a "safety valve" to release harmlessly the energy which might otherwise express itself in a large earthquake.

Strictly on an energy basis, such reassurance is doubtful, and it does not fit very well with the history of large and small earthquakes in areas where information is adequate. The energy relations suggest that major strains are released only in major earthquakes, and that minor earthquakes are rather incidental indications of the accumulation of regional strain. This fits with the geographic evidence that the larger earthquakes occur chiefly in association with the principal faults and active structures, while earthquakes of lower magnitude are mostly associated with minor tectonic features.

Dr. Benioff points out that if strains are compared directly, instead of by way of the energy, this no longer necessarily holds. The strains are taken as proportionate to the square roots of the energies; if these quantities are then combined linearly, the total contribution from smaller shocks is often of the same order as that of the comparatively few large shocks. This implies coherence of strain; if applied to southern California, for example, it would demand that all the shocks, large and small, are incidents in the release of a regional strain which accumulates in a more or less uniform manner over the entire area. This is a geologically justifiable assumption, and Dr. Benioff's results with curves of accumulated strain tend to support it (Chapter 6). Simple physical considerations lead to one more qualification with reference to the "safety valve." While small shocks of the observed frequency of occurrence relative to the large ones may release strain sufficiently to delay a major event, nevertheless once a major strain has accumulated it can only be relieved by a major earthquake or by a highly abnormal number of small shocks. The latter process may be favored by exceptional tectonic conditions, which may account for the prevalence of earthquake swarms in particular areas such as the Imperial Valley.

GLOBAL AND REGIONAL STRAIN RELEASE

Dr. Benioff's method of adding the square roots of the computed energies to exhibit strain release as a function of time has been applied to the year-by-year seismicity of particular regions and even to the world as a whole. This method presupposes mechanical coherence in the strains represented by the earthquakes which are being studied. For the world as a whole, such an assumption has far-reaching implications and is extended only to earthquakes of the largest magnitudes (8 or over).

The plot in Fig. 22-4, representing strain release in shallow earthquakes, has striking features which are less easily read from an energy-release plot. Intervals of high seismicity and large cumulative strain release are separated by intervals of lower activity; but the totals reached in the active periods successively decrease. These bursts of high activity were during 1896(?)–1900, 1904–1907, 1917–1924, 1931–1935, 1938–1942, possibly 1945–1948, and 1950. The initial date of the first period is indicated as 1896 with question, because before that year the necessary records did not exist. Guten-

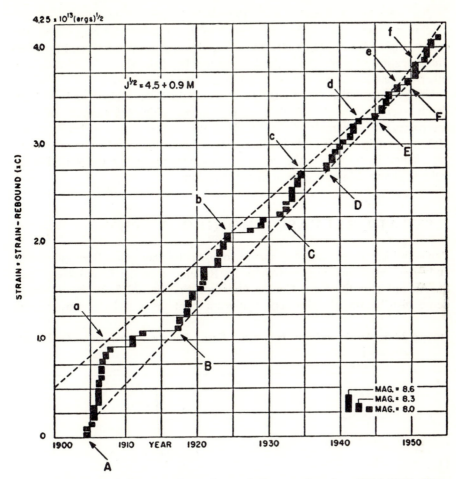

FIGURE 22-4   *Global strain release, large shallow earthquakes, 1904–1955.* [*Ben-ioff.*]

berg's latest work leaves little doubt that this active period was even more remarkable than the following one (which includes the extraordinary year 1906). The successive decrease of definiteness in the active cycles reaches its limit for 1945–1948, which hardly represents more than a return to normal activity after interruption. The year 1950, although not comparable with 1906 (still less with 1897), occupies a special and extraordinary position (especially when deep and intermediate shocks are considered). In some way it marks a turning point; the six years following showed generally low activity.

The first sign of a new pattern was a remarkable swarm of earthquakes in the Aleutian Islands (near 168° W) on January 2, 1957, including two of magnitude 7. On March 9 an earthquake of magnitude 8 or more originated farther west in the Aleutian arc (175° W); its immediate aftershocks, including at least five of magnitude 7 or over, were distributed over the whole range 168°–175° W. High aftershock activity continued for several months. Simultaneously the number of major earthquakes (magnitude 7 and over) in other regions increased; ten occurred in the first half of 1957.

Similar cumulative strain curves for shocks in given regions and at given depth levels show a classifiable variety of forms, which Dr. Benioff has used for comparative tectonic study. Many of these curves show a break about 1922, a year in which there is other geophysical evidence of some far-reaching change in the earth; the polar displacement producing the variation of latitude changed in character at about that time.

STRAIN RELEASE AND GEOGRAPHY

Benioff's idea of accumulating square roots of the energy has been independently applied to mapping seismicity by Ritsema (for the East Indies) and St. Amand (for the Kern County aftershocks), Fig. 22-5. With any fairly dense distribution of epicenters significant contours can be constructed to

FIGURE 22-5   *Strain release contours, Kern County aftershocks, 1952.* [*St. Amand.*]

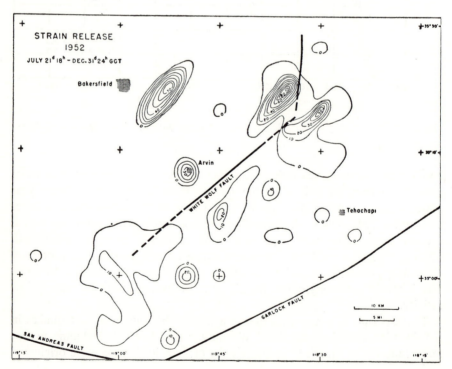

show the nominal distribution of strain release over the area. These contours are of interest in relation to faults and structures; with the lapse of time and the accumulation of data it may be possible to apply the method to seismic maps of the world.

Mapping in terms of energy release has been advocated by Båth and several other authors.

References

Development of the magnitude scale

Richter, C. F., "An instrumental earthquake scale," *B.S.S.A.,* vol. 25 (1935), pp. 1–32.

Gutenberg, B., and Richter, C. F., "On seismic waves (third paper)," *G. Beitr.,* vol. 47 (1936), pp. 73–131.

Gutenberg, B., "Amplitudes of surface waves and magnitudes of shallow earthquakes," *B.S.S.A.,* vol. 35 (1945), pp. 3–12.

———, "Amplitudes of *P, PP* and *S* and magnitudes of shallow earthquakes," *ibid.,* pp. 57–69.

———, "Magnitude determination for deep-focus earthquakes, *ibid.,* pp. 117–130.

Richter, C. F., "History and applications of the magnitude scale," *Publ. bureau central séismologique international,* Ser. A, vol. 17 (1948), pp. 217–224.

Båth, M., "Earthquake magnitude determination from the vertical component of surface waves," *Trans. Am. Geophys. Union,* vol. 33 (1952), pp. 81–90.

Gutenberg, B., and Richter, C. F., "Magnitude and energy of earthquakes," *Ann. geofisica,* vol. 9 (1956), pp. 1–15.

Other contributions

Hayes, R. C., "Measurement of earthquake intensity," *N. Z. Journ.,* Ser. B, vol. 22 (1941), pp. 202–204. (Magnitudes in New Zealand.)

Peterschmitt, E., *La magnitude des séismes. Comptes rendus des séances de la conférence réunie à Strasbourg en 1947,* Strasbourg, 1948, pp. 86–88.

———, "Etude de la magnitude des séismes," *Ann. inst. Physique Globe,* vol. 6 (1950), pp. 51–58.

di Filippo, D., and Marcelli, L., "La 'magnitudo' dei terremoti e la sua determinazione nella stazione sismica di Roma," *Ann. geofisica,* vol. 2 (1949), pp. 486–492.

———, "Magnitudo ed energia dei terremoti in Italia," *ibid.,* vol. 3 (1950), pp. 339–348.

Zátopek, A., and Vaněk, J., "On the regional distribution of magnitude differences between Pasadena and Prague, *Kartograf. prehled (Praha)*, vol. 5 (1950), pp. 41–55, 123–128 (in Czech).

————, "Les magnitudes de Praha et leurs relations avec les 'revised values' de Pasadena," *Publ. bureau central séismologique international,* Ser. A, vol. 18 (1952), pp. 137–151.

Vaněk, J., "Determination of earthquake magnitude from surface waves for the stations Hurbanovo and Skalnaté Pleso," *Trav. inst. géophys. Acad. Tchécoslovaque,* No. 6 (1953), pp. 83–89 (in Czech with extensive English summary).

Tsuboi, C., "Determination of the Richter-Gutenberg's instrumental magnitudes of earthquakes occurring in and near Japan," *Geophys. Notes* (Tokyo Univ.), vol. 4, No. 5 (1951), pp. 1–10.

Trapp, E., "Ableitung der Magnitudengleichung für die Erdbebenstationen Wien und Graz und allgemeine Bemerkungen zur Magnitudenberechnung," *Arch. Meteorologie, Geophysik, Bioklimatologie,* Ser. A, vol. 6 (1954), pp. 440–450.

Wadati, K., and Hirono, T., "Magnitude of earthquakes—especially of near, deep-focus earthquakes," *Geophys. Mag.* (Tokyo), vol. 27 (1956), pp. 1–10.

Vaněk, J., and Zátopek, A., "Magnitudenbestimmung aus den Wellen *P, PP* und *S* für die Erdbebenwarte," *Trav. géophys. Acad. Tchécoslovaque,* No. 26 (1955), pp. 91–107.

Tsuboi, C., "Energy accounts of earthquakes in and near Japan," *Journ. Physics of Earth* (Tokyo), vol. 5 (1957), pp. 1–7. (Revised formula for magnitude.)

Bonelli-Rubio, J., and Esteban Carrasco, L., "La magnetud de los sismos en Toledo," *Revista de Geofísica,* vol. 14 (1955), pp. 1–12. (Assignment of M for teleseisms.)

De Bremaecker, J.-Cl., "Determination des magnitudes des seismes en Congo belge," Acad. royale des sciences coloniales (Bruxelles), Bull. des séances, n.s. vol. 1 (1955), pp. 1043–1046.

Solov'ev, S. L., "O klassifikachii zemletryasenii po velichine ikh energii," Trudy Geofiz. Inst. Akad. Nauk SSSR, No. 30 (157), 1955, pp. 3–21.

Shebalin, N. V., "O svyazi mezhdu energey, ballnost'yu i glubinoy ochaga zemletryaseniya, Izvestiya Akad. Nauk SSSR, ser. geofiz., 1955, pp. 377–380.

Asada, T., Suzuki, Z., and Tomoda, Y., "On frequency distribution of seismic magnitude," *Publ. Bur. Central International,* ser. A, vol. 19 (1956), pp. 95–98.

Kawasumi, H., "Intensity and magnitude of shallow earthquakes," *ibid.,* pp. 99–114.

Båth, M., "The problem of earthquake magnitude determination," *Publ. bureau central séismologique international* (A), vol. 19 (1956), pp. 5–93.

Kárník, V., "Magnitudenbestimmung europäischer Nahbeben," *Trav. géophys. Acad. Tchécoslovaque,* No. 47 (1956), pp. 399–522.

Earthquake energy and statistics

Gutenberg, B., and Richter, C. F., *Seismicity of the Earth*, Princeton University Press, 2nd ed., 1954, pp. 16–25.

——, "Earthquake magnitude, intensity, energy and acceleration," *B.S.S.A.*, vol. 32 (1942), pp. 163–191; second paper, *ibid.*, vol. 46 (1956), pp. 105–145.

——, "Magnitude and energy of earthquakes," *Nature,* vol. 176 (1955), p. 795. (A preliminary announcement; see next reference.)

——, "Magnitude and energy of earthquakes," *Ann. geofisica,* vol. 9 (1956), pp. 1–15.

Tsuboi, C., "Isostasy and maximum earthquake energy," *Proc. Tokyo Acad.*, vol. 16 (1940), pp. 449–454.

——, "Earthquake energy, earthquake volume, aftershock area, and strength of the earth's crust," *Journ. Physics of Earth* (Tokyo), vol. 4 (1956), pp. 63–66.

——, "Energy accounts of earthquakes in and near Japan," *ibid.*, vol. 5 (1957), pp. 1–7.

Båth, M., "The relation between magnitude and energy of earthquakes," *Trans. Am. Geophys. Union*, vol. 36 (1955), pp. 861–865.

——, "The energies of seismic body waves and surface waves" (in press).

Honda, H., "Amplitudes of *P* and *S*, magnitude and energy of deep earthquakes," *Science Repts., Tôhoku Univ.*, Ser. 5, Geophysics, vol. 3 (1951), pp. 138–143.

Sagisaka, K., "On the energy of earthquakes," *Geophys. Mag.* (*Tokyo*), vol. 26 (1954), pp. 53–82.

Gutenberg, B., "The energy of earthquakes," *Quart. Journ., Geol. Soc. London,* vol. 112 (1956), pp. 1–14.

——, "Earthquake energy released at various depths," in *Gedenkboek F. A. Vening Meinesz, Verh. konink. Ned. geol.-mijnb. Genootschap,* The Hague, 1957, pp. 165–175.

Båth, M., "Erdbebenenergie," *Zeits. f. Bergbau usw.* (Freiberg), vol. 9 (1957), pp. 17–21.

Bune, V. I., "Ob ispol'zovanii metoda Golitsina dlya priblizhennoy otsenki energii blizkikh zemletryaseniy," *Trudy Inst. Seismol. Akad. Nauk Tadzhikskoy SSR*, vol. 54 (1956), pp. 3–27.

Nordquist, J. M., "Theory of largest values applied to earthquake magnitudes," *Trans., Am. Geophys. Union,* vol. 26 (1945), pp. 29–31.

Miscellaneous references

The Effects of Atomic Weapons, Gov. Printing Office, Washington, D. C., 1950. (See pages 13, 14, 111.)

Gumbel, E. J., "Statistical control-curves for flood discharges," *Trans. Am. Geophys. Union* (1942), pp. 489–509.

————, *Statistical theory of extreme values and some practical applications*, National Bureau of Standards, Applied Math. Ser., No. 33, Govt. Printing Office, Washington, D. C., 1954.

Gutenberg, B., "Great earthquakes 1896–1903," *Trans., Am. Geophys. Union*, vol. 37 (1956), pp. 608–614.

Richter, C. F., and Nordquist, J. M., "Minimal recorded earthquakes," *B.S.S.A.*, vol. 38 (1948), pp. 257–261.

Benioff, H., "Seismic evidence of crustal structure and tectonic activity," *Geol. Soc. Amer., Spec. Paper* 62 (1955), pp. 61–73.

Ritsema, A., "The seismicity of the Sunda arc in space and time," *Indonesian Journ. Nat. Science* (1954), pp. 41–50.

Benioff, H., Gutenberg, B., and Richter, C. F., "Progress report, Seismological Laboratory, California Institute of Technology, 1953," *Trans. Am. Geophys. Union*, vol. 35 (1954), pp. 979–987. (Reports part of the following.)

St. Amand, P., "Two proposed measures of seismicity," *B.S.S.A.*, vol. 46 (1956), pp. 41–45.

Båth, M., "A note on the measure of seismicity," *ibid.*, vol. 46 (1956), pp. 217–218.

Ritsema, A. R., "The seismicity of the Sunda arc in space and time," *Proc. 8th Pacific Science Congress*, vol. 2A, pp. 753–763.

CHAPTER 23

Microseisms

IN CHAPTER 12 microseisms were described as continuous natural disturbances recorded by seismographs. They are due to a variety of causes.
Some causes are local, minor, and easily identified; others are of a more
general character, disturbing seismograms nearly simultaneously at stations
over a wide area.

MINOR CAUSES

Volcanic tremor is a name given to more or less continuous vibration recorded by seismographs near a volcano when an eruption is going on. Since
such eruptions are accompanied by many small earthquakes of at least two
different types (Chapter 12), it has been thought that the tremor consists
merely of a succession of numerous, small, individual shocks. Recent discussion of tremor recorded on Hawaii has considered the possible oscillations of
laminae of volcanic rock set up by surging motions of lava.

Moving water is a common source of disturbance. Vibration due to local
surf is recorded at stations at or near the coast; its possible relation to the
typical microseisms discussed in the next section has been debated for many
years. Microseisms produced by running water, especially cascades and
waterfalls, are large locally but ordinarily are not recorded at a distance.
Water in reservoir spillways has caused trouble at nearby seismological stations.

Much microseismic motion is meteorological in origin. Certain irregular,
long-period disturbances have been successfully correlated with the effect of
frost on the local ground. Wind pressing upon and perhaps tilting the building housing seismographs has also been identified as the cause of long-period
disturbances (30 seconds to 1 minute). At Pasadena, similar effects due
to direct air currents are obviated by covering the pendulums. Strong local
winds often produce highly irregular short-period disturbances which make
it nearly impossible to use the seismograms of sensitive instruments with
pendulum periods near 1 second.

An interesting group of disturbances still under study has been observed at Pasadena and its auxiliary stations as loosely correlated with local rainfall. Sometimes, especially when there is small precipitation after a long dry season, rainfall at the station may be unaccompanied by microseisms. On the other hand, microseisms with the usual characteristics (periods near 1 to 1½ seconds, with some irregularity) frequently record when there is no precipitation near the station; in such instances there is usually a front not far away, and often rain begins to fall at the station some hours after the typical microseisms are noted. Comparison with a recording barograph shows that these oscillations definitely are not in the air where recorded, although certainly they originate in the atmosphere and must somehow be transferred to the ground.

TYPICAL MICROSEISMS

Of microseisms related to weather, those most widely recorded and most discussed commonly have periods near 6 seconds. With increasing amplitude their periods rise to 8 seconds or even more. On the seismogram they typically appear as more or less separate groups, in each of which the amplitude increases gradually to maximum and then falls off, suggesting some kind of resonance or beats (Fig. 23-1). Instruments with high magnification for periods near 6 seconds are almost useless during a "microseism storm"; since periods in the same range are characteristic of the S group for many teleseisms, certain details of teleseismic recording are lost at such times, although other important data (such as the time of arrival of P) can often be read with comparatively little trouble from the records of short-period seismographs.

Microseisms of this sort are set up by cyclonic storms over water, including both the normal "lows" of temperate latitudes and the violent cyclonic storms of the tropics, called hurricanes in the West Indies and typhoons in the Pacific. Discussion amounting to controversy has centered about the

FIGURE 23-1   *Seismogram showing earthquakes superposed on 6-second microseisms.*

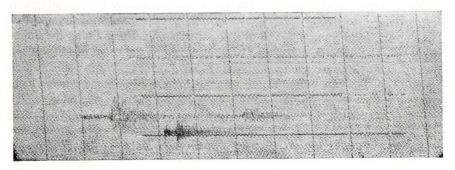

mechanism by which the storm energy is transferred from the atmosphere to the ground. Some misunderstandings have arisen from difference in detail between the conditions for microseism production by cyclones of temperate and of tropical types. Further confusion has arisen from misapprehension of the hypothesis which attributes the energy transfer to surf striking a steep or irregular coast; certain authors have confused this with the microseisms of shorter period which surf produces locally and have written papers on "the cause of microseisms" which cast no light on the origin of the 6-second oscillations.†

The prevalence of a particular period suggests resonance; it is likely that the 6-second period is associated in some way with the normal structure of the continental crust. Microseisms of this type are propagated with gradual loss of energy over wide areas of continental structure; those demonstrably due to storms off the coast of Norway are frequently observable at Irkutsk in Central Asia. On the other hand, they propagate only with considerable loss of energy across major structural boundaries and disturbed areas; thus they do not readily pass the Mediterranean and Alpide belt, or between Iceland and Scandinavia, and in North America they lose energy rapidly in crossing the Cordilleran zone between the Pacific and the great plains.

Microseisms associated with temperate-latitude "lows" are not propagated outward from the storm center, but from the peripheral zone of unstable air associated with strong winds when that is near the coast. As the storm passes inland, even though it remains intense, the corresponding microseisms decrease notably. It is possible, however, that the energy is transferred from the storm center to the coast by waves of disturbances which in propagation are controlled by a coupling between air, sea, and ocean bottom.

Tropical cyclones, on the other hand, appear to generate microseisms propagated directly from the center, whether on the continental shelf or on the continent itself. Such microseisms begin to be recorded at sensitive stations soon after the cyclone forms, and thus provide a useful means of detection. Moreover, by setting up a tripartite station, consisting of three seismographs at the corners of a triangle with sides a mile or so in length, the phase difference between the corners can be used to estimate the direction from which the disturbance is approaching, and so contribute to locating the storm center; but sometimes the indicated direction is considerably in error. When such installations are used to study the direction of approach for microseisms associated with ordinary temperate-zone storm centers, the directional indications are even less coherent. Indeed the observations often suggest a system of local standing waves rather than a progressing wave train.

----

† Eiby has aptly quoted from Tennyson:

"They take the rustic murmur of their bourg
For the great wave that echoes round the world."

References

PUBLICATIONS on microseisms are numerous, and too many of them are second-rate. The subject frequently attracts workers with poor critical judgment or inadequate training. There is a deceptive simplicity about the observations which makes it appear as if results of value might be reached with no real theoretical understanding. Some of the publications, indeed, represent work assigned to students. Occasionally a first-class worker gives his attention to this subject; the results have rarely been comparable in scientific value with those reached by the same man in other branches of seismology.

B. Gutenberg and F. Andrews, at the Seismological Laboratory of the California Institute of Technology, Pasadena, have edited a Bibliography on Microseisms listing 600 titles, and giving short abstracts of some of the papers (mimeographed, Pasadena; 2nd ed., first part, 1952; second part, 1956).

The following are useful discussions of recent date:

Coulomb, J., "L'agitation microséismique," *Handbuch der Physik,* Vol. 47, *Geophysik I,* 1956, pp. 140–152.

Gutenberg, B., "Observations and theory of microseisms," in *Compendium of meteorology,* American Meteorological Society, 1951, pp. 1303–1311.

———, "Untersuchungen zur Bodenunruhe in Südkalifornien," *Zeitschr. Geophysik,* Sonderband, 1953, pp. 177–189.

Båth, M., "The problem of microseismic barriers with special reference to Scandinavia," *Geol. Fören. Förhandl.* vol. 74 (1952), pp. 427–449.

———, "Comparison of microseisms in Greenland, Iceland, and Scandinavia," *Tellus,* vol. 5 (1953), pp. 109–134.

The student may derive profit from comparing the following:

Leet, L. D., "Microseisms in New England—Case history II," *B.S.S.A.,* vol. 38 (1948), pp. 173–178.

Katz, S., "Brief review and study of the microseismic storm of September 30–October 1, 1947," *ibid.,* vol. 39 (1949), pp. 181–186.

Volcanic tremor on Hawaii is discussed by:

Finch, R. F., "Volcanic tremor, Part I," *ibid.,* vol. 39 (1949), pp. 73–78.

Omer, Guy C., Jr., "Volcanic tremor, Part 2, The theory of volcanic tremor," *ibid.,* vol. 40 (1950), pp. 175–194.

CHAPTER 24

Earthquake Risk and Protective Measures

IN CHAPTER 11 the damaging effects of earthquakes were discussed as indications of intensity, which is the manner in which they directly concern the seismologist and geologist. However, in times of earthquake disaster —and sometimes, fortunately, before such disasters happen—the geologist or seismologist may be called into consultation with reference to risk and protection.

WEAK CONSTRUCTION

The Community Problem

Earthquake risk applies primarily to artificial structures. Except for the effects of great fires, over 90 per cent of the loss of life occasioned by earthquakes, and considerably more than half of the property loss, result from the failure of works of construction so weak that they should never have been permitted to be erected. From time to time such structures fail under stresses caused by wind and weather, overloading, traffic vibration, or settling of the ground; naturally, when large numbers of them are exposed to the unusual loads imposed by even a moderate earthquake, disaster follows.

For this reason, little of importance to the engineer is learned from damage in most earthquakes, even when studied with care and in detail. From Mallet's time to the present day, reports have confirmed, in melancholy repetition, the obvious fact that defective construction will not withstand an earthquake.

Lack of useful information is partly due to incomplete reporting. Attention is naturally attracted by the most spectacularly damaged structures. These, which are usually examples of the poorest workmanship, are most often photographed and examined. Only engineers, architects, and seismologists with previous experience of earthquake effects are likely to allot proportionate time and attention to the less conspicuous damage to better construction.

As noted in Chapter 8, many large populations customarily use weak types of masonry or substitutes for masonry (adobe, tamped earth); in earthquake regions such construction is fatal. Moreover, weak mortar is accepted passively, as if it were a law of nature; the fact that when masonry is damaged the bricks always separate and never break through is taken for granted. The proper use of reinforcement, in masonry and other types of construction, is ignored in many regions and progresses too slowly in others.

Countermeasures

In the absence of regulation, few structures are constructed to be earthquake-resistant; indeed under such conditions many structures of showy outward appearance are unsafe even for ordinary use. The appropriate community action for public safety is adoption and consistent enforcement of building codes. Even building regulations not including specific provisions for earthquake resistance usually set standards of design and workmanship which should eliminate the most dangerous weaknesses.†

Insurance companies and organizations have furthered safety measures greatly in contributing to the drafting and recommending the adoption of proper codes. A powerful factor is adjustment of insurance rates according to the degree of risk, which depends largely on the soundness of construction. This factor operates not only at the public but also at the individual level; a private owner will be more impressed by a high premium than by public discussion or official recommendations.

In many communities the setting up of disaster organizations has led to wider diffusion of earthquake risk information and has sometimes brought about better regulation or even definite action to remove specific risks.

THE HORIZONTAL COMPONENT

When we consider structures designed to meet specifications or regulations, rather than merely cast up according to the notions of an individual builder or the convenience of a general contractor, the outstanding problem is that of horizontal shaking.

As noted in Chapter 8, most ordinary works of construction—bridges, pipelines, dwellings, commercial and public buildings—are designed with a factor of safety against vertical overloading; but traditional practices do not provide a factor of safety against horizontal shaking. Indeed, a structure

---

† Not, of course, where regulations are poorly drafted or enforcement is perfunctory. As a distinguished Japanese remarked, "The fact that civil agencies issuing permits may approve a design does not necessarily mean that earthquakes will do so."

otherwise designed and erected with care may have almost no resistance to moderate lateral shaking; this is particularly true where severe storms and strong winds are uncommon.

Design for Horizontal Resistance

For an ordinary small structure of one or two stories, comparatively little additional expense in design and construction will result in greatly improved resistance to horizontal forces; it is much more difficult and expensive to introduce such resistance into an already existing structure. For small structures it is relatively easy to apply the rule-of-thumb method developed after 1906: to allow for horizontal acceleration of $0.1g$. Otherwise phrased, this calls for safe resistance to a steady horizontal pressure equal to one-tenth the weight. This is far from a correct representation of the forces likely to be applied during a strong earthquake; but experience has shown that the use of the rule, even for large construction, leads to reasonably good earthquake resistance.

Provisions for earthquake safety in building codes usually call for designing against lateral forces ranging from $0.05g$ to $0.25g$, the latter usually being specified for parapets, exterior ornaments, etc. The Uniform Building Code, now officially adopted in many American communities, offers such provisions in an optional appendix (so that in a majority of localities only the main body has been legally accepted and enforced and the earthquake regulations have been disregarded). Another code requirement is that all parts of the structure, including parapets and ornamentation, shall be bonded together into an integral whole or else separated into independent units with adequate clearance.

Reinforcing. The requisite strength in small structures, whether masonry or concrete, is generally attained by proper reinforcing. Steel frame or reinforced concrete can be made stronger than any masonry structure; however, masonry has many conveniences and is commonly retained in current construction as filler or facing, care being taken not to depend on it for bearing heavy loads or for horizontal strength.

Wood Frame. Resistance to earthquakes for ordinary small wood-frame structures depends simply on good workmanship, thorough bracing, and proper connection to the foundation (see Chapter 8). In weighing the relative advantages of wood construction and masonry, it is not possible to overlook the factor of fire risk.

Large Structures. In short, there are few real problems of design in small construction. Problems seriously in debate among architects and engineers arise almost exclusively in planning tall or complicated buildings to be earth-quake-resistant. For a brief discussion see Appendix II.

CUMULATIVE DAMAGE

Too often one hears, "This building is earthquake-proof; it stood through the last big shake." This may involve the usual confusion between magnitude and intensity, when the speaker has in mind a large earthquake originating at a distance, developing only moderate intensity at his locality. In any case it overlooks the weakening effect of repeated shaking on commonplace construction. Some spectacular failures of old buildings in recent earthquakes are attributable to progressive weakening in successive minor shaking, preparing the structure for serious damage when subjected to intensity VI or over.

FIRE RISK

Even in the most seismic regions, the long-term risk of fire is greater than that of earthquakes, and an earthquake may well bring fire in its train. The disasters at San Francisco and Tokyo are examples of the appalling consequences. Naturally, most large centers of population are exposed to the risk of fire starting from causes more common than earthquakes; fire departments and civic organizations are prepared to deal with such emergencies.

Water Supply

Fire following earthquakes presents special problems. The San Francisco conflagration was partly due to the rare and unfortunate circumstance of a water supply put out of action by actual displacement on a great fault; emergency water supply has since been provided which should effectively prevent repetition of that occurrence. Los Angeles and other large California communities are now getting much of their water supply by aqueduct from long distances. It is recognized that similar risk exists where these supply lines cross the major faults. Care has been taken to cross the faults directly, and not follow them for any appreciable distance; large storage reservoirs have been provided; where possible, to improve accessibility and the speed of repair, the crossing of the faults is made at the surface.

Even in regions as well investigated as California, unexpected events of this sort may occur. Thus major damage in Imperial Valley in 1940 was caused by large strike-slip on a fault previously only vaguely suspected to exist; and damage caused directly by faulting in Kern County in 1952 was equally unexpected, since the White Wolf fault, though known to exist, was not considered the probable seat of a major earthquake.

Pipelines; Shutoffs

Either shaking or faulting may open pipelines carrying gas, oil, or gasoline; moreover, tanks may spill or develop leaks. Minor occurrences of this sort in several California earthquakes have led to careful provision of shutoffs and to design and location of storage facilities with reference to earthquake risk. Shutoffs of electric supply are also provided to remove the fire hazard due to broken and fallen wires; usually care is taken not to cut off power from essential machinery such as pumps.

Alarm Systems; Sprinklers

The Long Beach earthquake drew attention to the need for safe housing of the fire alarm center; on that occasion the alarm system was put out of action, and fire risk was handled by each fire company continuously patrolling its district. Experience in the same earthquake also led to the recommendation that all kinds of shutoff valves and switches for buildings be located outside, so as to avoid the necessity of entering and perhaps searching a damaged structure to make a shutoff. Proper support for sprinkler systems is a necessity (see Chapter 8).

OTHER RISKS

Fissures

In the popular mind, the opening of fissures is a terrifying risk in earthquakes. This is either an effect in the zone of a great fault or a consequence of shaking and ground-water disturbance in heavily alluviated areas. Damage to property from this cause may sometimes be extensive, but the risk to life is small, in spite of the lamented cow in 1906 and the death of a Japanese housewife in 1948 (Chapter 30).

Landslides

A really terrible risk is that of great slides such as overwhelmed the valley of Pelileo, Ecuador, in 1949. The circumstances which produce such catastrophes do not come into existence suddenly at the time of an earthquake; such great slides are many years in preparation, and the risk could have been foreseen by proper geological study. Many similar occurrences in all parts of the world, some of them disasters to populated centers, have had no connection with earthquakes.

Supply and Communication Lines

The effect of faulting in interrupting water supply has been referred to under fire risk; but lack of water may be calamitous to agriculture, as in the Imperial Valley in 1940; on the same occasion the secondary effects on ground water resulted in costly damage in the Yuma Valley.

There is risk of interruption of long-distance supplies of gas and oil, and of communications by railway, highway, telegraph, and telephone. In California most private and public utility organizations have provided, by design of their lines and by emergency arrangements, against the consequence of a large displacement on one of the major faults. The plans include provision of shutoffs, automatic or manual, to reduce fire and other risks and avoid losses.

Utility, police, fire, and relief organizations in most larger communities have combined to work out a major disaster plan for their area. Since military disaster obviously has to be considered, this now usually centers about civilian defense provisions; but the possibility of earthquakes is almost always taken into account.

CHOICE OF GROUND

Seismologists and geologists are frequently consulted by those planning to build, for residence or business purposes, as to the safety of a given site with respect to earthquakes. The first inquiry usually refers to the location of active faults. It has to be explained that, in the long run, in a region like California the differences between two locations depend less on their distances from faults than on the character of the ground. Any given point is sooner or later going to be shaken heavily—either by a great earthquake far away or a smaller one comparatively nearby. In either case, intensity is going to be higher, other things being equal, on unconsolidated ground than on firm rock.

Strong shaking is especially likely where seismic waves emerge from basement rock into an alluviated area. Such danger spots are easily identified from experience in settled regions, and they should be avoided as sites for important construction.

Risk is particularly high on unconsolidated ground in a major fault zone; because of the broken character of the rock in the San Andreas and other major rifts, this combination is often to be found there. However, even adjacent to the major faults, investigation will identify individual small blocks of unfractured rock; structures can be founded on these with comparatively moderate risk. Such investigation has preceded location of new buildings on the campus of the University of California at Berkeley; the grounds are traversed by the active Haywards fault.

SPREADING OF POPULATION

In many regions increase of population has led to the settlement of areas which were practically unoccupied a generation or two ago. In such localities past seismic history does not exist, except as indicated by instrumentally recorded seismicity or, rarely, by geological evidence. This adds a new and difficult factor to insurance problems; and it renders exceedingly undesirable any current listing of large earthquakes based solely on macroseismic information, without regard to instrumental recording, magnitude determination, or geological evidence. Historical catalogues should be studied with due attention to their limitations.

SERIES OF SHOCKS

When a large shock has occurred, the public is anxious to learn whether more are to be expected. The continuance of even small aftershocks leads to disproportionate alarm, so that often reassurance is offered to the effect that "the worst is over," and no more serious shocks are to be expected. Careful study of Chapter 6 will show that such procedure may be dangerous. Particular caution is required in areas known habitually to be affected by earthquake swarms, or by series of nearly equally large shocks accompanied by small ones. Continuous instrumental recording extending over a long period of years is highly desirable for any region suspected of having such characteristics.

Due caution is also advisable in fostering the idea, usually born of wishful thinking, that a large number of small shocks may act as a "safety valve" to delay or even inhibit the occurrence of a destructive earthquake. The possibility exists; its realization is relatively rare (discussion in Chapter 22).

PREDICTION—A WILL-O'-THE-WISP

At present there is no possibility of earthquake prediction in the popular sense; that is, no one can justifiably say that an earthquake of consequence will affect a named locality on a specified future date. It is uncertain whether any such prediction will be possible in the foreseeable future; the conditions of the problem are highly complex. One may compare it to the situation of a man who is bending a board across his knee and attempts to determine in advance just where and when the cracks will appear.

Distant Hopes

Without specific prediction of date, there is some remote hope that it may be possible to detect the accumulation of strain toward a major earthquake in a given region or perhaps on a given fault. In the past, study of the pattern of occurrence of large and small shocks in an area has occasionally led to such forecasts; but the results have not been convincing. As noted in Chapter 6, an attempt by Davison to document a significant pattern in the seismicity preceding the Japanese earthquake of 1891 probably involves misinterpretation.

A direct approach to the study of strain accumulation is the resurvey of lines of triangulation and precise leveling, as carried out by the U. S. Coast and Geodetic Survey and by various agencies in Japan. Another is due to Dr. Benioff, who has lately initiated a program of setting up recorders to determine the progression of regional strain at a number of localities.

Amateurs and Astrologers

Amateur predictors are legion, and will continue to be, so long as claiming to predict earthquakes is an easy way to get one's name into the newspapers. Many of them are honestly self-deceived; they usually have (1) no conception of the frequency of small earthquakes (100,000 a year is a good figure to think of, although it needs definition in terms of the lower limit of magnitude included), (2) no means of knowing what earthquakes have occurred or how large they are, beyond mention in the press and the space allotted there, and (3) no effective training in scientific thinking. Some "predictors" select a large number of dates through the year and then claim as predicted anything which happens within a few days of any such date—so that the earthquakes of half the year or even more are called in as "verification."

Some predictors are astrologers and professional soothsayers, who claim to foretell future events of all kinds. Some are astrologers in disguise, who are actually basing predictions on the positions of the planets but justify them by some alleged mysterious radiation or modification of the law of gravitation. Predictions based on positions of the sun and moon have to be regarded a trifle more seriously, since there is evidence that tidal forces may occasionally act as triggers for earthquakes otherwise on the point of taking place; in this way the dates and hours of occurrence may show a slight statistical correlation with the tides.

A Case History

Perhaps the most "distinguished" of earthquake predictors became notorious in Italy early in the present century. We shall call him Graffiacane, and his city Malebolge. This man was an uneducated cobbler who began his career by foretelling local events; then, gaining the ear of newspaper correspondents, he began to predict earthquakes and claim verifications. The resulting furor in the press was too much for Agamennone, one of the most outstanding Italian seismologists. He published a paper pointing out Graffiacane's chicanery and general disregard for facts—his claiming minor events as verification of predictions of great earthquakes, surreptitiously changing his predictions to fit after the fact, even inventing reports of earthquakes which never occurred. The effect of Agamennone's taking notice of Graffiacane was to skyrocket the latter's popularity and set him up in the press as a "rival." Graffiacane found means to set up seismographs of a sort which recorded the earthquakes he predicted, whether others did or not. He issued dire predictions of destructive earthquakes for all parts of the world. He once created a local panic in Lima, Peru; the inhabitants took to sleeping out-of-doors and the authorities prohibited further publication of predictions.

During the Fascist regime Graffiacane curried favor with Mussolini; he invented new planets, one of which he named after Il Duce, and used their imaginary movements in earthquake prediction. To this day the Graffiacane Observatory exists in Malebolge; it apparently no longer issues predictions, but it often misinforms the press as to the nature and location of distant earthquakes (usually real).

All claims to predict the future have a hold on the imagination; it is not surprising that even qualified seismologists have been led astray by the will-o'-the-wisp of prediction. This has often been a factor in establishing seismological stations and research centers. The student should be aware of the little significance attempts at prediction have had in relation to the actual development of our knowledge and understanding of earthquakes.

RISK AND STATISTICS

Since specific prediction is at best a hope for the remote future, any study of earthquake risk has to be largely statistical. The general irregularity of earthquake occurrence and the relative rarity of large events are obstacles. Even in Japan, where records have been kept for centuries and seismicity is high, statistical risk is not easy to estimate. In California, with available history of less than 200 years, only the smaller shocks have been frequent enough to set up valid statistics. Yet within a century we know of three great earthquakes (1857, 1872, 1906), and several others have led to much property loss, which will probably be greater in the future because of the steady increase in population.

Insurance against earthquakes consequently presents an extremely difficult problem; and it is to the credit of insurance experts that so much progress has been made in the last 30 years.

EARTHQUAKE INSURANCE IN CALIFORNIA

Early History

In 1906 some insurance companies were not prepared for a major disaster, and failed to meet their obligations. In later years, earthquake insurance was regularly written, much of it without adequate attention to soundness; the total earthquake losses paid by the insurance companies for 1925, mostly at Santa Barbara, amounted to $666,265 (as reported by J. R. Freeman). This, together with dire predictions then in circulation, produced near panic in the headquarters of many insurance groups. Some arbitrarily increased rates, even trebling them; others temporarily discontinued writing earthquake insurance. Within a few years businessmen found that they could insure their property against earthquakes only at a high premium and only for losses above a high fraction. Desperate attempts to counteract this development included publication of a book, with the name of a well-known geologist on the title page, purporting to show that there was no serious risk of a destructive earthquake in southern California—or at least in Los Angeles. The disastrous Long Beach earthquake of 1933 relieved seismologists of the need to argue the matter.†

Organized Preparation

In the meantime the stock insurance companies had initiated organized and rational action. An important step was the setting up of the earthquake department of the Board of Fire Underwriters of the Pacific in 1926.‡ Their careful study showed the general inadequacy of existing building codes. Earnest efforts followed on the part of this and other insurance organizations, in cooperation with the California Development Association, the Pacific Coast Building Officials' Conference, and the California chapters of such organizations as the American Society of Civil Engineers, the American

---

† In 1933 the Los Angeles Chamber of Commerce printed and circulated the report of the Joint Technical Committee on Earthquake Protection, a fair and thorough account of the situation; it is now out of date only in so far as some of its conclusions have been further confirmed by later events.

‡ This work was transferred to the Pacific Fire Rating Bureau when that office was established in 1948.

Institute of Architects, and the Associated General Contractors of America; the Uniform Building Code, already prepared by the Building Officials' Conference (including its earthquake appendix), was studied and revised, and the general adoption of improved codes was strongly advocated.

The Zone Map

Historical and geological studies were used to distinguish regions at different levels of risk; the California-Nevada area, and finally the whole United States, were subdivided on this basis. A map zoning the United States at four levels was prepared by the U. S. Coast and Geodetic Survey with the cooperation of seismologists in all parts of the country. This map was incorporated in codes, such as the Uniform Building Code, of which it became legally a part. In the present state of knowledge, a map of this sort is necessarily imperfect. The chief deficiency of the 1948 map was its assigning probably too low a risk rating to areas with no known record of a strong shock, but tectonically similar to others where such shocks were known (which received higher risk ratings). The map was retired by the Survey in 1952 as "subject to misinterpretation and too general"; this action followed protests by groups of businessmen interested in getting lower risk ratings in their localities. A slightly revised form of the map still forms part of the Uniform Building Code.

It is geologically obvious that the earthquakes of two or three centuries do not provide an adequate basis for estimating the seismicity of any region. There is little to justify the common tacit assumption that the strongest shaking known to have affected a given point in the past will never be exceeded there. (See Chapter 6, under "Unusual Activity.")

Stratigraphic and geomorphic evidence should be used—but with judgment, since in many areas the tectonic features most obvious in exposures and in topography are not the most active.

In using macroseismic data there is difficulty in locating the probable epicenter and meizoseismal area, which may be distant from any of the places reporting an earthquake. Regional maps which ignore this usually exaggerate seismicity near the largest and oldest centers of population.

All these points have been under vigorous discussion during the last few years in the Soviet Union, where seismic zoning is an important public issue (see Chapter 33).

Sponsored Research

As noted in Chapter 8, a great contribution of the insurance industry, both in the direct public interest and in the furtherance of scientific understanding of earthquakes, has been the sponsoring and publication of factual reports on earthquake effects, particularly on damage. Other important results have followed from support of research in seismology—not merely of the engineering investigation aimed directly at decreasing damage, but of the physical and geological seismological work of cataloguing and analyzing earthquakes and their effects, on which any rational evaluation of risk must ultimately depend.

Present Problems

The insurance industry in the last generation has made enormous progress in efficiency and organization; it is now well prepared to cope with disasters of all kinds, including earthquakes. The panicky attitude of 1925 has been replaced by careful systematization. There is, naturally, still room for improvement. Earthquake insurance cannot be placed on as solid a basis as life insurance, nor can seismology be as well-organized as chemistry, so long as our knowledge of earthquakes remains comparatively imperfect and incomplete.

It is not intended to endorse the viewpoint of the customer who approaches insurance as if it were a public charity instead of a business, and grumbles at paying high premiums which result directly from his and others' neglect of safety precautions. However, there is still a need for some formula which will further the sale of earthquake insurance to small property owners on as generally accepted a basis as fire insurance. At present, small earthquake insurance is usually sold only in combined-coverage policies; otherwise the property owner does not demand it or the agent offer it. One reason for this is that premium rates are at a flat level high enough to take care of the weaknesses of common construction, with compensation beginning only above a percentage deduction. It is not easy to remove the latter obstacle, since it is necessary to restrict minor claims like those for cracked plaster, which may or may not be due to earthquakes; otherwise such claims tend to demand excessive time from investigators and adjusters.†

References

IN CONNECTION with this chapter the references for Chapters 8 and 11 and Appendix II may be consulted. For insurance and building code problems, the following may be noted:

---

† In 1956 standard premiums on earthquake insurance for average dwellings in the San Francisco and Los Angeles metropolitan areas were 15 cents per $100 insured, with 5 per cent deductible. Agents find that the average householder, insuring his property for $10,000, would rather take a chance on not having earthquake damage over $500 than pay $15 annually. The rate may go as low as 10 cents for structures of earthquake-resistant design; and equipment, stock, or contents may be insured. Ordinary hazardous structures may be rated at 25 cents; some companies insure average structures at that rate with a 1 per cent deduction.

Engle, H. M., "The earthquake resistance of buildings from the underwriter's point of view," *B.S.S.A.*, vol. 19 (1929), pp. 86–95.

Dewell, H. D., "The earthquake resistance of buildings from the standpoint of the building code," *ibid.*, pp. 96–100.

Chick, A. C., "Discussion of fundamental factors involved in the underwriting of earthquake insurance," *ibid.*, vol. 24 (1934), pp. 385–397.

Engle, H. M., "Earthquake provisions in building codes," *ibid.*, vol. 43 (1953), pp. 233–237.

Uniform Building Code, published for Pacific Coast Building Officials Conference, Los Angeles, California, 1955, Vol. 1.

Philbrick, F. P., "The effect of earthquakes on fire-alarm systems," *B.S.S.A.*, vol. 31 (1941), pp. 1–8.

Du Ree, A. C., "Fire-department operations during the Long Beach earthquake of 1933," *ibid.*, pp. 9–12.

Geography and Geology of Earthquakes

CHAPTER 25

Introduction to Geography and Geology of Earthquakes

IN PRECEDING CHAPTERS, earthquakes have been discussed rather generally, with details mainly for illustration. The better-known earthquakes individually call for more space, and many others deserve special notice, particularly those listed in Chapter 14 which give direct evidence of faulting. All this will be set forth in later chapters, in chronological order for each region. Part Two will be introduced by general discussion of the known geography of earthquakes and its geological implications.

HISTORICAL NOTE

Early studies of the geographical distribution of earthquakes were based on macroseismic data alone—accounts of destructive shocks or lists of shocks felt in a given locality. Information about small shocks was much influenced by accidents of location. Thus the community of Comrie in Scotland, where numerous slight shakes were felt, looms large in the annals of British earthquakes.

Some statistical studies were set up without reference to the magnitude of the shocks included, lumping important events and insignificant tremors together on an even basis. Still more commonly, the effects of density of population were neglected, so that the frequency and magnitude of European earthquakes were overrated.

Rudolph's studies of seaquakes showed that much imperfectly known seismic activity must be going on under the oceans. Here the major discovery is to the credit of John Milne; working through the British Association for the Advancement of Science, he caused seismographs of similar design to be set up all over the world. In consequence, rough epicenter locations were possible for all shocks of large magnitude—"world-shaking earthquakes," as Milne termed those which registered at all his stations. Milne's first correlations of these data showed a considerably larger fraction of

395

earthquakes originating under the oceans than under the continents. This finding was received with doubt at first, but it has been thoroughly confirmed by the later development of international seismology.

The logical extension of Milne's work is the *International Seismological Summary,* but it does not assign magnitudes, and false conclusions have often been drawn from the regional density of epicenters there catalogued. Maps drawn up from the *Summary* data show two conspicuous concentrations of epicenters, one in Europe, the other in Japan. The former is largely due to the number of stations in Europe and to the centralization of the international organization there; the seismicity of Europe, especially of western Europe, is low relative to that of many other areas. The seismicity of Japan, on the other hand, is really high, but it is accentuated in the *Summary* catalogues by the large number of stations operating there.

Milne, Turner, Tams, Sieberg, and others published critical studies of earthquake geography based primarily on microseismic data. By 1930 the main outlines were fairly well fixed, but important modifications in detail soon followed. The general recognition of deep-focus earthquakes led to verification of Turner's observation that they showed a significant geographical distribution, which soon proved to have an important relation to the varying depths of the shocks. However, when attempts were made to fit these new results into those considered established for shallow shocks, discrepancies arose. In using the *International Summary* for such investigation, three principal causes of confusion appeared.

(1) In spite of Turner's discovery of deep earthquakes, some such shocks were located as if they were shallow. In some instances this misplaced the epicenter by hundreds of miles; in others it placed a supposedly shallow shock in a region characterized by deep-focus earthquakes.

(2) Turner and his successors made praiseworthy attempts to locate all shocks recorded at a reasonable number of stations; but sometimes the data were misleading. Widespread microseisms confusing the records at many stations, or accidental errors in reporting the data for one or two key stations, might lead to a false epicenter.

(3) Many small shocks were listed, especially for Europe and Japan. Such shocks occur, roughly speaking, almost everywhere; their inclusion is no help in defining the main features of the seismicity of the earth, but rather tends to blur the picture.

Thorough revision, making use of all available instrumental data, was undertaken with these points in mind. The results are presented in publications by Dr. Gutenberg and the present writer, from which large parts of this and the next chapter are summarized.

SEISMICITY OF THE EARTH— MAJOR FEATURES

Epicenters occur chiefly in a few narrow belts or zones. Certain wider areas show fairly general moderate activity. Seismologically speaking, the most important subdivisions of the earth's surface are:

(1) The circum-Pacific belt, with many branches and complexities.

(2) The Alpide belt of Europe and Asia; this may be considered an extension of one of the main branches of the circum-Pacific belt.

(3) The Pamir-Baikal zone of central Asia.

(4) The Atlantic-Arctic belt.

(5) The belt of the central Indian Ocean, with branches.

(6) Rift zones (in the original sense), notably those of East Africa.

(7) A wide triangular active area in eastern Asia, between the Alpide belt and the Pamir-Baikal zone.

(8) Minor seismic areas, usually in regions of older mountain-building.

(9) The central basin of the northern Pacific Ocean; almost non-seismic except for the Hawaiian Islands.

(10) The stable central shields of the continents, also nearly non-seismic.

(The principal stable areas are the Canadian Shield; the Brazilian Shield; the Baltic or Fennoscandian Shield; the Angara Shield of northern Asia; the African mass, divided into a northern and a southern unit; western Australia; Antarctica; peninsular India. Arabia and Madagascar can be considered detached portions of the African stable mass, and Greenland a similar fragment of the Canadian Shield.)

The circum-Pacific belt accounts for a large majority of shallow shocks, a still larger proportion of intermediate shocks, and all but a very few of the known deep shocks. The Alpide belt and the Pamir-Baikal zone include most of the remaining shallow shocks, especially those of large magnitude; the Alpide belt accounts for all the remaining intermediate shocks and includes at least two important deep shocks (one under southern Spain and one west of Italy).

A globe with the seismic belts and areas indicated resembles a jigsaw puzzle. Figures 25–1 and 25–2 illustrate the individual pieces. Each conforms to a general pattern (except that which shows the Pacific Basin). The Australian piece, for example, centers naturally about the oldland of West Australia. The edges of the piece to the north and east follow the circum-Pacific belt through the East Indies and down to New Zealand; the other edges are the active zones of the Indian Ocean which separate this piece from those of India, Africa, and Antarctica. In eastern Australia is a region of minor seismicity corresponding to the older mountains of that

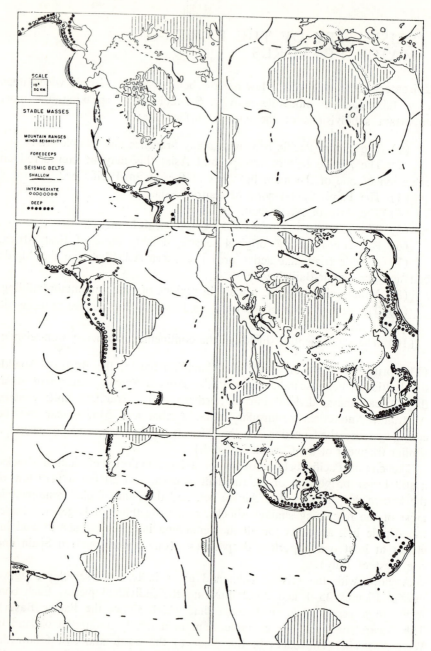

FIGURE 25-1   *Continental stable masses, and associated seismic belts.* [Seismicity of the Earth.]

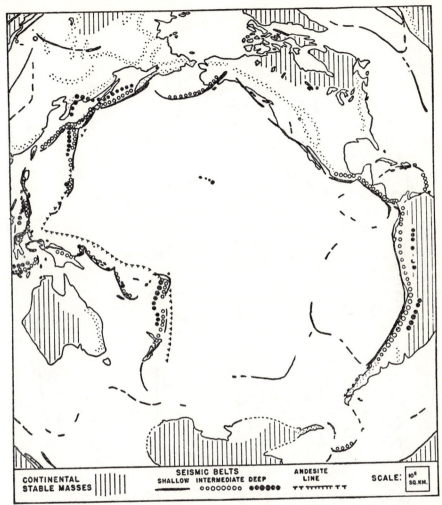

| CONTINENTAL STABLE MASSES | ‖‖‖‖ | SEISMIC BELTS SHALLOW INTERMEDIATE DEEP ⎯⎯ oooooooo ●●●●●● | ANDESITE LINE ⊤⊤ ⊤⊤⊤⊤⊤ ⊤ ⊤ | SCALE: 10⁶ SQ.KM. |

FIGURE 25-2   *Pacific stable area and surroundings.* [Seismicity of the Earth.]

area. In central Australia, at the margin of the stable shield, is the "shatter zone," in which some of the largest Australian earthquakes have occurred. Near the west coast, where the stable shield breaks down to the Indian Ocean, other epicenters are known. This includes the largest known Australian earthquake, on June 27, 1941. Similarly in North America we have the Canadian Shield with the circum-Pacific belt on one hand and the Arctic-Atlantic belt on the other; minor seismicity associated with the older structures of the Appalachians and Ozarks; and moderate seismicity along the rift zone of the St. Lawrence, marginal to the Canadian Shield.

The geography of intermediate and deep shocks is implied in their relation to the active arcs (Chapter 26), but the belts of deep shocks near Japan, originally described by Wadati, are still a little puzzling. One such belt,

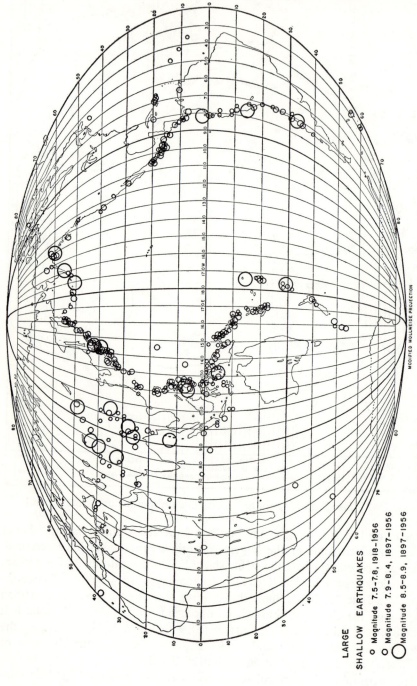

MODIFIED MOLLWEIDE PROJECTION

LARGE
SHALLOW EARTHQUAKES
○ Magnitude 7.5-7.8, 1918-1956
○ Magnitude 7.9-8.4, 1897-1956
◯ Magnitude 8.5-8.9, 1897-1956

FIGURE 25-3 *Large shallow shocks.* [*Data from Seismicity of the Earth, with additions and revisions.*]

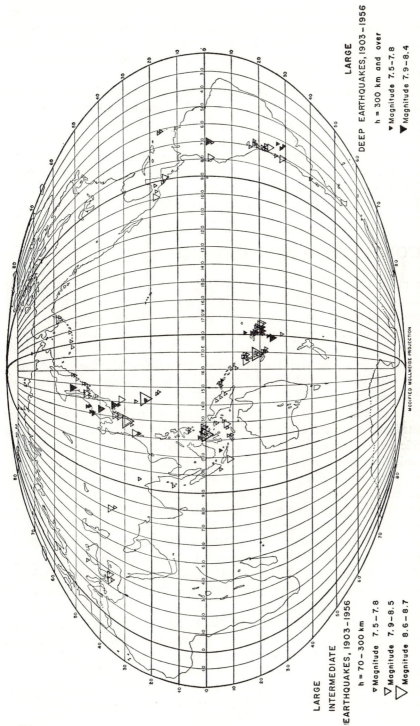

LARGE
INTERMEDIATE
EARTHQUAKES, 1903-1956

h = 70-300 km

▽ Magnitude 7.5-7.8
▽ Magnitude 7.9-8.5
▽ Magnitude 8.6-8.7

LARGE
DEEP EARTHQUAKES, 1903-1956

h = 300 km and over

▼ Magnitude 7.5-7.8
▼ Magnitude 7.9-8.4

MODIFIED WOLLWEIDE PROJECTION

FIGURE 25-4 Large intermediate and deep shocks. [Data from Seismicity of the Earth, with additions and revisions.]

continuing the zone of deep shocks associated with the Marianas arc, crosses Honshu transversely and continues across the Japan Sea to the vicinity of Vladivostok; there it intersects another belt, which follows the Siberian coast at first but then passes north of Japan and across the Sea of Okhotsk to Kamchatka. (Fig. 30-4.)

A curious fact is that the seismicity of the northern hemisphere is greater than that of the southern hemisphere. The assignment of magnitudes and the consequent cataloguing of all large earthquakes remove any possibility that this result is influenced by the greater number of seismological stations in the northern hemisphere. There is further decrease southward from the equator; less than 10 per cent of large earthquakes occur in the area beyond 30° S (a quarter of the earth's surface).

The epicenters of large shocks are mapped in Figures 25-3 and 25-4.

CONTINENTS AND OCEANS

The major seismic zones are conspicuously related to the chief units of the surface of the earth. The Pacific Basin occupies a unique position, separated from the continents by major seismic zones on all sides except the south. Thus most of the continental pieces of the jigsaw puzzle have the circum-Pacific belt as one boundary. Every piece is bounded in part by either this or the Alpide belt. The remaining principal boundaries, like the mid-Atlantic Ridge with its attendant seismic zone, separate the continental units from each other.

Some differences can be read from ordinary maps, for, as Suess long ago pointed out, there are two principal types of coast line, which he called the Pacific and Atlantic. The Pacific type of coast is more or less parallel to tectonic lines and is relatively unbroken; the Atlantic type cuts across structures, resulting in a ragged coastline with deep embayments and irregular headlands. The rule cannot be applied blindly, or in great detail; thus in Japan, while the east coast of Honshu is readily classifiable as of Pacific type, the coast more to the west and south is extremely irregular. The island arc of the Caribbean, fronting on the Atlantic Ocean, and that of the Sunda Islands in the East Indies, fronting on the Indian Ocean, are structurally of Suess's Pacific type. Also of Pacific type is the Alpide series of arcuate mountain systems extending across Asia from Burma and the Himalaya to the Mediterranean and the Alps.

On this evidence, such oceans as the Atlantic and the Indian are fundamentally different in location from the Pacific Basin, and it has often been suggested that they differ in underlying structure, perhaps representing an intermediate condition between the continents and the Pacific. Observational evidence for this in the past consisted chiefly of dispersion curves for seismic surface waves (see Chapter 17), which clearly show contrast between trans-Pacific paths and purely continental paths but seem intermediate for many Atlantic paths. An alternative explanation, always available for these data, was a complexity of the Atlantic paths, possibly traversing both oceanic and continental structure. This is the trend of the newer and more direct evidence now rapidly accumulating from refraction shooting and gravity measurement; except for the mid-Atlantic Ridge and the continental shelves, this work indicates structure of the same kind in the Atlantic as in the Pacific, with a general oceanic level of the Moho near 10 to 11 kilometers below the sea surface. Differences in detail are not excluded (compare Chapter 18); two regions in which the Moho is at the same level are not necessarily alike above it.

MARSHALL'S ANDESITE LINE

In the western Pacific one branch of the chief seismic belt follows the andesite line, first drawn by Marshall in 1911. This line has been taken by many authors as limiting the Pacific Basin proper; sea-covered areas outside it, such as the Philippine Sea between those islands and the Marianas, have been thought of as continental or subcontinental (though the latest data suggest that if there is any such difference it does not affect the level of the Moho). The Marshall line separates these outer regions, in which the younger eruptive rocks are largely andesitic although basalts occur, from the inner Pacific area in which andesites are not found. The possibility of such distinction, however, depends in part on the definition of andesite, which is rather elusive when the usages of different authors are compared. On the continental side of the line, some of the larger units, like the Fiji Islands, show their continental character by exposed granitic and dioritic rocks, while the Samoan Islands just across the line are of Pacific type, with large volcanic forms like those of Hawaii.

SEISMICITY AND GEOLOGIC HISTORY

Montessus de Ballore emphasized the geographical correlation between high seismicity and high relief, whether on land, associated with high mountains, or submarine, with the great trenches. Pursued further, this correlation indicates that present high seismicity is related to crustal disturbances of the younger geologic past. The highest mountains, the Himalaya and Andes, belong to this younger past; and so, presumably, do the deep trenches. Crustal irregularities even as great as these would be equalized by geological processes within a single epoch, unless stresses were continually at work to maintain them. The observed earthquakes are incidental to these stresses.

From the point of view of historical geology, our maps of seismic belts and zones are mere snapshots. For most of the world they depend wholly on instrumental epicenter locations, which are not accurate enough for even the roughest geological purposes prior to about 1903. Even in regions with both centuries-old culture and high seismicity, such as Japan, surprisingly few significant details can be added from macroseismic evidence. Thus our current picture of the seismicity of the earth may, for all we can tell, represent a temporary and unusual state of affairs. The record of the rocks leaves little doubt that tectonic geography differed considerably in the past, and vastly during the remoter eras.

We begin, accordingly, by correlating seismicity with the most recent occurrences in geologic history, and go on to consider successively older structures.

Seismicity and Quaternary Tectonics

We should expect close correlation between current observed seismicity and tectonic activity of Pleistocene or later age. Correlation exists but is not always conspicuous.

The writer believes that in many, if not all, active parts of the world the tectonic stresses are now changing rapidly, rather than slowly, when referred to the geological time scale. The basis for this conclusion is the production in earthquakes and otherwise of scarps and other tectonic effects which are not mechanically coherent with major structures of approximately Pleistocene age. In New Zealand, as indicated by Cotton, some of the small trace features attributable to Recent earthquakes appear discordant with the Pleistocene block structure. (See also Chapter 27.) In California, for example, Hill and Dibblee have suggested that the Garlock fault, with geologically young rift features, has been broken and offset along the San Andreas Rift; and it is difficult to maintain that the stresses now acting to produce displacements on the large faults of California are parallel to those under which they originated as fractures. The greatest fault scarp in California, the east face of the Sierra Nevada, shows no more than light seismicity, while in 1872 a great earthquake originated at the edge of the minor block of the Alabama Hills a few miles to the east in the Owens Valley graben. In Turkey, the rift zones associated with the earthquakes of 1939 and thereafter, at first supposed to separate the Tauride from the Anatolide structures, were later seen to cut into both. Similar discordances in the Tian Shan and other parts of the USSR have been noted by Petrushevsky and others (Chapters 13 and 33).

Present conditions are not necessarily exceptional in the history of the earth; it would be quite consistent with the geological record to suppose that rapid changes are normal. Intervals when the stresses shifted little would then be of relatively short duration on the total time scale; at such times displacements could accumulate coherently to large amounts, and these would be times of orogeny. However, H. Stille considers that the present time falls in the expiring stages of an orogenic period which reached a climax in the mid-Pleistocene; this he has made widely known under the name of the "Pasadena" orogeny.† Young thrusts and anticlines of corresponding age are fairly common in the active tectonic areas of California; the latter are frequently associated with oil production. Many producing oil fields, thus associated with very young orogeny, are aligned in the vicinity of the active faults.

Stille applies the "Pasadena orogeny" to the entire circum-Pacific active belt; this would logically associate it with the youngest tectonic features, such as the non-volcanic anticlines and the Meinesz gravity zones at the outer edge of the active arcs (see later subheads and Chapter 26).

Pleistocene and Pliocene

Going back in time, we encounter increasing divergence between current tectonics and the major land forms. The block structures now dominant in such regions as California and New Zealand were established largely in Pliocene and Pleistocene time; yet present seismicity is related to these structures only in general and not in detail.

Discussion is difficult and risky because of the changing picture due to the progress of investigation. Diastrophisms formerly described as regional and of brief duration have been localized and broken up into episodes distributed over entire epochs. Such are the Cascadian orogeny in North America and the Kaikoura orogeny of New Zealand. The principal uplift of the Cascade ranges, still assigned by many authors to the Pliocene-Pleistocene transition, is now often described as comparatively minor and episodic in an orogenic process considered to have begun with strong diastrophism even in the Miocene, continuing to the time of major elevation of the Sierra Nevada, which culminated in the Pleistocene. In the California coast ranges, Taliaferro and others have assigned the Tertiary orogenic climax to the Late Middle Pliocene. In New Zealand, the principal orogeny of the Kaikoura group is now considered definitely Pliocene; but diastrophism continued, and, especially in the North Island, there is indubitable evidence of major Pleistocene block faulting.

Present seismicity in the area of the Cascade Mountains is minor. In the California Coast Ranges and in New Zealand, it is only moderate as compared with highly seismic regions like Japan and the East Indies. There is frequent discordance between current tectonic movement and Late Tertiary

---

† This designation (which was not that originally attached by local investigators) arose out of the deep impression made on Stille during a visit to California in 1935, when he was shown Miocene and Pliocene sediments crumpled and thrust over horizontally bedded Pleistocene strata. Most of the localities are at considerable distance from Pasadena; some of the most striking exposures are in cuts along the coast highway between Ventura and Santa Barbara.

structures; the San Andreas fault cuts obliquely across the California Coast Ranges, and in the Kaikoura region active strike-slip faulting follows the line of old thrust faults bounding the conspicuous large blocks.

Tertiary and Cretaceous

Small-scale maps presented to show correlation of the present seismic belts with the younger mountains of the world usually indicate orogenies of Tertiary and Cretaceous date. The whole Alpide group, the younger circum-Pacific mountains, the Cordillera of western North America, and even the Andes are included. The general relationship is then established at the expense of detail. Thus the seismicity of the Rocky Mountain zone is much lower than that of the California coast ranges, and few important shallow earthquakes originate under the central Andes, the principal sources being farther west or deeper.

An example of loose correlation is that between present seismicity and Miocene orogeny. In the Alps and the Himalaya the climax of Tertiary mountain-building was in the Miocene. In the East Indies the interior was relatively stable at that time, but in the outer zones, including the Sunda arc, there was strong Miocene folding. In California, while the Miocene was a time of block displacement and volcanism, there was nothing mechanically resembling the Alpine orogeny (a point lately emphasized by Stille). Finally, in New Zealand the Miocene includes only a few of the moderate disturbances of the long interval called the Notocene by some authors (the Notocenozoic of Cotton; Chapter 27). Seismicity in the Sunda arc is now high; in California, New Zealand, and India it is moderate, and in the Alpine region it is low.

Older Orogenic Belts

Many regions now active were affected by a late Mesozoic (Jurassic or Cretaceous) orogeny involving folding and thrusting; in the Alpide belt this was the beginning of mountain-building, which increased in the Tertiary. However, these regions are now active largely because they were involved in the Pliocene-Pleistocene orogenic climax, although even the large block faulting of that period seems less closely related to current seismicity than the minor faulting and folding of later date.

Whatever the interpretation, there can be no doubt of the association of moderate to minor seismicity with certain Paleozoic orogenic belts far outside of the major seismic zones. A good instance is the Permo-Carboniferous Variscan belt, which strikes slightly south of east from the British Isles into the European continent. One of the largest of the rare English earthquakes, that of 1884, was destructive at Colchester on the northern edge of this zone. Minor activity occurs in the Appalachian region, and in the orogenic zones of comparable age trending westward in the United States.

The epoch of the Caledonian orogeny of northern Europe is so remote that when we encounter minor earthquakes in Scotland, for example, our judgment is that this is tectonic rejuvenation; minor displacements are being produced by stresses of modern origin along old lines of fracture, which, though ancient, may be much younger than the Caledonian mountains.

VOLCANISM

The comparatively minor earthquakes directly associated with volcanic processes (Chapter 12) may be neglected in mapping seismicity. Correlation of volcanic lines with the principal seismic belts, particularly those for shallow earthquakes, then appears as a loose parallelism like that with Cretaceous and Tertiary orogenic lines—and frequently coincident with it, for especially in the circum-Pacific belt active volcanic vents are set directly along the younger island arcs which indicate these lines. The close correlation of volcanic lines with epicenters of intermediate earthquakes will be discussed with the other characteristic arc features (Chapter 26). On the other hand, in the Atlantic and Indian oceans active volcanoes are located directly on the ridges associated with shallow earthquakes, forming the boundaries between the pieces of the jigsaw puzzle. Volcanoes also appear along active rifts, as in Africa.

Extinct volcanoes, like older orogenies, show remoter correlation with present seismicity, although extinct vents in some regions fill gaps in active lines, and in others form lines of their own which show correlation with shocks at depths of about 200 kilometers.

MEINESZ GRAVITY ZONES

Many valuable data on the internal constitution of the earth come from the precise measurement of gravity. Such measurements are made to about 1 milligal (since 1 gal is 1 cm/sec^2, this means measurement to one part in a million), since differences of the order of 10 milligals are of significance. Until lately good gravity data could be obtained only at land stations. In 1923 F. A. Vening Meinesz initiated an epoch with new apparatus enabling him to determine gravity at sea; a submarine was used for observation in order to get below the disturbed water near the surface. Among his first results was discovery, in the East and West Indies, of long, narrow belts over which the gravity anomalies (see Chapter 2) were negative, often as much as −200 milligals. Such Meinesz zones have been verified for most of the active arcs, where they closely follow belts of shallow earthquakes. In the same belts, usually where the gravity anomalies are less, there are young anticlines. These narrow belts represent the most disturbed parts of the earth's crust at the present epoch. Many authors have interpreted them as axes of orogeny now in progress; this is in accord with their high seismicity. Stille would place them in the latter part of his "Pasadena orogeny."†

PERIDOTITES AND SERPENTINITES

In most orogenic zones there are belts and aligned localities where ultra-basic rocks high in olivine—peridotites, dunites, etc., and the serpentinites into which they tend to be gradually transformed—are exposed. As investigation goes forward, this kind of evidence is becoming more and more applicable to the localization of orogenies. Fortunately this is independent of the details of interpretation; for, as Hess has lately put it:

> For most of this century a magnificent argument has gone on between field geologists who have worked on the peridotites of alpine mountains and laboratory investigators of their chemistry (particularly Bowen). The former stoutly maintain that the evidence indicates that they were intruded in a fluid state, as magmas; and the latter equally forcefully has proved to his own satisfaction that such magmas are not possible.‡

In certain active arcs the evidence from serpentines is of particular interest to the seismologist and geophysicist. Such an application was made by Hess to the loop of the West Indies, which branches from the Pacific belt. This loop has a Meinesz zone of negative gravity anomalies around the outside, with numerous shallow earthquake epicenters; but at the northwest, near the east point of Cuba, the Meinesz zone ends, and seismicity continues past Jamaica along the tectonic line of the Bartlett Trough, which appears to be a rift structure. In the Antilles east from Cuba serpentinites are found associated with the tectonic zone, and these rocks continue through Cuba. Hess proposed that this westward continuation may formerly have been associated with gravity anomalies, but that the tectonic strains have shifted, the pressures required to maintain gravity anomalies and seismicity have relaxed, and only the ultrabasic rocks remain as evidence. A similar application has been made to the peridotites of southeastern Celebes, which again suggest

---

† The writer cannot judge the somewhat controversial details of the system by which Stille and his followers have organized the whole history of the earth with reference to world-wide periods of orogeny. Especially in the remote past, the questions raised concern the seismologist very little, but the terminology and classification used by Stille often prove convenient for stating the facts of seismic geography (references at end of this chapter).

‡ "Serpentines, orogeny, and epeirogeny," *Geol. Soc. Amer., Special Paper* 62 (1955).

the course of a former tectonic line in which the strains have now largely relaxed. This relaxation may be related to the peculiar double curve of the indicated line in the Banda Sea and Celebes.

That the peridotite evidence indicates "fossil" rather than currently active tectonic belts must be remembered by the seismologist. Thus the peridotites of New Caledonia represent a possibly Mesozoic† orogeny which is only distantly related to the active arc of the New Hebrides north and east of it. In New Zealand, if Wellman is right, we have an ancient ultrabasic zone which has been cut and offset about 300 miles by motion on the Alpine fault since the Jurassic.

References

FOR MATERIAL on crustal structure and the Mohorovičić discontinuity refer to Chapters 17 and 18. Recent critical references on the structures underlying the Atlantic Ocean are:

Bullard, E. C., et al., "A discussion on the floor of the Atlantic Ocean," Proc. Royal Soc. (London), Ser. A, vol. 222 (1954), pp. 287–407. (Contributions from 17 authors; recommended to the student.)

Rothé, J. P., "Hypothèse sur la formation de l'Océan Atlantique," Comptes rendus, vol. 224 (1947), pp. 1295–1297. (A controversial suggestion, developed further in the following reference; the eastern Atlantic basin taken as continental, the western as oceanic.)

———, "The structure of the bed of the Atlantic Ocean," Trans. Am. Geophys. Union, vol. 32 (1951), pp. 457–461.

———, "La structure de l'Atlantique," Ann. geofisica, vol. 4 (1951), pp. 27–41. (Italian translation, ibid., pp. 118–125.)

Caloi, P., Marcelli, L., and Pannocchia, G., "Sulla velocitá di propagazione delle onde superficiale in corrispondenza dell'Atlantico," ibid., vol. 2 (1949), pp. 347–358.

Caloi, P., and Marcelli, L., "Onde superficiali attraverso il bacino dell'Atlantico," ibid., vol. 5 (1952), pp. 397–407. (Evidence against Rothé's hypothesis.)

Berckhemer, H., "Rayleigh-wave dispersion and crustal structure in the East Atlantic ocean basin," B.S.S.A., 46 (1956), pp. 83–86.

The student may trace the discussion further through many references given in the papers cited here.

---

† Hess remarks, "The New Caledonia serpentines are considered to be Eocene by most authors" [this includes Routhier, who has lately investigated them in detail] "but the present writer has rejected this opinion in favor of an older age which would be consistent with the tectonic setting. Admittedly the grounds for doing so are weak." The student should note this as an unusually good example of boldness in hypothesis.

Geography of earthquakes

Tams, E., "Die Seismizität der Erde," *Wien-Harms Handbuch der Experimentalphysik,* Vol. 25, 1931, pp. 361–437.

Sieberg, A., "Erdbebengeographie," *Handbuch der Geophysik,* Berlin, vol. 4, 1932, pp. 685–1005.

Gutenberg, B., and Richter, C. F., "Seismicity of the earth," *Geol. Soc. Amer., Spec. Paper 34,* 1941. (Preliminary form of the following reference.)

————, *Seismicity of the Earth and Associated Phenomena,* Princeton University Press, 1949; 2nd ed., 1954.

History of earthquakes

(The following are the two largest modern collections. For individual earthquakes consult the chronological bibliography, Appendix XVI.)

Davison, C., *Great Earthquakes,* Murby, London, 1936, 286 pp.

de Montessus de Ballore, F., *"La Géologie sismologique,"* Armand Colin, Paris, 1924, 488 pp.

Marshall line

Marshall, P., "Oceania," *Handbuch der regionalen Geologie,* Carl Winter, Heidelberg, vol. 7, part 2, 1912, 36 pp.

Bridge, J., "A restudy of the reported occurrence of schist on Truk, Eastern Caroline Islands," *Pacific Science,* vol. 2 (1948), pp. 216–222. (This documents drawing the andesite line west of Truk.)

Tectonic changes of recent date

Cotton, C. A., "Revival of major faulting in New Zealand," *Geol. Mag.,* vol. 84 (1947), pp. 79–88.

————, "Geomechanics of New Zealand mountain-building," *N. Z. Journ.,* Ser. B, vol. 38 (1956), pp. 187–200.

Richter, C. F., "Seismicity and structure of the Pacific region of North America," *Proc. 7th Pac. Sci. Congress,* vol. 2 (1953), pp. 671–681.

Hill, M. L., and Dibblee, T. W., Jr., "San Andreas, Garlock, and Big Pine faults, California," *Bull. Geol. Soc. Amer.,* vol. 64 (1953), pp. 443–458.

Stille, H., "Recent deformations of the earth's crust in the light of those of earlier epochs," *Geol. Soc. Amer., Spec. Paper 62,* 1955, pp. 171–191.

Petrushevsky, B. A., "Znachenie geologicheskikh yavleniy pri seysmicheskom rayonirovanii," *Trudy Geofiz. Inst. Akad. Nauk SSSR,* No. 28 (155), 1955, pp. 1–59. (Consult also appropriate references in Chapter 33.)

General tectonic history

Stille, H., "The present tectonic state of the earth," *Bull. Amer. Assoc. Petroleum Geologists,* vol. 20 (1936), pp. 849–880.

————, *Grundfragen der vergleichenden Tektonik,* Borntraeger, Berlin, 1942, 443 pp.

Lotze, F., ed., *Geotektonisches Symposium zu Ehren von Hans Stille,* Enke, Stuttgart, 1956, 483 pp.

Umbgrove, J. H. F., *The Pulse of the Earth,* 2nd ed., Nijhoff, The Hague, 1947.

Gilluly, J., "Distribution of mountain building in geologic time," *Bull. Geol. Soc. Amer.,* vol. 60 (1949), pp. 561–590 (criticism of Stille's theories). German translation, with discussion by Stille and Gilluly, in *Geol. Rundschau,* vol. 38 (1950), pp. 89–107; also the next reference.

Stille, H., "Nochmals die Frage der Episodizität und Gleichzeitigkeit der orogenen Vorgänge," *Geol. Rundschau,* vol. 38 (1950), pp. 108–111.

Eardley, A. J., "The cause of mountain building—an enigma," American Scientist, vol. 45 (1957), pp. 189–217.

Meinesz zones

Vening Meinesz, F. A., Umbgrove, J. H. F., and Kuenen, Ph. H., *Gravity Expeditions at Sea, 1923–32,* Netherlands Geodetic Commission, Delft, Vol. 2, 1934.

Vening Meinesz, F. A., *Gravity Expeditions at Sea, 1923–1938,* Delft, 1948, Vol. 4.

Ewing, M., Worzel, J. L., and Shurbet, G. L., "Gravity observations at sea . . . ," *K. Ned. geol.-mijnb. Gen., Gedenkboek F. A. Vening Meinesz,* The Hague, 1957, pp. 49–115.

Peridotites

Benson, W. N., "The tectonic conditions accompanying the intrustion of the basic and ultrabasic igneous rocks," *Mem. Nat. Acad. Sciences,* Vol. 1 (1926), pp. 1–90.

Hess, H. H., "Gravity anomalies and island arc structure with particular reference to the West Indies," *Proc. Am. Phil. Soc.,* vol. 79 (1938), pp. 71–96.

————, "Serpentines, orogeny, and epeirogeny," *Geol. Soc. Amer., Special Paper 62* (1955), pp. 391–408.

Routhier, P., *Proc. 7th Pacific Science Congress,* vol. 2 (1953), pp. 62–71. (New Caledonia.)

Roever, W. P. de, *ibid.,* pp. 71–81. (Kabaena, off Celebes.)

CHAPTER 26

Seismic Geography. Arc and Block Tectonics

RELATION OF ARC TO BLOCK STRUCTURE

In the chief seismic zones, shallow earthquakes occur in two different environments which which may be termed conditions of arc and block tectonics. Arcuate structures are dominant in most of the circum-Pacific and Alpide belts; in the western Pacific they generally take the form of island arcs, while those of the Alpide belt are chiefly mountain arcs like the Himalaya, the Alps, and the Carpathians. Historical geology indicates existence of similar arcuate structures in many regions in the remote past.

Block faulting is dominant in certain parts of the circum-Pacific belt, as in California and central New Zealand. This is probably true of most regions where only shallow earthquakes are known, like the Pamir-Baikal zone, and may apply to most of the mid-Atlantic belt. In some areas both tectonic types occur, as in Japan, Peru, the Philippines, and the North Island of New Zealand.

Although earthquakes in regions of arc structure are in the majority, most of our information bearing on their mechanism is microseismic, derived from the properties of surface waves, the azimuthal distribution of recorded initial compressions and dilatations, etc. On the other hand, most of our macroseismic observation of earthquake mechanism, including the occurrence of faulting, comes from areas of block tectonics (including those associated with arc structures, as in Japan); ·our ideas of the nature of earthquakes are based primarily on data from such sources.

Detailed geophysical study of the active arcs often runs into difficulty. Many of them, like the Aleutian and Kurile arcs, are submarine except for a chain of minor islands. Others are in regions where major stratigraphic and tectonic problems are yet unsolved. Some of the arcs of the Alpide belt, such as those of Italy and India (Chapters 4 and 5), are in well-investigated regions; but caution is needed in transferring conclusions drawn from these

412

less active structures to the Pacific arcs. Regions like Peru and northern New Zealand, where the tectonic types overlap, are difficult to interpret in the present stage of knowledge. In the East Indies and Japan, where seismicity is high and geological data have been collected by many investigators, tectonic relations are highly complex.

FEATURES OF PACIFIC-TYPE ARCS

The active arcs are three-dimensional in nature, requiring both a map and a profile for study (Fig. 26-1). They are normally associated with a group of typical phenomena and features. In their usual order, beginning at the outside or convex side of the arc, and passing toward its interior, the principal features are:

A. A deep oceanic trench (foredeep).

B. The principal tectonic line, with epicenters of shallow earthquakes, negative gravity anomalies, and non-volcanic anticlines (which may form a submarine ridge or emerge as a chain of small islands).

C. A belt of generally positive gravity anomalies with epicenters of earthquakes originating at depths of about 60 kilometers.

D. The principal structural arc, of Late Cretaceous or of Tertiary age, often consisting of large islands. Active or recently extinct volcanoes. Earthquakes at depths near 100 kilometers.

E. An inner structural arc. Volcanism usually older, and now in a late stage or extinct. Earthquakes at depths of 200 to 300 kilometers.

F. A belt of epicenters of deep earthquakes (hypocenters at 300 to 700 kilometers).

FIGURE 26-1   *Typical active arc features; gravity anomalies, idealized plan and profile. (Surface relief exaggerated; hypocenters on equal horizontal and vertical scale.)*

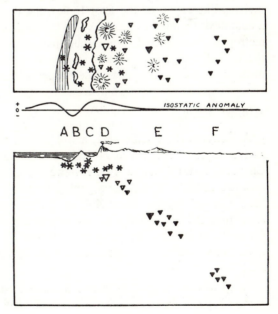

Active Surfaces

As Figure 26-1 shows, when hypocenters associated with a typical active arc are plotted in profile or contoured on a map, they are seen to define a surface which dips under the arc, in the majority of cases under a continent, at an angle not far from 45°. While the epicenters of the large, shallow earthquakes (B) are generally on the landward side of the foredeep, those of the deepest shocks are hundreds of miles (or kilometers) distant; thus in South America many of these epicenters are well to the east of the Andes. It is important that this relation of dip to continent, while prevailing, is not universal. Thus the structural arc of the New Hebrides is convex westward, and the surface defined by earthquake hypocenters there dips eastward toward the Pacific.†

Especially since the time of Suess, geologists have often interpreted the Pacific arcs as the surface traces of great thrusts. The effect of the location of deep shocks has been to identify the thrust surfaces with the surfaces defined by hypocenters, a point of view vigorously championed by the writer's colleague, Dr. Benioff. Thus we have identifiable tectonic structures extending to depths of the order of 700 kilometers; this contrasts strongly with earlier theories. Before 1930 it was generally believed that tectonic processes could not go on below about 100 kilometers, which was taken to be the level at which isostatic compensation is complete. Moreover, these large-scale structures must persist through geological periods.

The student should not think of the active surfaces as simple, sharp breaks. The range in hypocentral depth in a given locality indicates appreciable thickness, and generally epicenters in a given depth range cluster along relatively narrow belts, so that the dipping surfaces are traversed by narrow zones along which the seismicity is higher than elsewhere. In certain regions, like Japan or the East Indies, the three-dimensional pattern reaches a complexity almost as great as that of the freeway intersections in a modern metropolis. Dr. Benioff points out that in many areas there is a change in the dip of the general surface at about the level of transition from intermediate to deep earthquakes, the deeper surface dipping more steeply.

Alpide Arcs

Aside from the Pacific area, active arcs of this type are found particularly in the Alpide belt. Those of Italy and India have been noticed briefly in Chapters 4 and 5 and will be reviewed in Chapter 31. Although the Alpide arcs are roughly of the same geologic age as those of the Pacific, they exhibit a lower degree of activity, as if they had aged more rapidly. Gravity anomalies, active volcanism, and seismicity are indications of continuing processes maintained by progressive strains in the earth. If the strains relax, these more active manifestations decrease, although volcanism appears to

---

† Most of these shocks are shallow or intermediate, but one deep shock is known: 1949 July 18, 08:28:17, 12½ S 171½ E, depth 610 kilometers.

be more persistent than shallow earthquake activity. The arcuate structures and the evidence of past volcanism remain into later geological epochs.

The Spanish Deep-Focus Earthquake of 1954

Most of the arcs of the Alpide belt are associated with intermediate earthquakes like those of the Pacific arcs. Until March 29, 1954, it was largely a matter for speculation whether the Alpide structures should be considered as extending to greater depths than those of intermediate hypocenters. On that date an earthquake of magnitude 7 took place at a depth of 640 kilometers under southern Spain. In its implications, this is one of the most important earthquakes ever recorded; the microseismic data are excellent and leave no doubt of the facts. The Alpide structures, accordingly, must be considered as extending to great depths. This epicenter, however, is within the Alpide surface tectonic belt, instead of being horizontally at great distance; therefore the deep structure must here dip steeply. This contrasts significantly with the Sunda arc in the East Indies; that arc is on the western branch of the Pacific belt, but it can also without distortion be considered an eastern extension of the Alpide belt. Associated with the Sunda arc are many deep shocks, including some of the deepest known; but their epicenters lie well within the arc (northward), and the active surface dips in that direction in the usual fashion.

The occurrence of the Spanish deep shock indicates that in time deep shocks may be observed where they are to be expected but have not yet been located. Thus no shocks deeper than about 200 kilometers are known to be associated with the highly seismic arcs of the Aleutian Islands and Mexico, but it would be unwise to take this negative evidence as representing a structural difference between these and other Pacific arcs.

On February 17, 1955, a shock of magnitude 5½ originated at a depth of 470 kilometers north of Sicily; this extends the Italian arc structure into deep-focus levels.

REGIONS OF BLOCK TECTONICS

The active arcs do not account for all the seismicity, even of the Pacific belt. There are several areas where shallow earthquakes are frequent, from which the arc features are nearly absent, and where the chief activity appears to consist of relative displacements of blocks separated by large faults, which are often nearly straight and dip almost vertically. Among such regions are two discussed in the next chapters: California, with the Basin and Range province east of it; and the central seismic area of New Zealand, involving parts of both main islands. Regions of this type provide almost all of the authentic instances of surface faulting tabulated in Chapter 14.

Blocks Within Arcs

Regions of block faulting do not merely alternate with the active arcs; in some parts of the world they occur within the arcs. An example is the interior of Peru, which is almost the only part of the Andean region where shallow earthquakes associated with block faulting are frequent. (However, in the provinces of Mendoza and San Juan, northern Argentina, structures associated with shallow earthquakes lie directly above the hypocenters of deep shocks typically associated with the Andean arcs.) Another instance is the block faulting along the Japan Sea coast of Honshu.

Master Faults

The block faulting of several areas is primarily strike-slip in character and is dominated by a single master fault. The type example is the San Andreas fault of California; in New Zealand the corresponding feature is the Alpine fault of the South Island. In the Philippines such a strike-slip fault passes across Luzon, Leyte, and Mindanao (Fig. 26-2). A similar fault, whose rift is partly filled by volcanic deposits, passes longitudinally through Sumatra. The relative displacement on the Philippine fault is left-hand, that on the Sumatra fault right-hand. Vening Meinesz has interpreted these movements as lateral sliding incident to an eastward displacement of Borneo and Celebes, mechanically consistent with eastward thrusting along part of the Sunda arc and in the Banda Sea. There is a striking parallel in the West Indies, where the north limb of the active arc is cut by the Bartlett trough, which is a rift associated with left-hand strike slip; the south limb in Venezuela includes similar right-hand rifts. Both features are mechanically coherent with eastward thrusting of the West Indies arc.

The master fault of Anatolia, so plainly indicated by the earthquakes in Turkey beginning in 1939, is notable for its very curved course. Its mechanical relation to the Alpide arcs of that region is not simple, but it is plain that the block motion is chiefly subsequent to thrusting and folding. In Japan the great strike-slip fault associated with the earthquake of 1891 has many of the characteristics of a master fault, but it strikes across Honshu transverse to the arcuate structures; the great transverse structural break called the Fossa Magna must also be considered (Chapter 30).

In most of these areas there is a network of faults tending roughly parallel and perpendicular to the master fault. The relations can be shown as follows.

Region	NW-SE	NE-SW
California	R.H.*	L.H.
New Zealand	L.H.	R.H.*
Japan	L.H.	R.H.
Philippines	L.H.*	. . . .

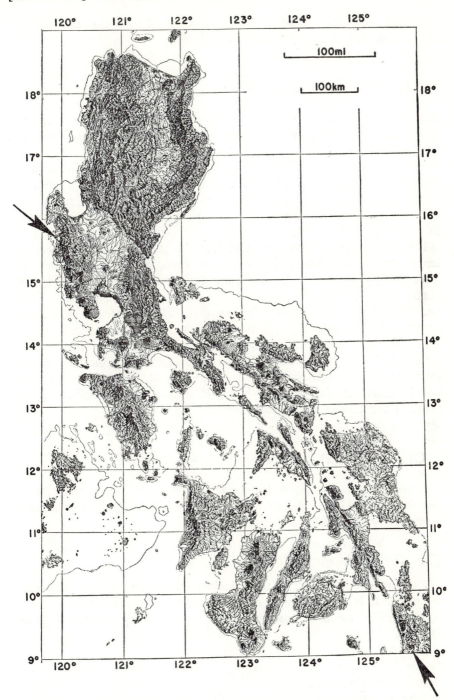

FIGURE 26-2 *Philippine Islands, tectonic relief.* [*King and McKee.*] *Arrows indicate the conspicuous rift of the master fault.*

Here R.H. and L.H. indicate the opposite senses of strike-slip, and an asterisk indicates the character of the master fault.

RIDGES AND RIFTS

The seismicity of certain oceanic ridges may be assignable to block tectonics; but it may represent a third principal type, coordinate with the active arcs and with the block tectonics of Pacific areas.

The typical specimen of this third group is the mid-Atlantic Ridge, which continues northward into the Arctic Ocean. Along the Ridge there occur many minor to moderate earthquakes, all shallow. Active and recently extinct volcanoes are numerous, but, although parts of the Ridge are curved, there are none of the other arc features.

Heezen and Ewing point out that the mid-Atlantic Ridge has a central longitudinal depression—a rift or graben, strongly suggesting vertical displacements, continuous with the central volcanic depression of Iceland. Similar rifts are probably associated with other active oceanic ridges. Ewing and Heezen class them with rifts like those of East Africa; indeed, they represent the African rift system as branching from rifts associated with ridges in the Indian Ocean. This classification cannot easily include all the large seismic rift structures. Strike-slip rifts, like that of the San Andreas fault, belong to Pacific-type block structure; the St. Lawrence rift is an internal fracture of the Canadian stable shield, and the Hawaiian rift is a break in the floor of the Pacific basin.

CHIEF SEISMIC BELTS AND AREAS

This section and those following should be read with reference to the appropriate maps (Figs. 26-3 to 26-15). Additional details may be found in *Seismicity of the Earth.*

The Circum-Pacific Belt

The circum-Pacific belt is the principal seismic and tectonic feature of the globe. It is complex, with several main branches, including arc structures, areas of block tectonics, and at least one example of the ridge and rift type.

Pacific Active Arcs. The arcs of the belt differ in detail, and few of them show all the characteristic arc features. Feature A, the foredeep, is always present; so are active volcanoes (feature D). Negative gravity anomalies have been demonstrated for nearly all the arcs, and probably exist in the

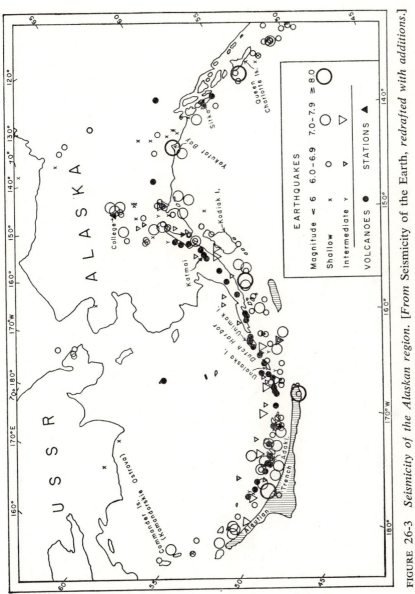

FIGURE 26-3 *Seismicity of the Alaskan region.* [*From Seismicity of the Earth, redrafted with additions.*]

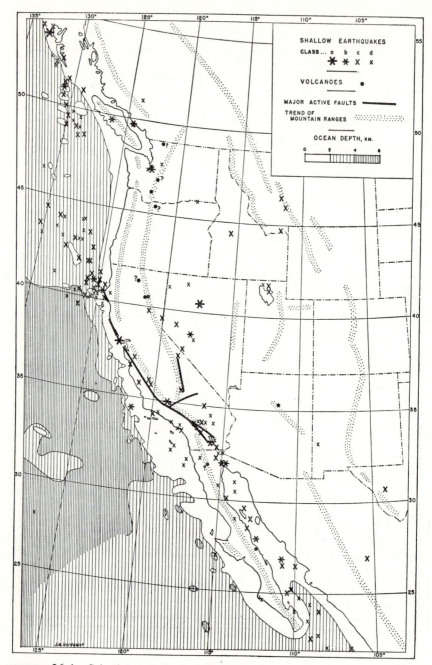

FIGURE 26-4 *Seismicity of the California region.* [Seismicity of the Earth.]

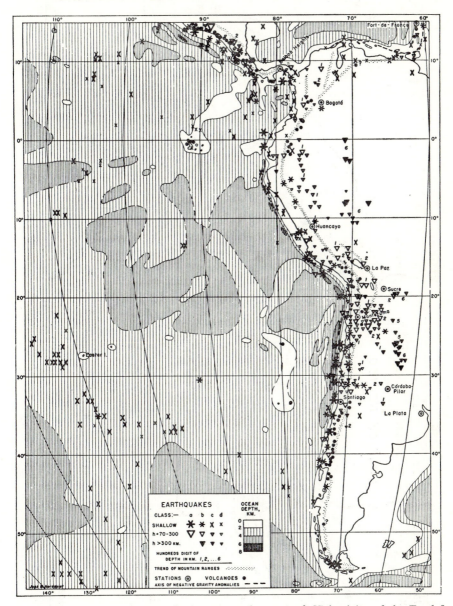

FIGURE 26-5   *Seismicity: South America and westward.* [Seismicity of the Earth.]

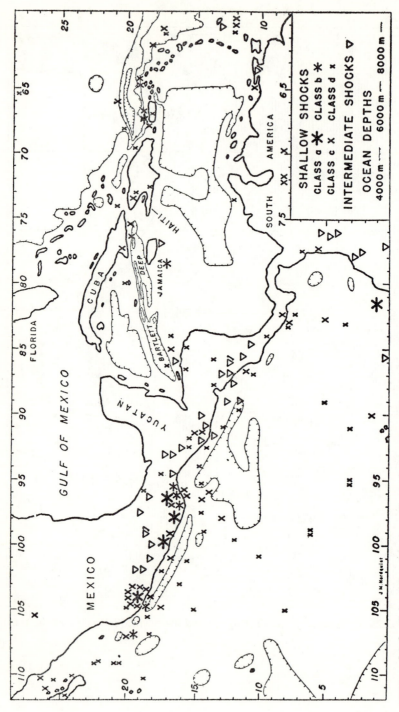

FIGURE 26-6   *Seismicity: Mexico, Central America, West Indies.* [*G.S.A. Special Paper 34.*]

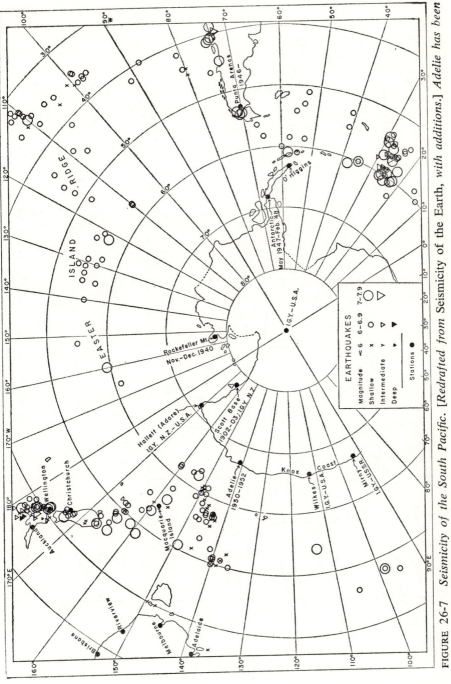

FIGURE 26-7 *Seismicity of the South Pacific. [Redrafted from Seismicity of the Earth, with additions.] Adelie has been reoccupied for the IGY.*

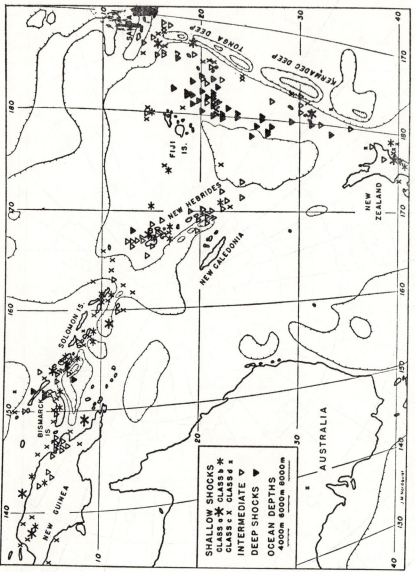

FIGURE 26-8 Seismicity: New Zealand to New Guinea. [G.S.A. Special Paper 34.]

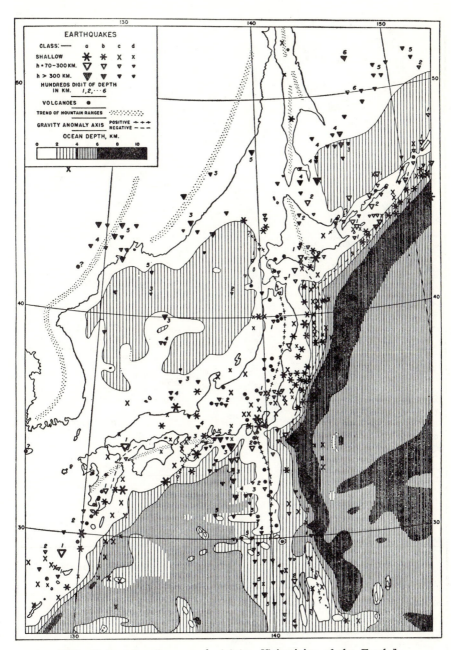

FIGURE 26-9 *Seismicity: Japan and vicinity.* [Seismicity of the Earth.]

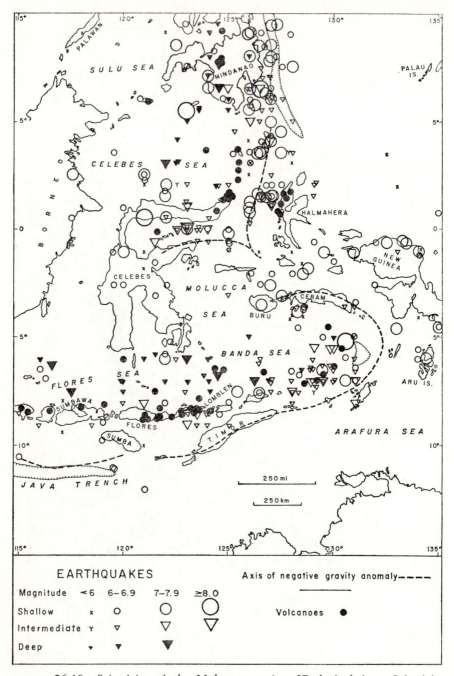

FIGURE 26-10   *Seismicity of the Moluccan region.* [*Redrafted from* Seismicity of the Earth, *with additions and minor changes.*]

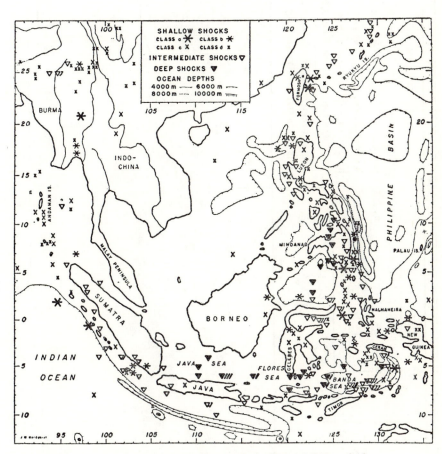

FIGURE 26-11   *Seismicity: East Indies.* [*G.S.A. Special Paper 34.*]

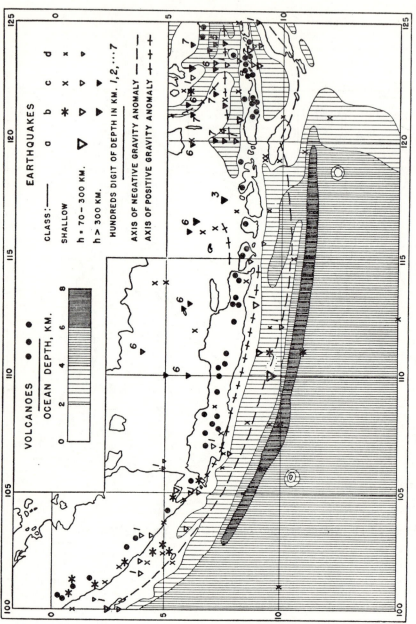

FIGURE 26-12   *Seismicity of the Sunda arc.* [Seismicity of the Earth.]

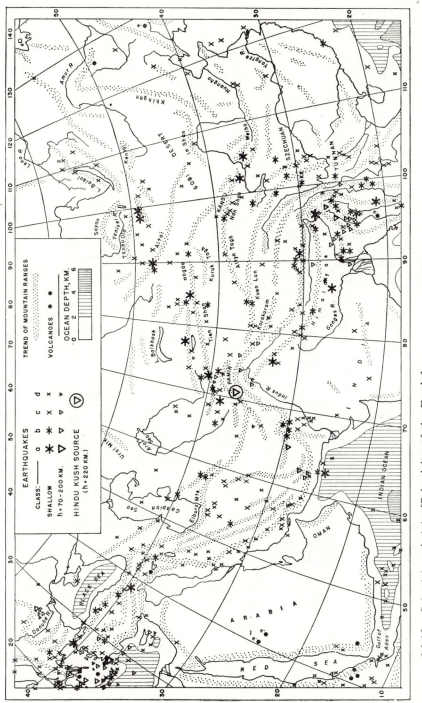

FIGURE 26-13 *Seismicity of Asia.* [Seismicity of the Earth.]

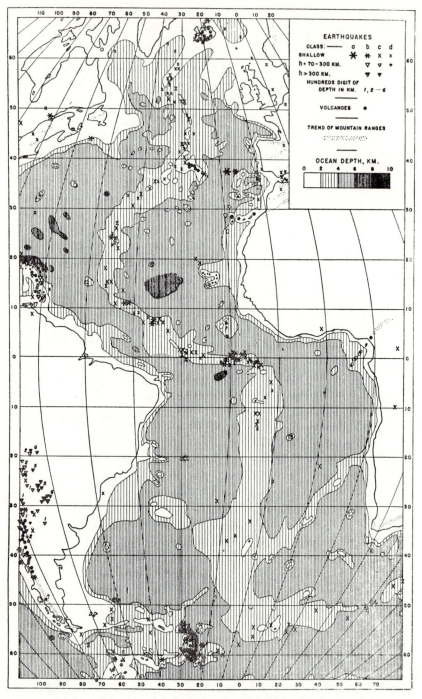

FIGURE 26-14 *Seismicity of the Atlantic.* [Seismicity of the Earth.]

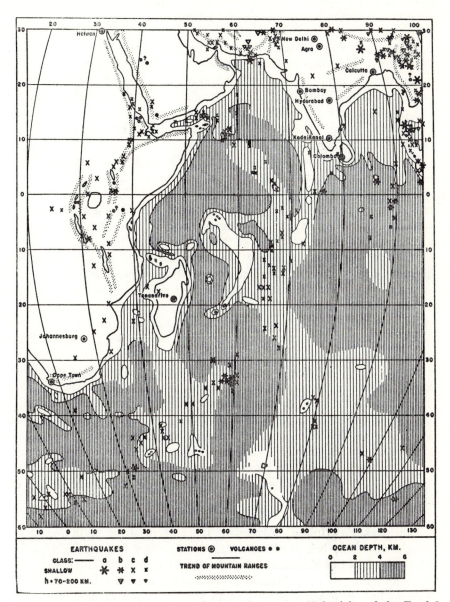

FIGURE 26-15 *Seismicity: Indian Ocean and environs.* [Seismicity of the Earth.]

others. Shallow and intermediate earthquakes are found in all the arcs, but for some of them no deep shocks are known.

The American Arcs.   Arcs fronting westward on the Pacific Basin are the Aleutian-Alaskan, the Mexican, the Central American, and the Andean arcs (roughly, Peruvian and Chilean). Two arcs, one on the West Indies and one in the South Atlantic, front eastward on the Atlantic Ocean. An exceptional feature of the Mexican arc is that feature B, with shallow earthquakes and young anticlines, runs on land instead of offshore or through a chain of small islands.

In the Americas shocks deeper than 300 kilometers are associated only with the Andean arcs; there they are infrequent, but some are large. None are known there between depths of 300 and 550 kilometers. In the Andes one of the best examples of feature E is provided by the Puna de Atacama, where a row of nearly extinct volcanoes is associated with epicenters of numerous earthquakes at depths of 200 to 300 kilometers.

In South America, as usually elsewhere, the active surface defined by hypocenters dips under the continent, suggesting westward overthrusting, or eastward underthrusting. Many geologists find this discordant with evidence of eastward overthrusting in the Andean region. Probably there is superposition of the effects of tectonic episodes of different ages. One may compare these thrusts with the great flat eastward overthrusts (as usually interpreted) in the Rocky Mountains and the Mojave Desert.

Seismicity falls off rapidly south of 37° S, but resumes near the Strait of Magellan, whence epicenters located from the records at distant stations mark the course of the South Atlantic tectonic arc (the Scotia Arc), by way of South Georgia and the South Sandwich Islands to the South Shetland Islands. The last group lies off the coast of Palmer Peninsula. Although there is geological evidence that structures similar to those of the Andes extend along the peninsula into and perhaps across Antarctica, seismological evidence is disappointingly lacking. No earthquake is yet known to have originated on the Antarctic continent.

Asiatic Arcs, East Indies, Polynesia.   If we begin at the north and follow the circum-Pacific belt southwestward, we encounter first the highly active and typical arcs of Kamchatka and the Kurile Islands. In Honshu the belt separates into two main branches. The eastern branch follows Marshall's andesite line, in a series of arcs (with some segments of block tectonics) from the vicinity of Tokyo by way of the Marianas to Guam, and thence southwestward through the Carolines. In Halmahera it swings sharply around to run eastward through New Guinea, the Solomon Islands, and the New Hebrides, then passes between the Fiji and Samoan groups, and finally runs southward past the Tonga and Kermadec Islands to New Zealand.

The western branch passes from Honshu through Kyushu, the Riukiu

Islands, Formosa and the Philippines to northern Celebes, where there is an east-fronting arc closely opposing the west-fronting arc of Halmahera. From Celebes the belt swings round the Banda Sea, and along the line of the lesser and greater Sunda Islands to Sumatra, the Nicobar Islands, and the Andaman Islands.

The Kamchatka arc has an especially well-marked feature E, similar to that of the Puna de Atacama in South America. The arcs of Luzon and the New Hebrides front anomalously west and southwest, away from the Pacific basin. In the Solomons shocks at depths from shallow to 300 kilometers have their epicenters nearly on the same line; consequently the active surface there dips almost vertically. The small arc of New Britain stands athwart the general trend.

The complex trends of the East Indies are associated with the first established and best known of the Meinesz gravity zones. (For further discussion of the East Indies see Chapter 31.)

Only shallow and intermediate shocks are known from Kyushu to Luzon; deep shocks occur under the central Philippines, in association with the Mindanao arc. In the region of the Fiji, Tonga, and Kermadec Islands deep earthquakes are more frequent than anywhere else. The deepest known earthquakes occur in the East Indies under the Flores Sea. The complicated pattern of deep-focus earthquake belts in the vicinity of Japan has been noted (Chapter 25).

Block Tectonics. The longest sector of the circum-Pacific belt characterized by block tectonics to the exclusion of arc features extends from southeastern Alaska into northern Mexico. A similar sector includes most of New Zealand and extends southwest to Macquarie Island. Block faulting occurs in the interior of arc structures, as in Peru and Japan; the central Philippines are in an area of block structure with generally northeast-southwest trends. Northern New Guinea is often described as a northward-fronting tectonic arc; but the frequent large shallow shocks and the absence of deep earthquakes suggest block structure. Strike-slip faults cutting the edges of active arcs were remarked on earlier in this chapter.

Easter Island Ridge. In the latitude of Colima, Mexico (about 19° N) the eastern circum-Pacific belt divides into three branches. One continues southeastward into the Central American arc; the second strikes across Mexico through Yucatan and around the loop of the West Indies, returning through Venezuela to rejoin the first branch in the Colombian Andes. The third branch, associated only with shallow shocks, trends almost due south past the Galápagos Islands, and follows the trend of the Easter Island Ridge. This feature resembles the mid-Atlantic Ridge; it probably has a central rift, associated with volcanism as in the Galápagos, or with islands of volcanic rock like Easter Island. The ridge trends gradually more westward, touches the Antarctic Circle, and extends to the vicinity of Macquarie Island; the accompanying epicenters complete the enclosure of the Pacific Basin by seismic belts.

The Alpide Belt

The Alpide Belt, in the sense here used, begins in Burma. Topographically, geologically, and seismologically, there is an approximate but not an exact connection between the tectonics of Sumatra and of Burma. From the present point of view, this break separates highly active structures of Pacific type from a series of arcs in which activity has to a certain extent relaxed. In its origins, the Alpide belt may perhaps be best considered continuous with the western branch of the Pacific belt; its present condition differs.

The Alpide belt can be traced westward as a series of arcs with generally southward front, in Burma, the Himalaya, Baluchistan, Iran, Anatolia, and the eastern Mediterranean (Crete). Recent authorities (Stille and others) continue the tectonic zone with sharp bends, reminiscent of the zone of the East Indies, around the Adriatic into the Alps and back by way of the Apennines; here we have the Italian arc of Chapter 4. The continuation is then through Sicily and the north coast of Africa, and into the Atlantic as far as the Azores, where the Atlantic belt is reached.

The Alpide belt has a northern front, characterized by minor to moderate seismicity, which involves the Pyrenees, the Carpathians, the Caucasus, the Crimea, and the Kopet-Dagh.

Most of the Alpide shocks are shallow. Intermediate shocks are fairly frequent in Burma; and in the Hindu Kush, near 36.5° N, 70.5° E, is an epicenter from which there has been a remarkable and persistent repetition of earthquakes (over 70 recorded at distant stations since 1905) at depths near 220 kilometers, some of them of magnitude 7 or larger. There is similar but less frequent repetition from a source about 150 kilometers deep under the Carpathians north of Bucharest. Several shocks at depths approaching 300 kilometers, and one near 470 kilometers, have been identified off the west Italian coast near the Lipari Islands. On March 29, 1954, a shock occurred at a depth of 640 kilometers under southern Spain (see discussion above).

Other Regions

The shocks of the remaining regions to be mentioned are all shallow; that is, their depths do not exceed 60 kilometers.

The Pamir-Baikal Zone.   Most of the largest earthquakes outside the Pacific and Alpide belts occur in the zone extending from the Pamir plateau to Lake

Baikal, along the southern margin of the Asiatic stable mass. Between this and the Alpide belt is a broad triangular area, including most of China, Tibet, and the Gobi Desert, with a series of ranges most of which are associated with known earthquakes. This is probably the broadest of known seismic regions; some of its earthquakes, such as those of Kansu in northwestern China, are very large.

Ridges and Rifts. The narrow belts which complete the divisions between the separate pieces of the global tectonic jigsaw puzzle mostly belong to the ridge and rift group. The most important of these is the mid-Atlantic Ridge; its Arctic continuation, after passing Iceland and Spitzbergen, turns to meet the Siberian coast near the mouth of the Lena River. The ridge and the seismic zone can be traced from the South Atlantic round Africa into the Indian Ocean; there the main feature trends northward to the Gulf of Aden, and a branch connects with the Easter Island Ridge at Macquarie Island.

Of the active continental rift zones, those in East Africa and along the St. Lawrence have been noted above. At the eastern margin of the West Australian stable mass is the "shatter zone" with the largest known Australian shocks. The Samoan Islands represent an intra-Pacific rift like that of the Hawaiian Islands, but associated with much lower seismicity.

Margins and Troughs. Earthquakes also occur about the continental margins, and in the troughs separating major and minor continental masses. Examples of the latter are the Gulf of Aden and the Red Sea, between the African and Arabian masses; the Mozambique Channel between Africa and Madagascar; and Baffin Bay between the Canadian mass and Greenland.

A few shocks occur in other parts of the generally non-seismic Pacific Basin. On December 10, 1949, there was one at 4° N, 129° W (*Seismicity of the Earth,* 2nd ed., addenda); and on November 22, 1955, one of magnitude 6¾ occurred near 24° S, 122½° W, in the Marquesas Islands.

MINOR SEISMIC AREAS

Some of the seismicity noted in the preceding section might be listed here. There remains a group of small regions characterized by moderately frequent minor earthquakes (rarely exceeding magnitude 6), associated with pre-Cretaceous mountain structures in continental areas between the chief active belts and the stable shields. Among these are the Appalachian area of eastern North America, and geologically similar localities in eastern Australia, central Europe, and South Africa.

NON-SEISMIC REGIONS

It is probable that no large area is permanently unaffected by earthquakes, but there are many to which no epicenters can yet be assigned, and others where only the most insignificant local seismicity is known. Sensitive seismographs, operating over an extended period of years, would probably record at least a few small local shocks if installed anywhere on the globe.

The least seismic regions, as noted in Chapter 25, are the Pacific Basin (excluding the Hawaiian Islands) and the continental stable shields (excluding internal rift zones). In the Atlantic region, seismicity is very low in the basins east and west of the mid-Atlantic Ridge. The same is apparently true of similarly placed areas in the Indian and other oceans, but the relevant observations are neither so complete nor so accurate as for the Atlantic region, where detection and location are facilitated by the numerous seismological stations in North America and Europe (and of late years by several excellent stations in Africa).

References

READING on the material of this chapter will be found in the general references following Chapter 25; details for specific areas appear in the text and references of later chapters. For major strike-slip faulting consult Chapter 13 and references. Reading on Andean problems may be found in:

Oppenheim, V., "Structural evolution of the South American Andes," *Am. Journ. Science,* vol. 245 (1947), pp. 158–174.

Jenks, W. F., ed., "Handbook of South American geology," *Geol. Soc. Amer. Mem.,* No. 65, 1956.

Gerth, H., *Der geologische Bau der südamerikanischen Kordillere,* Berlin, 1955.

The following are recent discussions bearing on arc and block structure.

Benioff, H., "Seismic evidence for the fault origin of oceanic deeps," *Bull. Geol. Soc. Amer.,* vol. 60 (1949), pp. 1837–1856.

————, Orogenesis and deep crustal structure; additional evidence from seismology," *ibid.,* vol. 65 (1954), pp. 385–400.

————, "Seismic evidence for crustal structure and tectonic activity," *Geol. Soc. Amer., Spec. Paper 62* (1955), pp. 61–74.

Gilluly, J., "Geologic contrasts between continents and ocean basins," *ibid.,* pp. 7–18.

Wilson, J. Tuzo, "The development and structure of the crust," Kuiper, G. P., ed., in *The Earth as a Planet*, Chicago, 1954, pp. 138–207.

Gutenberg, B., and Richter, C. F., "Structure of the crust. Continents and oceans," Chapter 12, pp. 314–339, in *Internal Constitution of the Earth*, 2nd ed., New York, 1951.

Hess, H. H., "Major structural features of the western North Pacific, an interpretation of H. O. 5485, bathymetric chart, Korea to New Guinea," *Bull. Geol. Soc. Amer.*, vol. 59 (1948), pp. 417–446.

Hess, H. H., and Maxwell, J. C., "Major structural features of the southwest Pacific; a preliminary interpretation of H. O. 5484, bathymetric chart, New Guinea to New Zealand," *Proc. 7th Pacific Science Congress, 1949*, vol. 2, pp. 14–17.

Richter, C. F., "Seismicity and structure of the Pacific region of North America," *ibid.*, pp. 671–681.

Ewing, M., and Heezen, B., "Mid-Atlantic Ridge seismic belt" (abstract), *Trans. Am. Geophys. Union*, vol. 37 (1956), p. 343.

Heezen, B. C., "Deep-sea physiographic provinces and crustal structure" (abstract), *ibid.*, vol. 38 (1957), p. 394.

CHAPTER 27

California and New Zealand

CALIFORNIA AND NEW ZEALAND are regions where block fault-ing occurs within the circum-Pacific belt. There are many interesting analo-gies between the two, and existing differences are highly instructive. During the Seventh Pacific Science Congress in 1949 the author had an opportunity to become acquainted at first hand with some of the outstanding features in New Zealand; observations and discussion at that time provided much of the foundation for the present chapter.

STRATIGRAPHY

The Basement

In both regions (Figs. 27-1, 27-2) the basement rocks† consist, over large areas, of more or less metamorphosed marine sediments, Mesozoic or older, of the type commonly believed to accumulate in geosynclines, largely classifiable as sandstones and particularly as greywackes, with typical minor constituents such as cherts, pillow lavas, and spilites. In California such rocks compose the Franciscan formation, whose date has been debated for many years owing to the rarity of fossil material. Evidence once generally accepted as restricting the Franciscan to the upper Jurassic now requires reinterpreta-tion as the result of discovery of positively identifiable Cretaceous ammonites in a typical Franciscan section. Informed opinion now dates the Franciscan as extending across the Jurassic-Cretaceous boundary, and contemporaneous at various horizons with formations of non-geosynclinal facies which are as-signed other names.

In New Zealand, fossil dating for the geosynclinal rocks is relatively good in the South Island. In the North Island, pre-Cretaceous fossils are scarce and often poorly preserved, so that dating has depended on stratigraphic and lithologic analogy with the South Island rocks. However, fossiliferous lenses of limestone in the greywackes of northern Auckland province have in recent years allowed close correlation of parts of the greywacke series with Permian

---

† New Zealand authors, following W. M. Davis, write of "the undermass."

438

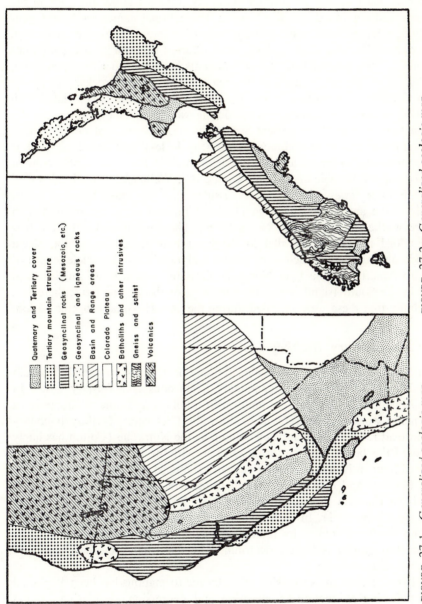

Quaternary and Tertiary cover
Tertiary mountain structure
Geosynclinal rocks (Mesozoic, etc.)
Geosynclinal and igneous rocks
Basin and Range areas
Colorado Plateau
Batholiths and other intrusives
Gneiss and schist
Volcanics

FIGURE 27-2 Generalized geologic map, New Zealand.

FIGURE 27-1 Generalized geologic map, California region.

stages known in the South Island. The geosynclinal deposition, in both islands at least locally of Permian antiquity, reached great thickness in the Triassic and continued into the Jurassic. Formational names are in use, generally transferred from contemporary rocks of different facies, but correlation is so difficult that the whole series is often referred to, from its dominant component, as "the greywackes." Later sediments have been derived from these, so that New Zealand in much of its area is a land of grey rocks.

In California, deposition in the Franciscan geosyncline is correlated with the late Jurassic Nevadian orogeny, during which fold mountains were developed on the site of the present Sierra Nevada. In New Zealand the similar Hokonui (or rather post-Hokonui) orogeny is less well dated, since it is represented chiefly by an almost complete break in the stratigraphic column between upper Jurassic and Middle Cretaceous. Wellman has lately identified a locally thick series with few fossils, assigned the new name Urewera series, as of Aptian and Neocomian age (Lower Cretaceous); these are underlain by greywackes, at least in the North Island. The igneous rocks of New Zealand afford no evident parallel to the great batholiths, that of the Sierra Nevada among them, which were intruded along almost the whole length of western America in the Late Mesozoic. Granitic batholiths of smaller scale exist in the South Island, west and northwest of the Alpine fault; there is a possibility that these were intruded during the post-Hokonui orogenic period.

The Sedimentary Blanket

In New Zealand as in California, the basement rocks are largely overlain by a blanket of principally sedimentary Cretaceous and Tertiary series. In New Zealand this blanket is so uniform in lithology and other characteristics that many workers have refused to draw the Cretaceous-Tertiary boundary and instead refer to the combined column as the Notocene (Cotton has lately proposed the better term Notocenozoic), separating it naturally into formations without insisting on correlations with standard sections elsewhere.

In Pliocene and Pleistocene time there occurred what in North America is called the Cascade Revolution, and in New Zealand the Kaikoura orogeny (but see Chapter 25). In both regions, previously folded areas were broken up into block mountains by faulting with dominantly vertical dip, and the present topography is largely determined by these block displacements and accompanying warping.

In New Zealand the succession of Notocene formations has been attributed to a pulsating alternation of elevation and depression, decreasing from a maximum at the Hokonui period to nearly zero in the middle of the interval, then increasing progressively to Kaikoura time. No such clear or simple statement appears possible for California, at any rate not for the Coast Range province, which has been most thoroughly investigated for the Cretaceous and Tertiary. There is evidence of many orogenic periods in this interval, dated differently and considered of differing importance by the various investigators; almost no one suggests that these disturbances were of equal significance even in all parts of the Coast Range province—certainly not outside it. Actual complexities in New Zealand are probably quite as great.

TECTONICS; THE MASTER FAULTS

Trends and Comparisons

In California the major tectonic trend is roughly northwest-southeast, in New Zealand northeast-southwest. These are the trends of the two master faults: the San Andreas fault, and the Alpine fault of the South Island. Both are right-hand strike-slip faults, and both have developed major rift features with abundant evidence of geologically recent activity. Two great earthquakes have originated on the San Andreas fault in the short historical period (1857, 1906), but no such event is known for the Alpine fault, although accumulated displacement has offset postglacial streams by as much as half a mile; there is a possibility that the Alpine fault was at least partly involved in the earthquake of 1848.

In both regions (Figs. 27-3 and 27-4) faults with trends nearly parallel to the master fault show evidence of right-hand displacement, while other faults with trends crudely at right angles to the principal structures are of left-hand type. In California the chief left-hand fault is the Garlock fault, which resembles the Alpine fault in strong topographic evidence of activity without clear indication of displacement in historical time. The White Wolf fault, associated with the 1952 earthquake (Chapter 28), is roughly parallel to the Garlock fault; the evidence indicated principally dip-slip displacement in 1952 but also established some left-hand strike-slip.

FIGURE 27-3 *Fault pattern of the California region and chief Quaternary volcanoes.*

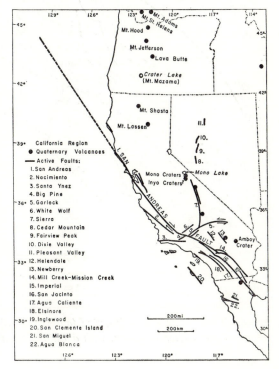

California Region
● Quaternary Volcanoes
— Active Faults:
1. San Andreas
2. Nacimiento
3. Santa Ynez
4. Big Pine
5. Garlock
6. White Wolf
7. Sierra
8. Cedar Mountain
9. Fairview Peak
10. Dixie Valley
11. Pleasant Valley
12. Helendale
13. Newberry
14. Mill Creek-Mission Creek
15. Imperial
16. San Jacinto
17. Agua Caliente
18. Elsinore
19. Inglewood
20. San Clemente Island
21. San Miguel
22. Agua Blanca

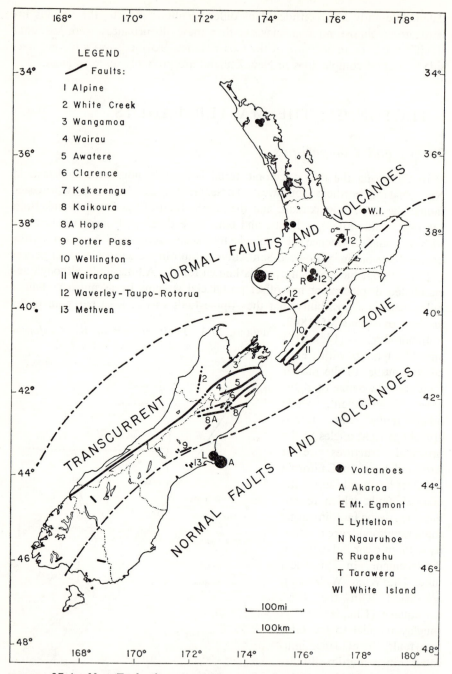

LEGEND

Faults:

1 Alpine
2 White Creek
3 Wangamoa
4 Wairau
5 Awatere
6 Clarence
7 Kekerengu
8 Kaikoura
8A Hope
9 Porter Pass
10 Wellington
11 Wairarapa
12 Waverley-Taupo-Rotorua
13 Methven

● Volcanoes
A Akaroa
E Mt. Egmont
L Lyttelton
N Ngauruhoe
R Ruapehu
T Tarawera
WI White Island

100mi

100km

FIGURE 27-4   *New Zealand: active faults, and volcanoes active since the Pleisto-*
cene. [After Wellman, with additions from other sources.]

The San Andreas Fault

The tectonic significance of the San Andreas fault is comparatively clear between 35° and 40° north latitude. Just beyond these limits the fault system is deflected and complicated by the intervention of two east-west structural features which recent oceanographic work has associated with the Murray and Mendocino escarpments; these are included in two great fracture zones given the same names.

Effect of the Transverse Ranges. The Murray fracture zone is aligned with an important structure of southern California, the east-west belt of Transverse Ranges. A western element in this belt is the line of islands on the parallel of 34° off Santa Barbara; these islands are traversed and their structures offset by large left-hand strike-slip faults. Faults with east-west trends occur in all parts of the Transverse Ranges; some of them appear as branches of faults with other trends, and some are characterized by north-south thrusting.

The San Andreas fault is deflected on approaching the Transverse Range belt from the north. Its trend curves almost to due east in Tejon Pass, where it meets or cuts the Garlock fault; thence it gradually turns more to the south and cuts through the transverse structural belt at Cajon Pass. Lawson's maps accompanying the report on the 1906 earthquake show the San Andreas Rift as continuing from Cajon Pass eastward on the north side of San Gorgonio Pass into Coachella Valley. This was based on field work by Fairbanks, which was extended by later investigators, and the interpretation remained unmodified for many years.

Very recently Dr. Clarence R. Allen, studying the critical area on the ground with the aid of air photographs, has demonstrated that the tectonic conditions there are much more complicated (Figs. 27-5 and 27-5A). The San Andreas Rift, with its characteristic features clear but diminishing, is traceable from Cajon Pass southeasterly across Pine Bench to Burro Flat, east and a little north of the town of Banning. Here it meets the east-west fault seen by Fairbanks. The California state geological map of 1938 correctly shows this fault as extending east and west through the vicinity of Burro Flat. Dr. Allen terms it the Banning fault, to distinguish it from the San Andreas fault as ordinarily described; he finds that it is a thrust fault dipping about 60° north. There is no evidence of fresh strike-slip, and none of the characteristic San Andreas rift features. Had Lawson been aware of these circumstances, it is doubtful whether he would have applied the name "San Andreas fault" beyond Burro Flat. The Banning fault converges in the hills north of Indio with the Mission Creek fault; this was long believed to be continuous with the Mill Creek fault, which branches eastward from the San Andreas fault in Cajon Pass. The Mill Creek and Mission Creek faults both are active earthquake sources; the large Desert Hot Springs earthquake of 1948 was

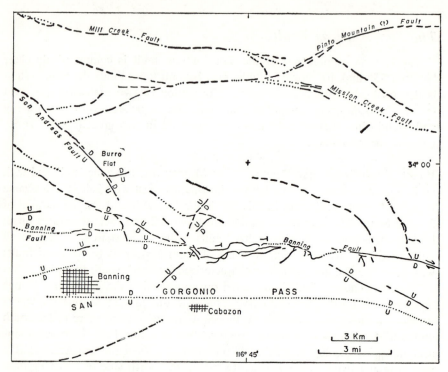

FIGURE 27-5 *San Andreas fault and related faulting near Cajon Pass and San Gorgonio Pass.* [*After C. R. Allen.*]

associated with the latter. It was accordingly suggested that these faults might represent the principal active continuation of the San Andreas system, leaving the fault ending at Burro Flat as a minor branch. However, Dr. Allen finds that the line of the Mill Creek and Mission Creek faults is not continuous but is offset by still another fault. Thus there appears to be no continuous surface fracture extending from Cajon Pass south of the transverse structural belt.

In discussion with the author, Dr. Allen has preferred to consider the present lack of continuity as resulting from complex processes acting on what may have been originally a simple fracture. Displacements coherent with the east-west structures of the transverse belt may have alternated with displacements of San Andreas type to produce the complex surface pattern. The Banning fault, for example, shows some evidence of ancient strike-slip; it is possible that it represents an ancient section of the San Andreas fault pushed out of its former position and now subject to thrusting. Hill and Dibblee have proposed an analogous history for the Garlock fault, which now breaks off southwestward at the San Andreas fault. They interpret the Big Pine fault, which trends southwest from a point on the San Andreas zone west of Tejon

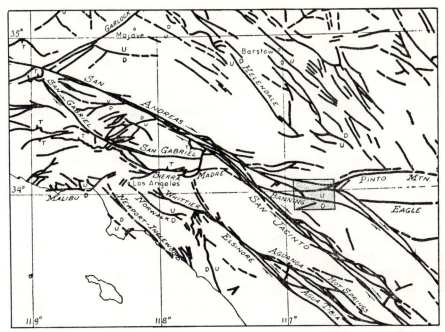

FIGURE 27-5A   *Fault pattern, Southern California, showing location of Figure 27-5.*

Pass, as the former continuation of the Garlock fault, now displaced by the San Andreas fault.

Southward Continuation.   Some writers suggest tracing the master fracture along the San Jacinto fault, which diverges from the San Andreas fault before it reaches Cajon Pass and extends as a more or less continuous broad zone with various echelon offsets and subsidiary faults through the mountains southeastward to Imperial Valley. However, there is no direct continuity between the San Jacinto and San Andreas zones; where they approach each other most closely, the San Jacinto fault shows chiefly dip-slip features, while the characteristic strike-slip features of the San Andreas Rift are strongly developed.

Thus there is no single surface fracture, breaking completely through the transverse belt, which has a claim to be considered the continuation of the San Andreas fault. There may be a continuity of fracturing at depth; but there is nothing to show which surface fracture farther south represents it. Rather than any one fault, there is a wide system of roughly parallel faults of which the continuation of the Mission Creek fault may be the easternmost. Westward from this in order are the Imperial fault, along which large strike-slip broke the surface in 1940 (Chapter 28); the San Jacinto fault; the Agua Caliente fault, which appears to break over at its northern end toward the next, the Elsinore fault; the Inglewood fault, associated with the Long Beach earthquake of 1933; and an offshore fault following the east face of San

Clemente Island, traceable northwestward and southeastward by submarine rift topography as well as earthquake epicenters.

The suggestion often seen in popular publications that "the San Andreas fault extends into Mexico and down the Gulf of California" is an oversimplification. The major seismic belt of the region does in fact extend down the Gulf of California (see Fig. 26-4), apparently along its western side rather than near the center, although epicenter location in that region is hardly good enough for such close distinction. The configuration of islands in the Gulf, and its bottom contours, strongly suggest faulting; but no one main fault is indicated. For such broad views, the term "San Andreas fault system" is recommended, but moderation in its use is advisable, or the incautious speaker will find he has equated the San Andreas fault with the whole circum-Pacific seismic belt.

Northward Continuation.   Related problems arise in tracing the San Andreas fault northwest. The San Andreas Rift strikes more westerly than the general trend of the Coast Range structures; so that the Rift, the main fault, and the visible trace of the 1906 earthquake disappear under the sea at Point Arena, roughly 100 miles northwest of San Francisco. The fault trace which appeared near the coast farther north in 1906 is not now regarded as a continuation of the main San Andreas fault (see Chapter 28).

Earthquake epicenters immediately northwest of Point Arena are few; but seismicity near and north of the Mendocino escarpment is higher than anywhere else in the California region (Fig. 26-4). However, there is no single narrow belt of epicenters suggesting a simple continuation of the San Andreas fault; rather there is a broad zone of seismicity. It has often been suggested that the Mendocino fracture zone deflects the San Andreas fault westward, just as the transverse structures in southern California deflect it eastward, and that north of the Mendocino escarpment the fault frays out into a broad band of faulting as it does in southeastern California. The only evidence to the contrary is that the bottom contours north of the escarpment, as now mapped, show alignments trending northeast rather than northwest. At about 45° N, 131° W, the seismic belt reaches the edge of the continent and ends. The northward continuation of the circum-Pacific belt begins farther east, off the coast of Vancouver Island, separated from the end of the San Andreas system by a large gap.

The Alpine Fault

Cook Strait.   The effect of the Transverse Ranges may be compared with the behavior of the New Zealand structures in the vicinity of Cook Strait, which separates the two main islands. Superficially at least, the North Island appears to be shifted eastward, or southeastward, with reference to the South Island. However, closer examination of the geology has suggested to some investigators displacement in the opposite sense, implying a left-hand fault in Cook Strait. Eiby has lately presented seismological evidence for this. As the Alpine fault approaches Cook Strait from the south, it curves eastward and frays out into several faults ("virgates") in the manner described above for the San Andreas fault. Wellman has identified some of these branches with active faults of the eastern North Island.

Southward Continuation. In the other direction the Alpine fault enters the fiord region on the west coast of the South Island. A moderately active seismic belt extends thence southwestward at least to the vicinity of Macquarie Island (54° 30′ S), but some evidence indicates that the Alpine fault follows the steep coast, which curves gradually eastward round to Foveaux Strait between the South Island and Stewart Island.

The exposed rocks of Fiordland are almost wholly schists. East of these, curving southeastward, is a large ancient syncline along the center of which are exposed ultrabasic rocks similar to those on the other side of the Alpine fault some 300 miles to the north.

Problems of the North Island

The principal faults on the North Island trend northeast, with evidence of strike-slip as well as dip-slip. The master fault, if it exists, may be associated with the central trough of the island, which includes several active volcanoes and is largely filled with volcanic deposits. In this it is comparable to the longitudinal rift in Sumatra, and perhaps to the Fossa Magna of Japan. Northward it continues as a submarine trench. It is not a highly seismic belt; shallow shocks are mainly volcanic in character. Intermediate shocks occur at depths ranging from 100 to 200 kilometers under the volcanoes, as in other Pacific regions. The principal belt of shallow seismicity is adjacent to the east coast of the North Island. The northwesterly trend of the Auckland Peninsula (which is almost non-seismic) has caused many authors to treat it as a structural continuation of New Caledonia; peridotites and their usual associates occur in both, but there is a problem of relative age (Chapter 25).†

New Zealand Arcs?

Macpherson treated New Zealand as a double arc, first convex to the east as it descends from New Caledonia through the Auckland Peninsula, then convex westward as it swings east through the southern geosyncline north of Foveaux Strait. This can only represent the older structures; it denies the northeast trend of the active structures of the east coast of the North Island, and it overlooks an apparent great break along the Alpine fault.

† On January 13, 1956, a shock of magnitude over 6 occurred near 28° S, 167½° E, not far from Norfolk Island, and almost on the indicated line.

Large Strike-Slip?

Large-scale interpretations of New Zealand structure are affected if one accepts a bold suggestion by Wellman, which would call for accumulated right-hand shift along the Alpine fault of nearly 300 miles since the Mesozoic. This is indicated, among other evidence, by the remarkable belt of ultrabasic and metamorphic rocks which breaks off in the northern part of the South Island at the northwest side of the Alpine fault and appears to resume on the southeast side about 300 miles southward. In the north, near Nelson, this "mineral belt" includes Dun Mountain, the type locality of dunite. In southern California there is evidence for accumulated shift of about 25 miles along the San Andreas fault since the mid-Tertiary, but Mason Hill and T. W. Dibblee suggest extending this back in time so as to imply displacements of hundreds of miles since the Mesozoic.

STRUCTURAL PROVINCES OF CALIFORNIA

For our purposes we need consider tectonic provinces only in relation to the occurrence and interpretation of earthquakes. Some of the chief divisions have been indicated in the discussion of the master faults. For the California region (Figs. 27-6 and 27-7) we distinguish:

(1) The highly seismic area offshore north of the Mendocino fracture zone, representing the northwestern extension of the San Andreas system. This is in the southern part of the Ridge and Trough province of Menard and Dietz.

(2) The north coastal mountains, a geologically diverse group including the Klamath Mountains, the coast ranges of Oregon, and the Olympic Mountains of Washington. This is an area of low seismicity.

(3) A trough including the Willamette Valley (Oregon) and Puget Sound. Seismicity increases northward into British Columbia (but the principal seismic zone there lies farther offshore).

(4) The Cascade Mountains, with minor seismicity, some of it volcanic (especially in the vicinity of Mt. Lassen, near the southern limit).

(5) The principal Coast Range province between the Mendocino and Murray fracture zones, with moderate to major seismicity. The San Andreas fault does not bound this province or its seismic zone, but cuts across it at a low angle. It has one principal active branch, the Haywards fault passing northward east of San Francisco Bay. West of the San Andreas fault is the Nacimiento fault; between them lies a long, narrow slice differing in stratigraphy (for example, almost complete absence of Franciscan rocks) from the areas east and west of it. The Nacimiento fault is an earthquake source, possibly a major one.

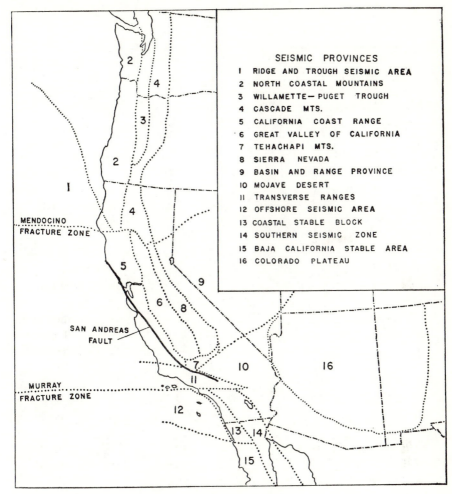

FIGURE 27-6 *Seismic provinces, California region.*

An observation significant in connection with the mechanics of large earth-quakes is the comparative lack of minor shocks along the sectors of the San Andreas fault which were displaced in 1857 and 1906. Earthquakes of a wide range of magnitudes have occurred on the adjacent faults; the major disturbance of 1952 in Kern County originated within 10 miles, and some of the aftershocks were even closer, yet activity did not extend to the San Andreas zone. Numerous shocks of magnitude up to 5 have originated along the sector southeast from that affected in 1906; in 1922 a shock of magnitude 6½ occurred at a position probably not far from the northern limit of fault-ing in 1857. East of Cajon Pass, probably beyond the southeastern limit of 1857, minor shocks have been frequent with epicenters between the surface

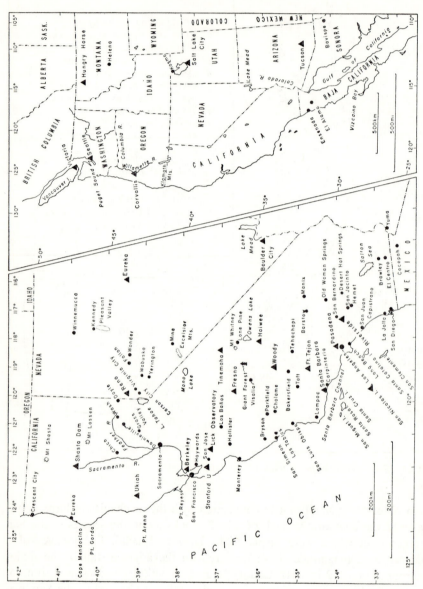

FIGURE 27-7   Location map, California region.

traces of the Banning and Mill Creek faults; these might be ascribed to either.

There are two almost opposite possible interpretations. The more obvious is that strain in the neighborhood of the master fault was completely relieved by fracture in 1857 and 1906 and has not yet begun to recover in those sectors. The writer prefers to consider that the fractures in 1857 and 1906 passed through masses of exceptionally competent rock, which yield only when strains have accumulated to a high level, while other parts of the fault zone can be displaced under smaller stresses.

(6) The Great Valley of California. Few, if any, shocks originate under this area. Those of the western margin belong to the Coast Range structures; at the east small shocks originate on minor faults in the foothills of the Sierra Nevada, as in many other areas of ancient mountain-building. Shocks under the southern end of the Great Valley belong with the following.

(7) The Tehachapi Mountains. Buwalda considered this region a part of the Transverse Ranges, with some elements of Coast Range structure. Its obvious northern boundary is the White Wolf fault, the seat of the major earthquake of 1952, but the aftershock activity following that event extended over an adjacent block to the north. Southward the Tehachapi Mountains extend to and across the Garlock fault; this is a major fault with evidence of geologically recent left-hand strike-slip, but showing in this vicinity no surface displacement of fresh appearance. It extends from the San Andreas fault in Tejon Pass northeast across the Mojave Desert.

(8) The Sierra Nevada. Marginal activity occurs on the west side of this block whose eastern edge has been tilted up in Quaternary time to form a scarp rising to 14,000 feet above sea level. Few earthquakes originate in its interior. A major north-south fault, first described by Lawson, follows the very straight course of the upper Kern River, but little, if any, present seismicity can be ascribed to it. Earthquakes near the east base of the Sierra Nevada belong largely to the next.

(9) The Basin and Range province. The Sierra Nevada is the westernmost of a series of generally north-south mountain blocks extending across a great area from California across Nevada into Utah. Much of the exposed rock is Paleozoic (rare elsewhere in the region). Many of the blocks, like the Sierra Nevada on its eastern side, are bounded by relatively fresh-appearing scarps; minor earthquakes are known associated with most of those in California and adjacent Nevada. Major earthquakes in Nevada in 1932, 1954, and 1915, which formed fault traces, had their epicenters along an alignment trending east of north; prolonged west of south, this would include the meizoseismal area of the great earthquake of 1872. The Basin and Range structures in California are sharply bounded on the south by the Garlock fault.

(10) The Mojave Desert area. The angle between the San Andreas and Garlock faults as they converge westward to Tejon Pass is almost non-seismic,

as indicated by instrumental recordings since 1929. Farther east, near the longitude of Barstow (117° W), seismicity is noteworthy and includes one known shock of magnitude over 6 (the Manix earthquake of 1947). In this area are many more or less parallel faults trending northwest, most of them demonstrably of strike-slip character, some cutting fans and showing other evidence of recent seismicity.

(11) The belt of Transverse Ranges. This is a broad zone; at the west it includes the mountains of Santa Barbara County, with the large east-west Santa Ynez fault, faults on the islands of San Miguel, Santa Rosa, and Santa Cruz, and faults in the Santa Barbara Channel put in evidence by located epicenters. Somewhere in this area originated the major earthquake of December 21, 1812. The transverse belt extends eastward through the San Gabriel and San Bernardino Mountains, with evidence of north-south thrusting as well as of east-west strike-slip, and including the complication caused by the partial breakthrough along the San Andreas fault. Immediately south of the Transverse Ranges the structure is dominated by the succession of northwest-trending faults already noted, but it may be separated as follows:

(12) The offshore area, possibly marked by the Inglewood fault zone as its eastern boundary.† It includes the fault following the coast of San Clemente Island. Bottom topography is extremely irregular; islands and banks alternate with deep depressions. This is generally considered a flooded part of the continental area, bounded on the west by a steep scarp descending to the floor of the Pacific, on the north by the islands of the transverse belt, and extending south to Vizcaino Bay on the coast of Baja California. The latest investigation suggests that the Moho may here stand at or near the oceanic level. Seismicity is moderate; the largest known shock, near the south tip of San Clemente Island, was of magnitude 5.9.

(13) A mostly non-seismic block, included at the north between the Inglewood and Elsinore faults, bounded on the south by a major and roughly east-west fault, with abundant evidence of right-hand strike-slip extending from south of Ensenada nearly across the peninsula of Baja California. Not far north of this conspicuous tectonic feature is a parallel and active fault which broke the surface in the earthquake of February 9, 1956.

(14) The chief seismic zone included between the Elsinore and Mission Creek faults, trending southeast into Imperial Valley and thence southward down the western part of the Gulf of California.

(15) Baja California south of the fault noted under (13) appears to be nearly non-seismic.

(16) The Colorado Plateau, cut locally by faults with geologically recent displacement, but showing only minor seismicity.

---

† A more important structural boundary may be the fault at the north base of the San Pedro Hills, extending west into Santa Monica Bay.

CONTINENTAL LIMITS

There has never been much dispute as to where to look for the boundary between the continental and oceanic regions off California. The bottom level descends rather suddenly to a depth of about 2000 fathoms (or 4000 meters) along a fairly straight scarp. This is ordinarily taken as the margin of the continent; the offshore island area (12 above) lies east of it. Nevertheless, Dr. Press's late results, cited in Chapter 18, suggest an oceanic level for the Moho in area (12).

On the east coast of New Zealand critical problems arise. Marshall's andesite line, the petrological representative of the Pacific boundary, is ordinarily drawn as descending from the vicinity of the Kermadec Trough and passing off the east shore of the North Island, then, at about 43° S, leaving the coast of the South Island in a great sweep eastward round the Chatham Islands, which are generally considered continental in character. Since 1950 this has been further supported by investigation of Mernoo Bank, on the Chatham Rise 90 miles east of the South Island, which proves to be non-volcanic; sample pebbles include argillites and sandstones like those of the Paleozoic to Mesozoic rocks of New Zealand.

Such a course for the Pacific boundary leaves Banks Peninsula, which projects eastward from the coast not much farther south, definitely on the continental side. Yet the peninsula consists of two calderas, flooded to form the harbors of Lyttelton and Akaroa, which in form and rock type are closely similar to the volcanic structures of the Hawaiian Islands and Tahiti—definitely Pacific to all appearances. Nothing closely resembling them occurs elsewhere in New Zealand. The eruptives of Lyttelton crater, the westward one, are in part andesitic; those of Akaroa are basaltic. It is clear from his publications that Marshall originally proposed to draw the andesite line between the two craters, a dubious procedure.

In other parts of the Pacific the principal seismic zone for shallow earthquakes accompanies the andesite line. In New Zealand the active eastern margin of the zone follows the east coast of the North Island, then crosses Cook Strait to enter the South Island in the region of the Kaikoura mountain group, thence southwest through the hills at the base of the Southern Alps. Between this zone and Banks Peninsula lies the wide Canterbury Plain, practically non-seismic and believed to represent a geosyncline—a conclusion not inconsistent with gravity observations. One suggestion is that the geosyncline continues into Bounty Basin, south of the Chatham Rise. Whatever the ultimate interpretation, it is clear that continental and oceanic structures are entangled here in a complex way.† A structure intermediate between typical

---

† The point is important because Turner and Verhoogen have included the rocks of Banks Peninsula among continental volcanics, thus blurring the distinction between Pacific and continental petrology.

continental and oceanic would fit the shallow level for the Moho in the New Zealand region (17 to 20 kilometers below the surface, as lately reported by Eiby and by Officer), and suggest comparison with the indicated shallow level off southern California.

STRUCTURAL PROVINCES OF NEW ZEALAND

For seismological purposes we distinguish, beginning at the south (Figs. 27-8 and 27-9), the following.

(1) The active belt extending from the fiord region at the southwest of the South Island almost to the Antarctic Circle. Seismicity apparently increases to the south; the two largest known shocks here, with magnitudes near 8, were at 56° and 53° S (1924, 1943).

(2) The southern transverse belt. This includes the region of Foveaux Strait. There is minor seismicity which may belong to structures following the suspected extension of the Alpine fault round the south coast of the South Island, and perhaps to internal faulting in the southern syncline. Thus in the Taringatura Hills, about 45.7° S, 168.3° E, Coombs has described a northeast-trending scarplet 4 to 10 feet high (with southeastward downthrow) cutting both Mesozoic rocks and recent alluvium.

FIGURE 27-8 *Seismic provinces, New Zealand.*

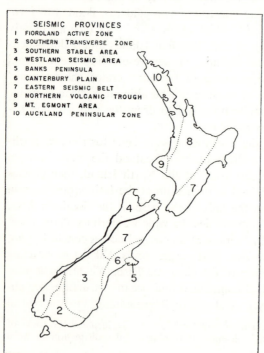

SEISMIC PROVINCES
1 FIORDLAND ACTIVE ZONE
2 SOUTHERN TRANSVERSE ZONE
3 SOUTHERN STABLE AREA
4 WESTLAND SEISMIC AREA
5 BANKS PENINSULA
6 CANTERBURY PLAIN
7 EASTERN SEISMIC BELT
8 NORTHERN VOLCANIC TROUGH
9 MT. EGMONT AREA
10 AUCKLAND PENINSULAR ZONE

A much more striking example is that reported by B. L. Wood just below the outlet of Lake Hauroko (also spelled Hauroto), about 46.1° S, 167.3° E. An east-west scarplet was discovered in 1948 as a straight feature passing through the forest, appearing on air photographs. Ground investigation showed a downthrow of 2½ to 3 feet southward. The main geological trends in the vicinity are north-south. The scarplet appears in fine sandstones; Wood suggests that it represents overthrusting of the Fiordland metamorphic complex, exposed a few miles to the north. From saplings growing on the scarplet, Wood gives its age as about 60 years. He estimates that a smaller motion of 2 to 3 inches took place within 4 to 5 years. This points with some probability to an earthquake of magnitude 6 on April 21, 1939 (see Table 29-1), which was widely felt in Southland province. Our tabulated epicenter is some 30 miles south in Foveaux Strait, but epicenter locations in this area are difficult and inaccurate, owing to its distance from the principal group of recording stations. The small motion may have been precipitated by shaking; or it may have occurred on some other occasion.

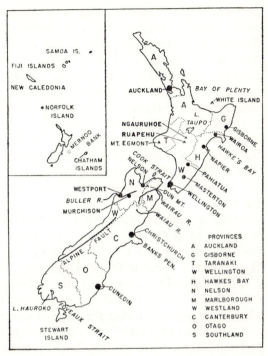

FIGURE 27-9   *Location map, New Zealand.*

(3) The southern stable region. This includes a large non-seismic area, almost entirely in the province of Otago, in which the country rock is prevalently schist, with patches of mostly Quaternary sediments, and an extremely complex group of volcanics surrounding Dunedin on the coast. The schists extend in a narrowing belt northward along the east side of the Alpine fault; next eastward is a wider belt where the surface rocks are of the greywacke group, which appears to be non-seismic north to about the latitude of Christchurch (between 43° and 44° S). Practically the whole southern stable region is broken up into blocks along geologically young fractures. In some localities there are scarplets of comparatively fresh appearance. Cotton has compared the block structures to those of the Basin and Range province of western North America (9 in our list); see Chapter IX of *New Zealand Geomorphology*.

(4) The Alpine fault and the area northwest of it. Lack of any but geological evidence for activity on the Alpine fault probably represents a temporary circumstance, like that noticed for parts of the San Andreas fault since 1906. Cotton has lately suggested that the Wairau fault, which may be the active continuation of the Alpine fault, may have been involved in the earthquake of 1848.

The area north of the Alpine fault in the northwestern part of the South

Island is largely highland, with a block structure trending slightly east of north. At least some of the faults bounding the blocks are active, as was demonstrated by vertical displacement in the earthquake of 1929. This and the southern geosyncline are the Paleozoic areas of New Zealand; Paleozoic sediments here underlie a generally thin Notocene blanket. This compares interestingly with the western part of the Basin and Range province in California, north of the Garlock fault.

(5) Banks Peninsula. This volcanic region shows minor seismicity; local shocks are recorded at Christchurch, and sometimes felt there; normal faults like those of the northern volcanic trough occur west of Christchurch.

(6) The geosyncline of the Canterbury Plain. This, like the Great Valley of California, is largely non-seismic; shocks on its western margin belong to the next.

(7) The eastern seismic belt of the two main islands. This perhaps should be separated at Cook Strait. In the South Island it runs through the highlands just west of the Canterbury Plain, where a series of depressions with Quaternary fill occur in the Mesozoic (greywacke) terrain. At the northeast it enters the region of the Kaikoura and more or less parallel block mountains. Active faults like that of the Awatere Valley follow these blocks, but often show upthrow facing the large blocks, a feature which Cotton at one time attributed to reversal of motion during late geological time. The interpretation may be correct, but Cotton himself has noted that similar effects may be produced by large strike-slip (discussion in Chapter 13 and at the end of this chapter).

In the North Island the principal seismic belt (for shallow earthquakes) includes the east coast; a few epicenters are offshore, and beyond them is the depression considered the southern extension of the Kermadec Trench. Several authors lately have regarded the geosyncline under the Canterbury Plain as the farther southward extension of this depression, which would accord with the connection of the seismic belts of the two islands across Cook Strait. The Strait itself represents a tectonic depression transverse to the main trends. Epicenters are often located in it; a series of shocks, two of which were strong enough at Wellington to cause slight damage, took place there in 1950. The epicenters were north of the projected line of the Alpine fault, and might be assigned to our province (4).

Seismic maps of New Zealand show what appears like a single principal seismic zone in the form of an elongated rectangle; it includes all the epicenters under the present number as well as (4), those in the central trough of the North Island, and a few still farther west. This results from the superposition of block and arc structure. The North Island includes a typical Pacific arc, less active than those to the north, but with most of the characteristic features; geophysical study there may provide the key to many of the problems raised in the preceding chapter. The offshore trench and the coastal belt of shallow earthquakes correspond to A and B in the classification of arc features.

(8) The central trough of the North Island. This is feature D of the arc classification. It includes the volcanoes of Ruapehu, Tongariro, Ngauruhoe, and White Island, and the hot spring and geyser areas near Lake Taupo and Rotorua. As the White Island Trench, it extends northward under the Bay of Plenty. Shallow earthquakes are common, but many of them are clearly volcanic in character. Some, like the Taupo earthquakes of 1922, have been accompanied by the subsidence of superficial blocks with formation of scarplets. Scarplets of the same general type, attributed to normal faulting, are found in the whole area of the trough, including its prolongation to the vicinity of Wanganui on the north coast of Cook Strait.

This belt includes the epicenters of many deep-focus earthquakes. Most of them are small, and their determination depends exclusively on the data of the local stations (which often record ScS sharply; this fixes the depth closely). To 1948 about seven have been listed in the *International Seismological Summary* with adequate data for confirming depth and epicenter from readings at distant stations. The results for these shocks accord well with those for the smaller ones worked out at Wellington.

Of these epicenters, those more accurately located are aligned along the west margin of the central trough; the line can be traced north to the Kermadec Islands and south into Cook Strait. A large majority of depths are in the range 150 to 200 kilometers. A few shocks have depths from 280 to 360 kilometers; these are mostly northward, the two largest being under the Bay of Plenty. One earthquake of magnitude about 5½ on March 24, 1953, had an epicenter near 39° S, 174.5° E, in the same general belt as the others, but with depth determined at Wellington from the data of eight local stations as 570 kilometers. The result has been further confirmed by the time of P' as recorded at Kiruna (Sweden), distant 149°; no other distant station has reported the shock. The depth is not likely to be in error more than 50 kilometers, nor the epicenter more than 1 degree. This minor shock thus assumes great significance, showing that the New Zealand active arc is of the general Pacific type, extending to great depth like the Kermadec and Tonga arcs. The location of the epicenter comparatively near those of shocks at intermediate depth assigns a very steep dip to the active surface comparable with the New Hebrides rather than with the Kermadec arc.

(9) The vicinity of Mt. Egmont (Taranaki province, the west-central coast of the North Island). Mt. Egmont is a well-preserved Pleistocene volcano, comparable with those of the North American Cascade Mountains. In location it corresponds with feature E of the typical arcs. However, the seismicity in this vicinity consists neither of superficial volcanic earthquakes nor of shocks in the lower range of intermediate depth; the latter occur in the belt discussed under (8). Instead, shallow earthquakes of magnitude up to 6 at least originate off the coast north of Mt. Egmont; one of magnitude

over 5 occurred there on October 18, 1953. Eiby has mapped a northeast-southwest line of epicenters passing nearly through Mt. Egmont; he believes that it represents a major line of tectonic weakness.

(10) The Auckland Peninsula. The structural axis here consists largely of igneous rocks and of Mesozoic rocks of the greywacke group. The boundary faults of the parallel graben to the east, in which lies the Firth of Thames, are probably active, but seismicity generally is low. This point has been stressed in connection with the safety of the metropolis of Auckland. Reports that an earthquake about 1834 was violent in the Auckland area have generally been discounted as involving confusion of place. (See Eiby's discussion.) However, on April 7, 1956, a shock from some nearby source caused minor damage at Auckland.

COMPARISONS

It is impossible not to be struck by the resemblance of individual features of New Zealand tectonics to individual features in California, but attempts at orderly matching, point by point, end in confusion. It is as if the same elements had been used to build up the two groups of structures, but in entirely different arrangement. Moreover, California almost completely lacks the Pacific arc features, which are recognizable, though not conspicuous, in New Zealand. We have just seen that the superficial correlation of the Cascade volcanoes with the active volcanoes of New Zealand is misleading; the proper comparison is with Mt. Egmont. Deep-focus earthquakes are wholly lacking in California; to find them, as well as the other arc features, we must travel south to Mexico or north to Alaska. Even the South Island presents, in Banks Peninsula, a feature wholly strange to western North America.

The eastern coastal belt of the North Island resembles the California Coast Ranges in structure and seismicity; but the Alpine fault is separate from the former, while the San Andreas fault cuts through the latter. The Canterbury geosyncline, with seismicity on its western margin, and the Alpine fault, still farther west with the mass of the Southern Alps between, may be compared with the Sacramento Valley (the northern part of the Great Valley of California), which has Coast Range seismicity at its margin and the San Andreas fault to the west.

The points of resemblance between the northwest section of the South Island and the California Basin and Range area have been noted. Cotton's parallel in central Otago and the surrounding area is locally very striking.

RECENT TECTONIC CHANGES?

In New Zealand there is particularly impressive evidence that present displacements are not coherent with the larger topographic features. (See Figs. 27-10 and 27-11.) Cotton has emphasized the occurrence of small fresh

FIGURE 27-10 *Rift of the Hope fault (on the South Island, New Zealand), associated with the earthquake of 1888, rising diagonally across terraces. [Crown copyright. Permission to publish granted by Land Survey Department, Wellington.] Compare Figure 29-4.*

FIGURE 27-11   *Reversed scarplet, Ruahine Range, the North Island, New Zealand.*
[*After a sketch by C. A. Cotton, from a photo by R. J. Waghorn.*]

scarps along the Awatere fault and elsewhere which face against the large
scarps at the front of the major blocks (Fig. 27-11). This does not have to be
interpreted as a direct reversal of movement, which would imply a reversal
in regional strain. Here as elsewhere in the region, evidence suggests strong
strike-slip; the reversal might exist only in relatively small and incidental dip-
slip. Or, if there had been a long period of very little dip-slip while strike-slip
accumulated to an amount measured in miles, scissor points might well be
shifted enough to produce a reversal facing the former topography. There is
clear evidence in New Zealand that at least some recent fracturing is on lines
at an angle to the older structures. Thus in the North Island a young scarp,
first figured by Waghorn, obliquely descends the face of a large old scarp. In
the Wairau valley a long fresh scarp, possibly connected with the 1848 earthquake and perhaps a branch of the Alpine fault, crosses the valley from side to
side at a low angle. Two miniature strike-slip rifts in the Waiau-Hope valley,
at least one of which was the scene of displacement in 1888, ascend obliquely
from river level up to that of a high terrace (see Fig. 27-10). Everywhere in
the principal active area of both islands are scarplets of the right height and
extent to have originated in single seismic events, without indication of accumulation or repetition, as if the locus of fracture were constantly shifting.
(See also Fig. 27-12.)

There are similar, but generally less conspicuous, indications in California.
Along the San Andreas fault there are small young scarps running at the
base of larger and presumably older ones, so that here and in many other
localities the most recent movements appear coherent with those of the

FIGURE 27-12   *Strike-slip faulting at foot of old scarp, New Zealand.* [*After a drawing by C. A. Cotton, from a photo by Charles Rich.*]

immediate geologic past. However, the lack of evidence of very recent movement along the Garlock fault has been noted, whereas in 1952 the White Wolf fault, not strongly expressed in the topography, was the seat of a major earthquake. Again, there is little current seismicity at the base of the great scarp of the Sierra Nevada, while the Owens Valley earthquake of 1872 broke the surface at the base of the smaller block of the Alabama Hills. The relations of the Imperial fault of 1940 to the San Jacinto fault and of the Agua Caliente fault to the Elsinore fault suggest a shifting fracture pattern. For the San Andreas fault Taliaferro has pointed out that in certain sections the recent fracture zone cuts at a small angle through the older elongated block and trough structures which are sometimes taken as evidence for the antiquity of the faulting.

This general lack of exact coherence of current tectonic activity with obvious structural features appears in other regions; the earthquakes in Anatolia from 1939 provide another example (Chapter 31). Thus it is all the more necessary for the field geologist and seismologist not to draw conclusions too rapidly as to the interpretation of seismicity, especially in the absence of seismographs.

References

PAPERS dealing with specific earthquakes are listed in the chronological bibliography, Appendix XVI. References on strike-slip faults follow Chapter 13. Those on crustal structure in both regions are given with Chapter 18.

California region

Important references on California tectonics and general geology are included in the following publications of the California Department of Natural Resources, Division of Mines, San Francisco.

Bull. 118 (1943); includes Taliaferro, N. L., "Geologic history and structure of the central Coast Ranges of California," pp. 119–163. (Gives data on the San Andreas Rift referred to above.)

Bull. 154 (1951); Jenkins, O. P., ed., "Geologic guidebook of the San Francisco Bay counties."

Bull. 170 (1954), Jahns, R. H., ed., "Geology of southern California," 392 pp., 34 map sheets; additional maps and charts. (Includes some references given in Chapter 13; note also map sheet 11, Geology of the Owens Valley region, and Hewett, D. F., "A fault map of the Mojave Desert region" in Chapter IV.)

Geological Map of California, 1938. (Scale 1:500,000. Shows faults and stratigraphy. Inset, Geomorphic map of California, scale 1:200,000, showing faults with surface and submarine contours, also issued separately.)

Geological Map of California. (30 sheets, scale 1:250,000; preliminary uncolored edition; 8 sheets available 1955.)

See also: *Tectonic map of the United States,* scale 1:2,500,000. Amer. Assoc. Petroleum Geologists, Tulsa, 1944.

San Andreas fault or Franciscan formation

Lawson, A. C., "Geology of the coast system of mountains" in: *The California earthquake of 1906, Report of the State Earthquake Investigation Commission,* 1908, vol. 1, pp. 5–24.

———, *San Francisco Folio, U. S. Geol. Survey,* 1914; geological atlas, folio 193.

Weaver, C. E., "Geology of the Coast Ranges immediately north of the San Francisco Bay region," *Geol. Soc. Amer.,* Mem. 35 (1949). Revised and abridged as "Geology and mineral deposits of the area north of San Francisco Bay," *Calif. Dept. Nat. Resources, Div. Mines Bull.* 149 (1949).

Louderback, G. D., "Characteristics of active faults in the central Coast Ranges of California, with application to the safety of dams," *B.S.S.A.,* vol. 27 (1937), pp. 1–27.

Taliaferro, N. L., "Geological history and correlation of the Jurassic of California and southwestern Oregon," *Bull. Geol. Soc. Amer.,* vol. 53 (1942), pp. 71–112.

———, "Franciscan-Knoxville problem," *Bull. Am. Assoc. Petroleum Geologists,* vol. 27 (1943), pp. 109–219.

———, "Geology of the San Francisco Bay counties," *Calif. Dept. Nat. Resources, Div. of Mines Bull.* 154 (1951), pp. 117–150.

Schlocker, J., Bonilla, M. G., and Imlay, R. W., "Ammonite indicates Cretaceous age for part of Franciscan group in San Francisco Bay area, California," *Bull. Am. Assoc. Petroleum Geologists,* vol. 38 (1954), pp. 2372–2381.

Important general discussions (now obsolete in many particulars)

Reed, R. D., *Geology of California*, American Association of Petroleum Geologists, Tulsa, 1933.

Reed, R. D., and Hollister, J. S., "Structural evolution of Southern California," *Bull. Am. Assoc. Petroleum Geologists*, vol. 20 (1936), pp. 1529–1704. (Also published separately, and later reprinted under one cover with the above.)

Offshore topography

Shepard, F. P., and Emery, K. O., "Submarine topography off the California coast," *Geol. Soc. Amer., Spec. Paper* 31 (1941).

Menard, H. W., and Dietz, R. S., "Submarine geology of the Gulf of Alaska," *Bull. Geol. Soc. Amer.*, vol. 62 (1951), pp. 1263–1285.

——, "Mendocino submarine escarpment," *Journ. Geology*, vol. 60 (1952), pp. 266–278.

Menard, H. W., "Deformation of the northeastern Pacific basin and the west coast of North America," *Bull. Geol. Soc. Amer.*, vol. 66 (1955), pp. 1149–1198.

——, "Fractures in the Pacific floor," *Scientific American*, vol. 193, No. 1 (July, 1955), pp. 36–41.

Other tectonic data

Hewett, D. F., "Structural features of the Mojave Desert region," *Geol. Soc. Amer. Spec. Paper* 62 (1955), pp. 377–390.

Curry, H. D., "Strike-slip faulting in Death Valley, California" (abstract), *Bull. Geol. Soc. Amer.*, vol. 49 (1938), pp. 1874–1875.

Allen, C. R., Silver, L. T., and Stehli, F. G., "Agua Blanca fault—a major transverse structure of northern Baja California, Mexico" (abstract), *Bull. Geol. Soc. Amer.*, vol. 67 (1956), p. 1664.

Nolan, T. B., "The Basin and Range province in Utah, Nevada, and California," *U. S. Geol. Survey Prof. Paper 197-D* (1943), pp. 139–196.

Longwell, C., "Tectonic history viewed from the Basin Ranges," *Bull. Geol. Soc. Amer.*, vol. 61 (1950), pp. 413–434.

New Zealand region

General geology and tectonics

Officers of the New Zealand Geological Survey, *The Outline of the Geology of New Zealand* (to accompany the 16 miles to 1 inch geological map), Wellington, 1948; geological map, 2 sheets, scale 1:1,013,760.

Fleming, C. A., "The geological history of New Zealand (with reference to the origin and history of the fauna and flora)," *Tuatara*, vol. 2 (1949), pp. 72–90. (Informal and readable.)

Finlay, H. J., and Marwick, J., "The divisions of the Upper Cretaceous and Tertiary in New Zealand," *Trans. Royal Soc. N. Z.*, vol. 70 (1940), pp. 77–135.

———, "New divisions of the Upper Cretaceous and Tertiary," *N. Z. Journ.*, Section B, vol. 28 (1947), pp. 228–236.

Cotton, C. A., "Review of the Notocenozoic, or Cretaceo-Tertiary, of New Zealand," *Trans. Royal Soc. N. Z.*, vol. 82 (1955), pp. 1071–1122.

Macpherson, E. O., "An outline of late Cretaceous and Tertiary diastrophism in New Zealand," *N. Z. Geol. Survey, Mem.* 6 (1946). (An important publication, although its chief thesis has not been accepted generally.)

Lillie, A. R., "Notes on the geological structure of New Zealand," *Trans. Royal Soc. N. Z.*, vol. 79 (1951), pp. 218–259. (A thoughtful discussion, which brings out the deficiencies in Macpherson's treatment.)

———, "The geology of the Dannevirke subdivision," *Bull. N. Z. Geol. Survey,* vol. 46 (N.S.) (1953). (This and the following reference are sumptuous monographs on small areas; the authors discuss many of the critical points in New Zealand geology, particularly of North Island.)

Fleming, C. A., "The geology of the Wanganui subdivision," *ibid.,* 52 (N.S.) (1953).

Wellman, H. W., "Structural outline of New Zealand," *Bull. N. Z. Dept. of Scientific and Industrial Research,* 121 (1956), pp. 1–128.

Cotton, C. A., *New Zealand Geomorphology,* New Zealand University Press, Wellington, 1955. (Reprints papers originally dating from 1912 to 1925; doubly valuable to the student because of numerous annotations indicating later revision and progress of investigation.)

Tectonics and faulting (see also Chapter 13)

Wellman, H. W., "Data for the study of Recent and Late Pleistocene faulting in the South Island of New Zealand," *N. Z. Journ.*, Section B, vol. 34 (1953), pp. 270–288.

———, "Active transcurrent faulting in New Zealand," (abstract), *Bull. Geol. Soc. Amer.*, vol. 65 (1953), p. 1322.

———, New Zealand Quaternary tectonics," *Geol. Rundschau,* vol. 43 (1955), pp. 248–257.

Cotton, C. A., "Tectonic relief; with illustrations from New Zealand," *Geographical Journ.,* vol. 119 (1953), pp. 213–222.

———, "Rejuvenation of the Awatere fault cicatrice," *T. ans. Royal Soc. N. Z.,* vol. 77 (1949), pp. 273–274.

———, "Fault valleys and shutter ridges at Wellington," *N. Z. Geographer,* vol. 7 (1951), pp. 62–68.

———, "Tectonic relief features in and around Wellington, *Proc. 7th Pacific Science Congress,* vol. 2 (1953), pp. 5–6 (Correction: for Pahautanui read Pauatahanui.)

————, "Submergence in the lower Wairau valley," *N. Z. Journ.*, Section B, vol. 35 (1954), pp. 364–369.

————, "Geomechanics of New Zealand mountain-building, *ibid.*, vol. 38 (1956), pp. 187–200.

————, "Geomorphic evidence and major structures associated with transcurrent faults in New Zealand," Revue de géographie physique et de géologie dynamique, series 2, vol. 1 (1957), pp. 16–30.

Waghorn, R. J., " 'Earthquake rents' as evidence of recent surface faulting in Hawke's Bay," *N. Z. Journ.*, vol. 9 (1927), pp. 22–26.

Fleming, C. A., "Earthquake traces near Waverley and their tectonic setting," *Trans. Royal Soc. N. Z.*, vol. 77 (1949), pp. 274–275.

————, "The White Island trench; a submarine graben in the Bay of Plenty, New Zealand," *Proc. 7th Pacific Science Congress*, vol. 3 (1953), pp. 210–213.

Munden, F. W., "Notes on the Alpine fault, Haupiri Valley, North Westland," *N. Z. Journ.*, Section B, vol. 33 (1953), pp. 404–408.

Coombs, D. S., "The geology of the northern Taringatura Hills, Southland," *Trans. Proc. Royal Soc. N. Z.*, vol. 78 (1950), pp. 426–448.

Wood, B. L., "A recent fault scarplet at the outlet of Lake Hauroko, Southland," *N. Z. Journ.*, Section B, vol. 30 (1948), 173–176.

The Pacific boundary

Speight, R., "The geology of Banks Peninsula," *Trans. N. Z. Inst.*, vol. 49 (1917), pp. 365–392.

Benson, W. N., "Cainozoic petrographic provinces in New Zealand and their residual magmas," *Am. Journ. Science,* vol. 239 (1941), pp. 537–552.

Hatherton, T., "Gravity profiles across the Canterbury Plains," *N. Z. Journ.*, Section B, vol. 34 (1952), pp. 13–20.

Fleming, C. A. and Reed, J. J., "Mernoo Bank, east of Canterbury, New Zealand," *ibid.*, vol. 32 (1951), No. 6, pp. 17–30.

Gerard, V. B., "Aeromagnetic observations over the Banks Peninsula area and the Mernoo Bank," *ibid.*, vol. 35 (1953), pp. 152–160.

Crustal structure

Note references for New Zealand at end of Chapter 18.

Auckland region

Eiby, G. A., "The seismicity of Auckland city and Northland," *N. Z. Journ.*, Section B, vol. 36 (1955), pp. 488–494.

CHAPTER 28

California Earthquakes

HISTORICAL NOTE

Earthquakes Before 1880

In both California and New Zealand knowledge of seismicity is limited by the relatively late dates at which they were opened to scientific investigation.

The earliest earthquake reported for California was on July 28, 1769; it was felt strongly by the exploring expedition of Gaspar de Portolá, in camp on the Santa Ana River near the present site of Olive (about 30 miles southeast of the center of Los Angeles); aftershocks were felt for days as the party traveled northwest. Something is known of earthquakes during the active period of the Franciscan missions; thus in 1812 on December 8, the church at San Juan Capistrano was destroyed, with 40 fatalities, and on December 21 a major earthquake destroyed Purisima mission, damaged those at San Fernando, San Buenaventura, Santa Barbara, and Santa Ynez and produced a tsunami in the Santa Barbara Channel. From 1830 to 1850 documentary evidence is fragmentary, so that little is known of the two large earthquakes in central California in 1836 and 1838. With the discovery of gold and the influx of population beginning in 1849 more records are in existence, particularly for central California.

Not enough is known of the important earthquakes of 1857, 1868, 1872. For the first, information is from non-scientific sources; for the second, a scientific report was lost. A report on the 1872 earthquake in Owens Valley was published by Whitney, then head of the state geological survey; but the meizoseismal area was vast and sparsely inhabited, and investigation was necessarily incomplete.

Earthquake Catalogues

When Lick Observatory was established at Mt. Hamilton in 1887, the best seismographs then available were installed there and at Berkeley. The director, E. S. Holden, published catalogues of known Pacific Coast earthquakes; the final inclusive edition covered the years 1769 to 1897. A catalogue for 1897–1906 was published by McAdie; it is less detailed than Holden's, partly because of loss of material in the fire of 1906. In 1939 Townley and Allen published a catalogue revising those of Holden and McAdie, and extending the record through 1927.

Under the auspices of the U. S. Coast and Geodetic Survey, pamphlets have been issued which summarize information on the larger earthquakes of California and western Nevada, occasionally supplement the Townley-Allen catalogue data from other sources, and extend the record (through 1950 in the latest revision, edited by Wood and Heck). The Survey has published more detailed annual data in the serial, *United States Earthquakes;* other records are preserved as papers and short notes in the *Bulletin of the Seismological Society of America.* Epicenter determinations, with some macroseismic information, are issued from Pasadena and Berkeley.

Progress Since 1906

For our scientific knowledge of the great earthquake of 1906 we are heavily indebted to Andrew C. Lawson, who prevailed upon the Governor of California to appoint a State Earthquake Investigation Commission. Funds for the operation of the Commission and publication of its findings were provided by the Carnegie Institution of Washington. Lawson, as chairman, organized the work of investigation; he acted as general editor for the report, to which he contributed much of the material.

In many countries destructive earthquakes have led to the foundation of permanent government-sponsored institutions for research in seismology. In California the immediate good effects ceased with the publication of the report of the Investigation Commission. A public "earthquake psychology" developed; the press was asked to suppress or minimize news of earthquakes. Scientific men were told they should not discuss or even investigate earthquakes because it was bad for business; if direct approach did not work, indirect pressures were applied. The Seismological Society of America, founded in California in 1911, had to meet this kind of opposition from the very first; one of its assets was the vitriolic personality of Lawson, who stated the case in no uncertain terms in the paper leading off Volume One, Number One, of the society's bulletin.

Seismological Stations

Scientific interest aroused after 1906 was too strong to be suppressed. The Seismological Society continued publication. Improved instruments were acquired for the two University of California stations at Berkeley and Lick Observatory; published readings begin in 1910 and 1911, respectively.

In 1921, on the initiative of Mr. H. O. Wood, the Carnegie Institution of Washington set up a program for earthquake registration and research with headquarters at Pasadena. The torsion seismometer was developed for this work by Anderson and Wood. By the summer of 1927 four stations were recording continuously in southern California. Public interest aroused by the damaging earthquake at Santa Barbara in 1925 also led to the installation of new equipment at Berkeley and Mt. Hamilton, as well as additional stations at Stanford University and San Francisco.

In 1937 the California Institute of Technology assumed administration of the southern California program; in 1956 there were sixteen stations in operation. Bulletins with readings for teleseisms and the larger local earthquakes were issued from 1931 and lists giving epicenters and origin times for the smaller local shocks from 1934. The northern California group, reporting from Berkeley, had eleven stations in 1956, including Reno (Nevada) and Corvallis (Oregon).

Stations operated by or in cooperation with the U. S. Coast and Geodetic Survey have contributed to the study of earthquakes in this region. In California these were at the International Latitude Observatory at Ukiah, and at Shasta Dam (the latter has been transferred to the Berkeley group). The station at the Magnetic Observatory, Tucson, Arizona, has operated since 1910 (and with improved instruments since 1937). A group of three stations in the vicinity of the artificial Lake Mead, in Nevada and Arizona, began recording in 1930; lately they have been reduced to one station at Boulder City, but some of the equipment has been transferred to Eureka in Nevada, where it is recording very effectively.

Work of the U. S. Coast and Geodetic Survey

The valuable and extensive program of recording strong motion with triggered low-sensitivity instruments, also operated by the Coast and Geodetic Survey, has been mentioned in Chapter 15; Chapter 11 has noted the equally important Survey program of collecting information on perceptible earthquakes by means of questionnaire cards (supplemented by personal field investigations). The seismological contribution of the Survey in its traditional function, establishing displacement of lines of leveling and triangulation, has been of critical importance; the results are implied in discussion of earthquake mechanism on many pages of this book.

Engineers in Seismology

Much of the progress of seismology in California, as elsewhere, has been due to the interest and support of public-spirited engineers. Their contributions have been made through the Seismological Society and various engineering organizations. Some of the engineers most actively concerned in furthering our factual knowledge of earthquakes have been consultants for

Table 28-1   Shocks of Magnitude 6 and Over—California–Nevada Region—Excluding those off the North Coast*

Date	G.C.T.	Lat. N	Long. W	M
1903 Jan. 24	05:	31½	115	7+
1906 Apr. 18	13:12.0	38	123	8.3
1906 Apr. 19	00:30	32.5?	115.5?	6+
1907 Sept. 20	01:54	34.2?	117.1?	6
1908 Nov. 4	08:37	36?	117?	6½?
1910 May 15	15:47	33.7?	117.4?	6
1911 July 1	22:00.0	37¼	121¾	6.6
1915 June 23	03:59	32.8	115.5	6¼
1915 June 23	04:56	32.8	115.5	6¼
1915 Oct. 3	06:52.8	40½	117½	7.6
1915 Nov. 21	00:13.7	32	115	7.1
1916 Oct. 23	02:44	34.9	118.9	6
1918 Apr. 21	22:32:25	33¾	117	6.8
1922 Mar. 10	11:21:20	35¾	120¼	6½
1923 July 23	07:30:26	34	117¼	6¼
1925 June 29	14:42:16	34.3	119.8	6.3
1926 Oct. 22	12:35:11	36¾	122	6.1
1926 Oct. 22	13:35:27	36¾	122	6.1
1927 Sept. 18	02:07:07	37½	118¾	6
1927 Nov. 4	13:50:43	34½	121½	7.5
1931 Oct. 1	11:45:38	30	114½	6
1932 Dec. 21	06:10:05	38¾	118	7.3
1933 Mar. 11	01:54:08	33.6	118.0	6.3
1933 June 25	20:45:27	39¼	119	6.1
1934 Jan. 30	20:16:31	38	118½	6.5
1934 June 8	04:47:45	35.8	120.4	6.0
1934 Dec. 30	13:52:14	32¼	115½	6.5
1934 Dec. 31	18:45:45	32	114¾	7.1
1937 Mar. 25	16:49:03	33.5	116.5	6.0
1940 Feb. 8	08:05:59	39¾	121¼	6
1940 May 19	04:36:41	32.7	115.5	7.1
1941 Sept. 14	18:39:12	37.6	118.7	6.0
1942 Oct. 21	16:22:14	33.0	116.0	6½
1946 Mar. 15	13:49:36	35.7	118.1	6¼
1947 Apr. 10	15:58:06	35.0	116.6	6.4
1948 Dec. 4	23:43:17	33.9	116.4	6.5
1948 Dec. 29	12:53:28	39.5	120.1	6.0

Table 28-1   Continued

Date	G.C.T.	Lat. N	Long. W	M
1952 July 21	11:52:14	35.0	119.0	7.7
1952 July 21	12:05:31	35.0	119.0	6.4
1952 July 23	00:38:32	35.4	118.6	6.1
1952 July 29	07:03:47	35.4	118.9	6.1
1952 Nov. 22	07:46:38	35.8	121.2	6
1954 Mar. 19	09:54:29	33.3	116.2	6.2
1954 July 6	11:13:20	39.5	118.5	6.6
1954 July 6	22:07:40	39.5	118.5	6.4
1954 Aug. 24	05:51:32	39.5	118.5	6.8
1954 Aug. 31	22:20:35	39.5	118.5	6.3
1954 Oct. 24	09:44:08	31.5	116	6.0
1954 Nov. 12	12:26:47	31.5	116	6.3
1954 Dec. 16	11:07:10	39.3	118.1	7.1
1954 Dec. 16	11:11:29	39.5	118.3	6.8
1956 Feb. 9	14:32:38	31.7	115.9	6.8
1956 Feb. 9	15:24:26	31.7	115.9	6.1
1956 Feb. 14	18:33:34	31.5	115.5	6.3
1956 Feb. 15	01:20:38	31.5	115.5	6.4

* See Figure 28-2.

the insurance industry. Among them may be mentioned the late J. R. Freeman whose large volume, published in 1931, includes material derived from widely scattered sources and some not available elsewhere;† and Mr. H. M. Engle, consulting engineer for the Pacific Board of Fire Underwriters, who has published, edited, and contributed to many valuable reports on the effects of earthquakes on engineering structures. The Pacific Board established its earthquake department in 1926. (See also Chapter 24.)

LARGER EARTHQUAKES OF THE REGION

Lists of the larger located earthquakes in the California region during the 50 years for which instrumental magnitudes can be assigned appear in Tables 28-1 and 28-2. (See also Figs. 28-1, 28-2, and 28-3.) These tables do not completely represent the local seismicity as indicated by historical data. The summary catalogue by Wood and Heck lists the following as "great shocks":

† Freeman's interpretation of his data, as well as his presentation of the facts of instrumental seismology, is deficient in many particulars.

Table 28-2   Shocks of Magnitude 6 and Over—California
North Coastal Area*

Date	G.C.T.	Lat. N	Long. W	M
1909 Oct. 28	06:45	40.2?	124.1?	6+
1910 Mar. 19	00:11	40?	125?	6+
1910 Aug. 5	01:31.6	42	127	6.8
1914 Aug. 22	05:28.3	44	129	6¾
1915 May 6	12:09.0	39½	126½	6¾
1915 Dec. 31	12:20.0	41	126	6½
1917 June 10	04:32.4	44	129	6½
1918 July 15	00:23:00	41	125	6½
1922 Jan. 26	09:31:20	41	126	6
1922 Jan. 31	13:17:22	41	125½	7.6
1923 Jan. 22	09:04:18	40½	124½	7.3
1925 June 4	12:02:52	41½	125	6
1926 June 5	19:50:24	43	127½	6
1926 Dec. 10	08:38:53	40¾	126	6
1928 Sept. 11	12:36:19	43½	130¼	6.3
1932 June 6	08:44:22	40¾	124½	6.4
1934 July 6	22:48:52	41¼	125¾	6.5
1936 Sept. 25	12:53:35	42½	128	6.2
1938 May 28	10:14:01	42¾	126	6
1941 Feb. 9	09:44:04	40½	125¼	6.6
1941 May 13	16:01:45	40	126	6
1941 Oct. 3	16:13:08	40¾	125	6.4
1945 May 19	15:07:04	40¼	126½	6.2
1945 Sept. 28	22:24:10	42	126	6.0
1949 Mar. 24	20:56:56	42	126	6.2
1951 June 17	09:40:17	44½	130	6
1951 Oct. 8	04:10:36	40¼	124½	6
1952 Aug. 20	15:25:04	43¼	126½	6.5
1954 Nov. 25	11:16:36	40.5	126	6.5
1954 Dec. 21	19:56:25	41	124	6.6
1955 Aug. 23	15:32:40	43½	128	6¼
1956 Oct. 11	16:48:46	40½	126½	6

* See Figure 28-2.

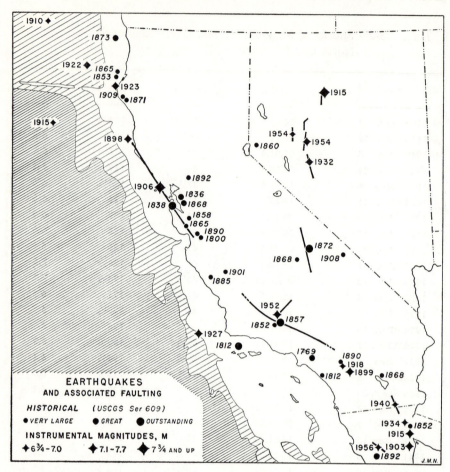

FIGURE 28-1   *Larger earthquakes of the California region. Extent of faulting indicated.*

1769, July 28	1857, January 9	1906, April 18
1790?	1868, October 21	1915, October 2
1812, December 21	1872, March 26	1922, January 31
1836, June 10	1873, November 22	1932, December 20
1838, June	1892, February 23	1940, May 18

Those of 1812, 1838, 1857, 1872, and 1906 are further qualified as "outstanding." Other shocks marked "strong" number 20 for 1800–1903 and 16 for 1909–1950. (Those so marked to which magnitudes can be assigned with any confidence are of magnitude 6 or over.)

Of the "outstanding" earthquakes, that of 1812 has already been described in a few words which cover most of what is known; its origin was in the province of the Transverse Ranges, and probably offshore. Those of 1838,

1857, and 1906, associated with the San Andreas fault, are discussed together in the next section; that of 1872 is the Owens Valley earthquake, to be discussed with those of Nevada. Of the remaining "great" shocks, that of 1769 may possibly have originated on the San Andreas fault, but other origins are equally likely; that of 1790 represents a vague Indian account of an earlier great shock in Owens Valley. Those of 1836 and 1868 were associated with the Haywards fault, discussed with the San Andreas group. That of 1873, known as the Crescent City earthquake, was one of the largest known shocks in the highly seismic offshore area north of the Mendocino escarpment; but it may not have exceeded that of 1922 in the same sub-

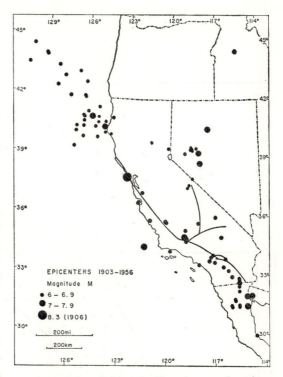

FIGURE 28-2 *California–Nevada earthquakes, magnitude 6 and over, 1903–1956.* [*From Tables 28-1 and 28-2.*]

province. That of 1892, originating in Baja California, will be discussed very briefly toward the end of this chapter with the earthquake of 1956 in the same region. Nevada shocks showing evidence of faulting in 1915, 1932, and 1954 will be discussed together. The Imperial Valley earthquake of 1940 will be taken up with the San Andreas fault group. The Kern County earthquake of 1952, here also assigned a section, ranks in magnitude with those listed above.

SAN ANDREAS FAULT—1838 AND 1857

Violent earthquakes occurred in the San Francisco Bay area in June, 1836, and June, 1838. Documents are few, and they tend to confuse the two events. The late Professor G. D. Louderback, by careful study, disentangled them. The 1838 earthquake is described as having opened a great crack many miles in length; the imperfect references to location would fit fairly well for a part of the San Andreas Rift along which fault traces appeared in 1906. Otherwise, details, for our present purpose, are lacking. Many such occurrences are

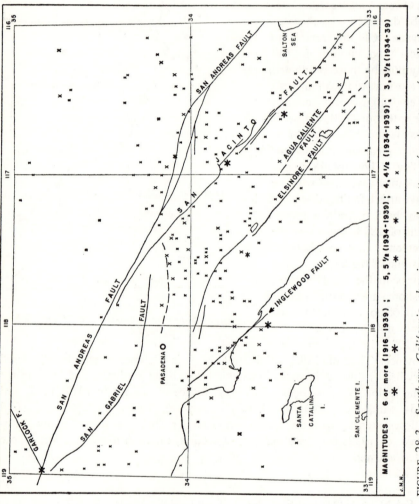

FIGURE 28-3 Southern California, showing general scatter of epicenters of small shocks.

listed in the chronicles of other regions, to be interpreted properly only by combining geological knowledge with historical scholarship.

The earthquake of January 9, 1857, was similar to that of 1906. In both there was displacement along the San Andreas fault, but the two traces were separated by about 150 miles. For 1857 our reports come mostly from untrained observers; the area over which shaking was felt, in spite of its thin population at the time, was comparable with that in 1906, and various intensities were reported at roughly equal distances from the fault on the two occasions. The magnitudes of the two events cannot have differed greatly.

Most accounts give little detail of faulting in 1857; they state that the ground opened in a great crack 40 miles long in the vicinity of Fort Tejon. This was an army post about 4 miles from the San Andreas fault, on the route of one of the present main highways (U. S. 99) between Los Angeles and San Francisco; official reports show that adobe buildings there were badly damaged. One observer, referring to the long "rent," writes: "This rent closed up immediately, but the loosened earth thrown up would not fit back into it, and therefore left more or less of a ridge which marked the line of eruption . . ."—a fair description of the mole-track effect characteristic of strike-slip faulting. At one point a round sheep-corral is said to have been changed to S-shape; this would result from right-hand strike-slip along a trace passing through the corral.

Fortunately a "History of Tulare County" published serially in a newspaper (the Visalia *Iron Age*) in 1876 contains the following.

The line of disturbing force followed the Coast Range some seventy miles west of Visalia and thence out on the Colorado Desert. This line was marked by a fracture of the earth's surface, continuing in one uniform direction for a distance of two hundred miles. The fracture presented an appearance as if the earth had been bisected, and the parts had slipped upon each other. Sometimes the earth on one side would be several feet the highest, presenting a perpendicular wall of earth or rocks. In some places the sliding movement seems to have been horizontal, one side of the fracture indicating a movement to the northwest, the other to the southeast. The fracture pursued its course over hill and hollow, and sometimes this sliding movement would give to the points of the hills and to gulch channels a disjointed appearance.

This is an unmistakable description of faulting involving strike-slip. Had it been published after 1906, invention might be suspected; but in 1876 such effects were not even known to specialists. The writer of this account probably did not clearly distinguish the actual displacement of 1857 (which he misdates 1856, but this is almost certainly a mere slip of memory) from the topographic features of the San Andreas Rift. This may account in part for the extent of 200 miles and the mention of the Colorado Desert. Other reports place the southeastern end of visible displacements near San Bernardino; the corral referred to appears to have been on the Carrizo plain about 150 miles to the northwest, and a point 70 miles west of Visalia would add over 100 miles more.

HAYWARDS FAULT EARTHQUAKES— 1836 AND 1868

The Haywards fault is the principal active branch of the San Andreas fault in central California; it diverges eastward from the master fault south of Hollister and runs northward at the base of the hills on the east shore of San Francisco Bay.† It has not been traced definitely north of the eastern branches of the Bay, which represents a tectonic feature that may offset or distort the fault lines. In part it shows strike-slip rift features comparable to those of the main fault.

The earthquake of October 21, 1868, caused extensive damage at San Francisco, particularly on filled ground, and until 1906 it was locally referred to as "the great earthquake." The most destructive effects were in the vicinity of San José and at Haywards (now officially Hayward), directly on the fault line. Scientific details have been lost. There was a large fissure following the fault which remained open and was found by the usual popular methods to be "bottomless." Contemporary accounts mention no lateral displacement; but, since most of the area was unoccupied, this is not significant. At one locality Professor Louderback has found some evidence still remaining for such displacement. Triangulation by the U. S. Coast and Geodetic Survey was going on in the region at the time; discrepancies in the position of monuments, especially of Mt. Tamalpais north of San Francisco, were later analyzed and they indicate considerable motion at or about the time of the earthquake.

The earthquake of June 10, 1836, was a similar but possibly greater event. Large fissures appeared, in all probability on the Haywards rift. Shaking was strong and damaging as far away as Monterey. The documents were carefully studied by Professor Louderback.

THE CALIFORNIA EARTHQUAKE OF APRIL 18, 1906

The heading is the designation used in the official report of the Earthquake Investigation Commission—and rightly so, for a large part of the state was seriously affected. In spite of this, ordinary discussion will probably continue to refer to "the San Francisco earthquake."

---

† Another active branch runs a few miles farther east, nearly through the town of Walnut Creek, where there was much minor damage in an earthquake of magnitude 5.4 on October 23, 1955.

Loss and Casualties

The Commission report does not explicitly discuss the earthquake as a disaster. As such, that of 1906 ranks below many others, including the catastrophe which befell Tokyo in 1923. Nevertheless, it involved the destruction by earthquake and fire of a large part of a great city, with a loss of the order of 400 million dollars, and possibly 700 lives.

Since, with the passage of years and repetition from book to book, there

FIGURE 28-4 *Isoseismals, California earthquake of April 18, 1906. Rossi-Forel scale. [Redrafted, after the State Commission Report.]*

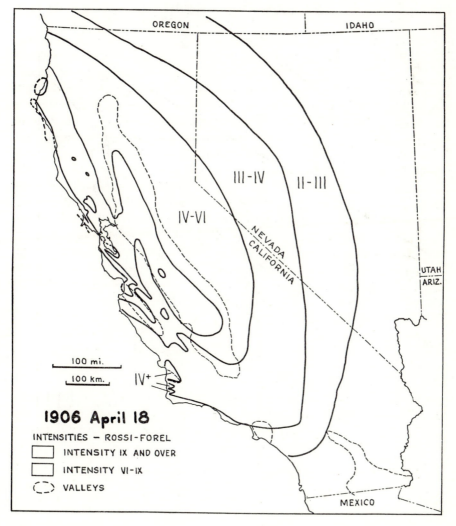

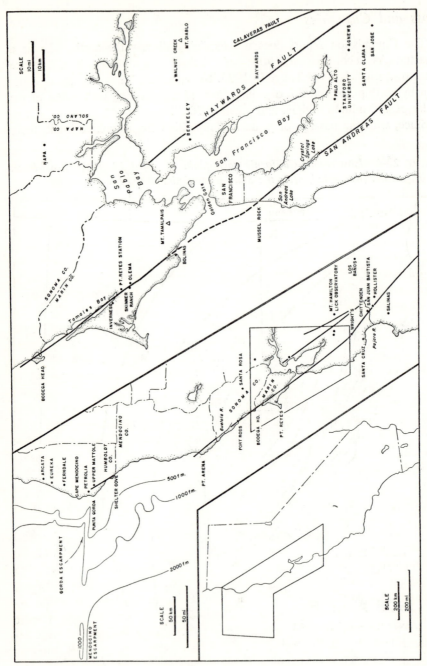

FIGURE 28-5 *Earthquake of 1906, location map.*

appears a tendency to reduce the casualty figures, it is worth remarking that no one will ever know how many persons perished in San Francisco. A number commonly quoted is 390; this is certainly very conservative and is based in part on guesswork. Contemporary writers all remark that many bodies were buried in haste without official record, and many structures were so com-- pletely consumed by fire that no trace could have been found of persons who were trapped there. In 1906 there was no complete census or directory of permanent residents, particularly in certain populous areas south of Market Street which were swept by fire; and the fate of the large number of transients can only be guessed.

Outside of San Francisco, casualty lists are more reliable. The Commission report notes 189 known deaths. This includes the outrageous number of 112 (which may not be a total) at the state insane asylum at Agnews, near San José. There were 61 identified dead at Santa Rosa, 9 at the Loma Prieta Mill not far from Santa Cruz. The report does not speak of fatalities at San José; contemporary newspaper accounts mention 19. San José and Santa Rosa were the two communities most spectacularly damaged apart from San Francisco. Destruction was particularly thorough at Santa Rosa, where the business section consisted wholly of brick and stone masonry structures which were almost totally demolished.

A chain of damaged towns stretched from Eureka on the north to Salinas and beyond on the south; the earthquake was perceptible well into Oregon and Nevada. Isoseismals (Figs. 28-4 and 28-5) show a great elongation; this is evidently connected with the extent of faulting, but it probably has a connection with geological structure as well.

The earthquake was regrettably destructive at Stanford University, near Palo Alto. It was noteworthy that the worst damage was to the outer quadrangle and to other structures erected after the death of Senator Stanford, probably under less careful inspection. The University of California at Berkeley suffered no serious damage.

Intensity at San Francisco

The distribution of intensity in the city of San Francisco is one of the best-known examples of the influences of ground and geological conditions on earthquake manifestations (Fig. 28-6). Every built-up block in the city was inspected by Mr. H. O. Wood, and the observed intensities mapped in detail. At its nearest point the San Andreas fault is well outside the city limits, and about 8 miles distant from the business center. Within the city intensity varied more obviously with ground than with distance from the fault. The highest intensity, which would rate IX or slightly over on the Modified Mercalli scale, applied to areas of fill or "made land" reclaimed from San Francisco Bay over many years. The most conspicuous of these areas, because it was heavily built up, was about the foot of Market Street. Several

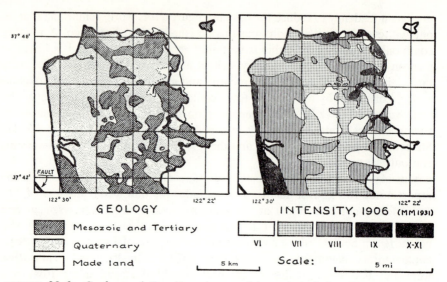

FIGURE 28-6   *Geology of San Francisco and intensity of the 1906 earthquake.*
[*After Lawson and Wood.*]

notable structures, including a new post office building, were just outside this area of fill; although badly damaged, they were repaired and restored to use. Locally throughout the city even small areas of fill due to grading showed higher intensity than their surroundings, while many residential areas, located on hills of firm rock (chiefly of the Mesozoic Franciscan formation), were relatively undamaged.

Fire

These earthquake effects were in part obscured by those of the great fire. Because of the effect of ground, the proportion of earthquake damage to fire damage was higher in the Market Street area than generally. Mr. Wood was of the opinion that the total earthquake damage was of the order of 20 per cent of the whole. Mr. J. R. Freeman concludes, in his volume repeatedly cited in this book, that the actual figure may have been much lower. Although this is a judgment based on considerable experience in estimating losses, it is probably a mistake. Mr. H. M. Engle, than whom no one is better informed in this connection, believes that the earthquake damage may well have ex-ceeded 20 per cent. Freeman's opportunity for investigation and study was limited, he was unaware of the existence of some documents then available, and he had no access to information developed after his death.

The destructiveness of the fire was itself a consequence of the earthquake, since displacement on the San Andreas fault put the pipe line carrying the main water supply out of action and the effects of shaking disabled the city's local water distribution. The fire was finally halted by dynamiting structures in its path, but only after 3 days of ineffectual efforts; success followed only on expert direction of the placing of charges.

"The Typical Earthquake"

Because so much is known about it, the event of 1906 often figures in general discussion as a typical earthquake. However, it must not be forgotten that there are large classes of earthquakes of which this one is not representative. Such are volcanic shocks and deep-focus earthquakes; but even the majority of shallow shocks appear to differ from that of 1906. Most of the larger earthquakes, and a corresponding fraction of the smaller ones, are associated with the great arcuate structures of the circum-Pacific and Alpide belts (Chapter 26). In those structures folding and large-scale thrusting appear to be the dominant processes, and block faulting is incidental; while in certain areas, such as California and part of New Zealand, block faulting is dominant. Moreover, the block faulting represented by the 1906 earthquake is primarily strike-slip, whereas, even in California and New Zealand, many earthquakes have shown at least a large proportion of dip-slip. In addition, the linear extent of faulting in this earthquake is the longest on record for a single event (a minimum of 190 miles or 300 kilometers, and possibly nearer 270 miles or 430 kilometers).

Because seismograms at stations all over the world were collected and published in the Commission report, the magnitude 8.3 assigned to this earthquake is exceptionally reliable for a shock of so early a date. (See Chapter 22.)

Among the most significant observations on this earthquake were the survey triangulations which provided the basis for Reid's theory of elastic rebound (Chapter 14). The generalization thus made needed to be supported by similar observations on other earthquakes; as pointed out in Chapter 14, such information has been forthcoming, notably in Japan. A feature not at first clearly formulated was the continuing of the displacement of the main earthquake during aftershocks, in the form of elastic afterworking. As shown in Chapter 6, Dr. Benioff's discussion has integrated these observations with Reid's conception.

Strike-Slip

Because of its great extent and general accessibility, the San Andreas Rift provides many of the type examples of topographic features associated with strike-slip faulting; such features have been discussed very briefly in Chapter 13. In 1906 actual strike-slip was observed along 190 miles of the Rift from Point Arena southeastward. It is probable that in 1857 there was a nearly equal extent of faulting, presumably terminating at San Gorgonio Pass; Allen's findings cast further doubt on the questionable references to displacements on that occasion extending farther east. That part of the Rift which was probably unaffected in either 1857 or 1906 has features clearly indicating large displacements in the geologically very recent past.

Largest Displacements. In 1906 the most significant feature of the fault trace was consistent right-hand strike-slip. Because the region was generally populated, this could be established almost everywhere by reference to such artificial markers as roads and fences. The largest displacements were in the area most visited and reported on, in Marin County north of the Golden Gate. The offset had a general maximum of about 15½ feet; however, on the soft alluvial ground approaching Tomales Bay it attained 21 feet, as observed at the much-photographed locality on the road west of Point Reyes Station. This increase on soft ground was attributed to lurching.

The Skinner Ranch. One of the best localities for observing the 15-foot displacement was that referred to in the report as the Skinner ranch. (It appears that Skinner was the name of tenants, not of the property owner.) See Figure 28-7.

Gilbert states that a fence south of the barn was offset 15.5 feet, and continues:

The barn, beneath which the fault trace passed, remained attached to the foundation on the southwest side, but was broken from it on the northwest side and dragged 16 feet. A path in the garden, originally opposite steps leading to the porch, was offset 15 feet. A row of raspberry bushes in the garden was offset 14.5 feet.

Jordan writes:

The eucalyptus trees in front of the dairy moved on to a position opposite the barn, and one detached from the others and to the westward of the crack was left near the head of the line instead of at its foot . . . Under each of the east windows of the barn stood a pile of manure. Each pile is intact, sixteen and one half feet south of the window to which it belongs.

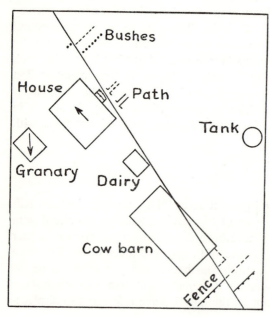

FIGURE 28-7 *Displacements at the "Skinner" ranch, 1906. [After Gilbert.]*

Not far from this was the Shafter ranch. Here the fault produced large cracks, which were seen to open and close by the alarmed men who had gone out for the morning milking. A cow fell into one of these cracks; she was injured and could not be extricated, so that the ranchers filled the impromptu grave, leaving only the tail visible. This story has sometimes been told in exaggerated form, which leads to questioning; but the general facts were well attested by the Shafter family, who were of reliable character, and the body of the cow was seen in place by many witnesses, among them G. K. Gilbert. This is a far cry from the legendary tales of earthquakes opening vast chasms into which persons, houses (and whole cities if the audience does not seem sufficiently impressed) are swallowed. The reader may compare the account of the Fukui earthquake (Chapter 30).

The Fault Trace Northward

Northward from this area the Rift is flooded by the sea, forming the long and narrow inlet of Tomales Bay. That fault displacement continued here was indicated by, among other things, the distortion of a pier at Inverness. Still farther north the trace cut across the sandy neck of Bodega Head; then after another short marine segment the Rift and the fault trace reach the coast of Sonoma County near Fort Ross. From here northwestward the features were very clear; the Rift partly follows the nearly straight valley of the Gualala River. In this area offsets of 15 to 16 feet were measured at fences and roads. Part of the trace here runs through the coastal forest of California redwood (*Sequoia sempervirens*). Some of the larger trees were split; and so many large branches were broken off that the resulting covering of the ground by fallen timber made it difficult to follow the trace.

The Humboldt County Trace. Northwestward the Rift goes out to sea at Point Arena; beyond that point the course of the San Andreas fault, and what happened there in 1906, are partly conjectural matters. However, in 1906 a fault trace appeared near the coast of Humboldt County, trending roughly parallel to the main trace but far out of line to the northeast. This was in thinly settled ranch country north of Shelter Cove. The evidence of faulting was positive; that of strike-slip was inferential. To quote the report:

While it has been found impracticable to demonstrate by actual measurement the existence of a horizontal displacement along any of these new fissures—in the absence of fences or other objects of sufficiently defined outline—yet it has seemed warranted to regard them as true fault or shear fractures, to be classed in the same category with those found farther south, merely on the strength of their superficial resemblance.

The effects of a horizontal shear on thick grass sod in open country, as observed in a number of localities along the zone of faulting in Sonoma and Mendocino Counties, are as follows: On fairly level ground, where conditions are simplest and no vertical movement is evident, the sod is torn and broken into irregular flakes, twisted out of place and often thrust up against or over each other. The surface is thus disturbed over a narrow belt, whose width apparently varies with the magnitude of the displacement. Along the main fault, where the throw amounts to 10 feet or more, a width of 5 or 6 feet is not uncommon; on the secondary fractures, where the throw does not exceed a foot, the belt is generally only a foot wide. Whatever the width of the belt, the sod within it, as well as the unconsolidated material underneath, appears loosened up and not compact. It consequently takes up more space than before it was disturbed, and the surface of the belt is therefore slightly raised above the level of the ground, from an inch to a foot or more, according to the magnitude of the disturbance. Within such a belt there is seldom, if ever, a well-defined, continuous, longitudinal crack, the toughness of the sod precluding a clean shear fracture. Rather, there is a marked predominance of diagonal fractures resulting from tensile stress.

This is an example of the sort of close examination necessary to interpret trace phenomena in open country. Relatively few field investigations come up to this standard; undoubtedly we have lost much valuable evidence of strike-slip which might have been observed. Even the careful account quoted fails to name the expected orientation of diagonal tension cracks, although elsewhere in the report orientation indicating right-hand strike-slip is commented on and shown in photographs. There is no statement as to such cracks or their orientation applying specifically to the trace in Humboldt County.

A Separate Fault?   Presence of an active fault in this Humboldt County area has been confirmed in later years by the occurrence of many local earthquakes, some of them damaging at Upper Mattole and Petrolia, which are near the northwestern projection of the Shelter Cove trace. Relation of this fault to the San Andreas Rift is still unclear. For the Commission report Lawson decided provisionally to accept a direct connection; his maps show the fault line with a submarine segment curving from Point Arena to Shelter Cove. At present (1956) most students incline to an opposite conclusion; nothing significantly new has been learned about the events of 1906, but seismographs at Ferndale and Arcata in Humboldt County have improved the location of epicenters. Few of these are off the coast between Point Arena and Shelter Cove; most of the earthquakes are of relatively small magnitude, and hence not a reliable index to the location of major faults. A possible exception is a shock of magnitude 5½ on December 20, 1940 (23:40:54 G.C.T.), near 39¾° N, 124½° W, a point almost in line with the San Andreas Rift as projected northwest from Point Arena.

A little farther north there is an important and sudden change. Point Arena is just south of 39° N latitude, Shelter Cove just north of 40°. At about 40.4° N, in the latitude of Point Gorda, soundings show a remarkable north-facing submarine escarpment running nearly due west. This feature is termed the Gorda escarpment. Westward it aligns with the Mendocino Escarpment, which has been traced into the Pacific for over 1400 miles; the scarp faces southward and is from 3300 to 10,500 feet high. For further discussion refer to Chapter 27.

Since the fault at Shelter Cove lies eastward, it points to branching of the San Andreas fault, or a complication of the fault system to which it belongs, beginning farther south. There is a major branch of the required character, the Haywards fault, on which the earthquake of 1868 originated. It has not been traced northward; but the Coast Range structures, with the same general trend and with some known faults, continue northwestward east of the San Andreas fault. The apparent high intensity at Santa Rosa in 1906 might be due to secondary action on some fault nearby; but the alluvial character of the ground and the poor quality of local masonry are quite sufficient to explain the destruction. A similar question arose with respect to an area of high intensity including Los Baños, in the western San Joaquin Valley; here the deep alluviated area, having a relatively high permanent water table is often flooded.

Related Shocks

Acceptance of the Shelter Cove fault segment as part of a structure lying east of the San Andreas fault is made easier by the occurrence of notable earthquakes not much farther north, with epicenters on the coast and even somewhat inland. (The shock which was damaging at Eureka on December 21, 1954, had an epicenter definitely on land.) The immediate result is to make the earthquake of 1906 more complex in that more than one fault was active. It is interesting that on the afternoon of the same day, April 18, 1906, there was a strong shock in Imperial Valley, which was then rather thinly settled; nevertheless, there was damage at Brawley, and a water tank was thrown down at the railroad station of Cocopah (Mexico).

Aftershocks of the 1906 earthquake were numerous; there is no means of placing them individually, although a rather large one on April 23 was strong in Humboldt County and less noticed southward, while one on May 17 was felt from Napa to San Luis Obispo.

The Fault Trace Southward

Southward from Tomales Bay the Rift and the 1906 trace cross land as far as Bolinas; then the fault is submarine, passing west of the Golden Gate and returning to land at Mussel Rock southwest of San Francisco. Here a great slump, which had been in motion for years, slid out and completely exposed its fracture surface.† Southeastward the trace enters the cultivated hill country

---

† This slide is pictured on Plate 12 of the Commission report, where the captions of the two photographs A and B should be interchanged.

of the San Francisco peninsula, where offsets measured on fences and other artificial features decreased gradually from about 15 feet to about 8 feet. Traversing the long valleys used for the San Andreas and Crystal Springs reservoirs, the faulting did not destroy the dams. The surface break passed at a distance of about 200 yards from the concrete structure of the Crystal Springs dam. Years later, when this reservoir was at a low level, an old small earth dam within it was found neatly sheared and offset along the fault line. The pipeline which connected the reservoirs with San Francisco crossed the fault repeatedly and was put completely out of service; whole sections were crushed, and others were torn apart. Southward the trace passed within 6 miles of Stanford University (the severe damage there has already been mentioned) and then through the Santa Cruz Mountains. A frame house near Wright's Station was torn apart along a branch fracture, but the two parts, though distorted, remained standing. At Chittenden the fault crosses the Pajaro River, and the movement was shown by disturbance of the concrete bridge piers. The highway along the river here has shown extensive cracking in many earthquakes since; this is due to slumping and settling in the unconsolidated material of the fault zone and does not indicate action on the San Andreas fault; some of the earthquakes producing such effects had their epicenters definitely on other faults. Southward from this locality the observable displacement gradually decreased, and the trace could not be followed farther than a point about a mile east of San Juan Bautista. The commonly quoted extent of 190 miles is between this place and Point Arena.

Vertical Displacements

Vertical displacements along the trace were, in general, relatively insignificant, never more than about 3 feet. There were reversals in the direction in which the scarp faced; it is not sure how far these represent true scissoring of tectonic significance. They were often connected with the passage between what Gilbert termed the ridge and trench phases of the trace. The ridge phase corresponds with the description already quoted at length; it does not consist of a high ridge of the mole-track type, but of a rather minor swelling of the surface. This may be connected with the generally grassy nature of the country; it certainly implies some difference in soil from Imperial Valley, where huge mole tracks appeared in 1940.

Minor Effects

Minor effects are discussed systematically in the report. The more important of these, including Lawson's discussion of slides and earthquake fountains, have been noted in Chapters 9 and 10.

Rebuilding

Rebuilt San Francisco has an emergency water system, capable of using sea water for fire control, and adequate to secure the city against any fire disaster like that of 1906. Improved building regulations call for some provision against lateral forces, so that the city is gradually being made safer against earthquakes, although few of the larger buildings have been designed to be earthquake-resistant.

THE IMPERIAL VALLEY EARTHQUAKE OF 1940

The most serious effects of this earthquake, and those of greatest geological interest, developed in the Imperial Valley (Fig. 28-8), extending across the international boundary from California into northern Baja California (then a territory, now a state of the Mexican Republic), although there was also

FIGURE 28–8 *Imperial Valley earthquake, 1940. Map showing location of fault, instrumental epicenter, etc.*

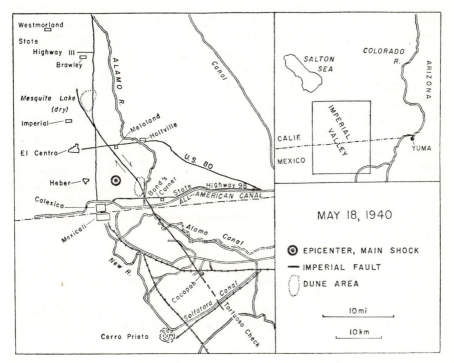

much damage in the adjacent Yuma Valley of southwestern Arizona. The region includes the delta of the Colorado River, and the depression north of it; most of the latter is below sea level and was dry until 1905, when irrigation projects brought about a flood that produced the Salton Sea (now a permanent body of water, since loss by evaporation is replaced by seepage).

Casualties and Damage

In the earthquakes of May 18, 1940, 7 persons were killed in the collapse of weak structures, 1 was burned to death, and 1 died a few days later of injuries. Property loss was estimated at 5 to 6 million dollars; only part of this was due to damage to buildings. There was loss of crops due to interruption of the water service, since the entire irrigation system, especially on the west side of Imperial Valley, was put out of order by breaks which necessitated costly repairs and reconstruction.

There was serious damage to structures in all the towns of the central and southern Imperial Valley: Brawley, Imperial, Holtville, El Centro, Heber, Calexico, and Mexicali (the two last are separated only by the international boundary). Most of the damage was confined to relatively old structures of weak masonry (masonry C and D in the notation of Chapter 11); however, two steel tank towers fell, a reinforced-concrete hotel in Brawley was badly damaged, and a concrete-framed warehouse in El Centro was seriously cracked. This earthquake provided the first good test of the effectiveness of the Field Act, enacted by the California legislature in 1933, which set improved standards for public construction. Schools in Imperial Valley built before 1933 were nearly all more or less damaged; but schools in the same communities, erected after 1933 and consequently under the Field Act provisions, showed no significant damage.

Large tanks were in use for storage by the city water systems. Those at Imperial and Holtville collapsed; at the latter place the distortion of the collapsing tank had sufficient force to shoot its rivets through the walls of the adjacent pumphouse like so many bullets. The tank at Brawley was on an elevated structure; after previous earthquakes it had been rebraced to be earthquake-resistant, following recommendations of the Board of Fire Underwriters of the Pacific.† This tank was uninjured in 1940.

The Imperial Fault

This earthquake was distinguished by the appearance of right-hand strike-slip effects, comparable with those of the 1906 earthquake but traced only for about 40 miles (although it may have extended farther south), along a previously unknown fault designated the Imperial fault by Buwalda. This

---

† These rules, still in force, were formulated by H. M. Engle and J. E. Shield, who specified and approved all details of the rebracing at Brawley.

fault is certainly part of the San Andreas fault system, in the same sense as the San Jacinto and Elsinore faults; but it has no traceable connection on the ground with the San Andreas and Banning faults in San Gorgonio Pass.

Magnitude

The published magnitude for this earthquake is 6.7; but recent revision, allowing for the latest data on magnitude determination, gives 7.1 as a more acceptable value (on the same magnitude system as that on which the figure 6.7 was originally reached). This value corresponds more nearly with experience as to magnitudes of other earthquakes developing comparable traces.

Epicenters and Northern Fault Trace

The instrumentally located epicenter is not very accurate, since all the nearer stations are to one side of it; it is roughly east of El Centro, toward Holtville, and on the surface trace within the limits of accuracy. There was an appreciable difference in character between the trace southeastward and northwestward from the vicinity indicated. Southeastward the trace was very nearly straight, trending about 30° east of south with displacements and trace phenomena increasing toward the international boundary. Northwestward from the indicated epicenter the strike-slip decreased, and the trace was no longer straight. The railroad between El Centro and Holtville showed displacement of about 18 inches, but where the trace crossed a road directly east of Imperial the slip was less than a foot. The surface break northwest of this was found curving around the base of an old erosional scarp bounding the dry bed of Mesquite Lake. Approaching the railroad line running from Imperial to Brawley, the trace frayed out into several branches with displacements of a few inches and finally disappeared.

The main shock was at 8:36 P.M., but the principal visible damage at Brawley took place during an aftershock at 9:53 P.M., which at that city was generally considered much stronger than the first shock. On the other hand, at Calexico, El Centro and Holtville, the shock at 9:53, although it added to damage, was generally considered much less strong than that at 8:36. It is a natural suggestion that the main earthquake fracture began near the instrumental epicenter at 8:36 and continued southeastward, while the small northwestward trace developed with the aftershock at 9:53. Because of the darkness and general confusion there were no observations which might confirm this directly. The curvature of the small trace along the older erosional scarp is explainable as the effect of topography on the surface expression of a comparatively small displacement.

There were no true foreshocks. On May 17 an earthquake of magnitude 5.4, originating north of Imperial Valley in the region of the Little San Bernardino Mountains, was felt sharply over the valley and as far as Los

Angeles. It had rather a long series of aftershocks which continued to record among those of the May 18 earthquake, from which they were easily distinguishable.

Unfortunately, the epicenter of the aftershock at 9:53 P.M. on May 18 cannot be located instrumentally with the desirable precision, nor can any other of the immediate aftershocks. Nearly all the large aftershocks were later members of bursts of successive earthquakes, of which the first were too small to be clearly recorded, while the later ones were large enough to confuse the recording of the following largest shocks in each group.

The Fault Trace Southward

The trace offset the painted center strip of the highway which runs between El Centro and Holtville just south of the railroad; one block of the concrete paving was rotated by the fault displacement (Fig. 28-9). Southeastward from here the trace ran mostly through a settled area divided by north-south and east-west roads (on quarter-section land survey lines), and offsets could be measured at intersections of the trace with these. Strike-slip increased gradually to 3 and 4 feet; mole-track features (Fig. 28-12A) became more and more pronounced. Before reaching the next main highway, the trace passed through a region of sand dunes; here it consisted largely of trenches traceable across the nearly level sand in spaces between dunes. At one point a small sand ridge had slumped where it projected across the trace, and was converted from a single to a double summit form.

FIGURE 28-9   *Imperial Valley earthquake, 1940. Strike-slip offsetting highway, indicated by painted center strip.* [*Photo by J. P. Buwalda.*]

Near Alamo River. One of the most visited localities was on the highway running eastward from Calexico toward Yuma, a short distance west of the Alamo River. The roadway was cut by a 4-foot scarp facing east. North-south fences were offset about 4 feet (Fig. 28-12B), east-west fences about 7 feet; the trace trended about N 33° W. The effect of drag in the fault zone was very clear, the posts nearer the trace being well out of line compared to those at a distance. North-south fence wires were slacked, east-west wires were tightened or snapped. Immediately south of the highway the trace ran through a citrus grove; the offset of the rows of trees showed the pattern of displacement very clearly, especially as seen from the air (Fig. 28-10).

The largest displacements measured in this earthquake were along small irrigation ditches and minor roads between this locality and the Mexican boundary; the greatest slip measured by the writer was 19 feet (right-hand strike-slip along the trace).

All-American Canal. At the international boundary the All-American Canal was approaching completion, but no water had yet been admitted. The earth banks were badly shattered in many places; there was large stepwise

FIGURE 28-10 *Imperial Valley earthquake, 1940. Strike-slip offsetting trees in a citrus grove. [U. S. Army photo, courtesy Mr. E. Marliave.]*

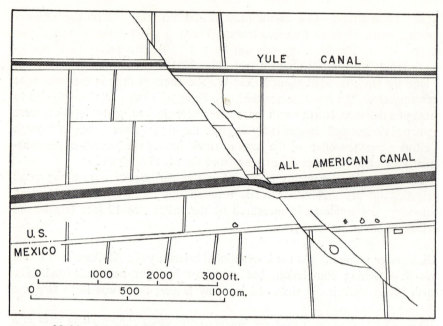

FIGURE 28-11 *Imperial Valley earthquake, 1940. Sketch showing fault traces near the All-American Canal.*

lurching, and slumping down into the excavations. The embankments were broad, with sufficient roadway for trucks; they were intersected and offset by the fault trace; the north-south offset on the northern embankment amounted to 14 feet 10 inches.

FIGURE 28-12 *Imperial Valley earthquake, 1940. Collapse of garage at Meloland. [Photo by J. P. Buwalda.]*

FIGURE 28-12A *Imperial Valley earthquake, 1940. Fault trace showing mole track appearance.* [*Photo by J. P. Buwalda.*]

There was a remarkable doubling of the trace on both sides of the canal excavation. On the north side, the more easterly of the two traces, with conspicuous mole tracks and offsets, died out rapidly within about half a mile; the western trace more conspicuously offset the embankment, extending northwestward continuously with the main trace already described. On the south side the western trace curved eastward and died out rapidly, while the eastern trace continued on into Mexico. (See Fig. 28-11.) The entire disturbance was so symmetrical with respect to the canal excavation that it is hard to dismiss the location as coincidence; it is at least plausible that the presence of a deep east-west trench in the alluvium here modified the surface expression of faulting.

Baja California. Canals providing the principal water supply for the west side of Imperial Valley were crossed by the fault south of the Mexican boundary. The most northerly was the Alamo Canal; this was offset, and the water created a local flood until it could be shut off and repairs started. Farther west, a flume in which the canal crossed a nearly dry stream course was completely collapsed by the shaking; this also required repairs before water service could be restored.

FIGURE 28-12B *Imperial Valley earthquake, 1940. Offset fence west of Bond's Corner, looking north. [Photo by J. P. Buwalda.]*

Twenty miles southeast the trace crossed a main canal, the Solfatara Canal, near Tortuoso Check. Strike-slip here was not over 2 feet, but there was considerable change of grade. Apparently the block west of the fault was tipped eastward against the flow of water from that direction; the resulting sudden rise in water ruptured the north embankment west of Tortuoso; the rush of water offset the bank over 10 feet, and this displacement was mistaken for fault motion by a number of observers. South of this canal observed displacement decreased to less than a foot within half a mile. However, there were reports of faulting extending still farther south; if this is true, there must have been another increase in displacement, or perhaps an echelon offset of the trace. Dr. C. R. Allen has found, on air photographs taken in 1941, a series of large north-south tension cracks for a few miles southeast from Tortuoso in the trend of the fault.

Between the two canals the trace crossed the railroad just west of Cocopah station. Here the east-west rails were offset about 7 feet by the right-hand strike-slip. No curvature of the rails other than that due to drag was detectable by sighting along the track for about a half mile, so that elastic rebound was not large enough to be seen in this way. South of the tracks the trace offset a small canal and passed through an adobe house, destroying it. Cocopah was the locality where a water tank was reported thrown down in the earthquake on the afternoon of April 18, 1906 (about 11 hours after the San Francisco earthquake; chimneys fell at Brawley).

Survey Work

Block displacement associated with right-hand strike-slip was partly confirmed by retriangulation by the U. S. Coast and Geodetic Survey. Among the monuments resurveyed were those used to mark the international boundary. However, as none of the available monuments was close to the actual trace, the shifts shown by survey were small compared to those near the fault.

The effects of this shock on ground water in Yuma Valley have been referred to in brief in Chapter 9.

DESERT HOT SPRINGS EARTHQUAKES—1948

On December 4, 1948, an earthquake of magnitude 6.5 originated near Desert Hot Springs, a newly developed resort community almost directly on the Mission Creek fault (the name Seven Palms appears here on some older maps). This is in the critical part of the San Andreas fault system east of San Gorgonio Pass. The occurrence bears on problems involved in field investigation by Dr. C. R. Allen (Chapter 27); he has carried the seismological

FIGURE 28-13 *Desert Hot Springs earthquakes, 1948–1953.* [*C. R. Allen.*]

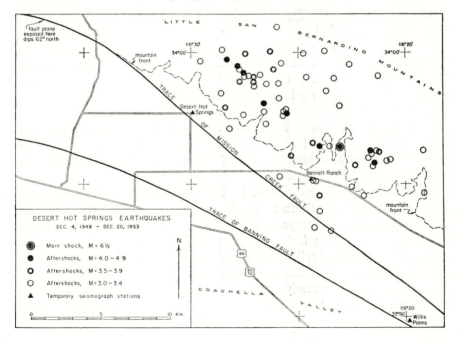

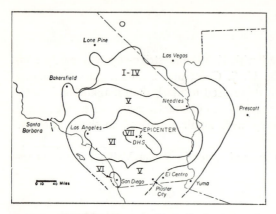

FIGURE 28-13A *Desert Hot Springs earthquake, 1948; isoseismals.* [*U. S. Coast and Geodetic Survey.*]

study to completion. Epicenters of the numerous aftershocks (Fig. 28-13) are concentrated on a line about 18 kilometers long, parallel to, but 5 kilometers north of, the surface expression of the fault. This suggests a northerly dip of about 73°, which is consistent with dips exposed a few miles west of Desert Hot Springs. The epicenter (Fig. 28-13A) of the main shock is near the southeast end of the active line, and aftershock activity is largely concentrated toward the two ends, as noticed in many other earthquake series. The investigation made use of seismograms of portable and other temporary stations operated in the meizoseismal area. Some of these records show the anomalous small apparent values of the *S-P* interval observed at short distances on other occasions. (Fig. 18-5.)

SAN JACINTO EARTHQUAKES— 1899 AND 1918

The San Jacinto fault, unlike the San Andreas fault, shows almost continuous minor seismicity from Cajon Pass nearly to the Mexican boundary. Moderately large earthquakes are not infrequent.

The earthquake of December 25, 1899, was destructive at San Jacinto, and even heavier at Soboba Hot Springs. It was felt over much of southern California, as well as in Nevada and Arizona, and must have had a magnitude comparable with that of 1918 in the same region. The most interesting account was published by Daneš, in a journal where most students of California earthquakes have overlooked it. The maximum intensity was reached along the San Jacinto fault near its highest level in the mountains; Daneš describes interruption of footpaths and development of large slumps. There may possibly have been fault trace effects, but interpretation is uncertain.

The earthquake of April 21, 1918, had an epicenter nearer San Jacinto and Hemet; both towns suffered serious damage. The meizoseismal area was closely examined, particularly near the fault, but no trace effects were seen. The magnitude was 6.8.

LONG BEACH EARTHQUAKE—1933

The destructive shock on March 10, 1933† (Fig. 28-14) was associated with the Inglewood fault, one of the more or less parallel faults representing the San Andreas system south of the Transverse Ranges (Chapter 27). Some principal facts have been noted in other chapters; preceding shocks, including one true foreshock, were cited in Chapter 6.

Epicenters

The instrumentally determined epicenter was not far from Huntington Beach, about 10 miles southeast of Long Beach; epicenters for immediately following aftershocks were scattered along the Inglewood fault zone into the city limits of Long Beach, near the Signal Hill oil field. After activity had nearly subsided, on October 2, 1933, a shock of magnitude 5.2 originated near this Signal Hill terminus, followed by many aftershocks in that vicinity. For several years, minor aftershocks continued to occur most frequently near the two ends of the disturbed fault segment.

FIGURE 28-14 *Long Beach earthquake, March 10, 1933, showing area of damage, and selected epicenters, 1933–1944.*

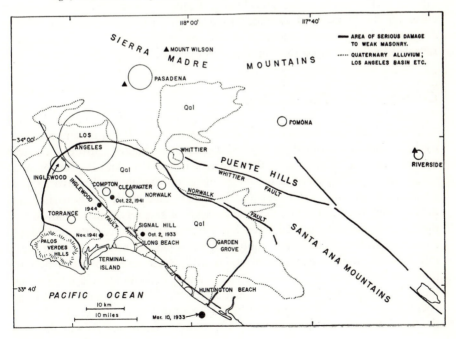

---

† Local date. March 11, G.C.T.

Epicenter determinations are fairly reliable, especially in terms of relative position for different shocks of the series. The active area was surrounded on three sides by stations of the southern California seismological network, and data were supplemented by those of a portable instrument operated successively near Laguna Beach, Huntington Beach, Santa Monica, and elsewhere, including one short run on Santa Catalina Island. Since all evidence indicates that this is an area of structure differing notably from that farther inland, future accumulation of local data on seismic wave propagation may lead to significant revision of the conclusions drawn in 1933.

Casualties and Damage

Though only a moderate earthquake, this one ranks as a major disaster. Its magnitude (6.3) was closely the same as that of the Santa Barbara earthquake of 1925, in which losses of life and property were much smaller; but in 1933 the meizoseismal area included heavily settled parts of the alluviated Los Angeles Basin, in the city of Los Angeles and its surrounding area as well as the secondary urban center of Long Beach.

Loss of life is commonly stated as 120, and property damage at 50 million dollars. Somewhat smaller figures are sometimes given for fatalities; but there is a little indefiniteness, since deaths as late as 1935 were attributed to injuries in this earthquake. The damage figure is conservative and does not cover all the "concealed damage" to business structures, especially in Los Angeles, where repairs undertaken because of apparently minor cracking revealed badly damaged interior walls and partitions.

Consequences—The Field Act

This calamity had a number of good consequences. It put an end to efforts by incompletely informed or otherwise misguided interests to deny or hush up the existence of serious earthquake risk in the Los Angeles metropolitan area. The appalling damage to school buildings, which would have resulted in great loss of life had the earthquake taken place a few hours earlier (the actual time was 5:54 P.M.), led to passage of the regulatory Field Act by the state legislature (see Chapter 8). Several excellent reports on damage and other circumstances were published which led to improvements in building codes, as well as adoption of codes with earthquake provisions by communities hitherto negligent. The strong-motion recording program, just initiated by the U. S. Coast and Geodetic Survey, was given impetus and support. Seismological work in California generally was furthered, although the good effects were hampered by the serious economic crisis at the time.

Transit Time

Another scientific contribution of the records of the Long Beach earthquake resulted from the unusually reliable epicenter and origin time available for a shock of this magnitude. The availability of times of arrival of *P* and other waves for stations as distant as Europe and even farther contributed significantly to the refinement of time-distance tables.

Later Shocks

The Inglewood fault has not been quiescent since 1933. Shocks in 1941 and 1944 were strong enough to cause minor local damage in the area, and apparently also triggered damaging subsurface displacements in the Dominguez and Rosecrans oil fields (Chapter 12). The earthquake of November 14, 1941, probably did not originate directly on the Inglewood zone; but, although only of magnitude 5.4, it was accompanied by serious damage to the business center of the town of Torrance. Probably nearly all of the loss was due to failure of walls and structures imperfectly repaired after the earthquake of 1933.

OWENS VALLEY EARTHQUAKE— MARCH 26, 1872

This and the earthquakes discussed in the immediately following sections belong to the Basin and Range province. That of 1872 (Fig. 28-15) is undoubtedly the largest known to have occurred there. It has generally been considered the largest known in the entire California–Nevada region, thus placing in magnitude above those of 1857 and 1906 on the San Andreas fault. Such judgment rests on the violence of effects over the large meizoseismal area, as well as perceptibility extending to great distances.

Documents

Our sources are limited. One important document is a rather sketchy report prepared by Professor J. D. Whitney,† then head of the now defunct California State Geological Survey. This report appeared in two parts in the *Overland Monthly,* a not widely distributed journal published at San Francisco; only the first part is of any present value, the second consisting of generalities and speculation. Whitney's meager data were reproduced in a

---

† Whose name is attached to Mount Whitney, highest summit of the Sierra Nevada and highest point in the United States, excluding Alaska.

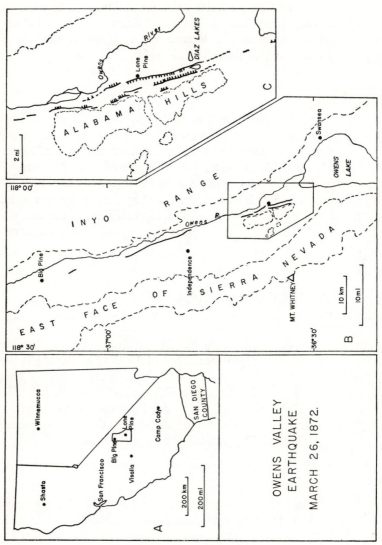

FIGURE 28-15   Owens Valley earthquake, March 26, 1872. A, location map; B, features of Owens Valley; C, faulting of 1872 as mapped by Johnson. [After Hobbs.]

paper by W. H. Hobbs, who added much important new material, including partial maps of the fault traces prepared in 1907 by W. D. Johnson of the U. S. Geological Survey from evidence then remaining on the ground. Holden's earthquake catalogue contains some information not found elsewhere; Townley, in preparing his revised catalogue, supplemented the material and discussed it critically. Knopf's geological reconnaissance of 1918 gives some further field data; the latest geological map accompanies *Bulletin 170* of the California Division of Mines.

Casualties, Damage, Intensity

The most disastrous effects of the earthquake were at the town of Lone Pine. Of a population between 250 and 300, 23 were killed and about 60 others were injured, of whom 4 subsequently died. Of 59 houses, 52 were destroyed; most of these were of adobe.

Considering the condition of settlement of the region in 1872, information about the extent of perceptibility is understandably incomplete. Whitney states that the shock was perceptible in California from Shasta on the north to San Diego County on the south; Holden quotes press reports that it was felt throughout the Sacramento and San Joaquin valleys. Townley points out that the best evidence is that the shaking was very slight or not felt at San Francisco. Perceptibility extended over most of Nevada, to Winnemucca at least, and probably into Utah and Arizona.

Holden states that at Camp Cady the shaking was strong enough to move heavy wagons, and Whitney says that mules were thrown off their feet (an odd observation, considering the hour, 2:30 A.M.). This locality was over 140 miles southeast of Lone Pine.†

Buildings were destroyed at Swansea, not far from Lone Pine; varying amounts of damage were reported from the few other centers of population in Owens Valley. Elsewhere, the greatest reported damage was at Visalia, at that time the only sizable town on the east margin of the San Joaquin Valley.

Throughout the Sierra Nevada large slides and rockfalls occurred. John Muir, in his book on Yosemite Valley, has left a graphic eyewitness account of the fall of one of the minor pinnacles there.

A large wave rose in Owens Lake. This may have been a seiche, although Hobbs and others, with reason, attribute it to tilting accompanying the vertical displacements next to be described.

Scarps

The faulting mapped by Johnson was chiefly in the vicinity of Lone Pine, near the east base of the Alabama Hills, an obviously tilted block internal

---

† Camp Cady was an Army post on the south bank of the Mojave River about 34.9° N, 116.6° W, not far from the epicenter of the Manix earthquake of 1947.

to the Owens Valley (which is a graben between the high scarps of the Sierra Nevada and the Inyo Range, rising to elevations over 14,000 feet above sea level to west and east). The features chiefly described are vertical scarps; the highest measured displacement on any of these was 23 feet. Johnson carefully studied the relation of these scarps of 1872 to older scarps of the same character; the new scarps generally appeared in extension of or in line with the old ones, while in the instances where the 1872 displacement followed an old scarp the additional new movement was relatively small. Nearly everywhere there are scarps facing in opposite directions, roughly east and west, with a depressed area between. The 23-foot east-facing scarp was thus fronted by a west-facing scarp over 10 feet high, leaving a net displacement not over 13 feet, which is more nearly representative of the observations.

It is open to question how far these scarps represent actual tectonic displacement. They may represent the local dropping of blocks in a fracture zone, due to shaking or to local redistribution of strains following the earthquake, as discussed in Chapter 7. Another consideration which gives pause is the fact that some of these scarps, near Diaz Lake for example, are conspicuous in alluvial material but die out on approaching basement rock. This adds importance to the evidence of scissoring; Hobbs emphasizes and figures reversal of throw on a trace about a mile south and a little east of Diaz Lake.

Strike-Slip

Fortunately for our doubts about the reality of faulting as such, there were many observations of strike-slip, a few of which are described clearly. Whitney, detailing one such instance, remarks:

> The same thing was noticed by us at Lone Pine and Big Pine, with regard to fences and ditches, the horizontal distance through which the ground had been moved varying from three to twelve feet. These are local phenomena, however, and not to be taken as indicative of a general motion of the valley in any fixed direction.

With this careless generalization we have lost the details of what would have been invaluable data on the mechanism of the earthquake.

The offset detailed by Whitney, taken from a diagram by Captain Scoones, is on an east-west road about 3 miles east of Independence, the west side having moved 18 feet relatively to the south; this is a left-hand offset. A similar left-hand offset of about 10 feet was described by Captain Keeler to Holden, as affecting east-west fences about 1½ miles south of Lone Pine. Right-hand offset is described by C. Mulholland as follows:

A half mile north from Lone Pine a row of tall trees extends westward at a right angle to the wagon road. About 100 yards from the road there is an offset in this row of trees. Beyond that point where the straight line is broken, the trees stand about 16 feet farther north than in the line from the same point back to the road.

Hobbs refers to what is probably this same offset, but without specifying its sense; he also reproduces a photograph taken by Johnson showing right-hand offset of 9 feet in a row of trees at an unspecified location.

Since both right-hand and left-hand offsets of the order of 15 feet occurred along faults of the same general trend, mechanical complexity is indicated beyond any possible analysis from the fragmentary data. Evidence of the offset east of Independence still remains; it is on a fault shown on current geological maps, extending northwest toward Big Pine. It is to this that the letter from Guy C. Earl, quoted on page 229 of Freeman's book, chiefly refers.

Whitney repeatedly states that features seen near Big Pine were comparable with those near Lone Pine; Hobbs seems sometimes to have confused the two names. Near Big Pine is a Pleistocene cinder cone cut by young faulting; at the base of a large scarp there is a smaller one which may date from 1872.

NEVADA AND UTAH EARTHQUAKES— 1915, 1932, 1934

Pleasant Valley—1915

The Nevada earthquake of 1915 is often overlooked, although it was described in detail by J C. Jones shortly after, and the fault scarps were restudied by B. M. Page in 1935. The meizoseismal area was in Pleasant Valley, some 40 miles south and a little east of Winnemucca; it included a small mining town, Kennedy, and some ranches.

Foreshock Activity. On October 2, 1915, at 3:40 P.M. a sharp earthquake was felt through northern Nevada, including Reno (125 miles from Pleasant Valley); at Kennedy there was no damage, but the shock was hard enough to alarm the inhabitants. At 5:49 P.M. there was a large shock; at Kennedy it was difficult to stand; at Reno clocks were stopped, and perceptibility extended for at least 200 miles from Pleasant Valley, into California. An observer at Kennedy writes (as reported by Jones):

From this disturbance on, it was an incessant continued trembling, the earth never appearing quiet. At about nine o'clock we retired for the night . . . one could shut his eyes and imagine he was occupying a berth in a Pullman car, accompanied with creakings and rattling of windows, to be abruptly awakened by outbreaks at intervals of twenty to thirty minutes, lasting from five to ten seconds. At 10:54 things had quieted, or perhaps we were unconscious in sleep, when without the slightest warning a great roar and rumbling was heard and we were thrown violently out of bed . . .

This was the major earthquake (magnitude 7.6, at 06:52.8 G.C.T.). At Kennedy some adobe houses were destroyed, and there was damage, soon repaired, to the mine and its mill. Adobe houses elsewhere in Pleasant Valley were also seriously damaged. Frame houses were racked and moved on their foundations. There was less damage to structures close to the actual fault trace than to others at some distance in the valley.

At Winnemucca many brick and adobe structures were damaged, and many chimneys were broken off at the roof lines. The earthquake was felt over a very wide area; Jones states that it was perceptible from Baker, Oregon, to San Diego, and from the Pacific coast to Salt Lake City. Similar general statements appear in various catalogues. As usual, there probably was no distinction between actual felt shaking and marginal effects like swaying chandeliers. The bulletin of the Berkeley station, for example, notes that "one newspaper stated the shock was felt in San Francisco." Jones draws isoseismals only for Nevada (see Fig. 28-16).

Faulting.   The fault scarp, from 5 to 15 feet in height, appeared on the east side of the south part of Pleasant Valley, opposite Kennedy (Fig. 28-17); it was traced for over 22 miles, along the base of the Sonoma Range, and extending about 2 miles farther south, where it fingered out into several branches which died out in a short distance. The Sonoma Range is one of

FIGURE 28-16   *Isoseismals for Nevada earthquakes of 1915 (R.F.), 1932, and 1954 (M.M.).*

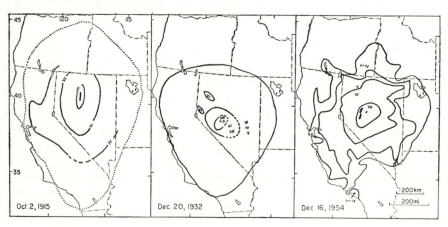

the typical north-south trending ranges of the Basin and
Range province; in the locality of the fault scarp the rocks
are chiefly dolomites capped
by rhyolites and basalts.
There is an older but post-
Pleistocene fault trace, 6 to
10 feet high, which is clear
in the northern part of Pleasant Valley and can be traced
southward into the line of the
1915 scarp. Jones writes:

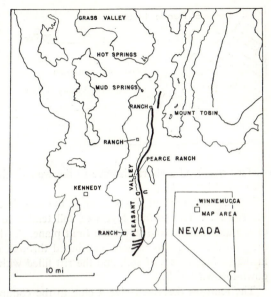

The recent fault scarp formed
at the time of the earthquake
follows the old fault very
closely. There is a decided tendency for the break to be somewhat further to the west as the
scarp crosses . . . the alluvial
fans, leaving a thin wedge of

FIGURE 28-17 *Nevada earthquake, October 2,*
1915. Meizoseismal area and faulting. [After
Jones.]

the alluvium still clinging to the rock; but where the rift crosses solid rock the old
and new fault traces are practically coincident.

This is further developed as follows:

The best and most accessible place to study the earthquake rift is at the ranch
of Mr. W. W. Pierce.† The ranch is located within a few hundred feet of the
rift . . . Beginning at the barn there is a fence that runs to the east across the
fault rift and up the slope of the mountain side . . . As closely as can be determined by eye . . . the fence was not laterally displaced. At the fault the fence
was broken, and an open pit fifteen feet in depth exposed the rock surface at the
bottom, giving an opportunity to determine the true dip of the fault . . . The
surface is striated and covered with a thin layer of clay gouge developed by the
movement along the fault. The striations incline less than five degrees to the north
from the vertical . . . The vertical displacement here, as measured, was 12 feet
and 5 inches. The dip as taken on the rock surface was 54° due west. The horizontal separation of the vertical walls of the alluvium covering the fault plane
was 9 feet, which corresponds very closely to the distance calculated from the
vertical displacement and the dip.
"The cause of the usual vertical face of the earthquake rift is very evident. As
the fault plane approached the surface . . . the thinner unconsolidated alluvium
offered less resistance to the movement than the longer contact between the
alluvial wedge above the break and the fault plane. As a result the break at the

---

† Properly spelled Pearce, as given by B. M. Page.

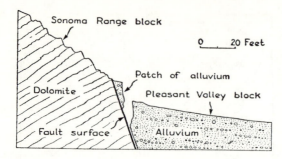

FIGURE 28-18

Nevada earthquake, 1915. Relation of new scarp to old fault surface. [After B. M. Page. From Principles of Geology *by James Gilluly, Aaron C. Waters, and A. O. Woodford. San Francisco: W. H. Freeman and Company, 1951.]*

surface occurred a short distance to the west of the outcrop of the old fault . . . [See our Fig. 28-18.]

The development of pits similar to the one described is exceedingly rare along the rift . . . The surface layers of soil tended to hang back as the deeper mass of the alluvium moved down the fault plane. At the surface the horizontal movement to the west was largely distributed through the secondary cracks developed in the surface layers of soil, and the rift filled with the caving of the disintegrated down-thrown block.

The features described give a clue to the common habit of the rift in dividing into two or more branches as it crosses the deeper alluvial deposits. Frequently one or more of the branches will die out, only to be repeated by another a short distance down the slope. It is only where the rift closely follows the outcrop of the old fault line that it has a single continuous face.

Jones gives a photograph showing a tree tipped by the relative motion of the deep and superficial layers.

An unusual amount of ground water came to the surface after the earthquake, considering that this is a rather arid region. At Mud Spring (a hot spring on the fault line in the northern part of the valley), "with every lurch of the ground the spring, which ordinarily has but a feeble flow, spurted water into the air to a height of two or three feet". This is one of the few clear reports of pulsation in the ejection of ground water.

Page, examining the features 20 years later, found them still fresh in appearance. He separates them into a northern scarp 4.6 miles long and a southern one about 2 miles to the west and 16.7 miles long. He found slickensides still remaining on the dolomites, and measured the dip as 49° due west. Page repeats a rumor that there was also faulting in 1915 at other places, including the Stillwater Range farther south; he did not investigate this point, which is of interest in view of the earthquakes of 1954. Other scarps farther north, also formed in 1915, appear on quadrangle maps lately issued by the U. S. Geological Survey (Mt. Tobin quadrangle, 1951; Golconda quadrangle, 1952). Several observers visited Pleasant Valley early in 1955, and found the scarps still conspicuous, though locally modified by slumping and washing.

Cedar Mountain—1932

An earthquake of magnitude 7.3 originated in west central Nevada on December 20, 1932 (December 21, 06:10:05 G.C.T.). It was felt over the whole of Nevada (intensity VI at Reno) and in adjacent states, as far as San Francisco, Los Angeles, and Salt Lake City (Fig. 28-16). The meizoseismal area (Fig. 28-19) was in one of the graben-type valleys of the Great Basin, flanked by fault-block mountains. It is one of the least inhabited parts of the United States; at the time of the earthquake hardly a dozen persons were in the region. A stone cabin was demolished, a mill at one mine was considerably damaged, and there was some underground sloughing in various mine workings.

Evidence of Faulting. The valley floor is a partially dissected lowland of Tertiary lake beds and volcanic rocks. Effects attributable to faulting were found in a belt 4 to 9 miles wide and 38 miles long, trending about N 21° W. Some sixty individual traces, from a few hundred feet to 4 miles in length, were found. The traces consisted largely of fissures, with occasional evidence

FIGURE 28-19 *Meizoseismal area, Nevada earthquake, December 20, 1932. [After Gianella and Callaghan.] Instrumental epicenters, and fault of January 30, 1934.*

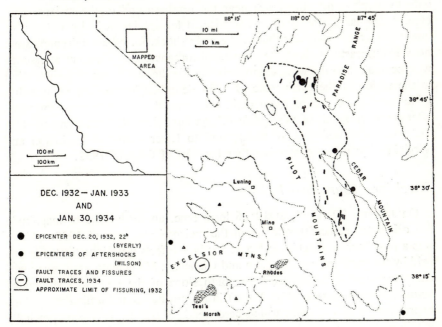

of vertical or lateral slip. At some points the traces were doubled, forming small grabens.

The field workers, Gianella and Callaghan, emphasized that these were not secondary features due to shaking. The breaks were traced across rock ridges and disappeared in alluvium; they often followed low scarps, or passed through depressions across ridges, suggesting earlier faulting along the same lines.

In general, there was an echelon pattern of these traces; the trends were mostly east of north, with a few running west of north. Along the individual traces there were also echelon offsets, the individual segments generally trending more easterly than the trace as a whole, corresponding to tension in right-hand faulting. Vertical throws were small, usually a few inches, although along one trace they reached 2 feet. Downthrows were both easterly and westerly; along some of the traces there were reversals, suggesting scissoring. One road showed a right-hand offset of 34 inches. Some of the traces included zigzag changes of trend; segments running northwest showed pressure ridges, while those running northeast remained as open fissures.

All these data are readily interpreted on the assumption of right-hand strike-slip on a fault underlying the valley but not reaching the surface as a continuous break. This is particularly plausible in view of the relatively incompetent character of the sedimentary rocks in which the traces were found.

Epicenter and Aftershocks. The epicenter as located from instrumental data by Byerly and others was near the northern end of the meizoseismal area. A few large aftershocks were selected for special study by Wilson; for one the epicenter was near that of the main shock, for others epicenters were placed southeasterly along the zone of disturbance. One apparent aftershock was located far to the west, completely outside the main zone of activity (but in the region of the earthquake of 1934 described in the next paragraph); minor earthquakes were frequent in this general part of Nevada for years afterwards. A shock of magnitude 6.1 on June 25, 1933, was damaging at the small community of Wabuska, and to lesser extent at Yerington and Virginia City; the epicenter is rather far west of those of 1932.

Excelsior Mountain—1934

On January 30, 1934, at 12:16 local time, an earthquake of magnitude 6.5 originated some 50 miles from the meizoseismal area of 1932. (There was a foreshock of magnitude 5.5 at 11:24.) Effects were investigated in the field by the same team (Callaghan and Gianella). Many features of secondary character were found; but on the south slope of the Excelsior Mountains actual faulting was seen (Fig. 28-20) in the crushed zone forming the surface expression of an old fault, which here trends N 65° E. This zone is wholly within metamorphic rocks, but a short distance northward there is a contact

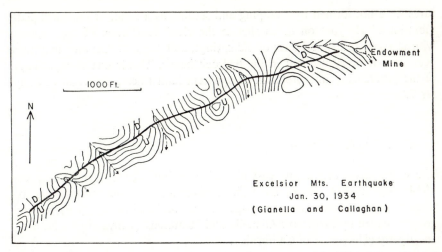

Excelsior Mts. Earthquake
Jan. 30, 1934
(Gianella and Callaghan)

FIGURE 28-20 *Fault trace, Nevada earthquake, January 30, 1934. [Gianella and Callaghan.]*

with volcanics. There was no old scarp or other surface feature indicating faulting prior to the earthquake.

The new scarp was followed for 4500 feet, with a maximum throw of only 5 inches. It was not completely continuous, but broken in part into series of echelon fissures trending more northerly. The trace curved northward into the gulches, indicating a northwest dip of about 73°. The most significant point, however, is the sense of downthrow, which is northwest—up slope. The feature thus cannot be attributed to slumping. However, with the given dip, it still constitutes normal faulting, which may perhaps be attributable to the action of gravity under shaking rather than to continuous fracturing extending to the surface from the hypocenter.

Utah—1934

The Utah earthquake of March 12, 1934, caused minor damage and some alarm at Salt Lake City. Effects perhaps attributable to faulting were observed near Kosmo on the north shore of Great Salt Lake. They followed the boundary of Hansel Valley, a typical Basin and Range feature. Although faulting had not here been established by geological field works, steep scarps exist which suggest it. Springs appeared in connection with a belt of fractures at least 5 miles long.

The field report was prepared by Philip J. Shenon of the U. S. Geological Survey. He notes vertical displacements up to 20 inches between the two sides of the fractures; evidence of horizontal offset was looked for and definitely not found. These fractures are all on salt flats or in poorly consolidated gravels. "Hence," Shenon writes, "it cannot be definitely stated that the fractures were not formed by slumping and settling as a result of the earthquake vibrations." He goes on to emphasize the close association of the fractures with terraced forms and lines of old springs, and suggests that they may have been caused by adjustment in the bedrock below. This is one more example of the problems of interpretation of apparent fault traces, discussed in Chapter 7.

NEVADA—1954

The earthquakes originating east of Fallon in July, August, and December, 1954 (Fig. 28-21), are of the highest importance for our purposes. Thorough field and microseismic investigation was undertaken jointly by research groups headed by Professors Gianella and Slemmons (University of Nevada) and Byerly (University of California).

Earthquakes of July and August

No notable foreshock activity preceded these events. (There was a probable foreshock of magnitude 3.) The first strong earthquake (revised magnitude 6.6), on July 6, was damaging at Fallon; it was followed by a long aftershock series, including a shock of magnitude 6.4 about 11 hours later. Another large shock (magnitude 6.8) on August 23 was again followed by many aftershocks. To quote D. Tocher, the earthquakes of July and August

> . . . were caused by movement along a fault extending along the eastern border of Rainbow Mountain (15 miles east of Fallon) northward to the southern edge of the Carson Sink. After the earthquake of July 6, this movement appeared at the surface as either a fresh, vertical scarp, from one to twelve inches in height, with the west side up relative to the east side, or as severe cracking with no apparent displacement, along a line or narrow linear zone running for eleven miles through the alluvial apron at the base of Rainbow Mountain, and north into the flat desert . . . Fault movement accompanying the earthquake of August 23 extended northward from the July 6 dislocation for an additional 14 miles, and evidence was seen of increased displacements at the northern end of the July 6 breakage. . . .
>
> Rainbow Mountain is a tilted block of Tertiary sedimentary layers which dips westward about 35° to 50°. The faulting along the eastern edge is everywhere found either in the bajada (alluvial apron) along the base of Rainbow Mountain, or in the valley floor to the east, which consists of recently deposited silts and gravel bars of lake Lahontan. North of the northern end of Rainbow Mountain the fault appears only in soft lake-bottom deposits. . . .
>
> The ground breakage . . . is all in surface material which can at best be called poorly consolidated . . . Nonetheless, the continuity of individual cracks, several of which can be traced for distances up to five miles, and the overall

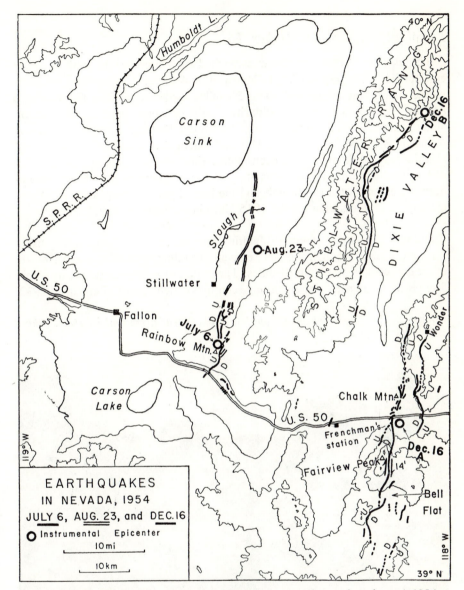

FIGURE 28-21   *Meizoseismal area and faulting, Nevada earthquakes of 1954.*

linearity of the zone of length 25 miles which includes these cracks, suggest that
the surface movements closely reflect movements of the bedrock, at least so far as
they indicate the strike of the Rainbow Mountain fault in the bedrock. . . .

The August 23 displacement was much more continuous than that of July 6,
probably as a result of the greater relative movement (30 inches as compared to
12 inches).

Principal Shocks—December 16

The climax of these events was reached on December 16 with a major earthquake of magnitude 7.1, followed after 4 minutes by a shock of magnitude 6.8. These originated some 30 miles farther east than the July and August shocks, in country ordinarily thinly occupied, and at that season almost deserted. Field investigators discovered faulting of such spectacular proportions (Figs. 28-21A, 28-21B) and extent that it was at first suggested that the magnitude must have been underestimated. However, seismograms at distant as well as nearby stations support the magnitudes just named; moreover, the extent of the isoseismals and the intensity at given epicentral distance are appreciably less than for the 1915 Nevada earthquake (Fig. 28-16).

Faulting.   The total extent of faulting was about 65 miles;† this consists principally of two roughly equal segments trending somewhat east of north, offset en echelon by about 10 miles. The southern segment extends to within 5 miles of the northernmost faulting seen by Gianella and Callaghan in 1932.

FIGURE 28-21A   *Faulting near Fairview Peak, Nevada, December, 1954. [Photo by Hugo Benioff.]*

---

† Much of the following information, including the quotes, is taken from preliminary material distributed at the Reno meeting of the Geological and Seismological Societies of America, April, 1956. Further details have been supplied by Professor V. P. Gianella.

FIGURE 28-21B *Faulting near Fairview Peak, Nevada, December, 1954.* [*Photo by Hugo Benioff.*]

Research by Romney on the instrumental records places the epicenter of the major earthquake on the southern segment, that of the strong shock 4 minutes later on the northern segment. It is believed that the two traces originated separately at those times. The traces are not connected by visible surface faulting or evident structures. The northern fault shows at least 7 feet of dip-slip and little or no strike-slip. The southern fault zone

. . . has up to 12 feet of vertical displacement and up to 12 feet of horizontal displacement; these maximum values are not at the same place. Where graben structures follow the main fault, the major scarp can be much higher than 7 or 12 feet; the highest scarp . . . is 23 feet in height. Vertical and horizontal displacements of two or three feet are common along the other faults of the area that were offset during the December 16, 1954 earthquakes. The faults are all of the normal type and commonly have dips of from 55° to 70°.

Most of the faults follow the boundaries between horst bedrock blocks and alluviated graben valley blocks. In places, however, the faults take short cuts across the irregular edges of the mountain or valley blocks. Most of the faults . . . follow one edge of the fault block, but in at least one area, the valley to the east of Chalk Mountain, a graben block has moved with simultaneous activity on faults bounding both sides of the block . . . The faults may be simple single fractures, particularly in bedrock, but may form branching zones or "fault line graben zones" . . . The structures here called fault-line grabens are gravity collapse structures which result from an increase in the angle of dip of the fault plane as it approaches the surface. The change in dip of the fault plane creates, with normal fault movement, a void in the upper part of the fault; collapse of the hanging block under the influence of gravity, perhaps assisted by vibration produced by the earthquakes. These collapse structures form a narrow graben zone that follows the edge of the downfaulted block. The graben structure accentuates the height of the main scarp . . . Earthslides from the top of the main scarp can further exaggerate the apparent displacement of the fault.

Compare this with Jones' description for 1915.

The scarps of the northern segment "seldom show appreciable lateral displacement, and where they do, the direction of lateral movement is inconsistent from place to place." Those of the southern segment "consistently show right-lateral components of movement." "The faults form en echelon patterns, or offset roads, stream lines or ridge lines, to verify the right-lateral nature of the displacement." Romney found from recorded first motions on seismograms that the initial movement of the main earthquake involved a right-lateral component, while that of the shock 4 minutes later did not. Right lateral deformation of the region was also indicated by remeasurement at triangulation stations in the area by the U. S. Coast and Geodetic Survey.

The southern segment passing Fairview Peak is a true fault trace, though modified by the minor graben development which the field workers describe. The northern segment differs in character; it is less linear, tending to follow contours on the west side of Dixie Valley. Here and there it cuts into the mountains, and occasionally crosses spurs in the manner regularly seen along the southern trace. There is at least one large scarp at a relatively high elevation. Although Romney assigns the second large earthquake of December 16 to an epicenter near this trace, he finds a hypocentral depth of 40 kilometers, as compared with 15 kilometers for the first large shock. Consequently, it is still more plausible that the northern trace is less directly connected with displacement at its corresponding hypocenter. It is not known whether any part of the northern trace features originated in the first earthquake, in which case some of them might be secondary, due to shaking. Crustal blocks reaching the surface have here dropped, creating a fault trace concerning which the only question is as to its continuity with the deep displacements. Apart from shaking, it is evident that since the first large shock was accompanied by regional distortion, this may also have resulted in block displacement along the northern segment.

Geodetic and Gravity Data. Evidence for regional distortion, as summarized by Whitten, is unusually good. Survey lines were run in the area after the earthquakes of July and August; this work was completed shortly before the December shocks. Triangulation and leveling were repeated early in

1955, so that displacements are more closely bracketed in time than on any former occasion. Right-lateral shearing is well determined, agreeing fairly closely with the results derived from seismograms by Romney (he finds a strike of N 11° W with a dip 62° eastward, with motion normal and right lateral, the latter about twice the former). The elastic-rebound pattern of displacement decreasing away from the fault was well established by the resurvey. An interesting observation by Professor Gianella accords with this; along a road running southeastward from Bell Flat south of Fairview Peak, for a distance of 3½ miles successive fractures were found with vertical and right-lateral displacements gradually decreasing away from the main fault. This evidence was destroyed by travel and weather in a few weeks.

G. A. Thompson has lately reported as follows.

Dixie and Fairview Valleys . . . are characterized by local negative gravity anomalies of more than 20 milligals. This indicates that these valleys contain several thousand feet of lightweight Cenozoic sediments and that their bedrock floor lies close to sea level. The 1954 faults . . . are thus only the latest of many movements that produced not only the visible topographic relief of 5000 ft. but also a buried relief of comparable magnitude.

An Earlier Event? Near Wonder, Nevada, which is south of Dixie Valley and nearly in the northeastward prolonged line of the Fairview Peak-Chalk Mountain fault, there are large features apparently due to very young faulting. There is a fugitive report that these features, or part of them, originated in an earthquake about 1906 or 1907. This was repeated by F. Schrader, in a report for the U. S. Bureau of Mines; it was also related to V. P. Gianella by an old resident. Whether the two reports are independent is unknown. No large earthquake which would fit is in the catalogues; the matter is so doubtful that it has been omitted from our tables and maps.

EASTERN CALIFORNIA—1875? AND 1950

The following was published by H. W. Turner in 1896.

"Faults with small throw in Pleistocene sediments were noted on the east side of the Middle Feather River, opposite Wash postoffice, and on the south bank of the river about one mile upstream from Wash postoffice. About a mile farther east . . . a fissure was formed in the tuffs and breccia beds at the time of an earthquake (about 1876)."

Residents stated to Turner that the fissure was about 2 feet wide. On the Downieville quadrangle the locality can be identified as near 39° 44′ N, 120° 34′ W. The earthquake intended may be that of January 24, 1875, strong in northeastern California and at Carson City.

Professor Gianella has lately reported on minor faulting in the Fort Sage

Mountains, near Doyle in southeastern Lassen County, California (not far from the Nevada boundary). This was associated with the earthquake of December 14, 1950 (magnitude 5.6; epicenter about 40.1° N, 120.1° W). On the west side of the mountains there were three fault traces, with a combined length of about 6 miles, and with scarps up to about 8 inches high.

MANIX EARTHQUAKE—1947

Manix, California, is a station on the Union Pacific railroad in the Mojave Desert, about 25 miles east and slightly north of Barstow. On April 10, 1947, an earthquake of magnitude 6.4 originated a few miles east of Manix (Fig. 28-22). The shock was felt over a wide area, including Los Angeles; structural damage was limited to widely scattered ranches, railway stations, and gasoline service stations on the adjacent trunk highway (U. S. Highway 91).

FIGURE 28-22   *Manix earthquake, April 10, 1947. Location map.*

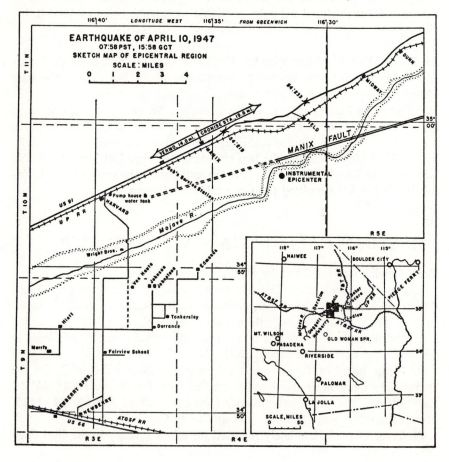

The railroad right of way in Afton Canyon was damaged by slides, and extensive repairs were necessary. Clouds of dust arose from slides on the Manix bluffs, along the south face of which the Mojave River runs for a short distance, at the time of the main shock and during many aftershocks. These bluffs trend a little north of east and south of west, marking the Manix fault, along which Pleistocene (or possibly Pliocene) lake beds have been uplifted against the fanglomerate material constituting the bluffs.

Fault Trace

Toward the west end of the bluffs there is a bench higher than the deeply dissected clays and sands of the Manix lake beds. After the earthquake a nearly straight trace was found cutting across this bench in line with the Manix fault; it consisted chiefly of open cracks. The trace crossed a number of shallow, dry gullies descending from the bluffs, and on most of them close inspection showed evidence of left-hand strike-slip of 2 or 3 inches. Westward the trace entered a dissected area of fanglomerate overlain by sand. Here the continuous crack was frequently accompanied or replaced by series of tension cracks, all striking roughly northeast, in the proper direction for tension connected with left-hand strike-slip. The total length for which this trace could be followed was not much over a mile; it lay wholly within the zone of the Manix fault as indicated by the larger features of displacement in the sedimentary rocks.

Epicenters

Permanent seismological stations were in operation in so many directions that the Manix area was well surrounded, but at distances all well over 100 kilometers. A portable instrument with good timing was operated for short runs at two locations about 22 and 60 kilometers distant, recording aftershocks which could be compared with the main shock at the permanent stations. In this way an epicenter was found for the main shock about 2 miles north of the Manix fault in the area of greatest disturbance. This position might easily be 2 or 3 miles in error.

Concealed Faulting?

An unexpected result appeared when the time differences between the arrivals of the first recorded waves at the permanent stations for the main shock were compared with the corresponding differences in the aftershocks. It is possible to determine the relative positions of the several epicenters with much less uncertainty than the placing of any one on the map, since the method of differences is less affected by the incompleteness of our knowledge of wave speeds and crustal structures. Epicenters were found aligned, not

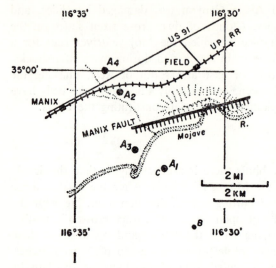

FIGURE 28-23 *Epicenters for Manix earthquake and aftershocks. A1, A2, A3, A4, B, C represent groups of 3 to 14 shocks. The main shock is assigned to epicenter A2. [Richter and Nordquist.]*

on the Manix fault, but along a line trending N 30° W, nearly at right angles to the fault (Fig. 28-23). No reasonable interpretation of the data will seriously modify this result.

The recording of initial compressions and dilatations at the several stations would be consistent either with left-hand strike-slip along the Manix fault or with right-hand strike-slip on a fault along the line of epicenters. The author favors the latter possibility. This means that the small displacement found on the Manix fault is an indirect consequence; presumably motion on the otherwise unknown fault led to a slight displacement along the old line of weakness. This is satisfactory mechanically; both motions, for example, would follow from relief, by shearing, of a north-south compression. An obvious objection is that no fault corresponding to the line of epicenters is known. This area is very incompletely investigated geologically, partly because good topographic contour maps are not yet available. The ridges are of igneous and metamorphic rocks, with their lower slopes disappearing under fan material and lake beds. Search on the ground disclosed nothing acceptable as faulting; the only significant observation was the existence of a line of demarcation, with the expected trend, west of which the exposed rocks are chiefly metamorphic, while to the east they are Tertiary volcanics (rhyolites and basalts).

There is no inherent impossibility in an active fault in basement rock which does not extend to the surface; this concept will be applied to some of the 1952 Kern County earthquakes.

WALKER PASS EARTHQUAKES—1946

On March 15, 1946, an earthquake of magnitude 6.3 originated under Walker Pass, near the eastern margin of the Sierra Nevada at about the latitude of transition to the Tehachapi Mountains. There was a large immediate foreshock (magnitude 5¾) and many aftershocks. Portable instruments were taken into the field. A synthetic study has been published by Chakrabarty and Richter; however, it is probable that many detailed results will need revision when the full consequences of the recording of earthquakes in Kern County in 1952 and thereafter are applied. The 1946 shocks are of particular interest because of probable tectonic connection with those of 1952.

No effects suggesting faulting were observed. Epicenters determined for aftershocks show a tendency, not well marked, to alignment somewhat north of east to south of west. They are definitely within the Sierra block, but the hypocenters can be placed on the boundary fault of the Sierra if that is assigned a dip of 60° to 70°. Some of the later shocks, considered aftershocks in the published study, are far to the west of the principal group of epicenters, and fall in the active area of the 1952 shocks, with which they provide a valuable link.

EARTHQUAKES IN KERN COUNTY—1952

The events next to be described afforded the first opportunity for detailed seismometric investigation of a major earthquake in California after the installation of modern seismographs. Many of the results obtained were highly significant; they have necessarily been referred to in their proper connection in other chapters (especially Chapter 8), but the discussion which follows presents all the principal points. Investigation is not yet fully completed.

FIGURE 28-24 *Isoseismals of the Kern County earthquake, July 21, 1952. Inset: isoseismals of the aftershock of August 22, 1952 (the "Bakersfield earthquake"). [After U. S. Coast and Geodetic Survey.]*

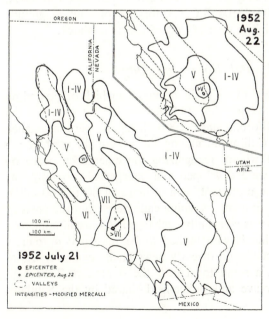

Tectonics and Past History

The principal Kern County earthquake on July 21, 1952 (Fig. 28-24), was in many ways an unexpected event. The White Wolf fault, on which it originated, had been described and named by Hoots, and it appeared on the official geological map of California, but there was no generally known reason to distinguish it from numerous other faults as the possible seat of a major earthquake. The fault had been mapped southwestward only to the point where it disappears under the alluvium of the southern San Joaquin Valley. Commercial geophysical investigation had shown that it continues and is marked by a very high step in the basement rock; but the details were not made public until after the earthquake.

Partly because of the accidents of population distribution, partly because of the late establishment of seismological stations in southern California, it is difficult to decide whether any of the moderate to minor shocks before 1929 could have been associated with this fault. An earthquake on October 22, 1916, which had a magnitude of about 6, was apparently strongest in the region of Tejon Pass and was attributed to the San Andreas fault on macroseismic evidence; but the data would not exclude an epicenter near that of 1952. A smaller but widely felt shock on February 16, 1919, is an even more likely instance.

During the years of slowly improving recording of minor shocks from 1929 to 1952, fairly numerous epicenters were located in the southern San Joaquin Valley and in the vicinity of the Kern River Canyon. In retrospect these appear scattered over the entire area which became active in 1952; but there is nothing to distinguish this from the peppering of epicenters over much of southern California.

In the first half of 1952 (and the first three weeks of July) general seismicity in southern California was lower than average and showed no distinctive pattern in relation to subsequent events.

The principal earthquake on July 21 was preceded by only one foreshock, 2 hours 9 minutes earlier (at 09:43:04 G.C.T.); this was of small magnitude (3.1) and had an epicenter slightly west of that of the main shock.

Events of July 21, 1952, and Following Days

The magnitude of 7.7 assigned to the main earthquake which originated on July 21 at 4:52 A.M. daylight-saving time (11:52:14 G.C.T.) makes it the largest with epicenter in California since 1906. Sleepers were awakened throughout most of southern California; among them were the staff of the Pasadena laboratory, most of whom, recognizing the emergency, were on duty by 7 A.M. Seismograms were developed and interpreted promptly; the epicenter was seen to be to the northwest, at a distance of 80 to 100 miles. Telephones were in constant use; many incoming calls were from the press or from officials, and information was constantly exchanged. As usual, numerous calls were from excited individuals who wished to be reassured or merely wanted to talk. Through the confusion of early news finally came reports of serious damage at Tehachapi, a small town near the summit of Tehachapi Pass, on the line between Mojave and Bakersfield shared by the Southern

Pacific and Santa Fe railways. Railroad and highway were reported closed by slides west of Tehachapi. An expedition with a portable seismograph left Pasadena about 7 A.M., and recording began at 11:57 A.M. at a point not far southeast of Tehachapi. At that location aftershocks were recorded in large numbers and in quick succession, many of them sharply perceptible to persons.

Eleven lives were needlessly lost at Tehachapi, the result of failure of the weakest type of outdated masonry. Many structures on the main street were of this character; they showed such damage as caused incautious observers to rate the local intensity at VIII, though it is doubtful whether more than VII was actually indicated. A structure of later date, originally a lodge hall, was undamaged and served as a shelter for the homeless and injured.†

From the vicinity of Tehachapi the portable instrument was taken to White Oak Lodge on the summit of the Tehachapi Mountains (and in the zone of the Garlock fault) and thence returned to Pasadena. Field observations along the Garlock fault zone showed no effects of high intensity; a phenomenon

FIGURE 28-25 *Investigation of the Kern County earthquakes, 1952. Map showing temporary and new permanent installations, with dates.*

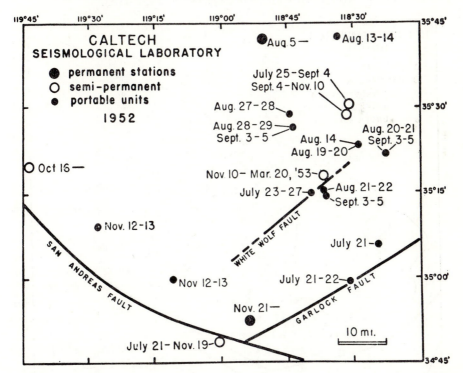

---

† For other details of damage in this earthquake see Chapter 8.

interpreted by some observers as faulting was examined and attributed to the dropping of a small block locally in the fault zone, due to shaking or to the general regional readjustment of strain (see Chapter 7).

A second party took a seismometer and recording unit to Chuchupate ranger station near Frazier Mountain on July 21 and arranged for its continuous operation there. This unit, later moved to Fort Tejon, has constituted a permanent recording outpost closer than any other station to the epicenter of the main earthquake of July 21.

Dr. Pierre St. Amand, at a time when news reports were incomplete and conflicting, located and studied the principal trace along the line of the White Wolf fault on the west flank of Bear Mountain. This trace passed a few miles east of the town of Arvin, where there was extensive damage (intensity VIII to IX).

Stations and Epicenters

At the time of the main earthquake the nearest established seismological stations were at Santa Barbara, about 90 kilometers southwest of the prin-

FIGURE 28-26   *Kern County aftershock epicenters, 1952–1953.*

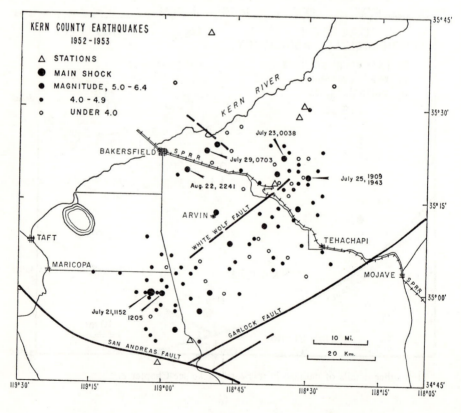

cipal epicenter, and China Lake, roughly the same distance northeast of the other end of the active area. In the following weeks the portable seismometer was operated at a number of points for a few hours to a few days each. Additional continuously recording stations were established at Havilah (on July 25; later moved to Knox Ranch, and in 1954 to Isabella), at Woody (August 5), and at King Ranch (November 4). These are shown on the map in Figure 28-25. A special program with three portable instruments in the field was operated on September 3–5.

Combination of the times recorded at all the stations, including the temporary locations, made it possible to determine epicenters for all the large shocks in the aftershock sequence. Those of magnitude 4 and over were

Table 28-3   Numbers of Aftershocks in the Kern County, California, Active Area

Date		Magnitude		Date		Magnitude	
		4.0 and over	3.0–3.9			4.0 and over	3.0–3.9
1952	July	135	—	1954	July	0	10
	Aug.	32	—		Aug.	0	7
	Sept.	12	—		Sept.	0	9
	Oct.	4	—		Oct.	0	6
	Nov.	5	—		Nov.	0	4
	Dec.	2	17		Dec.	0	2
1953	Jan.	2	26	1955	Jan.	1	4
	Feb.	1	30		Feb.	1	7
	Mar.	2	17		Mar.	0	3
	Apr.	1	12		Apr.	0	4
	May	3	31		May	1	4
	June	1	15		June	0	12
	July	0	10		July	1	5
	Aug.	2	10		Aug.	2	6
	Sept.	3	9		Sept.	0	3
	Oct.	1	9		Oct.	0	2
	Nov.	1	17		Nov.	1	3
	Dec.	1	12		Dec.	0	2
1954	Jan.	4	23	1956	Jan.	0	2
	Feb.	4	9		Feb.	0	1
	Mar.	0	7		Mar.	2	5
	Apr.	0	12		Apr.	0	4
	May	2	19		May	0	5
	June	0	8		June	1	7

investigated with especial care. Beginning with January, 1953, those of magnitude 3 and over were located in routine fashion, with systematic special checks.

Statistics and Mapping

Statistics for aftershocks of magnitude 3 and over are included in Table 28-3. The located epicenters appear on the map in Figure 28-26. They are distributed over a quadrilateral with sharp, straight boundaries on the south,

Table 28-4 Larger Shocks of the Kern County Series
(Magnitude 5 and Over)

Date			G.C.T.	Lat. N	Long. W	Magnitude
			h m s			
1952	July	21	11:52:14	35° 00′	119° 02′	7.7
	July	21	12:02			5.6
	July	21	12:05:31	35.0	119.0	6.4
	July	21	12:19:36	34 57	118 52	5.3
	July	21	15:13:59	35 11	118 39	5.1
	July	21	17:42:44	35 14	118 32	5.1
	July	21	19:41:22	35 08	118 46	5.5
	July	23	00:38:32	35 22	118 35	6.1
	July	23	03:19:23	35 22	118 35	5.0
	July	23	07:53:19	35 00	118 50	5.4
	July	23	13:17:05	35 13	118 49	5.7
	July	23	18:13:51	35 00	118 50	5.2
	July	25	13:13:09	35 19	118 30	5.0
	July	25	19:09:45	35 19	118 30	5.7
	July	25	19:43:23	35 19	118 30	5.7
	July	29	07:03:47	35 23	118 51	6.1
	July	29	08.01:46	35 24	118 49	5.1
	July	31	12:09:09	35 19	118 36	5.8
	Aug.	1	13:04:30	34 54	118 57	5.1
	Aug.	22	22:41:24	35 20	118 55	5.8
1953	None					
1954	Jan.	12	23:33:49	35 00	119 01	5.9
	Jan.	27	14:19:48	35 09	118 38	5.0
	May	23	23:52:43	34 59	118 59	5.1
1955	None					
1956	None					

west, and north; to the east a similar boundary is indicated, but straggling epicenters extend beyond it. The south and west boundaries are close to known faults; they are definitely short of the two major faults in those directions (Garlock and San Andreas). The north boundary corresponds to a structural trend of the region; it is nearly parallel to the Garlock fault. However, it passes through an area where the surface features have a different trend, more nearly paralleling the San Andreas fault. This suggestion of a deep, active fault beneath surface structures of different trend is reminiscent of the findings after the Manix earthquake of 1947.

Table 28-4 gives date, origin time, epicenter, and magnitude for all the shocks of the series which were assigned magnitudes 5.0 and over as originally published (listing complete to the end of 1956; some of the magnitudes near 6 may need revision when present restudy of the magnitude scale is completed). Most of these shocks have been referred to in connection with the discussion of secondary aftershocks in Chapter 6. Times are G.C.T., from which Pacific daylight-saving time, then in use in California, can be obtained by subtracting 7 hours.

Principal Occurrences

The series of tectonic events as indicated by these epicenters (and, for the main shock, by field evidence) may be outlined as follows.

July 21, 11:52. Main shock. Fault rupture began at a hypocenter about 16 kilometers below the tabulated epicenter and extended northeast along the White Wolf fault; the fault surface dips about 60° in this part of its course, but the dip probably decreases significantly northeastward. Epicenters of aftershocks at depth comparable with the main hypocenter hence fall southeast of the White Wolf fault surface trace and tend to cluster along a line passing not far from Tehachapi. Initial rupture probably followed this line at the corresponding depth, thus accounting in part for the observed intensities about Tehachapi. Fracture extended upward and westward, breaking the surface along the trace shortly to be described.

July 21, 12:05. Largest aftershock, with epicenter near that of the main earthquake.

July 23, 00:38. Large aftershock marking a tectonic change. Previous aftershocks were all located, so far as the sometimes difficult data permit determination, on the southeast side of the White Wolf fault; they belong to the upthrown block. No epicenters in the interval following the main shock up to July 23 00ʰ were placed on the downthrown side. This large aftershock, however, and others following it have epicenters on the northwest side of the fault and are logically attributable to the downthrown block. If these and all those following on the downthrown side are combined into a Benioff strain-release curve, the result is of the type originally attributed to shear; while those on the upthrown side, before and after the critical shock, combine into a strain-release curve of compressional type (Fig. 6-1B).

July 25, 19:09 and 19:43. Two strong aftershocks probably representing a slight northeastward extension of faulting, and establishing a terminal epicenter which continued very active in subsequent months.

July 29, 07:03. Extension of strain release to the vicinity of Bakersfield; series of shocks setting the northern boundary of the active area.

August 22, 22:41. The "Bakersfield" earthquake, with an epicenter only slightly different from that of July 29, but causing much more damage.

1954, January 12. Largest shock in the series since August 22, 1952; epicenter very close to that of the main shock.

Main Shock—Intensities and Isoseismals

Isoseismals for the principal earthquake of July 21, as constructed by the U. S. Coast and Geodetic Survey, are shown in Figure 28-24. Wisely, no attempt was made to draw detailed isoseismals in the meizoseismal area where intensities were VIII or over. Confusion in assignment of intensity arose from the destructive effects at the fault trace, which the M.M. scale would rate as XI, although there was no evidence of extremely violent shaking at points near the trace, as judged by damage to houses. Intensity IX was manifested over much of the area near the fault, and effects due to shaking assignable to X (such as large fissures and damage to underground pipes) were developed at many localities.

A striking feature of the map is the general elongation of the outer isoseismals in the northwest-southeast direction parallel to the regional structure, although the White Wolf fault trends almost at right angles to this.

The placing of the instrumental epicenter at one edge of the meizoseismal area, far from its center, is quite positive and conforms to experience in other earthquakes. It is, of course, correlated with the progression of faulting away from the epicenter. A natural suggestion is that much energy may have been released from the fault near the middle of its course. There is also the direct effect on wave propagation from the moving source; this will be described a little later.

Long-period effects were highly developed, and this accounts for apparently high intensities at surprisingly large distances; much of this evidence consisted of slides, and of cracks in soft ground.

The Fault Trace

The principal surface expression of faulting (Fig. 28-27) appeared on the west flank of Bear Mountain, as a scarp generally about 4 feet high, downthrown westward. The trace showed much curvature and many echelon

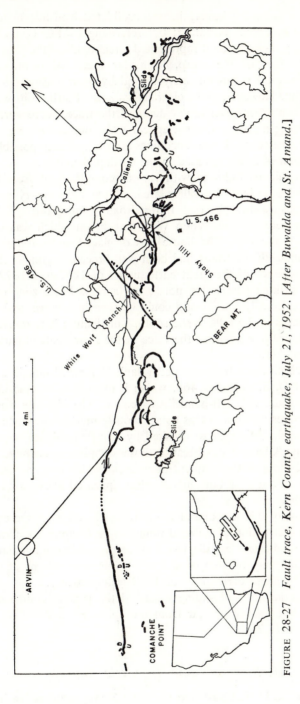

FIGURE 28-27 Fault trace, Kern County earthquake, July 21, 1952. [After Buwalda and St. Amand.]

offsets. Since it appeared largely in unconsolidated landslide material, there was initial doubt about its nature; but releveling by the U. S. Coast and Geodetic Survey showed that it represented a real uplift of the Bear Mountain block. At the foot of the scarp, triangular facets of disturbed ground, bounded by pressure ridges on one side and by open tension cracks on the other, stood out. However, these clear indications of strike-slip were neither coherent nor continuous over long extents of the trace. Moreover, a number of fences crossing the trace did not appear to be out of line, except for disturbances near the scarp. These fence lines are old and may possibly antedate the accumulation of strain released in the earthquake.

This trace could be followed with various complications from Comanche Point (a projection, into the valley area, of the hill country southeast of Arvin) northeastward to beyond the railroad passing through Tehachapi Pass. Halfway between was White Wolf Ranch, on which was one of the oldest houses in the area. The house was displaced on its foundations, but not greatly damaged; the frame had been well braced diagonally throughout, suggesting experience of the great earthquakes of 1857 and 1872. Past the ranch house, striking nearly due north, ran a remarkable zone of cracking and left-hand offset. The strike-slip could be seen in a fence near the house and also at fences in the hills to the south. This feature is reminiscent of the transverse strike-slip segment of the trace of the New Zealand earthquake of 1931 (Chapter 29).

A clear trace diverged from the north flank of Bear Mountain a short distance south of U. S. Highway 466, which it crossed as a zone of cracking about 20 feet wide, with left-hand displacement. The highway was soon repaired, but fissures and low scarps conspicuously marked the trace ascending a hill to the north. This spot was visited by crowds of sightseers, and it became known as "Shaky Hill" because those stopping to examine it usually noticed one or more of the frequent aftershocks which were perceptible over the entire area. A similar but smaller trace appeared about a quarter of a mile to the west.

Just northeast of this was one of the most costly effects of the earthquake, the offset and destruction of railroad tunnels (see Chapter 8). Cracking and other evidence indicated less displacement at the surface immediately above the tunnels. This may be compared with similar observations at the Tanna tunnel in the Japanese earthquake of 1930 (Chapter 30).

Spectacular cracking and sliding indicated the general course of faulting northeasterly, terminating not far from the epicenters assigned to the large aftershocks on July 25.

Work of the U. S. Coast and Geodetic Survey

Members of the U. S. Coast and Geodetic Survey were active in the field immediately. Among their results were the assembling of photographic and other documentary data on the effects of the earthquake, which was used in constructing the isoseismal map (Fig. 28-24). The strong-motion instruments operated by the Survey wrote records of special value on this occasion; the most interesting was that at Taft (Fig. 28-28), only about 45 kilometers (28 miles) from the epicenter of the main earthquake. Additional strong-motion instruments installed in the area during the aftershock period wrote many seismograms valuable for the study of earthquake motion at short distances.

Of great significance was the result of resurvey of precise levels and triangulation in the area; some of the lines repeated had been surveyed only a few months before the earthquakes. Whitten reports:

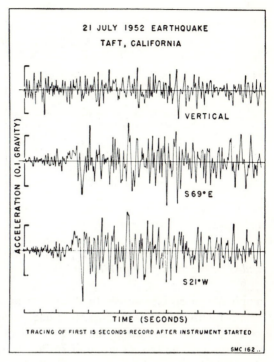

FIGURE 28-28 *Strong-motion seismograms of the major Kern County earthquake, July 21, 1952, recorded at Taft, distance 45 kilometers (28 miles) from the instrumental epicenter.* [*U. S. Coast and Geodetic Survey.*]

The Bear Mountain block, southeast of the fault, moved toward the north-northeast a distance on the order of one to two feet; but the southwest segment of that block . . . appears to have moved upward and toward the northwest over the valley. The one triangulation station on the valley floor suggests movement of the valley block a similar distance in a west-southwest direction. Greatest vertical movement, an elevation of 2 feet, appears to have taken place . . . with the Bear Mountain block elevated and tilted toward the east, but moved northwest. Depression of the valley side was on the order of a foot and a half, centering in the basin-shaped area southwest of Arvin.

These statements describe thrust (reverse) faulting accompanied by left-hand strike-slip; they agree with the field evidence and with the microseismic data (see Fig. 28-29).

Main Shock—Microseismic Data

Seismograms or copies from stations all over the world were collected at Pasadena, through the established courtesy of international exchange, and

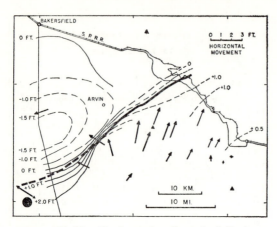

FIGURE 28-29   *Horizontal and vertical displacements shown by resurvey in Kern County after the earthquake of July 21, 1952.* [*U. S. Coast and Geodetic Survey; from Whitten.*]

were investigated by Dr. Gutenberg. The nature of first motion (compression or dilatation) at the various stations was observed and was used to infer the mechanism of origination of the earthquake. The data fit very well for thrusting on the White Wolf fault, with a dip of about 62° at the hypocenter. The evidence less positively indicates a further component of left-hand strike-slip which would make the exact direction of initial block displacement nearly due north-south.

In applying the maximum amplitude of surface waves to determine magnitude, it was found that the results for European recordings were unexpectedly high. However, the amplitudes recorded in Australia and New Zealand were correspondingly low; at those stations the second group of surface waves, which had traveled around the world by the major arc, was actually larger than the direct waves across the Pacific. These results could be explained in terms of piling of seismic wave energy in the direction in which rupture progressed from the instrumentally determined hypocenter along the White Wolf fault; the speed of progression of such rupture might be expected to be of the same order as that of surface waves (which is 3 to 4 kilometers per second). Something related to the Döppler effect would result, with increased concentration of energy in the direction of faulting; this might have contributed to the apparent intensity at Tehachapi.

AFTERSHOCK MECHANISM

Dr. Båth, when at Pasadena as a visiting research associate (1956–1957), investigated the initial motions recorded for 57 of the larger aftershocks. South of the White Wolf fault, left-hand strike-slip dominates, with strike approximately parallel to the fault. North of the fault, in the area of the large shock of July 23, right-hand strike-slip of various orientation is associated with dip slip. Of the shocks near Bakersfield, three show right-hand strike-slip roughly parallel to the White Wolf fault, while three others show dip slip, or right-hand strike-slip with different orientation.

The mechanical relations of the northern aftershocks, consequently, are not simple; this has important bearing on theories of regional strain.

MISCELLANEOUS RESULTS

The two principal reports on these earthquakes contain a wide variety of information contributed by many authors. The engineering report reviewed in Chapter 8 includes a valuable discussion of the geology and one of the two chief accounts of the fault trace. Other descriptions of faulting are included in the general publication *Bulletin 171, Division of Mines.* An unusual item is a detailed account of the remarkable system of large ground fractures developed in the alluvium of the San Joaquin Valley. These are shown to be similar in character and in trend with older fracturing for which evidence remains in and around the margin of the valley. There are also interesting reports of effects on ground water, including fluctuations of level in wells, of the type described in Chapter 9.

BAJA CALIFORNIA—1892 AND 1956

This section reports incomplete and, in large part, unpublished investigation. In 1954 three important shocks took place in Baja California (Fig. 28-30): on October 17, magnitude 5.7; on October 24, magnitude 6.0; on November

FIGURE 28-30   *Baja California earthquakes, 1954–1956.*

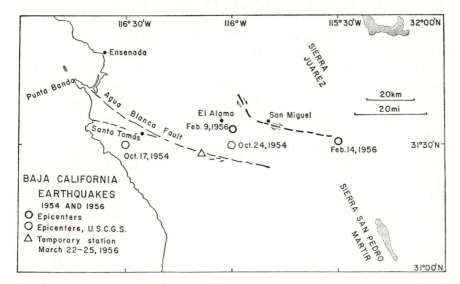

12, magnitude 6.3. Each was accompanied by many aftershocks. Epicenters assigned by the U. S. Coast and Geodetic Survey, good to about the nearest quarter degree, were 31.5° N, 116.5° W for the October 17 shocks, and 31.5° N, 116° W for those on October 24 and November 12. These data drew attention to an east-west structure, conspicuous from the air, cutting across the Baja California peninsula in that latitude.

On November 26, 1955, a shock of magnitude 5.4 occurred in the same area; and on February 9, 1956, there was an earthquake of magnitude 6.8 followed by a long and complicated aftershock sequence, with the development of visible fault traces and other circumstances forming an interesting parallel to the Kern County disturbances of 1952.

Field investigations and study of the fault trace were carried out by G. G. Shor (Scripps Institution of Oceanography, La Jolla) and independently by Baylor Brooks and Ellis Roberts (San Diego State College). The large structure mentioned above was investigated on the ground and from the air by C. R. Allen, F. Stehli, and L. Silver (California Institute of Technology). A portable seismometer was operated in the area for 60 hours on March 22–25.

Results may be summarized briefly as follows: Faulting began on February 9 at about 31.7° N, 115.9° W, not far from the mining town of El Alamo. New springs, hot at first, appeared in this vicinity. Maximum damage was observed a little farther east, at the village of San Miguel (then temporarily unoccupied). There was a fault trace with a sharp change of direction, or two fault traces meeting near San Miguel, the eastern trending N 60° W, the western N 40° W. There were vertical displacements, downthrown to the southwest, of 1 to 8 feet; in a few places it was possible to observe strike-slip, which was right-hand and amounted to at least 2 feet.

Probably the faulting on February 9 extended eastward to near 31.5° N, 115.5° W, in uninhabited country difficult of access. On February 14 a new burst of seismicity began in that vicinity, including two shocks of magnitudes 6.3 and 6.4, with many smaller ones. (This resembles the concentration of aftershocks at the terminal point of faulting in the Kern County earthquakes after July 25, 1952.) Strain-release curves indicate that the entire series of shocks form a single unit and do not support any hypothesis of independent events.

This faulting is north of the large feature studied by Allen and others, which proves to be a major fault (the Agua Blanca fault) cutting at least halfway across the peninsula, with spectacular evidence of geologically recent right-hand strike-slip. This is probably an important tectonic line of demarcation. Its relation to the lately active fault north of it is reminiscent of that of the Garlock to the White Wolf fault.

The large earthquake of February 23, 1892, originated in this area; it was reported as very strong at Ensenada and El Alamo. At San Diego it caused much minor damage, up to the level of cracks in masonry. It was felt as far north as Visalia. In magnitude it must have exceeded the 1956 shock; it is impossible to tell whether it originated on the same fault or on the Agua Blanca fault.

OTHER SHOCKS OF INTEREST

1868, September 4 and thereafter. Numerous shocks, locally severe and producing rock slides in the upper Kern River region of the Sierra Nevada; often referred to in connection with possible activity of the conspicuous north-south fault described by Lawson.

1885, April 11. A widely felt shock in central California, commonly attributed to the San Andreas fault. However, the strongest reported effect is the throwing down of chimneys at Las Tablas, 30 miles northwest of San Luis Obispo; this suggests a sizable earthquake on the Nacimiento fault. On November 21, 1952, a shock of magnitude 6 originated near the small community of Bryson, which is close to the Nacimiento fault in the same general area. This shock was fairly sharp at San Simeon, and it is noteworthy that damaging shocks at San Simeon are on record for 1852.

1892, April 19 and 21. Shocks damaging to towns in the valley area west of Sacramento and north of San Francisco Bay. The fault or faults responsible cannot be identified.

1899, July 22. A shock felt over nearly all southern California, with greatest apparent intensity in the vicinity of the San Andreas fault in Cajon Pass; many slides blocking the highway in the Pass; damage at San Bernardino and Patton, and as far away as Los Angeles.

1907, Sept. 19. Similar to the above: landslides in the mountains, damage at San Bernardino and San Jacinto.

1908, November 4. The strongest of a series of shocks in the Death Valley region; intensity VI at Lone Pine; prospectors frightened out of the meizoseismal area; widely recorded on seismographs.

1910, May 15. A shock of magnitude 6 or over; apparently originating on the Elsinore fault.

1915, June 22. Destructive in Imperial Valley; attributed to the San Jacinto fault (on field evidence only).

1920, June 21. The Inglewood earthquake; a minor damaging shock which drew attention to the existence and probable activity of the Inglewood fault, as later demonstrated in the Long Beach earthquake of 1933.

1922, March 10. Magnitude 6½; originating on or close to the San Andreas fault in the vicinity of Cholame (about 35¾° N, 120¼° W); fissures, probably of secondary character, in the unconsolidated material of the Rift zone. This epicenter may possibly indicate the northern limit of faulting in 1857. Southeast of it along the San Andreas fault even small shocks have been rare since the beginning of adequate instrumental recording about 1927, while northwest of it shocks of magnitude up to 5½ have occurred repeatedly as far away as Hollister, which is just southeast of the limit of faulting in 1906.

1925, June 29. The Santa Barbara earthquake; magnitude 6.3. The epicenter remains uncertain, but records of aftershocks written at Santa Barbara beginning in July, 1927, indicate an epicenter a short distance (not over 10 miles) west of the city, suggesting association with one of the known faults in the Elwood oil field at the coast. The chief importance of this damaging shock was in its effect on earthquake safety measures, in California and elsewhere; its exposure of improper practices led to preparation and adoption of improved building codes in many communities. Such action, however, progressed very slowly until after the Long Beach disaster of 1933; even since then, public indifference and official inertia continue to delay the general adoption of adequate regulations.

1927, November 4. Major earthquake; magnitude 7.5; originating at the edge of the continental shelf, on the north margin of the Murray fracture zone. There were extensive slides along the nearest seacoast (the west coast of Santa Barbara County), and service on the Southern Pacific Railroad, which follows the coast there, was interrupted. A sea wave rose to 6 feet on the same coast; it recorded on tide gauges in California and the Hawaiian Islands. Chimneys were thrown down or damaged at Lompoc and other towns near the coast.

1930, February 25 and March 1. Shocks locally damaging at Westmorland and Brawley in Imperial Valley; of interest in connection with the determination of the active fault pattern in that region.

1933, January 4. Magnitude 5½; epicenter 28° N, 126½° W, far outside the continental area.

1934, June 7. Damaging at Parkfield, on the San Andreas fault, northwest of the epicenter of 1922; magnitude 6. This, with its group of associated shocks, is of tectonic interest because of detailed study by Wilson. A large foreshock originated northwest of the epicenter of the principal earthquake, from which the epicenters of aftershocks extended southeastward for about 20 kilometers (12 miles).

1934, December 30 and 31. Shocks of magnitude 6½ and 7.1, originating south of Imperial Valley in Mexican territory, possibly along the fault system which showed displacement in 1940. Railroad bridges were damaged and tracks distorted in a manner attributed to faulting at the time.

1937, May 8. A locally damaging shock originating on the Haywards fault a few miles from the station at Berkeley (University of California).

1940, February 8. Damaging at Chico and Grass Valley; magnitude 6. The epicenter, on the western margin of the Sierra Nevada province, is difficult to locate, and there is evidence for unusual hypocentral depth, possibly 35 kilometers.

1941, June 30. Santa Barbara Channel, off Carpinteria; magnitude 5.9;

damage at Santa Barbara and Carpinteria, chiefly to structures weakened in 1925.

1951, December 25. Magnitude 5.9; epicenter near the southeast point of San Clemente Island, at 32.8° N, 118.3° W; largest known shock in the southern offshore province.

1954, December 21. Damaging at Eureka; magnitude 6.6; epicenter, determined at Berkeley: 40° 47′ N, 123° 52′ W; interesting because it is inland, although north of the Mendocino fracture zone.

1955, October 23. Shock of magnitude 5.4; VI at San Francisco; damage (VII) at Walnut Creek and a few other points east of San Francisco Bay; epicenter probably on the Calaveras (or Sunol) fault, which branches eastward from the Haywards fault.

1957, March 22. Magnitude 5.3; epicenter on the San Andreas fault, near Mussel Rock at the coast just outside the city of San Francisco. The first shock approaching this magnitude since 1906 on the sector of the fault displaced at that time. Consequently, the strongest earthquake shaking at San Francisco since 1906. Although this bore about the same relation to the earlier event as a firecracker to a cannon, it touched off a tremendous journalistic sensation and occasioned public alarm. Intensity barely reached VII near the coast; a section of highway cut in the bluffs north of Mussel Rock was closed by slides. In the urban area damage may have exceeded half a million dollars; this is the sum of many individual small items. Damage was heaviest in Daly City, a surburb south of San Francisco which includes the coast near the epicenter. Many chimneys were damaged; over forty were pulled down by the fire department. There was much plaster damage, especially to dwellings in a new development on unconsolidated foundation adjacent to the coastal bluffs.

LISTS OF SHOCKS, 1903–.
CALIFORNIA REGION

In Tables 28-1 and 28-2 (Fig. 28-2) attempt has been made to list all shocks of magnitude 6 and over in the region from 1903 to date; for the years before 1915 the listing is almost certainly incomplete. Table 28-1 lists shocks from 30° northward (between 30° and 32° there is increased likelihood that shocks of magnitude 6+ have been missed), except those near and off the north coast, from 124° westward which are given in Table 28-2. Data are chiefly from *Seismicity of the Earth*; one shock listed for 1903 is from a recent investigation by Gutenberg, and magnitudes for the larger shocks are as revised by him (with *m* replaced by *M* according to Table 22-4). A few shocks of dates 1909–1915 have been added on the basis of macroseismic data. For these, magnitudes are rough estimates; the question attached to coordinates indicate that these are little better than guesswork. Throughout these tables the smaller magnitudes have been re-examined by the writer, and in some cases slightly altered. Times are G.C.T.; to reduce to Pacific Standard Time subtract 8 hours.

References

PUBLICATIONS referring primarily to individual earthquakes (including that of 1906) are cited in the chronological bibliography (Appendix XVI), where notes are occasionally added to identify the source of material in this chapter.

Catalogues, etc.

Holden, E. S., "Catalogue of earthquakes on the Pacific coast 1769 to 1897," *Smithsonian Inst. Misc. Collections,* No. 1087 (1898).

McAdie, A. G., "Catalog of earthquakes on the Pacific coast 1897–1906," *ibid.,* No. 1721 (1907).

Townley, S. D., and Allen, M. W., "Descriptive catalog of earthquakes of the Pacific coast of the United States 1769 to 1928," *B.S.S.A.,* vol. 29 (1939), pp. 1–297.

Wood, H. O., and Heck, N. H., "Earthquake history of the United States. Part II, Stronger earthquakes of California and western Nevada," *U. S. Coast Geodetic Survey,* Ser. No. 609, rev. ed., 1951.

United States Earthquakes—serial published for individual years 1928 ff. by the U. S. Coast and Geodetic Survey.

Freeman, J. R., *Earthquake Damage and Earthquake Insurance,* McGraw-Hill, New York, 1931. (Reprints much source material on California earthquakes.)

Williams, J. S., and Tapper, Mary L., "Earthquake history of Utah, 1850–1949," *B.S.S.A.,* vol. 43 (1953), pp. 191–218. (On p. 212 by oversight the effects at Salt Lake City of the Nevada earthquakes of 1932 and 1934 are ascribed to a local source.)

Milne, W. G., "Seismic activity in Canada, west of the 113th meridian, 1841–1951," *Publ. Dominion Observatory,* Ottawa, vol. 18 (1956), pp. 119–146. (Among other shocks this cites a strong earthquake on Dec. 14, 1872, epicenter probably in British Columbia near 49° 10′ N 121° 00′ W).

History of seismology in the region

Most of the important developments after 1910 are reflected in the pages of the *Bulletin of the Seismological Society of America (B.S.S.A.).* Some historical details will be found in the following references.

Lawson, A. C., "Seismology in the United States," *B.S.S.A.,* vol. 1 (1911), pp. 1–4.

Louderback, G. D., "History of the University of California seismographic stations and related activities," *ibid.*, vol. 32 (1942), pp. 205–230.

Wood, H. O., "Earthquake study in Southern California," *Scientific Monthly,* vol. 39 (1934), pp. 323–344.

Day, A. L., "An adventure in scientific collaboration," *Carnegie Inst. Publ.,* No. 501 (1938), pp. 3–35. (History of the Pasadena seismological program through 1936.)

Ulrich, F. P., "The California strong-motion program of the United States Coast and Geodetic Survey," *B.S.S.A.,* vol. 25 (1935), pp. 81–95. See also: "Earthquake investigations in California 1934–1935," *U. S. Coast Geodetic Survey, Spec. Publ.,* 201 (1936).

Progress reports for the U.S.C.G.S. appeared approximately annually in *B.S.S.A.* and in *United States Earthquakes.*

Progress reports for the Pasadena program appeared in the *Year Book of the Carnegie Institution* of Washington through 1941 and, beginning with 1944, in the *Transactions of the American Geophysical Union.*

Other references

Wilson, James T., "Foreshocks and aftershocks of the Nevada earthquake of December 20, 1932, and the Parkfield earthquake of June 7, 1934," *B.S.S.A.,* vol. 26 (1936), pp. 189–194.

Eckart, N. A., "Development of San Francisco's water supply to care for emergencies," *ibid.*, vol. 27 (1937), pp. 185–204.

Byerly, P., "Earthquakes in the San Francisco Bay area," *Calif. Dept. Nat. Resources Div. of Mines, Bull.* 154 (1951), pp. 151–160.

———, "Earthquakes off the coast of northern California," *B.S.S.A.,* vol. 27 (1937), pp. 73–96.

———, "The earthquake of July 6, 1934: amplitudes and first motion," *ibid.*, vol. 28 (1938), pp. 1–13. (One of the shocks located in the preceding reference.)

Tocher, D., "Earthquakes off the north Pacific coast of the United States," *ibid.*, vol. 46 (1956), pp. 165–173.

Wood, H. O., "Earthquakes in southern California with geologic relations," *ibid.*, vol. 37 (1947), pp. 107–157, 217–257. (Has a good fault map. The maps and discussion of epicenters can be used for general purposes but are uncritical and untrustworthy in detail.)

Whitten, C. A., "Crustal movement in California and Nevada," *Trans. Amer. Geophys. Union,* vol. 37 (1956), pp. 393–398.

Thompson, G. A., "Gravity measurements between Hazen and Austin, Nevada," (abstract), *ibid.*, vol. 38 (1957), pp. 408–409.

CHAPTER 29

New Zealand Earthquakes

HISTORICAL NOTE

Lists of New Zealand earthquakes generally begin with 1834 or 1835, when shocks are stated to have been violent in the Auckland area. This occurrence has been questioned, with the suggestion that it possibly took place at some locality now considered distant from Auckland, such as the Lake Taupo area.

Bastings listed 69 New Zealand earthquakes for 1835–1934 reported to have reached intensity VIII or over on the Rossi-Forel scale. Especially for the earlier years, these constitute a more or less accidental sample, conditioned by the distribution of population and the circumstances of reporting.

FIGURE 29-1 *New Zealand earthquake epicenters. (Tables 29-1 and 29-2, plus 1848, 1855, 1888, 1901.)*

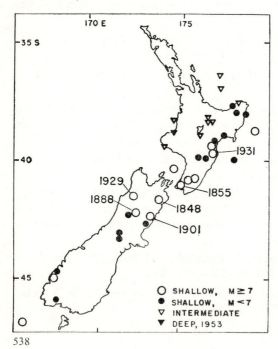

The earthquakes in New Zealand of greatest historical importance (Fig. 29-1) are the major disturbances of 1848, 1855, 1929, and 1931, all with evidence of faulting, and the somewhat smaller shock of 1888 which is exceptionally interesting as the earliest well-documented instance of strike-slip.

Catalogues were published for 1848–1890 by Hogben and for 1868–1890 by Hector; annual lists for 1869–

1902 appeared in the *Transactions of the New Zealand Institute*. For 1903–1920 material is on file at the Dominion Observatory, Wellington, but has not been published in detail. (Hayes has given a two-page summary.) Since 1921, annual summaries have been published by the Observatory.

Largely through the efforts of John Milne, stations were installed at Wellington (1898) and Christchurch (1901); but for more than 20 years there was no further installation, and few papers dealing with the seismology of the area were published.

New stations were set up after the earthquake of 1929; seismological investigation expanded and was given a great stimulus by the disaster of 1931. Valuable papers bearing on New Zealand seismology have appeared in considerable number, chiefly in the *New Zealand Journal of Science and Technology* and in the *Transactions of the Royal Society of New Zealand*. Many have originated at the Dominion Observatory, which continues to issue station bulletins. Readings are now reported from Wellington for 12 stations, including Suva (Fiji). Epicenters and origin times are given for local earthquakes. Many of the stations are equipped with torsion seismometers of the same standard type as used in California, and earthquake magnitudes are assigned from these on the basis of the tables developed in California for the first form of magnitude scale.

Numerous important papers bearing on faulting and on the geology of earthquakes come from the New Zealand Geological Survey, with headquarters at Wellington. The distinguished geomorphologist, C. A. Cotton, of Victoria University College, Wellington, has published and inspired many books and papers bearing on tectonic problems. The Department of Scientific and Industrial Research, with which the Dominion Observatory is now affiliated, has sponsored much seismological and geophysical investigation.

LARGER EARTHQUAKES SINCE 1914

Table 29-1 lists shallow earthquakes in the New Zealand region from 36° to 47° south latitude. It includes all those shown as of magnitude 6 or over in the second edition (1954) of *Seismicity of the Earth,* with additions for 1947 and 1948 derived from the *International Seismological Summary*.

Table 29-2 lists deep-focus shocks of all magnitudes from the same publication, with two important additions for 1953. The cataloguing here is far from complete, since it includes only shocks clearly recorded at distant stations. Bulletins from Wellington list many more, especially in the main belt of intermediate shocks (depths of 150 to 200 km), as noted in Chapter 27.

The epicenters in the two tables, with those of four earlier earthquakes, are mapped on Fig. 29-1.

Table 29-1   New Zealand Shallow Earthquakes*

Date (Local)	Lat. S	Long. E	M	Location
1914 Oct. 7	38	178½	6½	Gisborne province
1914 Oct. 28	40	178	6½	Gisborne province
1918 Nov. 3	47	165	6¾	
1922 Dec. 25	43	173	6¼	North Canterbury
1927 Feb. 26	38	178	6¾	
1929 Mar. 9	42½	172	6.9	Arthur's Pass
1929 June 17	41¾	172¼	7.8[1]	West Nelson (Buller)
1931 Feb. 3	39½	177	7.9[1]	Hawke's Bay
1931 Feb. 8	39½	177	6½	Hawke's Bay
1931 Feb. 13	39½	177	7.3[1]	Hawke's Bay
1932 Sept. 16	39	177½	6.8	Wairoa
1934 Mar. 5	40½	175½	7.6[1]	Pahiatua
1934 Mar. 15	40	176	6¼	Hawke's Bay
1937 Oct. 24	37¾	177¾	6¼	
1938 Dec. 17	45	167	7.0	Southland, Otago
1938 Dec. 30	40¼	176½	6¼	Hawke's Bay, etc.
1939 Apr. 21	46½	167½	6	
1942 June 24	41	175½	7.3[1]	Wairarapa
1942 Aug. 2	41	175¾	7.3[1]	
1943 Aug. 2	45	167	6¾	
1945 Sept. 2	46½	165½	7.2	
1946 June 27	43¼	171½	6½	Lake Coleridge
1946 June 28	43½	171½	6	Lake Coleridge
1947 Mar. 26	38¾	178½	7.1[1]	Damage at Gisborne; small tsunami
1947 May 17	39	179	6½	
1947 Aug. 28	39¼	179	6¾	
1948 May 23	42½	172¾	6¼	Waiau, Hanmer

* From *Seismicity of the Earth* and I.S.S., with Greenwich dates changed where necessary.

[1] Revised magnitudes, replacing those reported in Seismicity of the Earth.

AWATERE EARTHQUAKE— OCTOBER 19, 1848

Lyell published a fragmentary account of this great earthquake in *Principles of Geology*. It was perceptible over about half of New Zealand, and violent on both sides of Cook Strait; three deaths are recorded at Wellington. Lyell's

Table 29-2　New Zealand Intermediate and Deep Earthquakes
(Magnitudes for the smaller shocks are rough estimates)

Date (G.C.T.)	Lat. S	Long. E	Depth (km)	M
1914 Nov. 22	39	176	100	7
1921 June 28	38½	175½	140	6¾
1931 Sept. 21	37½	178	80	6¾
1938 Oct. 30	38½	176½	150	5
1938 Nov. 1	38½	176½	150	5
1939 May 14	36½	179	80	6
1940 Oct. 7	38½	176¾	170	5¾
1951 Mar. 28	36½	177	360	6¼
1949 Feb. 9	39¾	174	170	6¼
1953 Mar. 24	39	174½	570	5¼
1953 Sept. 29	37	177	300	6.8

informant, Sir F. Weld, described fissures extending many miles along the Awatere Valley, hence in the zone of the Awatere fault; these were found immediately after the earthquake. There is no way of knowing what part of them were secondary features due to shaking, nor whether the observers of 1848 took rift features of older date as having just been produced (they still preserve a relatively fresh appearance). Until lately, informed opinion in New Zealand favored these possibilities as against actual displacement on the Awatere fault in 1848.

Cotton (1954) has rediscussed the matter. He considers that a subsidence at the mouth of the Wairau River, often attributed to the 1855 earthquake, may have accompanied that of 1848; this suggests that fresh-appearing scarps and other trace effects on the Wairau fault may have originated in 1848. Cotton prefers to consider that both Awatere and Wairau faults were affected, with incidental subsidence of the block between them.

EARTHQUAKE OF 1855

This earthquake and its accompanying faulting were also reported by Lyell; it was the first instance of observed faulting to become generally known. Our knowledge of the event has been improved by field work reported in 1943 by Ongley, who extends and corrects the information published by Lyell. The fault trace of 1855 can still be followed on the ground (Fig. 29-2). It passes about 15 miles from Wellington, the capital, where there was serious damage, and strikes northeastward, conforming to the trend of the major structures of

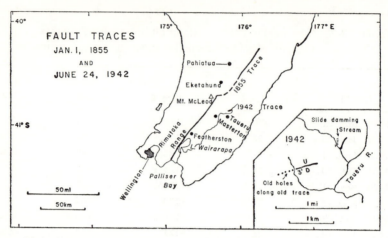

FIGURE 29-2   *Map showing fault location, New Zealand, 1855. Inset: faulting of the Wairarapa earthquake, 1942. [After Ongley.]*

that part of the North Island. The country to the west was generally uplifted; Ongley, following the trace, found indications of upthrow of from 3 to 10 feet in that sense, though oddly he found two places where the east side had been relatively raised by 6 to 10 feet (scissoring?). Lyell's informants described the trace as 90 miles long, and this is roughly confirmed by Ongley. At one point near the trace a large slide scarred a conspicuous mountain and produced a small lake by damming. This, as usual, gave rise to a report that the mountain (Mt. McLeod) had been "rent in twain."

Lyell was misled in one important detail. He published a schematic figure showing faulting of beds exposed in a sea cliff; but Ongley reports that this locality is on an old fault well to the east of the 1855 trace. He could not follow the actual displacements down to the coast; but the location was made quite clear by elevation to the west which affected the port of Wellington and the intervening coast. At the time of the earthquake a new road was being cut at the base of cliffs above tide level; after the earthquake there was at this point a shelf high enough to afford a dry route.

EARTHQUAKES OF 1888 AND 1901

New Zealand authors refer to the earthquake of September 1, 1888, by the name of the district, Amuri, in which the meizoseismal area was located; this is in the northern part of the South Island, near its center line. Right-hand strike-slip occurred on a few miles of a feature which Cotton has since named the Hope fault. Fences were offset 8 feet and more. The occurrence was described with care by A. McKay, one of the most distinguished New Zealand field workers (see Fig. 29-3). At that time strike-slip during earth-

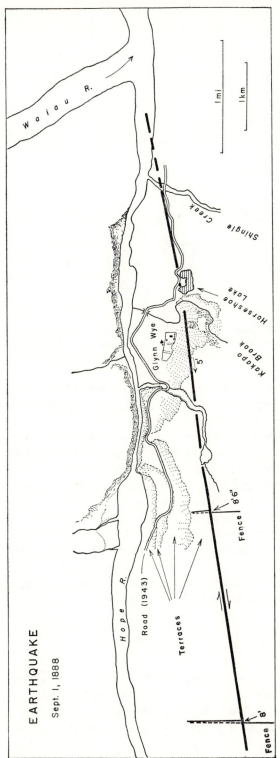

FIGURE 29-3 *Faulting in the Amuri earthquake of 1888, according to McKay. Topography from air photographs, 1943. [Crown copyright. Permission to publish granted by Land Survey Department, Wellington.]*

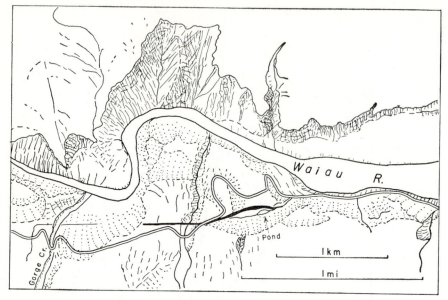

FIGURE 29-4   *Faulting in the Hope fault rift, east of that shown in Figure 29-3. Drawn from photographs. [Crown copyright. Permission to publish granted by Land Survey Department, Wellington.] Compare Figure 27-10.*

quakes was practically unheard of. It is not surprising to learn that other geologists, going over the ground and seeing the same effects, misinterpreted the displacements as a series of slides.

The present writer, on visiting the locality, was struck by the resemblance of features on the Hope fault to those of the San Andreas fault on a small scale. It is equally striking that the miniature rift (Figs. 27-10, 29-3, and 29-4) ascends obliquely from stream level to a high terrace. At one point there is a large pond which appears like a sag pond but is not; horizontal shift of a ridge has displaced the lower course of a stream, ponding it until it finds its way down. Small, true sag ponds occur nearby in the fault zone.

The large Cheviot earthquake of November 16, 1901, with epicenter in the northeast-trending structures of northern Canterbury province, is reported to have been accompanied by the formation of cracks and perhaps of scarplets; but McKay's description indicates only secondary effects.

EARTHQUAKE OF 1929

On June 17, 1929 (local date; June 16 G.C.T.) a major earthquake occurred on the South Island (Fig. 29-5). Publications have referred to this as the earthquake of West Nelson (the province), Buller (a county), and Murchison (the next county, and the town most affected). This was the first

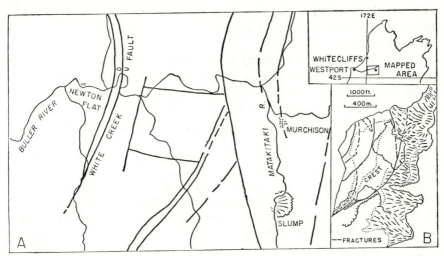

FIGURE 29-5 *New Zealand earthquake of 1929. A, meizoseismal area and fault-ing; B, Whitecliffs slide. [After Henderson.]*

large earthquake affecting populated areas of New Zealand since 1855; previously only 11 persons are known to have been killed during earthquakes in New Zealand, but 17 were killed in 1929.

Murchison at the time had a population of about 300. Most of its buildings were one-story wood structures; some of these were shifted or badly racked. "The most striking wreck was a two story concrete store . . . , which leaned dangerously to one side and later collapsed altogether under the aftershocks." At and near Westport, a town of about 4000 over 30 miles to the west, there were more brick structures and consequently more damage. Shaking was perceptible over a large area, including most of the South Island and extending far beyond Wellington into the North Island to distances of over 250 miles.

At this time a scarp was formed, 14 feet 9 inches in height where it crossed a main road which follows the Buller River, about 14 miles west of Murchison. This was on the line of the White Creek fault, the eastern one of an almost parallel pair, between which a slice of Tertiary limestones, sandstones, and other rocks about 7 miles long and 2000 feet wide is faulted down into granite. The bounding faults dip almost vertically and converge north and south to the ends of the slice. The scarp of 1929, downthrown to the west, was highest at the crossing of the river and road; H. E. Fyfe of the Geological Survey traced it 3 miles northward and 2 miles southward until the throw disappeared and other evidence like cracks and split trees faded out. (This was no small feat, as anyone who has encountered the New Zealand "bush" will appreciate.) A displaced fence south of the river indicated left-hand strike-slip, the eastern block having moved relatively 7.2 feet north and 5 feet west; the westward component would call for a 70° eastward dip. On the south side of the Buller River the displacement stepped up a river terrace which had previously been nearly level and showed no evidence of any older events of the same kind.

Releveling of monuments along railway lines in the area indicated uplift on both sides of the fault, that on the east being larger but decreasing more rapidly with distance from the fault. The observed points are few in number.

Landslides have figured prominently in the accounts of New Zealand earthquakes; those of 1929 were indeed spectacular. Henderson notes:

> The earthquake occurred in mid-winter when soil and subsoil were saturated, and throughout the severely shaken areas multitudes of slips descended from the slopes of terraces, hills, and mountains. It is to these that most of the fatalities and the greater part of the material damage were due. The main high road through the Lyell gorge and that between Seddonville and Karamea could not be reopened for wheeled traffic for many months.

Other roads were hastily repaired; a temporary bridge was built to cross one stream. The Buller Gorge, where the 14-foot scarp broke the road, was passable in a week; but two roadmen were killed on the highway in this area, and slips were large and numerous everywhere for 10 miles west from the White Creek fault. Three miles south of Murchison, where Tertiary sediments dip steeply into the Matakitaki valley,

> . . . the whole face of the hill to the top of the ridge came away, burying the valley bottom and low terraces—in places to a depth of 180 ft. or more—extending across to the 100 ft. terrace on the east side of the river . . . A very flat cone of debris a mile wide was formed and a lake three miles long and up to 80 ft. deep now fills the valleys . . . The Busch and Morel homesteads were destroyed, and five people killed . . .

Several large slumps are described, with the toe of the slide elevated well above its former level. The greatest of these was on the west coast at Whitecliffs, south of Karamea (see A, Fig. 29-5). It affected a lens-shaped area about a mile long and a little less than half a mile wide at the broadest point. The rocks are Tertiary sediments, chiefly sandstones but with some limestone, breccias, etc. The cliffs are about 1200 feet high; at their base before the earthquake there was a broad bench rising to about 300 feet above the sea. The slump broke primarily on a fracture at the back of this bench, concave toward the sea, where the material dropped as much as 120 feet. The elevated toe of the slump forms the western part of the lens shape; here a part of the former sea floor, over 300 feet wide and nearly a mile long, was raised to an elevation of about 40 feet. A known fault between the Tertiary rocks and granite passes through or near this area, and there was some suspicion that movement on this fault might have been a secondary part of the earthquake. This cannot be decided from the evidence; there is always the possibility, as in other cases of the kind, that the earthquake originating on the White Creek fault triggered the dropping of a block along a local fault at or near Whitecliffs, and that this in turn set off the slump.

HAWKE'S BAY EARTHQUAKE—1931

The earthquake of February 3, 1931 (local time; 22:46:42, February 2, G.C.T.) was the first great earthquake disaster in New Zealand; 255 were killed. Property damage was never reliably estimated. About 2½ million pounds were paid out in government relief and in insurance, but there is no doubt that total loss was much greater. Much of the loss represents destruction by earthquake and fire at Napier and Hastings, with populations near 16,000 and 10,000.

Napier

The earthquake is termed the Hawke's Bay earthquake, from the name of the province most affected. (The actual bay where Napier is located is more often referred to and mapped as Hawke Bay.) Napier is on a peninsula, the northern part of which is a limestone hill,† which was used for residences and public institutions, including the hospital. The principal business district was on the flat approach to the hill. Business blocks were nearly all masonry; many collapsed completely, and most were seriously damaged, choking the streets with fallen debris. A few better-constructed buildings survived without damage. There was not the difference in apparent intensity which might have been expected between the hill and the flat land. On the hill, residences, especially two-story structures, were seriously racked and distorted; the hospital was badly damaged, and its three-story nurses' home totally collapsed into a pile of rubble. Here the better conduction of seismic waves through the limestone may have compensated for the normal increase

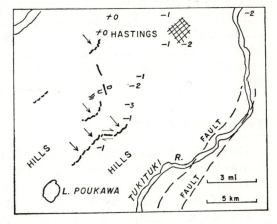

FIGURE 29-6 *Meizoseismal area and fault lines, Hawke's Bay earthquake, 1931. [After Henderson.]*

---

† This hill is actually a land-tied island, and is locally known as Scinde Island.

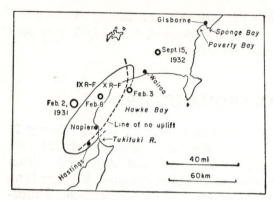

of amplitude on the softer ground of the flat. (Compare the discussion of Monghyr in the Indian earthquake of 1934, Chapter 5.) The area about Napier was uplifted, and former lagoons were turned into mud flats; the decay of dead marine organisms on these flats created a serious sanitary risk.

FIGURE 29-7   *Hawke's Bay area. Isoseismal of IX–X R.F., 1931, and line of no uplift. [After Henderson.] Instrumental epicenters. [Bullen.]*

Isoseismals

The meizoseismal area, with intensities of IX and over (Figs. 29-6, 29-7), was elongated, extending about 100 miles northeast to southwest, with a width of about 30 miles; Napier was near the center. All the isoseismals showed similar elongation, parallel to the general structure of New Zealand. Thus the earthquake was felt as far as Timaru on the South Island, 460 miles to the southwest, but not at Auckland, 200 miles northwest. Intensity was abnormally low (IV) in a small area near the south coast of Cook Strait, a so-called "earthquake shadow."

Microseismic Data

The earthquake was well recorded at distant stations. Seismograms of this (and also of the 1929 earthquake) were studied in detail for the information they give about the core of the earth; the European stations, in particular, were at great distances (the point antipodal to Napier is in central Spain), and waves reaching them from New Zealand pass near the center of the earth. Four stations were recording in New Zealand, two in the North Island and two in the South Island; of these two had only insensitive strong-motion instruments, so that the times of first motion there are less reliable than for the chief stations at Wellington and Christchurch.

Thrusts and Faulting

The careful field investigation revealed effects not closely like those of any previously well-described earthquake. These and many other details were strikingly similar to the phenomena of the 1952 California earthquake (Chapter 28). Taken together, they indicate seaward overthrusting of a large block.

The most noteworthy features were in the vicinity of the Poukawa Valley, which trends northeast-southwest a few miles southwest of Hastings. Here, except for small areas of alluvial cover, Tertiary limestones are exposed, dipping 15° to 20° northeastward. For about 6 miles along this valley there was a zone of ridges, rents, and cracks. Henderson reports:

The ridges are obviously due to the shortening of the surface, and resemble the pressure ridges formed at the toe of a surface slump. Like them also they usually occur at the foot of a slope, but their continuity in a general direction for miles, and the fact that in most places no gaping cracks occur higher up the slope, show that they are not mere surface slips. The ridges usually rise 3 or 4 ft. and in places 6 ft. to 8 ft. above the general surface. At some points the main rift branches, and at others there are two or three sub-parallel ridges. Occasionally the tough dry turf is folded into broken recumbent folds or masses of turf override the ground in front for several feet. In general, these folds and overthrusts indicate a movement of the country on the west side of the fracture relatively toward the east, and, as judged from the slackening of wires where fences cross, this movement is measurable in feet rather than in inches. The ground immediately west of the fracture is nearly everywhere much more disturbed than that east of it and, in general for a few yards from the fracture, the surface bulges and is seamed with gaping cracks as if the ground beneath had swollen and stretched the turf . . .

This description applies actually to two separate parts of the trace, both having a northeast trend and forming an echelon offset; the southwestern segment, about 3½ miles long, was connected with the northeastern one, about 1 mile long, by an east-west trace somewhat over a mile long. This connecting segment was very different in character, being a typical strike-slip trace. It crossed a railway and road with right-hand offset of 6 to 7 feet. Here

. . . cracks and gaping fissures are decidedly more common than pressure ridges. These are more or less parallel with the general course of the fracture, but there are important groups of fissures that, parallel among themselves, follow the fracture *en echelon* obliquely transverse to it. These fissures strike south-east, whereas the dislocation as a whole strikes east, and, as the ground on the north side has moved eastward relatively to that on the south, the fissures gape and the strips of turf between, compressed lengthwise by one component of the horizontal movement, are confusedly folded and overthrust . . .

A photograph of this part of the trace shows an unmistakable mole-track feature, with the earth piled up to heights of about 4 feet. Henderson writes:

The course of the major part of the fracture along the scarp side of two strike valleys suggests that movement occurred along weak layers between the thick beds of strong limestone; the east-west part of the fracture is then the outcrop of a rupture plane connecting these bedding planes. The country west of the fracture moved east in respect to the country to the east and along the strike valleys tended to override it.

There was a second northeast-southwest trace of pressure ridges about 1¾ miles long, located about a mile northwest of the Poukawa Valley feature; this line of disturbance at its northeast end turns nearly north, and after nearly 2 miles ends in a northwest-trending fracture with left-hand strike-slip of about 3 feet 9 inches as shown by fences. About 2 miles still farther north there was a third short northeast-trending trace of pressure ridges. Other minor ridges and cracks were found.

Changes of Level

Releveling was carried out over all available monuments; most of these were along the railroad, which skirts Hawke Bay from Wairoa down to Napier, and then runs southwestward, passing through the Poukawa Valley. Other bench marks were along the rivers southwest of Napier. A line of zero change of level was found extending from a point on the coast between 1 and 2 miles south of Napier southwesterly, parallel to the regional structures and the main earthquake traces. On the landward side of this line there were elevations generally up to 3 or 4 feet; on the coastal side subsidences, mostly about 1 foot, but some as much as 2 or 3 feet, were found.

Along the railroad round the coast northwesterly from Napier, uplifts at first amounting to 6 to 7 feet diminished, reaching zero at a point on the north shore of Hawke Bay which is west of the direct prolongation of the zero line from south of Napier. This can be represented by curving the line of no change; soundings showed a general rise of about 6 feet off Napier, without fixing its geographical limits.

Changes of level along the coast were confirmed by inspection. On the south side of the line of no change established by releveling, marshes and lagoons had increased in area, indicating subsidence of about 1 foot. Toward Napier uplift became evident; wave-deposited detritus was found well above the level reached by waves after the earthquake. The tide gauge at the Napier breakwater showed a land uplift of 6 feet. The wharf at Napier, however, was uplifted nearly 7 feet as measured by Professor Cotton from the evidence of upraised marine organisms (see Fig. 316 of his book, *Landscape*). Northward along the coast evidence of uplifts up to 9 feet were found, but observation was difficult because of large slides from cliffs.

Effects Near Gisborne

At Sponge Bay, a little over 2 miles from Gisborne on the north shore of Hawke Bay, on the afternoon of February 17 a boulder bank previously covered by 1 to 2 feet of water was seen to rise rapidly to 7 feet above water level. No earthquake was felt at the time; by this date the aftershock activity was subsiding. On the land side of the beach a corresponding area was depressed a few feet with much cracking. This displacement is on or near the line of a fault which passes close to Gisborne, but it is not likely that it represents faulting. It was evidently a large slump somewhat smaller than the one at Whitecliffs in the 1929 earthquake. On March 25, 1947 (March 26, New Zealand time), an earthquake of magnitude 7 originated not far from Gisborne. The town was damaged, and a destructive wave entered the adjacent Poverty Bay. This wave appears to have been rather local and may have originated in a slump. Two large aftershocks in May and August, 1947, are noted in Table 29-1; their instrumentally determined epicenters may be in error by half a degree.

Epicenters

For the principal earthquake of February 3, 1931, preliminary determinations of the epicenter placed it near Napier. Bullen, after careful rediscussion of the recorded times, including those at distant stations, derived the position 39° 20' S, 176° 40' E. (See Fig. 29-7.) This is well inland, even outside the meizoseismal area; however, it would not be inconsistent with the probable dip of the thrust fault, given a source at depth of about 20 kilometers.

The largest aftershock, on February 13 (magnitude 7.3), had nearly the same epicenter as the main earthquake. Some of the other aftershocks differed; three strong ones on February 20 and 24 were placed by Bullen nearer to Wairoa than to Napier, on the direct line between the two places, and they were reported felt more intensely at Wairoa than elsewhere. Many smaller aftershocks were felt, and recorded instrumentally, but the seismograms were not adequate for precise location. A valuable list of amplitudes recorded by the torsion seismographs at Wellington was published; from this Dr. Benioff plotted a strain-release curve (Fig. 6-1). This curve shows a definite break after 2 days 10 hours, interpreted as passage from the compressional to the shear phase of strain release. Whether this corresponded to a change in the geography of the epicenters, as in California in 1952, cannot be decided.

The Wairoa Earthquake

The Wairoa earthquake of September 15, 1932 (September 16, New Zealand time), was a remarkable case of the late occurrence of a large following shock not precisely to be considered an aftershock, since its energy must derive largely from the principal regional strain (Chapter 6). This shock, of magnitude 6.8, developed intensity IX in a small area including Wairoa. Considerable trouble was experienced in placing the epicenter instrumentally. By using the same methods as for the preceding earthquakes, Bullen found a reasonable interpretation of the readings with an epicenter near Wairoa.

About 3 miles northeast of Wairoa, Ongley identified a fresh fault trace, breaking a ridge transversely with trend east-northeast and downthrow northward. This parallels a fault previously mapped. Retriangulation of the Wairoa area showed displacements attributable to the earthquake, indicating right-hand shear with a roughly northeast strike.

WAIRARAPA EARTHQUAKE—1942

The earthquake of June 24, 1942, offered a bare minimum of data for establishing surface faulting. The meizoseismal area was on the North Island; it included Wellington, where there was more damage than at any time since 1855, but the town most damaged was Masterton. After careful search, probable trace phenomena were found along a fault roughly parallel to that of 1855, about 10 miles east of it. In the center of the area most heavily shaken and most affected by slumps and cracking, there were two linear traces made up of fissures and low scarps. These traces followed the bases of older ridges or crossed ridges through old notches, without much relation to the general topography. The better developed trace trended N 70–80° E, with a throw of 3 feet; the other trace more nearly followed the grain of the country, trending N 40° E. The combined length was not over half a mile. With features so small and of such short extent, it is hard to distinguish between primary and secondary effects; even the larger older scarps and notches might indicate simply the accumulated results of repeated shaking. Comparing with the other earthquakes discussed, one would expect more definite faulting from one of this magnitude (7.3); however, a nearly equal shock on August 1, 1942, with an epicenter a little more to the east, occasioned no serious damage, nor was a fault trace found (there is some evidence for hypocentral depth near 50 kilometers).

OTHER INVESTIGATED EARTHQUAKES

A number of other shocks have been studied specially or deserve special mention.

1921, June 29 (June 28, 13:58 G.C.T.). This strong shock, in the Hawke's Bay region, was felt over a wider area than the major earthquake of 1931. Bullen, carefully studying all the available readings, arrived at an epicenter near 39.3° S, 176.4° E, with a focal depth of about 80 kilometers; the latter result indicates why this shock was not more destructive (a later revision appears in Table 29-2). Bullen applied the same methods to the earthquakes of 1931 and 1934.

1922. The block displacements (normal faulting) accompanying a swarm of earthquakes in the Lake Taupo area have been described by Morgan. This is a typical phenomenon of the North Island volcanic zone.

1929, March 9. The large Arthur's Pass earthquake is interesting because it reached its maximum intensity in the vicinity of the Alpine fault, but other probably active faults are nearby, and neither instrumental readings nor the macroseismic observations available in the sparsely inhabited mountain region are adequate for decision.

1934, March 15. The Pahiatua earthquake was the first damaging shock in a populated part of New Zealand following the Hawke's Bay and Wairoa earthquakes of 1931 and 1932. Field investigation revealed no trace phenomena (hardly to be expected at magnitude 6¼). As in other cases, the epicenter originally derived from seismograms conflicted with macroseismic evidence; Bullen's revised result is in better agreement.

1939, April 21. This has been noted in Chapter 27 as possibly related to a recent fault trace.

1953, March 24. The significance of this small shock at a depth of about 570 kilometers has been noted in Chapter 27.

1953, September 29. This earthquake of magnitude 6.8, at a depth of over 300 kilometers under the Bay of Plenty, is notable for having been felt over most of the North Island (excluding the Auckland Peninsula), and south to latitude 43° in the South Island.

References

Publications on individual earthquakes are listed in Appendix XVI. The following are general.

Bastings, L., and Hayes, R. C., "Earthquake distribution in New Zealand, 1848–1934," *N. Z. Journ.*, vol. 16 (1935), pp. 308–312.

Hayes, R. C., "A summary of New Zealand earthquakes for the period 1903–1920," *ibid.*, pp. 361–363.

Bastings, L., "Destructive earthquakes in New Zealand, 1835–1934," *ibid.*, vol. 17 (1935), pp. 406–411.

Hayes, R. C., "The seismicity of New Zealand, *ibid.*, Section B, vol. 23 (1941), pp. 49–52.

Henderson, J., "Earthquake risk in New Zealand," *ibid.*, vol. 24 (1943), pp. 195–219.

Hayes, R. C., "On earthquake distribution in New Zealand," *ibid.*, pp. 236–238.

Annual summaries have been published as *Bulletins of the Dominion Observatory, Wellington,* and also in the *New Zealand Journal of Science and Technology.*

CHAPTER 30

Japan and Formosa (Taiwan)

THE REGION OF JAPAN—
PRINCIPAL FEATURES

For the seismologist, Japan has the advantages of a long recorded history, a
high density of population, and an urban culture with active research centers.
To quote Imamura:

Although the first recorded earthquake of authentic history bears date of A.D.
416, the number of those recorded is very small until the great Nankaidō earth-

FIGURE 30-1   *Large earthquakes of Japan.* [*After Imamura and Musha.*]

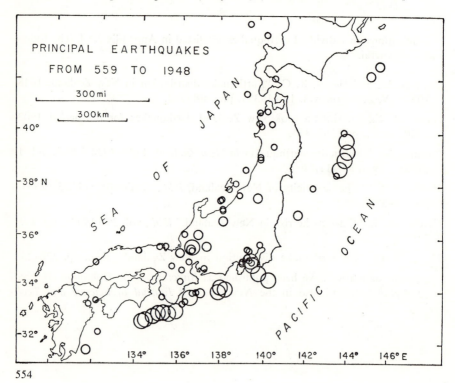

PRINCIPAL   EARTHQUAKES
FROM   559   TO   1948

300 mi

300 km

40°

38° N

36°

34°

32°

134°   136°   138°   140°   142°   144°   146° E

JAPAN

SEA   OF

PACIFIC   OCEAN

quake of November 29, A.D. 684. While even as early as A.D. 684, Central Japan was more or less cultured, and earthquake records are fairly comprehensive, this cannot be said of localities remote from the centre of culture; but from 1596 and onwards, the records for the whole country can be regarded as fairly complete.

Imamura lists and maps 66 destructive earthquakes from 1596 to 1935 (See Figs. 30-1 and 30-2), and goes on to say:

While admitting the futility of any attempt to deduce comprehensive seismic zones from a catalogue of earthquakes covering no more than 330 years, it may be pointed out that statistics also of older dates have been drawn upon.

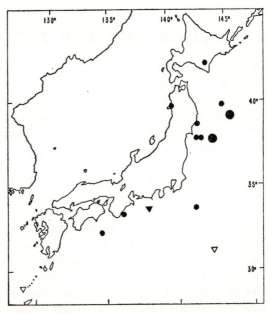

FIGURE 30-2 *Large earthquakes of the Japanese region, magnitude 8 and over, 1897– .* (*Table 30-1*). *Circles, shallow; open triangles, intermediate; solid triangles, deep.*

The literature of Japanese seismology is extensive. The large part of it published only in Japanese is inaccessible to most western readers; fortunately the most important materials have generally been republished in English or some other western language, at least in summarized form. The writer has occasionally had to seek help in getting a few critical passages translated.

Japan is located in the circum-Pacific belt just where an important branching occurs. There are several arcs of Pacific type with which are associated most of the larger, shallow earthquakes of the area, as well as the intermediate earthquakes and active volcanoes. These arcs front on the Pacific and on the Philippine Sea. The Japan Sea coast of Honshu is a region of block tectonics. Two belts of deep-focus earthquakes, named the transverse and the Sōya zone (from Sōya Strait north of Hokkaido) by their first describer, Wadati, take courses which appear at first unrelated to the surface structures. The principal features of Pacific arcs, including deep oceanic trenches (foredeeps) and belts of negative gravity anomalies, are all represented.

STRATIGRAPHY AND TECTONICS

In much of Japan, the older accessible rocks are igneous and metamorphic, overlain by volcanic formations of all ages down to the most recent. However, in many areas sedimentary rocks with adequate fossil material for dating are found. Enough is known to piece out a geological history in some measure comparable with that of California and New Zealand, in terms of a Mesozoic and a Pleistocene orogeny.

Principal Tectonic Divisions

The Mesozoic folding produced mountains in a zone crossing the islands of Kyushu and Shikoku and the large Kii peninsula. Associated with this is a major fault zone along which there was Cretaceous and Tertiary southward overthrusting; this is the Median Tectonic Line (Fig. 30-3), which divides that part of Japan into an inner zone extending to the Japan Sea and an outer zone facing the Pacific. The line terminates eastward at a still more important tectonic break which forms the western boundary of what Naumann named the Fossa Magna; this crosses Honshu and separates Japan into a southwest and a northeast tectonic division. It is the southwest division which is sharply separated into an inner and outer zone; a similar separation in the northeast division probably exists but is obscured by the prevalence

FIGURE 30-3   *Outline map of Japan, showing major tectonic lines. Shading, outer zone of northeast Honshu.*

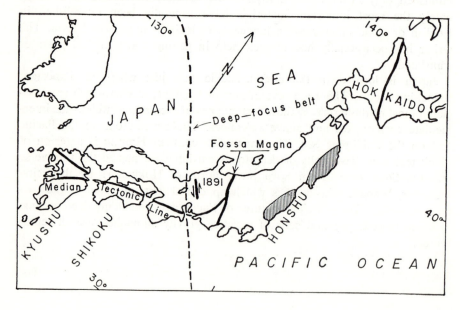

of young volcanic products at the surface. An important fault branching from the Median Line crosses northern Kyushu.

The Fossa Magna takes the form of a graben filled with volcanic material; from it rise several volcanic cones, including Fujiyama. Southward from Honshu the same structure expresses itself in a chain of volcanic islands extending to the Marianas and belonging to typical arcs.

Several authors have noted that the major tectonic lines of demarcation, with their striking expression in stratigraphy and relief, show almost no correlation with current seismicity. Comparison is suggested with the Garlock fault in California and the Alpine fault of New Zealand; both of these are major faults with abundant evidence of geologically recent activity, and both separate provinces of different stratigraphy and tectonic character, yet neither could be identified from a plot of located epicenters. As observed for other volcanic lines, a few shocks at depths of the order of 100 kilometers have epicenters in the vicinity of the Fossa Magna or its southern prolongation, but most of these would fit into a different representation of the three-dimensional seismic pattern.

Block Structure

As in other regions, the late Tertiary and Pleistocene orogeny broke up the region into blocks. This block structure applies over the whole of the Japanese main islands, but it is most evident in the inner tectonic zone and toward the Japan Sea. Several of the important observations of faulting during an earthquake belong to this inner zone and hence resemble the phenomena of California and New Zealand; the bearing of these observations on the mechanism of Pacific arcs is rather indirect.

Fault of 1891

A major tectonic feature is the strike-slip fault associated with the Mino-Owari earthquake of 1891, which crosses Honshu at its narrowest part. This is entirely within the inner tectonic zone, and far west of the Fossa Magna. The fault strikes roughly northwest, and the strike-slip is left-hand, as for most of the faults in Japan with the same general trend, while those trending northeast-southwest show right-hand strike-slip. These characteristics are general over the entire region; they would be mechanically coherent with general east-west compression, north-south tension, or both.

Direction of Displacements

Misunderstanding has arisen over statements that in the Japanese area the continental region is being displaced southward with reference to the Pacific Basin, "as in California." For the California region the reference is to the type of displacement represented by the San Andreas fault, which is roughly parallel to the coast. In Japan there is no such parallel feature, except for the Median Tectonic Line of west Japan; the corresponding displacements, if referred to this, would give northeastward relative motion of the continent. Those making the quoted statement had in mind dividing Japan at the Fossa Magna into a western continental and an eastern Pacific region, in which case the western region is in fact being displaced relatively southward (and slightly eastward). This formulation is consistent with observed faulting in earthquakes, with triangulation data, and with abundant observations of recorded initial compressions and dilatations.

Arc Structure

On small-scale maps it is easy to sketch a single arc including Honshu and Shikoku, but this ignores important details. The east front of Honshu conforms well to the typical arc description, with the Nippon Trench succeeded by shallow and intermediate earthquake belts, while gravity anomalies and active volcanoes appear in their normal order. On the south front, however, west of the Fossa Magna, the trench is less pronounced, and volcanoes and intermediate earthquakes are lacking. Instead of the constant occurrence of earthquakes of all magnitudes which characterizes the east front, the south shows rather low general seismicity, with short active epochs marked by great earthquakes (as in 684, 1605, 1707, 1854, 1944, 1946).

The arc of the Ryukyu Islands, from Kyushu to Formosa, in its slightly less pronounced characteristics, and in the presence of intermediate but absence of deep shocks, resembles the Alpide arcs. It may be that the branch of the Pacific belt of which this is the first clearly defined arc is classifiable as Alpide.

Formosa, or Taiwan, is marked by an intersection of structures where the arc just named meets another extending northward from the Philippines. It is of exceptional interest to seismologists because of two described instances of faulting (1906, 1935).

On the Pacific side of Honshu, most of the larger shallow earthquakes originate off the coast, conforming to the normal pattern of active arcs; but there is a conspicuous exception for the Kwantō district (in which Tokyo is located). Many large and destructive shocks have originated here in the neighborhood of Sagami Bay. Hypocenters here are often found to be deeper than average, in the depth range near 40 kilometers.

Deep-Focus Zones

Wadati's two main zones of deep shocks (Fig. 30-4) are not easy to relate to the structural arcs. South of Honshu the structures of the Fossa Magna have their logical prolongation in the volcanic islands of a normal east-fronting arc, and the deep shocks of Wadati's transverse zone appear there in normal relative position west of the islands. Northward toward Honshu, the transverse zone diverges westward and crosses Honshu (at depths chiefly in the range 350 to 400 kilometers) near and west of the Mino-Owari fault zone.† It extends across the Japan Sea to meet the Sōya zone in Manchuria. The northeasterly part of the Sōya zone is in the right place for deep shocks belonging with the Kurile arc, but it extends disproportionately far southwest. An epicenter which at present looks oddly isolated is that of a shock on May 17, 1950, at a depth of 580 kilometers off the Korean coast near 39¼° N, 130¼° E. This may represent the extension of the structures of the transverse zone to great depth.

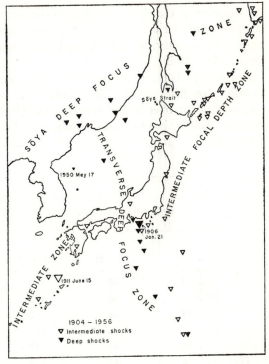

FIGURE 30-4 *Japanese region, showing belts of deep-focus earthquakes. Epicenters for shocks of magnitude 7 and over* [Seismicity of the Earth, *with later additions*], *and for May 17, 1950 (6.7).*

LARGER EARTHQUAKES OF JAPAN

Statistics

Seismicity of the region is so high that lists, even of large earthquakes, grow lengthy. The Central Meteorological Observatory published with its bulletin for 1950 a catalogue of 4195 earthquakes from 1885 to March, 1951, each of which was felt over an area of 30,000 square kilometers or more. Magnitudes were determined from macroseismic and microseismic data, independently by Kawasumi (using a scale of his own) and by Tsuboi

---

† Some authors working from small-scale maps have perpetrated absurdities by geographically identifying the Mino-Owari fault or the transverse zone of deep shocks with the Fossa Magna.

(conforming as closely as possible to the scale used in this book). Tsuboi's determinations yield magnitudes generally ranging down to 5.5, but in exceptional cases down to 4.5. In 1951 Kawasumi listed 342 large earthquakes for A.D. 599–1949.

Gutenberg's latest revision of magnitudes results in totals, for 1918–1954 inclusive, of 85 shallow, 20 intermediate, and 17 deep earthquakes of magnitude 7 or over in the general area (including Japan, Manchuria, the southern Kurile Islands, and Formosa). Table 30-1, also based on his figures, lists 27 shocks of magnitude 8 and over for 1897–1955 inclusive

Table 30-1 Large Earthquakes of the Japanese Region (Magnitude 8 and over, 1897–   )

Date (G.C.T.)			Lat. N	Long. E	Depth	Magnitude
1897	Feb.	7	40	140	shallow	8¼
	Feb.	19	38	142	shallow	8¼
	Feb.	19	38	142	shallow	8
	Aug.	5	38	143	shallow	8¾ ±
1898	Apr.	22	39	142	shallow	8¼
1901	Apr.	5	45	148	shallow	8
	June	24	27	130	shallow	8
	Aug.	9	40	144	shallow	8
1904	June	7	40	134	350 ± km	8
	Aug.	24	30	130	shallow	8
1905	June	2	34	132	100 ± km	8
	July	6	39½	142½	shallow	8
1906	Jan.	21	34	138	340 km	8.4
1909	Mar.	13	31½	142½	80 km	8¼
	Nov.	10	32	131	190 km	8
1911	June	15	29	129	160 ± km	8.7
1918	Nov.	8	44½	151½	shallow	8
1920	June	5	23½	122	shallow	8¼
1923	Sept.	1	35¼	139½	shallow	8.3
1927	Mar.	7	35¾	134¾	shallow	8.0
1933	Mar.	2	39¼	144½	shallow	8¾
1941	Nov.	18	32	132	shallow	8
1944	Dec.	7	33¾	136	shallow	8¼
1946	Dec.	20	32½	134½	shallow	8.5
1950	Feb.	28	46	144	340 km	8.0
1952	Mar.	4	42½	143	shallow	8.6
1933	Nov.	25	34	141½	shallow	8¼

(see Fig. 30-2). For 1897–1903 a great intermediate or deep shock may perhaps have escaped notice; thereafter, listing for these is probably as reliable as for the corresponding shallow shocks.

The remarkable series of shocks off the east coast in 1897 and 1898 is rendered yet more surprising by the fact that they followed a great shock in the same area, the Sanriku earthquake of June 15, 1896, with a devastating tsunami.

The earthquake of January 21, 1906, is the largest known in any region at a depth exceeding about 200 kilometers; that of June 15, 1911, has the maximum magnitude assigned to intermediate shocks (one or possibly two in other regions equal it).

Losses

To avoid repetition, Table 30-2 lists in summary form various elements of loss in the more destructive Japanese earthquakes since 1700 (and the

Table 30-2 Some Larger Losses in Japanese Earthquakes, 1700– *

Date (Local)	Deaths	Houses Destroyed	Houses Burned
1703 Dec. 31	5,233	20,162[1]	
1707 Oct. 28	4,900	29,000[1]	
1751 May 20	2,000	1,128	6,088
1847 May 8	12,000	34,000	
1854 Dec. 23	3,000	25,000[1]	600
1855 Nov. 11	6,757	17,444	33,000
1891 Oct. 28	7,273	142,177[2]	
1896 June 15	27,122	106,170[1]	
1923 Sept. 1	99,331	128,266	447,128
1925 May 23	395	3,333[2]	
1927 Mar. 7	3,017	10,633	4,961
1930 Nov. 26	259	2,142	
1933 Mar. 3	2,986	4,086[1]	
1935 Apr. 21	3,276	17,907	
1943 Sept. 10	1,190	7,485	254
1944 Dec. 7	998	22,563[1]	
1945 Jan. 13	1,901	5,539	
1946 Dec. 21	1,330	16,289[1]	
1948 June 28	5,386	63,000	3,960

* From Imamura and Japanese official sources.

[1] Includes tsunami effects.
[2] Includes fire losses with earthquake.

Formosa earthquake of 1935), as well as in a few less disastrous earth-quakes discussed later. This information is merely representative and incomplete; thus the official totals for the 1923 earthquake take the following form.

Deaths	99,331
Injured	103,733
Missing	43,476
Houses demolished	128,266
Houses partially demolished	126,233
Houses burned	447,128
Houses washed away	868

Figures such as these necessarily involve some uncertainty, and data for any one earthquake are often stated with slight variations by different authors.

Early Instances; Changes of Level

Earthquakes for which faulting is described or for which there is other important tectonic evidence will be taken up generally in chronological order, with incidental references to comparable shocks of earlier date.

Chapter 14 included brief reference to the numerous earthquakes of old date cited by Imamura as accompanied by changes of level at the seacoast. Among them are three instances (1793, 1802, 1872), all on the Japan Sea coast, of upheaval causing remarkable retreat of the sea preceding the strong shocks. The documents should be reliable, since they describe the alarm of the populace, expecting a tsunami—in place of which they were caught off guard by the following large and destructive earthquakes.

Attempts have been made, in Japan, California, and elsewhere, to detect and record such preseismic motions with tiltmeters. In California failure seems definite; small, irregular tiltings, due probably to local causes, are recorded but show no correlation with large shocks. In view of the much lower level of seismicity there than in Japan, the Japanese observations are in no way discredited. A number of published papers from Japan purport to show remarkable recorded tilting and other effects preceding large shocks, but most of them are scientifically invalidated by lack of proper control, there being no assurance that similar phenomena were not occurring at other times.

The earthquake of September 4, 1596, was followed by submergence of the island Uryū-jima, off the present city of Oita on the northeast coast of Kyushu. The island measured 4 by 2.3 kilometers and had a population of 5000. Flooding followed the earthquake and has continued until at the present day the water over the former island is 15 to 20 meters deep. The locality is on or close to the Median Tectonic Line; nevertheless, this may have been subsidence rather than true tectonic downwarping. During the earthquake of October 28, 1707, which Imamura considered the greatest on his long list of Japanese earthquakes, there were remarkable changes of level in the vicinity of the city of Kochi on Shikoku. Subsidences of nearly 6 feet near the city were matched by upheavals of 7 to 8 feet at points farther east. This is also close to the Median Line.

MINO-OWARI EARTHQUAKE—1891

On October 28, 1891 occurred the first really great earthquake to affect the main islands of Japan after the establishment of the active seismological program with headquarters in Tokyo (about 1880). This earthquake (Fig. 30-5) was investigated in the field chiefly by B. Kotō, professor of geology at the Imperial University. It is commonly termed the Mino-Owari earthquake, from the names of the two provinces which were chiefly affected. The following items are quoted from Kotō's principal publication.

The twin provinces are on three sides bounded by masses of mountains, and in their very centre lies an extensive populous plain. The general aspect of the plain, which inclines slightly toward the foot of the Yoro ridge, is monotonous and flat. It is covered with a network of rivers and artificial canals . . . The lowland of Mino and Owari is usually spoken of as having been once an im-

FIGURE 30-5   *Central Japan. Location map, showing epicentral areas of 1925, 1927, 1943, 1945, and faulting of 1891. Triangles, epicenters of earthquakes at intermediate depth.*

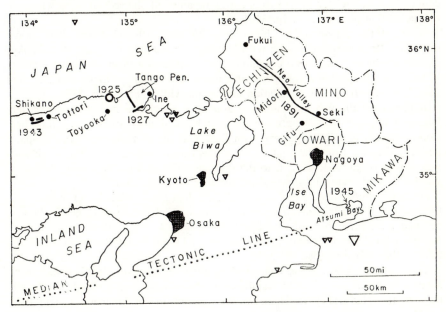

mense swamp, since converted to the present paddyland within historical times. . . . Although [this story] seems to be a great exaggeration, yet we see in it some germ of truth; for if all the artificial canals and dams were to be removed, the present fertile land would in a moment be nothing better than a moor. It is these dams and the drainage system that suffered the greatest damage in the last earthquake . . . [p. 307].

The north of Mino is a mountainous district, consisting mainly of Palaeozoic formations . . . Beyond the mountain-ridge lie the provinces of Echizen and Kaga . . . A plain of considerable extent . . . is the basin . . . in the centre of which lies the city of Fukui. This region felt severe shocks during the last earthquake, and suffered calamity surpassed in extent only by that felt in Mino and Owari [pp. 308–309].

Going northward from Nagoya to Gifu, we find a series of villages, one running into the next; that is to say, there is a nearly continuous street of . . . more than twenty miles in length. The rows of houses were thrown over by the shocks, and these twenty miles of road became simply a narrow lane between two interminable heaps of debris . . . Gifu, the provincial capital, was for the greater part overthrown, and then burnt by fire . . . [p. 314].

Total killed, 7,279; wounded, 17,393; buildings entirely destroyed, 197,530. . . . [p. 317].

. . . disturbances were distinctly felt from Sendai, in the north, to the west coasts of Kyushu in the south, or over an area of 134,722 square kilometres . . . [p. 319].

Generally speaking, the frequency of shocks was high during the month of October, 1891, and the earth was by no means tranquil then at any time, especially in the provinces of Musashi and Shimosa. Seventeen separate quakes had been already recorded by the seismographs, previous to the 28th, all however confined to the vicinity of Tokyo. Central Japan itself showed no signs whatever of the coming disaster . . . [p. 319].

Amongst the extraordinary things done by the earthquake, one that always drew my attention was the earth-rent. It strikes across hills and paddy-fields alike, cutting up the soft earth into enormous clods and raising them above the surface. It resembles the pathway of a gigantic mole more than anything else . . . Indeed it is known by this appellation among the villagers . . . [p. 328].

It goes right through the hamlet of Jōbara . . . where, in a front garden adjoining a farmer's house, there are two stately persimmon trees . . . which had stood time out of mind in an east-west line. The line of fault traverses the space between the two from the south-east to north-west, and as usual the north side was shoved north-west, so that, to the great astonishment of the owner, they now stand in a north-south line instead of east-west . . . [p. 337].

South of Midori . . . a fine new road . . . had been obliquely cut in two, and the lower end with the surrounding fields had sunk about 6 metres below the upper end . . . That the east half had been pushed 4 metres northwards, in conformity with the general rule, is well seen . . . by an abrupt change in the direction of the displaced road [p. 337].

This unique track . . . is characterized by constancy of direction and regularity of course. Starting from Katabira near the Kiso-gawa, it runs up its length through the Neo valley to Haku-san (a distance of 64 kilometres), and then seems to proceed north-west to the city of Fukui, for the extraordinary distance of 112 kilometres . . . the ground on the left side had subsided from ⅔ to 6 metres, and at the same time been horizontally shifted for 1⅔ to 2 metres in a northwesterly direction. The only exception to the general rule was the fault at Midori, where the land was lower on the west than on the east . . . [p. 349].

The sudden elevations, depressions, or lateral shiftings of large tracts of country which take place at the time of destructive earthquakes are usually considered as the effects rather than the cause of subterranean commotions; but in my opinion, it can be confidently asserted that the sudden formation of the "great fault of Neo" was the actual cause of the great earthquake . . . [p. 352].

The first extract is quoted to show the heavily alluviated character of the plain which constituted the meizoseismal area. Naturally any isoseismal map shows the limits of the alluvium, as the effects fell off rapidly in the surrounding hilly area. Note that, while Kotō remarks on an increase in earthquake activity in Japan in the preceding month, he points out that all this was at a distance from Mino and Owari. This bears on the discussion in our Chapter 6 of Davison's publications on the foreshocks.

Kotō's photograph of the fault near Midori is one of the most reproduced documents in the geological sciences; no seismological text is complete without at least a small version of it (Fig. 30-6).

Later investigation has left no doubt that the fault trace continued to the

FIGURE 30-6 *Faulting at Midori. Mino-Owari earthquake, October 28, 1891.* [*Kotō.*]

vicinity of Fukui, right across the island of Honshu. However, it was not one continuous trace but consisted of three or four separate segments offset en echelon. All of these show left-hand strike-slip. There were reversals in the vertical throw; Kotō was apparently puzzled by finding the downthrow of the Midori scarp opposite to that at most points. This is simply the scissoring characteristic of strike-slip faults. A scissor point was probably not far from the celebrated pair of persimmon trees, for Kotō remarks that the trace was so inconspicuous there that the peasants were completely mystified by the displacements.

The last extract from Kotō's paper shows that he was quite clear as to the causal relation between faulting and shaking; but it took many years for this view to be generally accepted.

Imamura notes that

. . . the 1891 earthquake being the first great earthquake since Western culture was introduced into Japan, all constructional work such as bridge-piers, buildings, and other structures of brick patterned after those of the Occident without regard to seismic stability were completely wrecked . . . in the succeeding year the Japanese government, which now became interested in this subject, created the Earthquake Investigation Committee as a national institution . . .

FAULTING OF 1894 AND 1896

Two earthquakes accompanied by faulting took place on October 23, 1894 and August 31, 1896, not far from the Japan Sea coast of Honshu (Fig. 30-7). In the earlier earthquake the fault was traceable for about 25 km., with a northeast trend and a downthrow toward the coast. The later earthquake developed two faults, trending more nearly northward on opposite sides of a mountain block, and downthrown away from it, as in Dutton's description of the Sonora earthquake of 1887 (Chapter 31). They were traced for distances of about 50 and 80 kilometers. The western of these two is probably the same fault as that of 1894, although the connection was not traced across the intervening gap of 25 kilometers. Both earthquakes were destructive, although that of 1896 had a larger meizoseismal area and was evidently of greater magnitude. Imamura's notes on the remarkable foreshock series of 1896 have been cited in Chapter 6.

FIGURE 30-7 *Faulting in the Japanese earthquakes of 1894 and 1896.* [*Redrafted from de Ballore, after Yamasaki.*]

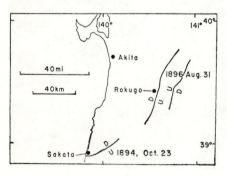

KAGI, FORMOSA (TAIWAN)—1906

The earthquake on March 17, 1906, in Formosa, was investigated by Fusa-kichi Omori, for many years the most distinguished of Japanese seismologists. His investigations had to be limited to the settled, and policed, western coastal plain; he would have found the central mountains very unhealthy,

FIGURE 30-8

Formosa earthquake, March 17, 1906. Faulting according to Omori.

because the Japanese never fully subdued the native tribes. The faulting in the accessible area (Figs. 30-8 and 30-9) trended roughly east-west, with down-throw to the north and right-hand strike-slip; it was traced for about 8 miles, including one short southeastward branch with the same type of displacement. The strike-slip reached 8 feet, but the vertical throw did not exceed 4 feet. Omori published a good photograph of a road offset 6 feet. Near the east end of the observed trace was a scissor point, beyond which the throw was down to the south, the strike-slip still being right-hand. In this disturbance, which is called the Kagi earthquake, 1258 were reported killed.

FIGURE 30-9   *Index map of Taiwan (Formosa), showing locations for earthquakes of 1906 and 1935.*

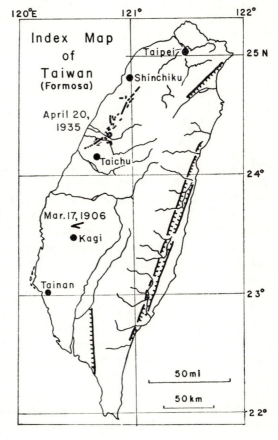

KWANTŌ EARTH-QUAKES—1923 AND EARLIER

Catastrophe

The great disaster of September 1, 1923, which devastated Tokyo, Yokohama,

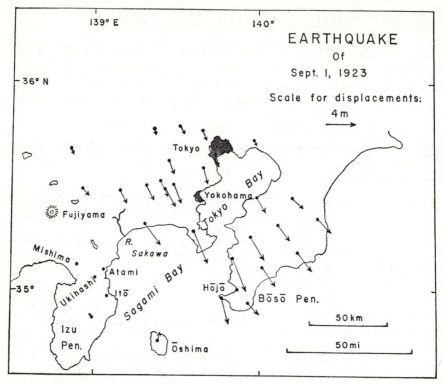

FIGURE 30-10   *Kwantō earthquake, September 1, 1923. Location map showing also horizontal displacements found by geodetic triangulation, as revised by K. Muto, 1932.*

and environs, is referred to by Japanese authors as the Kwantō earth-quake, the provincial name. The population of the meizoseismal area was of the order of 2 million; according to official figures, 99,331 lost their lives. This includes 38,000 who were burned to death in an open area of Tokyo where they had congregated, supposedly for safety, and were overwhelmed by one of the fiery whirlwinds which often originate in large conflagrations. Much that was destroyed was irreplaceable: libraries, museums, art collections, archives, scientific data.

In the seismological station at Tokyo University, the staff, headed by Professor Imamura, were at work interpreting the seismograms within 2 minutes; in half an hour reasonably accurate statements were being given out to the press. Although all the seismographs were damaged, six of them were promptly repaired and recording continued effectively without interruption. All this while violent aftershocks were occurring, and fire was spreading and destroying some of the University buildings.

To western readers the most accessible general account of this earthquake is the book published by Davison. Unfortunately, it is unreliable in details, so the serious student can only use it as a general guide. The book is much disliked in Japan because of Davison's flair for the sensational; he has treated hair-raising press reports on the same level as official documents and has often deliberately used language tending to give exaggerated impressions, both of the disaster and of the earthquake phenomena.

Epicenter and Displacements

All authorities place the epicenter of the principal shock in Sagami Bay, which is southwest of Tokyo and is separated from the bay of Tokyo by the Miura peninsula. Imamura interpreted the seismograms as due to three shocks: (1) a small one under Sagami Bay; (2) a very large one under the land to the north; (3) the main shock, again under Sagami Bay.

At least 15 lines of vertical faulting or sharp warping were found on land, and checked by releveling, to the north of Sagami Bay, and on the Bōsō peninsula to the east. Displacements ranged from about 1 to 6 feet. These are clearly minor fractures incidental to the displacement of large blocks. Releveling established large uplifts along the coast; on the north shore of Sagami Bay they approached 2 meters (6 feet 8 inches), and on the Bōsō peninsula they reached 1.5 meters.

Retriangulation showed even larger horizontal displacements; the largest was that found for a point on the island Oshima, south of Sagami Bay, which the first reduction indicated had shifted nearly northward by 3.8 meters (12 feet 5 inches). This reduction assumed that a primary triangulation point about 50 miles north of Tokyo had remained fixed, and that the direction of the line from this point to another eastward from it was unaltered. In 1932 K. Muto rediscussed the data; the observed apparent displacements were found to be affected by systematic error dependent on position of the triangulation points. The very slight change in reduction required to remove the error led to a significant change in the pattern of the displacements (Fig. 30-10). What had previously, after the first reduction, seemed like a peculiar rotation of the whole area about Sagami Bay now was replaced by a general southeasterly shift of the Tokyo area, with displacements decreasing from the coast inland. The northward motion of Oshima remained but was decreased to 0.9 meter (3 feet). However, the displacements on the Bōsō peninsula, large before, were now increased to a maximum of 4.55 meters (15 feet).

Soundings in Sagami Bay

After the earthquake, preliminary soundings for navigational purposes indicated extraordinary changes of depth in parts of Sagami Bay. The hydrographic department of the Japanese navy then organized a sounding program with four ships and four additional parties operating in shallow water. This continued into the following January; over 83,000 soundings were taken. Comparison with hydrographic surveys made between 1912 and the date of the earthquake resulted in a map of Sagami Bay showing nearly incredible changes of level. A large part of the bottom in the central part of the bay north of Oshima is shown as having subsided between 100 and 200 meters; elsewhere are smaller areas of subsidence in one of which the depth is mapped as increased by 400 meters! North of the principal area of subsidence is shown one of uplift, reaching as high as 250 meters.

Reactions to this publication have ranged from the extreme of rejecting the whole as "obviously" wrong to the opposite extreme of acceptance. (Davison unhesitatingly discusses the origin of the earthquake in terms of crustal blocks being displaced vertically by hundreds of meters.) Trying to discredit the observations of 1923 gets nowhere. The work was so careful and extensive that errors of the order of 100 to 200 meters, even though the greatest depths measured were near 2000 meters, must have been fairly well eliminated. It is easier to attack the soundings taken before the earthquake; by comparison, these must have been carried out in routine fashion, certainly with no such elaborate precautions as the later work.

The chief source of error in such work is usually not in the soundings but in fixing the position of the vessel.†

As Sagami Bay is not large, positions determined with reference to points on land should not be greatly uncertain. However, we have seen that during the earthquake points on the coast were displaced horizontally and vertically by several meters. Such motions might in special circumstances introduce even larger errors into positions referred to these points. Now the bottom of Sagami Bay is extremely irregular; not far from the point where a subsidence of 400 meters was reported in 1923, charts show the bottom descending northward by 400 meters in less than 2 kilometers. Here relatively small errors in position might result in large apparent changes in depth. Terada attempted to meet this type of objection by analyzing the data statistically and showing that there was no correlation between the reported changes in depth and the steepness of the bottom; however, this statistical argument applies properly only to the larger areas and has no force where great apparent changes were found at points close together.

One would like to explain the results as due to submarine sliding; but the principal closed area of subsidence, north of Oshima, is also the deepest part of the bay, into which material should have slid. This is also the vicinity of the most reliable instrumental locations for the epicenter. Under violent shaking the bottom may have compacted and settled, but 100 meters seems

---

† A chart showing contours off the shore of southern California, drawn up from echo sounding in 1922, showed two peculiar parallel alignments of ridges and other features trending southwest away from the vicinity of San Diego. These were due to error in position of the ships making traverses along those lines; they disappeared from charts when the work was repeated with better control.

excessive. Nevertheless, such an effect, combined with the minor sources of error discussed, seems at least more plausible than enormous motions of crustal blocks, greater than those known for any other earthquake.

The deep part of Sagami Bay has the form of a submerged trench which extends northwestward toward the coast, where it appears to be continuous with the valley of the river Sakawa, a graben bounded by probably active faults. The horizontal displacements as revised by Muto strongly suggest right-hand strike-slip on the east side of this trench. The coastal uplifts and the throws on the small faults found on land cannot be so simply interpreted.

Whatever occurred under Sagami Bay was doubtless connected with the tsunami waves which rose on its shores. These were not exceptionally large; heights of the order of 10 meters (30 feet and over) were reached at only a few points. The waves were traced by tide gauges along the coast of Honshu, decreasing rapidly away from Sagami Bay.

Ground-Water Effects

Ground-water effects were many, and some were of particular interest. The raising of pillars of an old bridge and the remarkable observation of pulsating earthquake fountains have been cited in Chapter 9.

Earlier Events

Imamura compared the changes of level in 1923 with those, naturally less well documented, which accompanied the earthquake of 1703 in the same area and concluded that they were closely similar. Raised strand lines with perforations of the boring mollusk *Lithophaga* are interpreted by him to indicate elevations in 1703 up to 2½ times those in 1923. Imamura correlated a higher strand line with a great earthquake in 818; there is evidence of a fourth similar earthquake in the region in A.D. 33.

NORTH TAJIMA EARTHQUAKE—1925

On May 23, 1925, a locally destructive earthquake occurred on the Japan Sea coast of Honshu, in the province of Tajima (see Fig. 30-11). This is described as accompanied by faulting; but the evidence is at the lower limit of observation. The principal line was examined by several investigators, one of whom was B. Kotō, the veteran of the 1891 earthquake.

The 1925 epicenter, as determined from all available data, was near a small inlet, Tsuiyama cove. On the east shore was a village, Tai, north of which a coastal hill rose to 231 meters. Part of the summit of this hill is nearly level, and was in cultivation for lilies and for willow (used for wicker baskets). Two nearly parallel lines, about 400 meters apart and trending southwest-

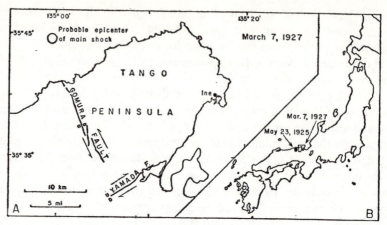

FIGURE 30-11   A, *faulting of the Tango earthquake.* [*After Yamasaki and Tada.*]
B, *location map; North Tajima earthquake, 1925; Tango earthquake, 1927.*

northeast, were traced for about 1500 meters. By including more or less discontinuous fissures and slides, Kotō gives the total length of this feature as 5.6 kilometers (3½ miles). Vertical displacements were generally from a few centimeters to 20 centimeters, with one throw of 1 meter, possibly a slide. Downthrows were mainly northwestward, toward the coast, although some reversals are reported where the trace was multiple. These might represent scissoring; and this is consistent with the observation of left-hand strike-slip, indicated by offsets of rows of lilies and of young willow shoots. At the best-observed locality, this shift was 6 centimeters (2½ inches); this was the maximum. Kotō reports and gives a photograph of a shift of 2.5 centimeters (1 inch)!

If it were not for the strike-slip, the evidence could be quickly dismissed as secondary—a series of lurches, slides, and fissures. Some Japanese observers did so dismiss it. The strike-slip demands more coherent block motion; but even this is so small that it is not necessarily correlated with any tectonic displacement of basement rock. The material in which offset is observed is relatively unconsolidated volcanic breccia and agglomerate, chiefly rhyolitic.

The earthquake was moderately large (magnitude 6¾); at the time no adequate seismographs were operating in that part of Japan, and a precise epicenter is not available. However, the principal meizoseismal area was roughly semicircular, about 12 miles in diameter. There was also considerable damage at Toyooka, about 7 miles south on unconsolidated ground, where the earthquake was followed by a destructive fire. The official death list was 395.

This shock received special attention because it was the first serious earthquake on Honshu after the great earthquake of 1923, after which the seismological service was reorganized and a new Earthquake Research Institute was added which gradually replaced the former Imperial Earthquake Investigation Committee.

TANGO EARTHQUAKE—1927

The first major earthquake to affect Japan after 1923 was the Tango earthquake on March 7, 1927 (Fig. 30-11). The new organization went vigorously into action. Other Japanese workers also took the field, with the result that this was one of the most lavishly investigated of all large earthquakes. The effort was repaid, since the faulting and other phenomena were of unusual interest.

Epicenter and Time Data

The Tango earthquake was well recorded at the better seismological stations all over the world. After the 1925 Tajima earthquake a station had been established at Toyooka, about 30 kilometers from the epicenter of the Tango earthquake. The time recorded for the main shock at Toyooka, together with records there and at other points for the aftershocks, determines the epicenter more closely than for most large earthquakes. The point found is at least 15 kilometers offshore (see Fig. 30-11). The recorded times were employed by E. A. Hodgson for a revision of the standard time-distance tables for seismic waves; his results were included in more extensive work by other seismologists, who made use of the large body of well-observed times which rapidly accumulated in the following years.

Magnitude and Intensity

The magnitude determined instrumentally is 8.0 (revised). The macroseismic effects were violent and extensive. "Considering the proportion of damage suffered by its people," wrote Imamura in 1937, "the town of Mineyama can probably lay claim to having established a record in Japan. Out of a total of 998 dwellings, 988 were either shaken down or reduced to ashes, while the killed numbered 1122 . . ." He went on to point out that much of the property loss at Mineyama was due to lack of fire control measures; there was no organized emergency action.

Eastward the earthquake was felt to and beyond Tokyo, over 400 kilometers distant; to the southwest perceptibility extended farther, to Kagoshima near the south end of Kyushu. The meizoseismal area surrounded the Tango (or Oku-Tango) Peninsula on the west and south.

Faulting

Two faults, the Gōmura and Yamada faults, appeared along the structural boundaries of the Tango block (Fig. 30-11). The Japan Sea coastal region of Honshu is divisible largely into tectonic blocks with boundaries trending roughly northeast and northwest. Some of these boundaries are active faults, like those which broke the surface in 1894 and 1896. The Tango block is roughly rectangular, with sides measuring about 40 kilometers (or 25 miles). It rises to 600 meters and consists of a granitic basement overlain by volcanics with limited deposits of shale, sandstone, and conglomerate.

To quote Yamasaki and Tada:

Across the neck of the Oku-Tango peninsula there is a remarkable rift valley . . . The fault scarp along the southwest side of this valley makes step faults and is well dissected by the young stage of erosion. Along each one of these steps the following five new seismic faults or rifts have been formed in the recent earthquake. . . .

Four of these five segments, with minor echelon offsets, make up the Gōmura fault or fault zone, which has a strike about N 30° W, and extends about 18 kilometers inland from the Japan Sea coast. The four segments all show downthrow to the east (maximum 79 centimeters) and consistently left-hand strike-slip, with a maximum of 281 centimeters (9 feet 2 inches). The fifth trace is less than 1 kilometer long. It is east of the principal Gōmura feature, with a small westward downthrow (40 centimeters) and no strike-slip; it could be a large slump or lurch. Especially where the main trace is not continuous, there are left-hand tension cracks; these are especially noticeable on hill ground. In the valleys mole-track traces were produced. Near the middle of the Gōmura trace a slickenside was produced in granite; the dip was measured as 80° NNE, with strike N 19° W. A similar slickenside at another locality showed dip 65° WSW, with strike N 10° W.

The Yamada fault is found just along an old fault scarp in the southeastern side of the Oku-Tango peninsula . . . Its general trend is S 55° W to N 55° E. In the village of Yamada the fault appears across a highway and cultivated lands forming a remarkable flexure instead of a steep rift . . . It seems that the land on the southeastern side of the fault has been depressed and shifted to the southwest, with the vertical and horizontal displacements of 0.7 meters and 0.8 meters respectively. The fault, however, makes a fracture cut deep into granite, which forms the ground rock and is exposed near a railway tunnel . . . Many other huge cracks have been formed on the flank of the hill. The main fault extends then northeastwards . . . until it reaches the Bay of Yosa . . . Very remarkably some tracts of land in the southeastern side of the fault in Otokoyama are partly submerged into the bay . . . On the other hand the southern extension of the Yamada fault passed through the village of that name crushing all houses and cottages into pieces and reappears in the mulberry fields in the west end of the village, where it makes many large rifts . . . Further westsouthwestwards it extends to a large village of Ichiba, which was totally destroyed by shock and fire. The total length of this fault measures 7.5 km.

The left-hand strike-slip on the Gōmura fault and the right-hand strike-slip on the Yamada fault are mechanically consistent; that is, they could have been produced by the same stress system, for example, east-west compression. Downthrows respectively eastward and southward indicate a tilting of the Tango block. There were elevations as large as 80 centimeters (30 inches) along the coast westward from the Gōmura fault, and slight elevation just to the east of it. Soundings showed no great changes of level comparable with those reported in 1923; but the evidence indicated that dip-slip continued along the line of the Gōmura fault northward off the coast, and apparently with increasing throw.

Geodetic Results

An elaborate program of leveling and triangulation was begun; levels were run five times, from immediately after the earthquake until 1930. The leveling routes crossed both Gōmura and Yamada fault lines. The first releveling showed that the dip-slip on both faults included both upthrow and down-throw decreasing away from the faults, as the elastic-rebound theory would require. The block between the two faults (part of the Tango block) showed uniform tilting. Successive repetitions showed displacements continuing in the same sense; but, whereas those on the Yamada fault continued throughout, those on the Gōmura fault decreased rapidly and had practically ceased by 1929 (Fig. 30-12). Naturally the later displacements amounted to only a few centimeters, but the work was done with great care. The plots against time correspond strikingly to Benioff's two types of strain-release curve (Chapter 6).

Similar repeated triangulations were carried out. The results are much less simple than those of the leveling. The general displacement of the Tango block corresponds to that found along the Gōmura fault, but between the later repetitions it shows local irregularity and even reversal. The effect of the Yamada fault is not so clearly shown. Tsuboi divides the area into a number of blocks, of which the Tango block is one, such that points

FIGURE 30-12 *Displacements on the faults associated with the Tango earthquake of 1927, during the following three years.* [*Tsuboi.*]

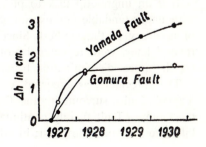

on each one move nearly parallel and simultaneously. These blocks apparently were differently affected at different times. He notes the strong possibility that the measured horizontal displacements on the Tango block were complicated by its tilting. Analysis of the first triangulation strongly suggested the elastic-rebound pattern for horizontal displacements on the two sides of the Gōmura fault.

All evidence, including that of aftershocks, places the epicenter of the principal earthquake off the coast on the prolongation of the Gōmura fault. If this actually does represent the point of initial rupture, it means that the displacement on the Yamada fault occurred as a consequence; either it was initiated by the arrival of the fracture progressing along the Gōmura fault, or it was triggered by the shaking from the elastic waves arriving a little sooner. In fact, there are reports to the effect that the Yamada scarp was seen forming after the first violent shaking in its vicinity.

Investigation of Aftershocks

A practically new technique was applied to the aftershocks. In 1924 workers in Switzerland set up a small instrument at Visp, near the epicenter of a strong local earthquake, to record aftershocks simultaneously there and at Zurich; but in 1927 three portable instruments were sent into the Tango meizoseismal area, two of which began recording on March 11 (4 days after the main earthquake), and the third on the following day. With these and the station at Toyooka, a close-range determination of epicenters, and to some extent of hypocentral depth, was practicable for a large number of aftershocks. In this pioneer undertaking no attempt was made to solve the awkward problem of providing time control for emergency instruments in the field. The rate of rotation of the drums was verified by using a marking clock, and determination was based on the measured time intervals between S and P at the stations. The velocities of P and S were at first assumed constant; this would make the straight-line distance from hypocenter to station equal to $k(S\text{-}P)$, where $S\text{-}P$ is a time interval in seconds and tenths, and k can be evaluated from the data if enough earthquakes and stations are used. Variation of velocity with depth was partly compensated by allowing k to vary in different shocks.

It is an ungrateful task to pick flaws in this work, but the investigators, with understandable enthusiasm, carried their calculations beyond the limits of accuracy of the method. Since some of these first results, even though revised later, were repeated by other authors and tend to become fixed in the literature, discrimination becomes necessary.

The epicenters, as expected, clustered in the vicinity of the Gōmura and Yamada faults, surrounding the Tango block. The belt of epicenters following the Gōmura fault plainly continues northward in direct line for at least 20 kilometers beyond the coast; in other words, it extends to the vicinity of the probable epicenter deduced for the main shock from the data at distant stations (although Hodgson, following Imamura, adopted an epicenter close to the coast). For the largest aftershock, on April 1 (Japanese time; March 31, 21:08, G.C.T.), which was recorded at distant stations, especially in Europe, the epicenter found from the Japanese stations is south of the Yamada fault. This had its own series of aftershocks of the second order. Nasu remarks that "the active centre changed its position from place to place in an oscillatory manner"; that is, normally no considerable succession of shocks followed one another in the same vicinity. (Compare the Kern County shocks of 1952, Chapter 28). The shocks appeared to be of two types, differentiated largely by the amplitude ratio between recorded P and S waves. Type A has a relatively small S and long-period motion attributable to surface waves; type B has large S and less evident surface waves. Imamura attributes type A to the Yamada fault and type B to the Gōmura fault.

Sources of error may be listed as follows:

(1) The velocities vary with depth, perhaps even slightly in different directions. Varying the constant k is not a proper theoretical correction for variation of velocity with depth.

(2) Nasu chiefly used $k = 8.4$ kilometers per second. This calls for measuring S-P to the tenth of a second. The exact time at which the S motion begins often is hard to read with accuracy; and if P is small, it is likely to be read late. Moreover, a high standard of performance is demanded for uniform rotation of the recording drum; this is difficult to provide in a field installation.

(3) Evidence from many regions indicates that seismograms at short distances frequently show an apparent S-P interval of about 1 second, which is definitely too short when correlated with other data. This circumstance is nearly impossible to detect when working with S-P time intervals only, without reduction to standard time.

From these causes it develops that slight errors in measurements or in velocities may lead to errors of several kilometers in the epicenter, and uncertainties as large as 10 kilometers in the depth found for the hypocenter. This means that very little weight can be given to the depths found by Nasu and others. These were generally in the quite reasonable depth range from 10 to 20 kilometers; but large systematic error is not excluded. Consequently it is not legitimate to use these depths for accurate geological purposes like determining dips of the faults. As placed by Nasu, the epicenters are almost all outside the Tango block, beyond the Gōmura and Yamada faults. This would suggest that these faults dip outward; the result is not inconsistent with field observation, but is in itself very uncertain. Much depends on the one station, Ine, which was on the Tango peninsula (near its east coast). A small decrease in the coefficient k, by which S-P at this station is multiplied to give distance, would shift all the epicenters toward Ine and remove the apparent

FIGURE 30-13   *Itō and Izu earthquakes, 1930. Location map, showing faulting of the Izu earthquake of November 26. [After Otuka and after Kunitomi.]*

effect of dip. It would not, however, remove the scattering of epicenters westward from the Gōmura fault; this scatter appears to have extended with the passage of time (to August, 1928).

Nasu set up an elaborate stereographic investigation of his hypocenters, placing them on the surfaces of a three-dimensional lattice system. The spacing of this lattice is from about 5 kilometers down to about 1 kilometer. Now an error of one tenth of a second in the *S-P* read for any one station changes the distance by nearly 1 kilometer; it is difficult to believe that the three-dimensional distribution represents anything but normal scatter, with alignments representing constant *S-P* at one or more of the stations used. In a later paper Nasu recognized these limitations and placed a more conservative interpretation on the data.

IZU EARTHQUAKES—1930

The earthquake of November 26, 1930, was accompanied by faulting on the Izu (Idu, Idzu) peninsula (Fig. 30-13), which bounds Sagami Bay on the west. Because it was preceded by remarkable swarms of small earthquakes, the pattern in time is highly abnormal. The town of Itō is on the east coast of the peninsula.

Itō Earthquake Swarms

In the words of Nasu (and others):

. . . Some 9 months before the occurrence of the destructive Idu earthquake . . . the town of Itō and vicinity . . . with one short period of intermission was, figuratively speaking, bombarded daily for more than two months by showers of earthquakes, the number of sensible ones of which at the end reached the total of more than 4880 shocks. Beginning in February 13, 1930, they increased progressively in both intensity and frequency until the culmination came as the big shock of March 22, at 17h 51m; after which they gradually declined in severity and in numbers, but on May 7 renewed their activities. In August, however, things had quieted down.

To the amazement of all in the town of Itō, the swarm of earthquake shocks began on November 7, and as on the previous occasion, daily increased in number and severity. On November 25 as many as 690 shocks were recorded at Itō, and on the morning of the 26 came the main destructive shock . . .

Microseismic Investigation

Temporary stations were installed at Itō and four neighboring points, beginning March 6, 1930. The diameters of the pentagon were little more than 10 kilometers; and apparently the epicentral area of the Itō swarm was surrounded. There was no time control at these stations, and locations were made on the basis of the S-P time intervals. These ranged from 1 to 2 seconds at all five stations, so the epicenters were necessarily found in a small area inside the pentagon. Question arises as to whether some of the waves taken for S were not of the unexplained type occasionally observed at short distances. This is particularly serious because the interpretation required unusually small velocities for P waves (about 3 kilometers per second), and the depths found were only a few kilometers.

The Izu peninsula consists almost wholly of volcanic rocks; it lies in the volcanic zone of the Fossa Magna. It is natural to interpret a swarm of small shallow earthquakes in such a locality as volcanic. For the largest shock of the first swarm, that on March 22, Kunitomi collated types of first motion (compressions and dilatations) from many seismological stations on Honshu. There was a nodal line trending north-northwest–south-southeast with compressions in the northeast quadrant and dilatations in the northwest. This of course excludes any purely explosive process as the source, but it is quite consistent with shallow faulting caused by volcanic processes. The entire series of events, including the subsequent swarm and major tectonic earthquake, strongly suggests such occurrences as have been described for Hawaii in Chapter 12.†

The foreshock swarm beginning on November 7 was concentrated in a small region, differing from that of the March swarm. They were recorded at Itō, where the station had continued in operation, and at Ajiro, reoccupied on November 25, as well as at Mishima, to the northwest. Rival investigators drew slightly different conclusions, but there is little doubt that the epicenters were all near Ukihashi, about 10 kilometers northwest of Itō.

---

† This and other earthquake swarms were discussed by Terada in a paper which included the charming and characteristically Japanese idea of comparing the time pattern of swarm earthquakes with that of the fall of camellia blossoms.

Principal Earthquake—November 26

The major earthquake on the morning of November 26, Japanese time (19:02:47, November 25, G.C.T.) certainly originated near Ukihashi. Imamura placed the initial epicenter about 4 kilometers north and slightly west of that locality. From analysis of the Tokyo seismograms he concluded that this was followed within 8 seconds by three larger shocks originating at different points on the Izu peninsula. It is doubtful whether these records are adequate for such detailed conclusions; however, from the field evidence there is little doubt that there was immediate extension of the disturbance, with faulting over at least 30 kilometers.

On November 26 two new special stations were initiated at points westward from Ajiro and Itō, giving a quadrilateral with a spread of about 15 kilometers. The records showed that many aftershocks were originating in the same general area as the foreshocks, near Ukihashi. First determinations placed most of these epicenters a little to the east, toward Ajiro; but this need not be significant. By no means all the well-recorded aftershocks originated in this limited vicinity; epicenters were scattered over a large part of the northern Izu peninsula.

Faulting.   The principal macroseismic event was displacement on the Tanna fault, which had been known since 1924, when it was encountered in excavating a railway tunnel. The evidence of faulting during the earthquake was greatest in this tunnel, with a left-hand strike-slip of 2.4 meters (7 feet 10 inches) and a westward downthrow of 60 centimeters (not quite 2 feet). On the surface above the tunnel the displacements were "much less" (strike-slip at one point was 70 centimeters).

The trace was followed for roughly equal distances north and south from the Tanna tunnel, over a total length of about 17 kilometers, largely by low scarps, landslides, and secondary cracks, but with consistent horizontal offsets where they could be checked by artificial features, as in rice fields. The course was nearly straight north and south; but south from Ukihashi, which is about 8 kilometers from the tunnel, there is a westward trend, separate secondary faults appear, and near the southern limit there was even a nearly east-west fault with small right-hand strike-slip.

Kunitomi mapped initial compressions and dilatations at the Japanese stations for the principal earthquake. The distribution is clearly in quadrants, with the nodal lines nearly east-west and north-south. The northeast and southwest quadrants show compression, the others dilatation, in conformity with the observed left-hand strike-slip on the Tanna fault.

Meizoseismal Area.   The principal earthquake was felt over most of Honshu, to distances of about 500 kilometers. The meizoseismal area extended about

50 kilometers from north to south, and 30 kilometers from east to west, including the narrowest part of the peninsula and centering in the vicinity of Tanna and Ukihashi. Within this, Imamura maps four smaller areas of high intensity, as estimated by the percentage of structures destroyed or damaged. One of these is along the central part of the Tanna fault, and here the damage was caused chiefly by the faulting itself, with accompanying cracking, fissuring, and slumping. The other areas show evidence of violent shaking; at many points relatively rigid structures slid over the ground for 70 centimeters or more. They are near the north and south ends of faulting, and west of Ukihashi. Imamura correlated the four areas of high intensity with the four separate shocks that he believed were identified on the Tokyo seismograms; this is probably going too far.

Geodetic Data. Rerunning of triangulation over the area showed northward displacement east of the Tanna fault, southward displacement west of it. There can be little doubt that the strike-slip on this fault represents the major tectonic process; the other faulting is mechanically consistent and might be interpreted as a relatively superficial complication due to the necessity for the fracture to reach the surface through a highly complex and probably incoherent mass of volcanic rocks.

Lights. The Izu earthquake is remarkable for one of the most extensive investigations of reported "earthquake lights" ever undertaken. K. Musya collected about 1500 observations of this kind from a large part of the shaken area; they are sufficient in quantity and character to eliminate thunderstorms and accidents to electric power supply as general explanations. What remains represents real phenomena, still very incompletely understood. It is no wonder that Terada, who studied and discussed the reports, reached no definite conclusion. In fairness to Terada, and as an excellent example, some of his preliminary remarks are here quoted:

With regards to all these testimonies of witnesses, it must be always kept in mind that people are naturally alive to all kinds of phenomena observed at the time of a severe earthquake and apt to regard them as something connected with the catastrophal occurrence, while they forget to consider that the same phenomena are frequently observed on many other occasions not at all connected with earthquake. On the other hand, we learn from the results of investigations by psychologists in what a ludicrous manner the testimonies of people, otherwise quite normal in mentality, may appear distorted when compared to the bare truth.

Lights were seen over a large part of the shaken area, to distances as great as 110 kilometers. Since most of the observers were awakened by the earthquake (at 4:03 A.M.), the reports of lights seen before the shock are very few. Most of the reports refer to flashes of light in the sky resembling sheet lightning, but longer in duration, and repeated several times according to some accounts. From the numerous reports at Tokyo, Terada concludes that the whole sky, including some low clouds, was more or less illuminated. Among those who saw flashes was Dr. Wadati, later Director of the Central Meteorological Observatory. In the meizoseismal area, fancier things were reported seen—balls of light, streamers like the aurora.

It was definitely established that no thunderstorm was anywhere in the area at the time. Some electrical transmission lines in the meizoseismal area were affected; some of the light seen may have been due to arcing. Terada was satisfied that this could account for only a small part of the reports; and it must not be forgotten that similar descriptions are on record from times long before the use of electric power.

Directions in which light was seen yielded no pattern capable of consistent interpretation. For Tokyo, Terada thinks some light was probably seen in whatever direction one happened to be facing.

While the Izu earthquake was investigated almost as much as the Tango earthquake, the literature dealing with it is more difficult to study. Publications are scattered widely in time, and those in English are in part incomplete. Some of the difficulty seems to have originated in work being done by two independent organizations, the Earthquake Research Institute and the Central Meteorological Observatory. Research and publication were carried out in a spirit of competition rather than cooperation. While this may have led to a little more critical caution, there seems to have been a good deal of duplication on one hand, and of failure to share information on the other.

FIGURE 30-14   *Faults of the Formosa earthquake of April 21, 1935. [After Nishimura, with data from Otuka and others.]*

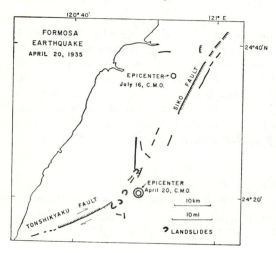

FORMOSA EARTHQUAKE—1935

Although the earthquake in Formosa, or Taiwan, on April 21, 1935 (Figs. 30-9, 30-14) (local date; April 20 G.C.T.) was of magnitude 7.0 only, it was accompanied by extensive faulting and constituted a major disaster. The western reader has the usual difficulty of finding most of the material published only in Japanese, with short English summaries; moreover, he has to reconcile two independent reports issued by competing organizations. The catastrophic consequences were largely due to the weak native construction on Formosa, which consists of sun-dried clay blocks analogous to the adobe of California and Latin America. Wooden houses of Japanese type, themselves no models for earthquake-resistant construction, nevertheless performed much better.

All place names here are transliterated in the form used by the Japanese, which frequently differs from that preferred by the Chinese. (Both groups write the names in the same Chinese characters.) The meizoseismal area was in the two prefectures of Taityu (or Taichu) and Sintiku (or Shinchiku); these are also the names of the corresponding principal towns.

Faulting

There are two principal fault traces, separated by about 30 kilometers. Both are close to known geologic faults. The northern or Siko fault extended for 15 kilometers with a trend N 30° E. Horizontal displacement, if any, was insignificant; but there was downthrow eastward, reaching a maximum of 3 meters. The dip was between 70° and 80° westward, and accordingly the trace curved westward in valleys and eastward on ridges. Temporary lakes were formed where drainage was dammed east of the scarp.

The southern, or Tonshikyaku, fault trace was about 12 kilometers long, trending N 60° E. The downthrows were from 60 centimeters to 1 meter and they reversed at a scissor point, east of which the downthrow was southward, while west of it the downthrow was northward. There was consistent right-hand strike-slip of 1 to 1.5 meters, offsetting roads, fields, footpaths, and the like. At one point the trace passed directly through a circular flower bed, deforming it accordingly. The trace showed echelon offsets in the right-hand sense (that is, in the same sense as tension cracks accompanying right-hand strike-slip).

Otuka, in describing these features, includes the following noteworthy statement:

> During this earthquake of Central Taiwan, Mr. D. Ho and K. Kwo observed at Sintakusan and Roppun in Sitan-syo, respectively, that the Siko earthquake fault formed after their houses were destroyed by the earthquake shocks, and not simultaneously with the initial shock.

This led Otuka to far-reaching speculation on the nature of earthquakes. However, such observations, though interesting and important, are easily reconciled with ordinary ideas of earthquake mechanism. Elastic waves from the point of initial rupture might well be expected to reach the surface earlier than the rupture itself, which must almost certainly propagate at less than the velocity of longitudinal waves, and in many cases even at less than that of transverse waves. In this particular case there is a further consideration: the instrumental epicenter is on the southern fault trace, not on the northern one where these observations were made.

Instrumental Epicenters

Main Shock. This epicenter accordingly is of considerable importance in the interpretation of the event. There is, fortunately, little difference among authorities. The following determinations have been published:

	Lat. N	Long. E
Central Meteorological Observatory, Tokyo	24° 21'	120° 49'
Kawasumi and Honma	24° 19.6'	120° 37.6'
International Seismological Summary	24.0°	121.0°
Gutenberg	24¼°	120¾°

Of these, the first was a preliminary determination at Tokyo. The second was derived by taking the first as an approximation and applying a careful reduction by the method of least squares, using transit times determined for the Japanese region by Matuzawa. The *International Summary* epicenter is admittedly rough; to save labor the coordinates were rounded to the nearest whole degrees, giving a position from which distances had been calculated for an earlier earthquake. Gutenberg's determination used the *Summary* epicenter and distances as first approximation, applying a rapid method of reduction which is good, as indicated, to about the nearest quarter degree. The resulting epicenter, and that of Kawasumi and Honma, are close to the Tonshikyaku fault, and definitely not to the Siko fault.

Aftershocks. For an aftershock of magnitude 6, which followed 24 minutes after the main earthquake, Gutenberg, revising the *Internaional Summary* result, finds an epicenter near 25° N, 120½° E. This is too far north for the Siko fault, and it suggests a wide extent for the total seismic disturbance. The workers at Tokyo, as well as the *International Summary,* refer the following shocks to many different epicenters; but the circumstances of registration make all these determinations relatively uncertain. For a comparatively large aftershock on July 16 (July 17, local time) the epicenter determined at Tokyo was 24.6° N, 120.9° E. This is accepted in the *International Summary,* where it is shown to fit the data of both near and distant stations fairly well. The location is practically on the Siko fault.

From August to December, 1935, a set of four stations was operated near and to the west of the Siko fault. Using the same methods as on previous occasions, Nasu determined epicenters for 55 aftershocks, all at short distances from these four stations and all close to or west of the Siko fault. Neither from this nor from other instrumental data is there good evidence of aftershocks associated with the southern fault.

It is unsafe to draw conclusions about the events of April from the large shocks in later months. Formosa is a region subject to bursts of relatively high seismicity, and such a burst occurred in 1935. Shocks of magnitude 6 and over were scattered over a much wider area than that of the April earthquake; on December 17 there was one of magnitude 7.1 far off the east coast. As another example of what this region is capable of, note that within 9 hours on October 21–22, 1951, there were three shocks with magnitudes over 7 originating close together near the northeast coast.

TOTTORI EARTHQUAKES—1943

These earthquakes, originating near the Japan Sea coast of Honshu, belong to the same group as the Tango earthquake of 1927; the epicenters are about 100 miles farther west. On March 4 and 5, 1943 (local time), there were two nearly equal shocks (magnitude about 5¾), separated by 10 hours, which heavily shook Tottori and environs. There were no deaths, but 11 persons were injured and about 70 Japanese houses collapsed. There was no observed faulting. These were foreshocks, for on September 10, 1943, there was a shock of magnitude 7.2 with nearly the same epicenter. Over 1000 were killed, and 7500 houses were destroyed. Two fault traces (Fig. 30-15) were found in the country south of west from Tottori, trending nearly parallel about west-southwest–east-southeast, about 3 kilometers apart. Both show consistent right-hand strike-slip up to 1 or 2 meters. The southern, or Shikano fault, showed scissoring, being downthrown southward in the east half and northward in the west half; its length was about 8 kilometers. The northern,

FIGURE 30-15   *Faulting of the Tottori earthquake, September 10, 1943.*

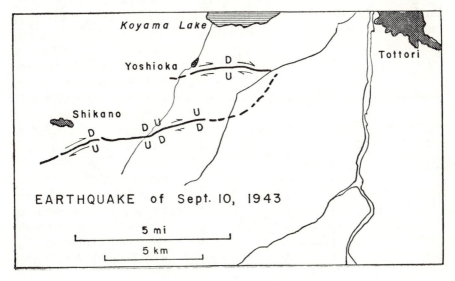

or Yoshioka, fault showed a trace about 4 kilometers long, with northward downthrow.

On this occasion occurred the remarkable displacement of stone dogs referred to in Chapter 3.

Aftershocks were studied with a large number of temporary stations, using *S-P* intervals for location. The evidence clearly shows extension of activity westward from the principal epicenter, near Tottori, to points considerably beyond the observed fault traces.

TONANKAI AND NANKAIDO—1944, 1946

On December 7, 1944, and December 20, 1946, great earthquakes originated at not widely separated epicenters off the Kii peninsula and Shikoku, in a part of the Pacific seismic belt which had been almost completely quiet during the rise of international seismology (although two great earthquakes had taken place there in 1854, less than 24 hours apart and not far from the same points; these, as well as the two later shocks, were accompanied by destructive tsunamis).

After each of the earthquakes of 1944 and 1946, aftershock activity extended over large areas, on land as well as offshore. Releveling showed that extensive deformation of the land surface occurred on both occasions; these events thus belong with those described in Chapter 14 as accompanied by regional distortion.

MIKAWA EARTHQUAKE—1945

On January 13, 1945 (Japanese time; January 12, 18:38:26, G.C.T.), an earthquake of magnitude 7.1 violently shook the coastal district of Mikawa

FIGURE 30-16   *Faulting of the Mikawa earth-quake, January 13, 1945.*

(Fig. 30-16), southeast of Nagoya. As this occurred within the meizoseismal area of the Tonankai earthquake of 1944, the Mikawa earthquake is regarded as an aftershock. The 1944 epicenter was offshore to the southwest. The Mikawa earthquake may indicate a northeasterly extension of faulting, but the most probable interpretation is somewhat different. A fault trace

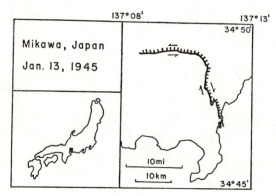

was indeed formed; Tsuya, who describes it under the name of the Fukōzu fault, quite properly calls it remarkable. Part of the trace trends nearly east for about 5 kilometers; then it turns abruptly south-southeast for another 4 kilometers, which brings it to the coast of Mikawa Bay (or Atsumi Bay). The trace skirts a mountain block of Paleozoic metamorphics; this block has been overthrust northward and eastward, and the blocks in those directions show relative downthrow of as much as 2 meters. Tsuya was able to measure dips directly at the trace, and he found the fresh fractures dipping 50° to 70° south and west; except that occasionally, where basement rock was not exposed, so that the fracture emerged through thin subsoil or through Quaternary deposits, the dip at the surface was reduced to 15° to 20°. There is topographic evidence of older faulting along the same lines. The mechanics of the motion is further confirmed by strike-slip, not exceeding a meter, left-hand along the northern east-west trace, but right-hand on the eastern trace. It appears that the crustal readjustments following the December earthquake set off displacement on an old thrust surface.

FUKUI—1948

While no surface faulting was observed for the Fukui earthquake of June 28, 1948, some important data were obtained. The magnitude was 7.3. In the meizoseismal area, which included most of the alluviated Echizen plain, over 75 per cent of the structures were demolished. Over 5000 were killed, about 900 in Fukui city alone. Intensity was markedly lower on adjacent bedrock.

Evidence for Faulting

Fukui is in the line of the Mino-Owari fault zone, and it was violently shaken in the 1891 earthquake. Compressions and dilatations in the first motion recorded at Japanese seismological stations for the 1948 earthquake show a clear quadrantal distribution, consistent with left-hand strike-slip on a fault trending N 20° W; this would fit the facts of 1891. Triangulation surveys confirmed this, indicating that the fault passes east of Fukui. No clear surface indication of faulting was found, although cracks and pressure ridges were found which might have had some tectonic significance.

An Unusual Tragedy

Large fissures of secondary character opened in the alluvium. To quote an official report:

A tragic event frequently described in fiction but rarely occurring in actuality was verified. The party learned of the fact that a young woman had been crushed to death in a fissure in the near neighborhood. The site was visited and the witnesses interviewed. The victim, Mrs. Sadako Nankyo, age 37, was working in a rice paddy, located close by the house at 33 Shissaku-machi, Fukui City, when the shocks were felt. She started out of the paddy but fell into a fissure which, it was said, opened to about 4 feet in width. It closed upon her to the chin, instantly crushing her to death. The body was immediately dug out of the ground (only a faint remainder of the course of the fissure could be traced) by neighbors and the woman's husband. Seven people saw the woman in the ground after the quake and several of them were eye-witnesses of the happening.

This is the only properly documented case of the kind in recorded history, even in Japan, except for that of the cow in the California earthquake of 1906. There is nothing incredible in the account; from the nature of fissuring and cracking described as seen in many earthquakes, such accidents must occasionally happen. The literature contains many vague and doubtful or incomplete references to such instances; in Japan, as in other countries, the exaggerated fear of being "swallowed up" during an earthquake is widespread. The risk, though real, is much less than that of many other earthquake dangers.

Aftershocks

Ten temporary stations were set up by S. Omote and collaborators of the Earthquake Research Institute, and aftershocks were located, as before, by the method using S-P time intervals. For the first time in Japan an attempt was made to determine the actual instants of arrival at the temporary stations; a broadcast radio time signal (once an hour) was received at five of these stations for use in correcting the clock marks. The drum rates were so irregular that this method was applicable only in the event of getting a shock recorded close to the time of the radio signal. One fortunate case of this kind was recorded, and the data were used for determining the hypocenter and studying velocities. The epicenter is about 25 kilometers north and slightly east of Fukui, near one of the temporary stations. Japanese workers generally place the epicenter of the main shock nearer Fukui, but the readings at distant stations fit better with the epicenter given by Omote for this aftershock. Omote maps epicenters scattering southward from this point as far as Fukui, and fairly well covering the Echizen plain. None is shown outside his network of stations.

IMAICHI EARTHQUAKES—1949

On December 26, 1949 (local date), two earthquakes of magnitude 6 and 6¼ originated near Imaichi, which is about 110 kilometers north of Tokyo and 7 kilometers east-southeast of the celebrated shrine at Nikko. Thorough investigation was carried out by the Earthquake Research Institute staff. Epicenters were determined as close to 36° 42′ N, 139° 39′ E, with origin times 23:17:29 and 23:24:52, December 25, G.C.T.

Intensities and Accelerations

The meizoseismal area was small, with relatively high intensity (M.M. VIII, probably reaching IX locally). This and the microseismic data suggest a rather shallow hypocenter; volcanic origin is possible, but normal tectonic causation is at least as probable. At and near Imaichi 299 dwellings collapsed, large landslides occurred even on gentle slopes, and 10 lives were lost. At Ochiai village, in the center of the meizoseismal area, there was evidence of prevailing vertical motion; houses were damaged as by vertical overloading, loose objects remained on shelves, and tombstones were not overthrown. Yet at Imaichi, about 7 kilometers distant, hardly a tombstone remained standing. Ikegami and Kishinouye, studying the overturning moments, concluded that accelerations reached 0.9g. This is consistent with recording of transient impulses of high acceleration in other minor earthquakes (Chapter 3).

Aftershocks

Aftershocks were recorded at 11 temporary stations; as many as 5 were in operation on some days. One was equipped for precise timing, with carefully regulated drive and marking clock, and time control from radio signals; but there was no such timing at the other temporary stations (most of which did not even have marking clocks). Unfortunately, no well-timed aftershock was large enough to record clearly at Tokyo or other permanent stations with precise time.

Epicenters determined from S-P intervals scattered along a narrow belt extending about 10 kilometers southward from a point just west of Imaichi. Depths determined by the same method (and subject to the usual doubts) ranged from zero to 10 kilometers. At one station three components were recorded. Azimuths of epicenters as determined from the horizontal components generally agreed with the locations found by using several stations, but sometimes diverged widely. The ratio of horizontal to vertical motion was used to determine the depth (with proper allowance for the effect of the free surface). The results were not satisfactory; apparently they indicated a discontinuity a few kilometers deep (which would call for modifying the epicenters as well).

Initial recorded compressions and dilatations, for the principal earthquakes at distant stations and for aftershocks at the temporary stations, were examined, but they fitted no simple pattern. Sensitive instruments were used to record micro-earthquakes among the aftershocks, the smallest of which appear to have been of magnitude zero or slightly higher (although the investigators estimate −0.8).

ARC AND BLOCK MECHANISM

Descriptions of faulting in the earthquakes of the Japan Sea coast and of Formosa, as well as in the great earthquake of 1891, present little that does not appear in observations in California or New Zealand. Forces operating in the continental crust of the same general character, though with different orientation, are suggested in these three parts of the circum-Pacific belt. A distinguishing characteristic of the Japanese region may perhaps be seen in the frequent development of evident multiple traces, as in the Tango earthquake of 1927 and the two Formosan earthquakes, although the indicated complexity is certainly no greater than for the Nevada events of 1954.

A different impression follows from the abundant data on the earthquake of 1923, supplemented by similar but more fragmentary results for 1944, 1945, and 1946, and by historical records like those for 1596, 1703, and 1707. There is suggestion that strike-slip is involved, but the indicated regional distortion is strongly reminiscent of the Indian earthquakes of 1897 and 1934 (Chapter 5), and this may be taken as indicative of the mechanism of shallow earthquakes directly associated with the active arcs. The great earthquake of 1933, best known for its destructive tsunami (Chapter 9) probably should be cited here, since it evidently belongs with the Honshu arc; but little is known about it that can be applied to the purpose, except the great number and wide geographic extent of its aftershocks.

JAPANESE GEOGRAPHICAL NAMES AND SPELLING

The names of the four chief islands, Honshu (or Honsyū), Shikoku, Kyushu (Kyūsyū), and Hokkaido appear repeatedly in the foregoing pages and on the accompanying maps.

The naming of local areas has a double aspect here, as in France. In attaching local names to earthquakes, the Japanese prefer to use the traditional names of provinces and districts, such as Kwantō, Mino, Owari, Nankaidō. The modern administrative districts differ from these. They correspond to the French departments and are known as prefectures (*ken*); their names commonly appear in postal addresses.

Spelling is often a stumbling block to the western reader. It is perfectly definite in the Japanese form; the ambiguities arise in attempts to represent the Japanese sounds in the western alphabet, which is rather inadequate for the purpose, especially to sensitive Japanese ears. A certain amount of arbitrary conventionality is unavoidable.

Spellings at first favored and long current were stabilized in a form reasonably comprehensible to English or American readers, but often awkward for users of other European languages, who sometimes introduced expedients of their own. Increasing nationalism in the 1930's led to gradual introduction of other usages, to some extent more convenient to the Japanese, but highly confusing to others. At present there is a tendency to return to the older Anglicized spelling; but in the interval many important publications were printed, and the reader must be prepared to deal with the equivalence of tsunami and tunami, or to recognize Fujiyama disguised as Mount Huzi. Current Japanese usage prefers Fujisan to Fujiyama. There was no authority to enforce uniformity; thus three successive papers in one issue of a journal in 1935 use the spellings Sizuoka, Siduoka, Shizuoka.

Japanese words are formed from a limited number of relatively fixed syllables, and problems of transliteration reduce to those of spelling these syllables individually. In the following little table, the column at the left gives the more familiar spelling, with the commonest alternative on the right.

chi	*ti*	*jo*	*zyo*
cho	*tyō*	*sha*	*sya*
zu	*du*	*shi*	*si*
fu	*hu*	*sho*	*syo*
ja	*zya*	*shu*	*syu*
ji	*zi*		

The greatest confusion results when these alternatives occur in the first syllable, particularly when the name is then widely displaced in alphabetical lists, as in replacing Fuji, Fukui, Fukuoka by Huzi, Hukui, Hukuoka.

The syllable *du, zu* appears by compromise as *dzu*. The long O syllable, which occurs often at the beginning of names, as in Osaka, Oshima, Otuka, is commonly printed Ō; occasionally the O is doubled instead, and we read Oosaka, Ooshima. The long mark over o and u often appears as a circumflex: ô, û.

Further complication occurs in the south, where both Japanese and Chinese spellings have gained international currency, so that we have to choose, for example, between writing of the Ryukyu or the Luchu islands. Such alternatives are the rule in Formosa; fortunately both Japanese and Chinese use the name Taiwan for the island itself, but the administrative center (and location of the chief seismological station) is known to Japanese as Taihoku and to Chinese as Taipei (or Taipeh).

References

FOR PAPERS dealing with individual earthquakes see Appendix XVI. On geology and tectonics it is difficult to find papers which are inclusive, up-to-date, and accessible to western readers. The following are suggested:

Otuka, Y., "Median dislocation line of southern Japan and the Nagasaki 'Dreiecke,'" *Bull. E.R.I.,* vol. 13 (1935), pp. 457–466. (In Japanese; a useful English summary on pp. 467–468.)

Collins, J. J., and Foster, Helen L., *The Fukui Earthquake, Hokuriku Region, Japan, 28th June 1948,* Vol. 1, *Geology,* Office of the Engineer, General Headquarters, Far East Command), Tokyo, 1949. (Contains much general information.)

Kobayashi, T., "The mountain systems on the western side of the Pacific Ocean classified from the standpoint of genesis," *Proc. 7th Pacific Science Congress,* Vol. 2, *Geology,* pp. 255–261.

Honda, H., and Masatuka, A., "On the mechanisms of the earthquakes and the stresses producing them in Japan and its vicinity," *Tôhoku Univ. Science Reps.,* Ser. 5, vol. 4 1952), pp. 41–60; vol. 8 (1957), pp. 186–205.

Imamura, A., *Theoretical and Applied Seismology,* Maruzen, Tokyo, 1937. (An invaluable general reference, the immediate source for much material in this book.)

Wadati, K., and Iwai, Y., "The minute investigation of seismicity in Japan," *Geophys. Mag.* vol. 25 (1954), pp. 167–173; vol. 27 (1956), pp. 11–15. First paper reprinted, *Proc. 8th Pacific Science Congress,* vol. 2A, pp. 775–782; second paper reprinted, *Publ. Bur. Centr. seismologique internat.,* Ser. A, vol. 19 (1956), pp. 261–265. (Many details of the distribution of small shocks.)

Kobayashi, T., "The mountain structure of the Japanese islands," *Proc. 8th Pacific Science Congress,* vol. 2A, pp. 743–751. "The insular arc of Japan," *ibid.,* pp. 799–807.

Murakoshi, T., and Hashimoto, K., (editors). *Geology and Mineral Resources of Japan.* Geological Survey of Japan, 1956. (Includes geological map, scale 1:3,000,000, dated 1953.)

Kawasumi, H., "Measures of earthquake danger and expectancy of maximum intensity throughout Japan as inferred from the seismic activity in historical times," *Bull. E.R.I.,* vol. 29 (1951), pp. 469–482. (Includes a catalogue of 342 large earthquakes in Japan, A.D. 599–1949.)

CHAPTER 31

Tectonic Earthquakes of Other Regions

THE SOMEWHAT MISCELLANEOUS materials of this chapter are arranged geographically. The order of treatment is: North and South America, East Indies, the Alpide belt from east to west, and Africa. Only earthquakes are discussed which furnish data bearing on the study of tectonic mechanisms or are of special interest otherwise.

MISSISSIPPI VALLEY, 1811–1812

Great earthquakes are commonest where the general level of seismicity is high, but occasionally they occur in other regions. Such were those of 1811–1812 under the central Mississippi Valley, near New Madrid (Missouri). There were three great shocks on December 16, January 23, and February 7. One of these must have been the largest known earthquake in the present territory of the forty-eight states. All of them, judging by their effects, exceeded magnitude 8; evidence for comparing them is in part conflicting and suggests that the epicenters differed. The first great earthquake on December 16, 1811, was perceptible over an area of at least a million square miles, extending from the headwaters of the Missouri River to the Atlantic seaboard (as far as Boston) and from Canada to New Orleans. (Fig. 31-1.)

In the heavily alluviated meizoseismal area, fissuring, lurching, slumping, emer-

FIGURE 31-1 *New Madrid earthquakes, 1811–1812. Meizoseismal area, shaded; points at which shaking was felt, dots. [After Fuller.]*

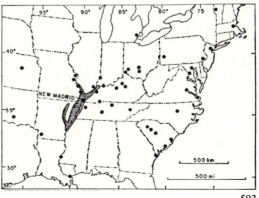

gence of ground water, and associated effects took place on a large scale. The investigators of the Indian earthquake of 1934 have compared these effects closely with those in the "slump belt" on that occasion. The numerous long, narrow areas of subsidence suggested the dropping of crustal blocks rather than slumping or warping. Such subsidence amounted in places to 15 feet and formed large new lakes, such as Lake St. Francis west of the Mississippi and Reelfoot Lake (Tennessee) to the east. There were also large uplifts; their exact extent is not known, since investigation many years later showed that the relief had changed since 1812; however, evidence of old trees indicated elevations of at least 10 feet at the time of the earthquakes. Small scarps were formed which produced temporary waterfalls; since these were not in rock, it is impossible to tell whether they were the result of slumping and fissuring, or whether they were related to displacement in the basement rock.

MEXICO—1887 AND 1912

On May 3, 1887, a great earthquake took place in the northern Mexican state of Sonora; a report on field investigation was published by J. G. Aguilera, chief of the Mexican Geological Survey. Effects were also investigated in the field by G. E. Goodfellow, of Tombstone, Arizona, who collected other information and sent a report to C. E. Dutton; Dutton published a large part of this material.

The largest town affected (1500 inhabitants, of whom 42 were killed) was Bavispe, which was situated on unconsolidated material and suffered heavy damage. Shaking was reported felt from Prescott, Arizona, to Mexico City, or to distances of over 400 miles from the probable epicenter. Faulting was reported on both sides of the Sierra Teras, a north-south range forming part of the Sierra Madre Occidental, which is continuous with the ranges of southeastern Arizona. The rocks are chiefly either crystalline metamorphics or Tertiary and younger volcanics.

Aguilera describes faulting only on the west face of the Sierra Teras, where the scarp (Fig. 31-2), somewhat less than 300 feet above the valley floor, followed a winding course over 35 miles long,† with maximum throw 26 feet. On crossing ridges, the curve of the trace was generally convex toward the valley, suggesting dip eastward under the mountains. Aguilera directly measured eastward dips of 75° at two locations at least.

Of the eastern faulting Goodfellow reports: "I have been told by Colonel Kosterlitzky, who has recently been there, that on the Chihuahua side of the

---

† Professor Gianella examined the area many years later, finding the trace of 1887 still clear. However, it was much longer than as described by Aguilera and by Goodfellow, who were probably deceived by the fact that the trace apparently dies out and then resumes farther on. Field investigators of other earthquake traces have been trapped into incomplete description in the same way.

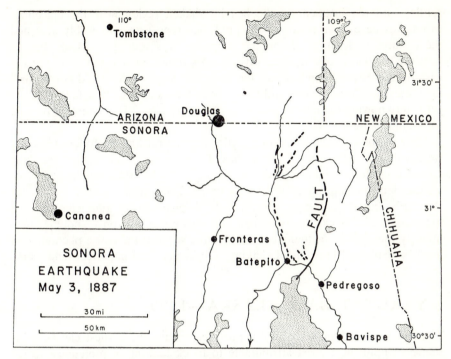

FIGURE 31-2 *Sonora earthquake, 1887. Faulting according to Aguilera. Shaded areas are at elevations over 6000 feet.*

Espuelas and Pitaicachi is a duplicate . . . fault. I have endeavored to confirm this, but without success." Dutton writes of this: "In other words, the range seemed to have been uplifted several feet between faults on either flank." It is not clear how anyone could be assured that the two scarps appeared simultaneously. There were many strong aftershocks.

Evidence for faulting in the Mexican earthquake of November 19, 1912, offers a fair example of the type which has to be rejected. This shock was very widely felt but was damaging only in a rather sharply bounded area measuring about 50 by 20 kilometers, centering about 100 kilometers from Mexico City. This was a highly unusual event, apart from its occurrence so far from the principal seismic zone of Mexico near the Pacific coast. The area is one of volcanic rocks, partly alluviated. What is described as faulting took place along three rather broken and irregular zones extending east and west. The traces consisted mainly of fissures; the sides showed differences of level, not exceeding about 60 centimeters, with apparent downthrows toward the center of the disturbed area.

These were almost surely secondary features due to shaking. They were found only in the meizoseismal area, as judged by damage to structures. They may have been mere lurching; but more probably they represented general settling and readjustment of the heavily shaken region near the epicenter. Montessus de Ballore reviewed the published report and ostensibly accepted the features as faulting due to the dropping of a four-sided block; but he remarked that it is odd that the investigators found no sign of faulting at the eastern or western boundaries of the disturbed area. "Why it did not occur," he wrote, "we do not know; and this impossibility shows us how far we are as yet from knowing the inmost processes of Nature . . ." He cannot be blamed for being puzzled.

Most of the earthquakes strongly affecting central Mexico follow the east-west volcanic belt from Colima past Popocatepetl to Orizaba; the best recorded of these originate at depths of the order of 100 kilometers. Instrumental data for epicenters and depths in 1912 are unsatisfactory by present standards, but they strongly suggest a depth of about 80 kilometers for the hypocenter of this earthquake. Since the magnitude was only about 7, it is unlikely that faulting actually extended to the surface; the entire phenomenon must be explained in terms of shaking, subsidence, and imperfect field work.

YAKUTAT BAY, ALASKA—1899

The Alaskan earthquakes of 1899 provide perhaps the earliest clear-cut instance of uplift at a coast attributable to block faulting in preference to regional warping; they also retain the record for the greatest known displacement in a single seismic event. There were two great earthquakes on September 3 and 10 (local dates). Seismograms, though few, were sufficient to place the epicenters on the coast of Alaska, and they have lately provided data for estimating the magnitudes as 8.3 and 8.5, respectively.

Macroseismic Data

Field investigation was not undertaken until 1905, but it was then continued for several seasons. The meizoseismal area was located adjacent to Yakutat Bay, especially in its arm, Disenchantment Bay. Both large shocks were felt to distances over 400 miles from Yakutat Bay; seiches and other marginal effects occurred much farther away. There is confusion where the date is not certain; however, there seem to be more observations for September 10 than for September 3. The only settlement in the meizoseismal area was Yakutat village, over 30 miles distant from Disenchantment Bay. There the shaking on September 3 was violent, and it was impossible to stand without holding onto something; on September 10 the shaking was at least as strong, but the observers chiefly describe great waves and disturbances in the harbor. Prospectors in Disenchantment Bay considered the shock on September 3 relatively slight in retrospect, since on September 10 they went through an appalling experience and barely escaped with their lives.

Eyewitness Accounts. Eight men in two parties were camped on the east shore of Russell Fiord near Disenchantment Bay. During the violent shock on September 10 one group at first failed to get out of their tent; two of the men held onto the tent pole to keep from being thrown down, while a third man was thrown over the camp stove into a corner. The face of the adjacent Hubbard Glacier was broken up so much that the ice ran out into the bay for half a mile. The waters of a small lake behind one camp escaped and flooded down toward the beach, carrying masses of rock with it. Immediately afterwards the water of the fiord rose in a wave the men describe as 20 feet high, which washed inland over the beach. The campers lost everything but a few provisions and one boat. By luck they found a native canoe afloat, and all managed to reach Yakutat on September 14.

The apparently lower intensity of the September 3 shock as felt near Yakutat Bay is significant. This earlier shock was reported as extremely violent at Cape Yakataga, about 100 miles west of Yakutat. The authors of the memoir incline to believe that the epicenter on that occasion was actually to the west; this would mean that the faulting in the vicinity of Disenchantment Bay occurred mainly during the great earthquake of September 10.

Faulting. Faulting is indicated by changes in level, chiefly uplifts, as shown in Figure 31-3. There is little doubt that these changes occurred at the time of the earthquakes. Photographs taken in 1895 show coasts and islands at their previously mapped levels; and three months before the earthquakes the bay was visited by G. K. Gilbert, a distinguished authority on elevated shore lines, who saw none of the spectacular examples observed by Tarr and Martin in 1905.

These investigators found the normal textbook evidences of elevation, wave-cut benches and caves well above tide level, in very perfect form. Not less remarkable was the botanical and zoological evidence, the decaying remains of tide-level animals and plants clinging to the rock, marking the former shore line in its raised position. Among these the shells of barnacles and mussels were especially common, and some of the more accurate measures of uplift were made from the level of these dead creatures down to that of the living specimens at the new tide line. In a few places submerged trees and the like indicated depression.

The sharp changes in these elevations across narrow and roughly straight channels like Disenchantment Bay and Russell Fiord are readily explained by faulting, while explanation in terms of warping or folding runs into trouble mechanically.

The largest uplifts, ranging from about 30 feet up to the observed maximum of 47 feet 4 inches, were found on the west coast of Disenchantment Bay; fortunately these changes are the best substantiated by evidence of all kinds. Uplift of the opposite coast was between 7 and 8 feet, while on the small Haenke Island, near that coast, it was near 18 feet. Since the bay is less than

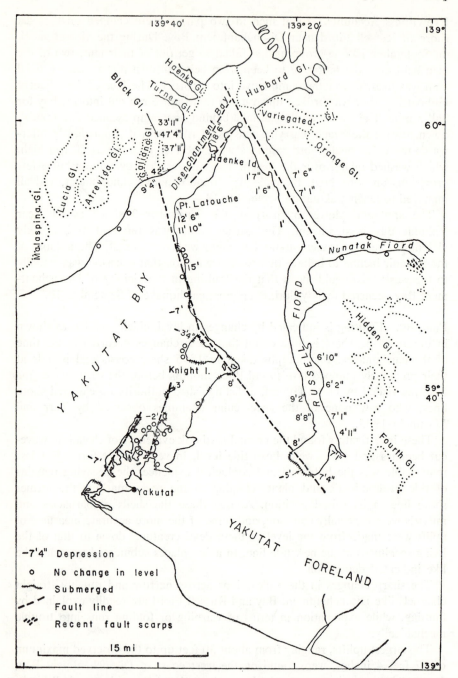

FIGURE 31-3   *Yakutat Bay earthquakes, Alaska, September 1899. Changes in level [after Tarr and Martin], and indicated faulting.*

5 miles wide, two faults are called for between Haenke Island and the coasts. Uplift of the block west of these will account for the major elevations found. Along the coast westward out of Disenchantment Bay there is a sudden drop in elevation as the Black Glacier is crossed, from an estimated 30 feet to a well-observed 9 feet 4 inches. This glacier and its stream were greatly changed between 1899 and 1905, and the description strongly suggests that the edge of the upraised block broke along a fault striking northward through Black Glacier.

The block bounded on three sides by Russell Fiord, Disenchantment Bay, and Yakutat Bay was uplifted about 12 feet at its northwest corner. On the west this block is bounded by one or two sharply defined faults, beyond which there was generally depression; this includes the depressed coast at the head of Russell Fiord. Uplift is less than 2 feet on the west shore of Russell Fiord, while on the east shore it is over 7 feet; this is the reason for drawing a fault down the fiord.

All these displacements fit reasonably well into the idea of a breaking up of the area in fault blocks under a coherent system of stresses; the opposition of uplift and depression on the western faults strongly suggests vertical elastic rebound. On the other hand, an area of remarkably fresh step faulting found near the head of Nunatak Fiord may be secondary in character, due either to violent shaking or to a comparative superficial local readjustment to the displacement of a large block. St. Amand and others have speculated whether this may not have been part of more extensive faulting inland.

Effects on Glaciers

These earthquakes are noted for their effect on the numerous glaciers in the vicinity of Yakutat Bay. Here, as in many other regions, the glaciers had been generally retreating since the mid-nineteenth century. After the earthquakes this particular group of glaciers advanced greatly for a few years each; the shorter and smaller glaciers first, then the larger ones. Eventually the general retreat was resumed. These changes are attributable to the shaking down of masses of snow and ice at the head of the glaciers; the additional supply would take longer to advance the fronts of the larger and longer glaciers. On the other hand, the Muir Glacier, 150 miles away, broke up and retreated abnormally in the years between 1899 and 1907; this effect might not have been due entirely to the earthquakes.

Structural Relations

The Yakutat Bay region is remote from the arcuate structures of Alaska; on seismic maps it appears near the northern end of the belt of epicenters off the coast of British Columbia. However, the great faults of central Alaska, which follow the general east-west trend of the ranges, strike southeastward toward this region. A fault at the base of Mt. St. Elias has been thought to follow the center line of the Miller Glacier into Disenchantment Bay and Russell Fiord.

CHILE—1822, 1835, 1906

Elevations on the coast of Chile accompanying the Valparaiso earthquake of 1822 and the Concepcion earthquake of 1835 have already been mentioned (Chapter 14) in connection with the refusal by Suess to accept the evidence. They probably do not directly represent faulting, as if the coast line were determined by an actual scarp. They belong in the class of regional warping, possibly associated with thrust faulting at depth. It is no wonder, therefore, that the reported uplifts were not uniform and that in certain localities none was noticed. Suess made much of this, and he rejected much detailed testimony, including that of Charles Darwin, whom the celebrated voyage of the *Beagle* brought to the coast of Chile at the time of the 1835 earthquake. The uplifts, some of which were measured and reported in detail, amounted to as much as 8 or 10 feet in the vicinity of Concepcion. A reported uplift near Valdivia in 1837 was of the same order.

A rather amusing situation arose after the Valparaiso earthquake of 1906, when changes of level at the coast were reported officially and unofficially. Montessus de Ballore, who investigated and published a report on the earthquake, tried earnestly to discredit these changes, but he was an avowed and enthusiastic disciple of Suess. Brüggen, in a critical review of the reports and of de Ballore's handling of them, leaves his reader in little doubt that some uplift actually took place. It is possible that in this instance, and perhaps in those of the previous century, the uplifts decreased with time; similar observations have been made, with greater precision, in the course of releveling on the Japanese coast, particularly after the earthquake of 1923.

ARGENTINA—1944

Before 1944, no fault trace phenomena associated with earthquakes had been reported from South America. This is remarkable in view of the high seismicity of Chile and Peru. Instrumental seismology suggests a partial explanation: a high proportion of the larger earthquakes originate at depths greater than those normal for other regions. Exceptions occur (1) off the coast, (2) in central Peru, (3) in northwestern Argentina. In both land areas there is good geological evidence of block faulting.

San Juan, Argentina, was devastated by an earthquake of magnitude 7.8 on January 15, 1944. The microseismic epicenter, though not very closely fixed, is near San Juan, in the immediate vicinity of a fault trace described by Castellanos (Fig. 31-4).

The publication in which this valuable information is almost concealed represents a course of four lectures given under the auspices of a cultural association at Rosario. They are less scientific than popular, with the laudable intention of informing the public and the authorities about the nature of earth-

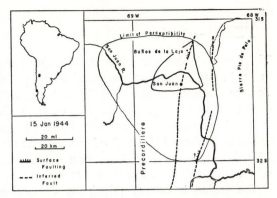

FIGURE 31-4 *San Juan, Argentina, 1944, showing faulting.* [*After Castellanos.*]

quakes and earthquake risk and the proper measures for earthquake safety. The first two lectures are elementary and general, based largely on handbooks which were then twenty years out of date. The second two, by Castellanos, deal specifically with the 1944 earthquake and its implications; they are remarkably thorough for a nearly single-handed undertaking, but the treatment suffers in places from lack of correlation with modern seismological investigation. Very annoying are the absence of any index or detailed table of contents and the representation of geography on figures which are intended to be maps but are not.†

Castellanos observed the principal trace feature crossing a road 3 kilometers south of Baños de la Laja, a locality north of San Juan; its strike was N 20° E, while the road trended N 35° W. When seen in February, 1944, there was westward downthrow of about 60 centimeters. This was in a dry river bed, but the trace could be followed for several kilometers in both directions, and southward it passed into rocky hills. A photograph shows the crack leaving the road and ascending a bank diagonally. According to Castellanos, it is a bedding-plane fracture in the sands and clays of the Pliocene Calchaquenos formation, which here dips 42° southeast. This means that the downthrow is opposite to the dip, which rules out any explanation in terms of secondary effects like slumping. Castellanos has the excellent judgment to say that he considers this fracture a result of the earthquake, and not its cause. In his interpretation, the origin of the earthquake was a displacement at depth on the boundary fault of the Pie del Palo range to the east. In consequence of elastic rebound the corresponding block rose relatively to that under the plain of San Juan, and this upward motion of the eastward block led to up-dip fracturing on the Calchaquenos bedding planes. This interpretation certainly fits all the available data; and it compares interestingly with Tsuya's observations on the Japanese earthquake of 1945 (Chapter 30).

---

† "When is a map not a map? When it has neither scale nor coordinates."

PERU—1946

The first well-observed instance of major faulting accompanying an earthquake in South America was provided by the Ancash earthquake in central Peru on November 10, 1946. This shock, of magnitude 7.4, disturbed an enormous area (Fig. 31-5), and extreme violence was reported in the meizoseismal region near the fault. Structures were demolished, and the death toll reached about 1400, which is remarkable for so thinly populated a region. This devastation was in large part due to great landslides, one of which covered the village of Acobamba to a depth of 20 meters.

The region is one of principally Cretaceous rocks dissected into deep canyons by the Marañon river and its tributaries. The principal fault scarp appeared on the relatively level summit of a ridge which rises over 1000 meters above the almost completely demolished town of Quiches (Fig. 31-6). This scarp, 5 kilometers long, has an average trend of N 42° W, paralleling the ridge, but it is downthrown southwestward, away from the canyon, and the fracture surface dips 58° southwest. This surface in places shows clean slickensides in limestone, with striation indicating that the entire displacement was dip-slip, with no sign of strike-slip. The manner in which this trace cuts

across the topographic irregularities of the elevated surface is so characteristic that anyone with a little experience will recognize it as faulting from the photographs (Fig. 31-7). The throw reached a maximum of 3.5 meters. Northward the trace disappears on approaching a deep canyon, but 10 kilometers beyond it resumes with similar characteristics for a length of 3 kilometers. West of the main fault is another fracture, paralleling it with opposite dip (30° northeast) and striking N 40° W; this is downthrown a maximum of 1 meter northeast, so that there is a graben about 2 kilometers wide.

The country is not easy to travel over under any condi-

FIGURE 31-5 *Ancash (Peru) earthquake of 1946. Faults and higher isoseismals. [After Silgado.]*

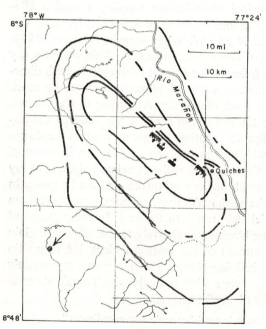

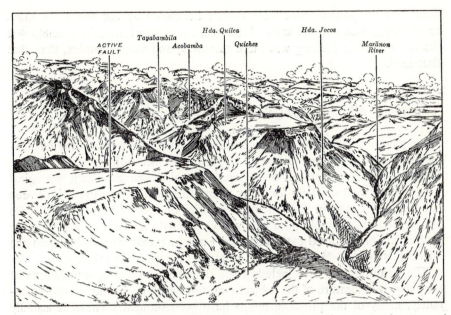

FIGURE 31-6 *Ancash (Peru) earthquake, 1946. Topography, towns, location of fault. [Courtesy E. Silgado.]*

FIGURE 31-7 *Ancash, Peru, earthquake, 1946. Fault trace crossing a saddle. [Courtesy E. Silgado.]*

tions, and the investigators were further hampered by the effects of slides and fissuring which closed some of the roads. It is therefore quite possible that the actual surface faulting was more extensive, and more complex, than that described.

THE EAST INDIES

The region of the East Indies, the greater portion of which now is the territory of the Republic of Indonesia, is important in seismology and tectonics. Largely because of interest in oil and mineral resources, much geological investigation went on there for many years. Although a great part of the area is under water, and in spite of the difficulties offered by its tropical climate and vegetation, valuable information accumulated and became available for synthesis. This involved extending field work as evenly as possible from the Nicobar Islands to New Guinea, a distance of over 3000 miles.

The writer cannot here include more than the outstanding conclusions of the extensive literature of the subject (most of which he has not read). Fortunately, many of the principal points have been brought forward repeatedly in the lively discussions centering about Meinesz' gravity observations.

Tectonic Lines

Fifty years ago, tectonic maps commonly showed the belt of Alpide structures as extending across Asia and the East Indies to New Guinea and the Solomon Islands. Field study in the Moluccas gradually rendered this interpretation unacceptable. Halmahera and the northern arm of Celebes were found to constitute two opposite arcs, and logical continuation of the structural trends led to the concept of the Pacific belt outlined in Chapter 26: the major branches diverging in the region of Honshu approach each other in the Moluccas, but separate again, the eastern through New Guinea and the western by way of Celebes and the Banda Sea to the Sunda arc. The sharply winding courses of both lines marked the East Indies as one of the most disturbed areas of the earth's surface. It was felt that here, if anywhere, the orogenic processes which played so large a part in the past history of the earth could be observed in full action at the present time.

The Sunda Arc. The Sunda arc, although it fronts on the Indian Ocean, is a nearly perfect example of a Pacific-type arc. All the characteristic features, from an oceanic trough (in this case the Java Trench) to deep-focus earthquakes (including some of the deepest known) are found here, especially if one combines features over its whole extent from Flores past Java and Sumatra. It was here that Vening Meinesz, with his apparatus installed in a submarine, established the first and best known of the "Meinesz zones" of negative gravity anomalies. In a narrow belt following the tectonic arc off the shore of the large islands, observed gravity was 200 milligals or more below the calculated value. This could mean only that here was a belt of disturbance and instability in the crust of the earth, perhaps the actual seat of the orogeny in progress. In following years, as critical investigation of seismic geography proceeded, and the effect of confusing deep with shallow shocks was removed, the epicenters of almost all the large shallow shocks of the region were found to lie directly along this Meinesz zone.

Off the coast of Sumatra the depression of the Java Trench decreases and finally almost disappears; the gravity anomalies decrease somewhat, and here in the Meinesz belt are a series of small islands whose rocks show evidence of strong compression, folding, and thrusting in the upper Miocene. Apparently a later repetition of compression has now brought these rocks up in an anticline. Such young anticlines are common about the Pacific, even in regions of block faulting.

Banda Sea and Celebes. The Meinesz line, and the determination of earthquake epicenters, have fitted in well with the geological evidence, so that there is little hesitation about continuing the Sunda arc eastward and then northward in a loop round the Banda Sea to Ceram. There is at one point a disturbance or interruption represented by the anomalously placed large island of Sumba.

Meinesz also found large negative gravity anomalies between the opposed arcs of Halmahera and northern Celebes; the Meinesz belt then runs northward to the Philippines. The course of the tectonic belt from Celebes to Ceram has been variously interpreted, since the gravity data and most other lines of evidence are less definite there. Intermediate and deep earthquakes, as well as volcanoes, show that the northern arm of Celebes represents an arc fronting east and south. From the west end of this arc the most favored procedure is to run the tectonic line through the southeastern projection of Celebes and thence northeastward to Buru and Ceram, then to complete the circuit about the Banda Sea. The evidence of peridotites (see Chapter 25) supports this procedure.

Block Structures

Borneo and western Celebes are parts of a stable mass where rocks of continental character and considerable geologic age are exposed. The Strait of Macassar between them marks an internal rift in this mass which is associated with fairly frequent shallow earthquakes, some of them large. The southwestern tip of Borneo, a small part of Sumatra, and the Malay Peninsula were involved in a Late Mesozoic folding.

The only directly observed faulting in the region is described in the following paragraph.

SUMATRA—1892

On May 17, 1892, there was an earthquake in northwestern Sumatra which was particularly strong in the Residence of Tapanoeli, where secondary triangulation surveys were going on at the time. Cracks appeared near one monument, which was accordingly replaced, but in the course of fixing this it was found that other monuments had moved, and a resurvey was necessary. Elevations were unchanged; but from six monuments horizontal shifts of the order of a meter were found (Fig. 31-8). H. F. Reid, analyzing the observations long after their publication, found that they fitted well for elastic rebound of the type found in 1906; it was possible to determine the strike of the probable fault, the sense of the strike-slip (right-hand), and roughly its amount (3.5 to 4 meters). There is a rift feature extending most of the length of Sumatra, obscured by volcanic deposits and jungle, for which geological evidence indicates strike-slip of this kind.

FIGURE 31-8   *Sumatra earthquake, May 17, 1892. Displacement of survey monuments, and inferred faults. [After Reid. Data from Muller.]*

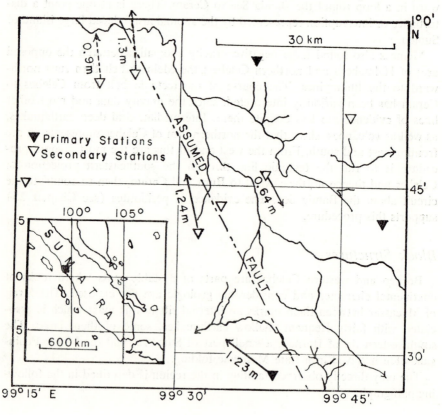

ALPIDE BELT—INDIA

Earthquakes of the Alpide arcs are of potentially great tectonic interest, since most of these arcs are continental and in some degree more accessible to all types of geophysical investigation than those of the circum-Pacific belt. Nevertheless, similar problems arise. Thus the large earthquakes of the Himalayan arc originate under or at the margin of the Ganges Plain, below a great and only vaguely known thickness of alluvium and sedimentary rock† which thoroughly blankets the surface manifestation of such earthquakes as that of 1934.

The great Arakan earthquake of April 2, 1762, is reported as accompanied by extensive changes of level on the Burmese coast. In 1878 a maximum uplift of 20 feet on the west coast of Ramree Island at 19° 10′ N was described as shown by recent remains of shells of the borer *Phola*. Farther north the vertical displacement reversed, and there was submergence in the Chittagong district.

How to interpret the faulting and regional deformation in 1897 is still a moot point; in view of the establishment of a first-class station at Shillong and expansion of seismological work in India and Pakistan, it is to be hoped that a comparatively few years may bring clarification. In Chapter 30 the complex manifestations of 1897 have been compared with the accessible surface effects of Japanese earthquakes in 1923, 1944, 1946. It is tempting to compare the Assam hills with the non-volcanic ridges appearing in the fronts of the Pacific-type arcs (in feature B of the classification in Chapter 26); this would mean that the surface effects of 1897 are of a type to be expected for a great earthquake originating under one of the small islands off Sumatra, or under the coastal mountains in the Mexican state of Oaxaca. This is tentative speculation, to be modified or abandoned in the light of new evidence.

Earthquake of Cutch—1819

The earliest clear and circumstantially described occurrence of faulting was that during the great earthquake in India on June 16, 1819. The fault scarp, locally called the Allah Bund (or Band) appeared in the Runn (or Rann) of Cutch (or Kacch); see Fig. 31-9. The general location is shown on our map of India (Fig. 5-1). The state of Cutch consists mainly of a so-called island on the west coast, a hilly area separated from the mainland by the remarkable muddy salt flat named the Runn of Cutch, frequently flooded and quite uninhabited, although at Sindri in the northern and relatively dry part of the Runn a fort had been established. Several rivers flow into

† For brief geological discussion and for descriptions of the earthquakes of 1897 and 1934, refer to Chapter 5.

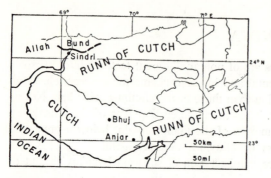

FIGURE 31-9   *Location map, Cutch earthquake, 1819.*

the arid area of Cutch, and the local potentates constructed dikes for irrigation purposes; such a dike is denoted by the word *bund*.

The earthquake of 1819 was not felt over so great an area in India as those of 1897 and 1934, although it was perceptible as far away as Calcutta. It was violent in all of Cutch, notably at Bhuj, the capital, and at Anjar. Loss of life was given as over 1500. The Allah Bund made its appearance at this time as a scarp about 5 miles north of Sindri, running roughly east and west for about 16 miles. It was downthrown to the south, toward the Runn, so that as seen from that direction it had the appearance of a *bund* about 10 feet high. The first part of the name duly credits its creation to Allah, in the fashion of our western legal lights who classify earthquakes as "acts of God." Simultaneously the ground about Sindri was depressed, and the river waters flooded in, so that the garrison had to make their escape by boat from the upper floors of the fort.

From the character of the flooding it was clear that the downthrow was greatest near the Allah Bund and decreased southward into the Runn. This was confirmed by surveys made in 1844. The disturbed area was then shown to be more than 40 miles from east to west. Moreover, the upthrow was greatest at the scarp and it decreased northward. This combination of upthrow and downthrow is exactly that to be expected on the theory of elastic rebound, as may be seen by turning Figures 14-1 and 14-2 to represent vertical instead of horizontal displacement.

Strong earthquakes have repeatedly affected Cutch; one of the most recent, on July 21, 1956, was reported as particularly severe at Anjar, with great damage and loss of life. This seismicity is not easy to refer to arc tectonics, because the area is far to the front of both the Himalayan and Baluchistan arcs, and has been interpreted as fracturing of the northern margin of the peninsular stable mass. Auden, however, has lately emphasized evidence of Mesozoic and post-Mesozoic folding in Cutch.

Baluchistan Earthquake—1892

The Baluchistan earthquake of December 20, 1892, disturbed a railway line near Old Chaman, between 40 and 50 miles from Quetta. This was due to left-hand strike-slip of at least 2 or 3 feet (the precise displacement

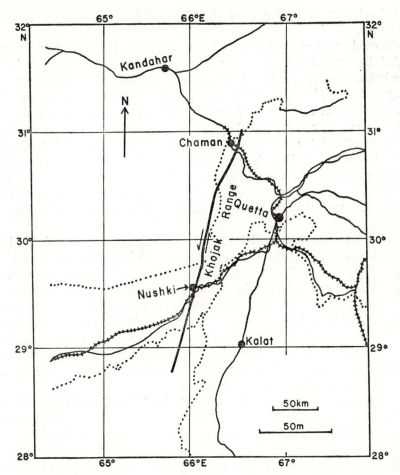

FIGURE 31-10 *Baluchistan earthquake, December 20, 1892, showing the location of the strike-slip rift zone and the point where the railroad was offset. [After MacMahon.]*

was not determined, since the fault cut the railway at a sharp angle). The new fracture was traced for only a few miles; however, it occurred on a fault zone with features remarkably resembling those of the San Andreas Rift, as has been described in Chapter 13. (Fig. 31-10.)

This faulting represents fracturing interior to the Baluchistan arc. The left-hand strike-slip can be thought of as representing the displacement of the southward front of this arc relative to that of the Himalaya.

CAUCASUS AND CENTRAL ASIA

In the Russian language, descriptions have been published covering macroseismic effects of some important earthquakes. In several instances the reported evidence strongly suggests faulting, but for only one (1911) is it conclusive.

The earthquake of 1902, destructive at Shemakha (Caucasus) was accompanied by extensive fissuring along a fault valley. Bogdanovitch in his description remarks that the principal fissures showed independence of topography and marked parallelism to tectonic lines, and consequently inclines to consider them as direct surface expression of dislocation in the underlying rock. They appeared not simply in alluviated valleys, but on limestone ridges.†

The effects of the earthquake of 1887, which was very destructive at Vyernyi (now Alma-Ata), indicate an epicenter in the northern Tian Shan, where vast slides and fissuring occurred. Some authors mention horizontal displacements caused by this earthquake, but the writer has not seen the original reference; the remark may merely refer to the lateral motion of great landslips.

Two great earthquakes originated in Mongolia on July 9 and 23, 1905 (see Table XIV-2). Each formed its own system of fissures, with a combined total extent of over 700 kilometers. The investigators spent almost their whole time in the field on the mere charting of these fissures, rendered still more time-consuming by lack of base maps. The lines of fissuring show distinct parallelism to tectonic trends, and faulting is certainly suggested. However, it is emphasized that the fissures were generally seen only in soft or marshy ground; the investigators note only one exception, which unfortunately they were unable to approach nearer than "a few tens of sazhen" (one sazhen is 7 feet, or 2.13 meters).‡

Another great earthquake on January 3, 1911 (again see Table XIV-2), originated in the Tian Shan region, but farther southward from Alma-Ata than that of 1887. Instrumental epicentral locations agree within their limits of accuracy with the field evidence that this originated in the valley of the Kebin river, between the Zailisk (or Trans-ilisk) and the Kungei Ala-Tau ranges. This earthquake is noted for enormous development of earth lurching, which broke the frozen and snow-covered ground into huge blocks which dwarf the men photographed standing on them. Fissures were seen cutting basement rock as well as alluvium, but the principal fault trace appears not to have been observed at the time. Later investigators found it conspicuously still in existence. A photograph published in 1954 shows a fault scarp cutting across the landscape with complete indifference to the minor irregularities of topography; the throw is stated to have reached 10 meters.

---

† Professor Gubin, who is thoroughly familiar with the area, inclines to doubt actual surface faulting in 1902; the features described by Bogdanovitch have now disappeared.

‡ Professor Gubin informs the writer that the features have been seen by recent geological field workers in Tannu-Ola, still nearly in the same condition as in 1905.

ASHKHABAD EARTHQUAKES

On October 5, 1948, an earthquake of magnitude 7.6 originated near 37½ ° N 58½ ° E, southeast of Ashkhabad (Turkmenian SSR). It caused heavy damage in a district not affected by strong earthquakes for several centuries, although many such occur in Iran not far south of Ashkhabad, especially near Kuchan.

The area was already under geophysical investigation in connection with a major canal project, but there was only one seismological station; it was at Ashkhabad itself.

The Geophysical Institute organized a complete special program. The field geology was intensively studied. No fault trace due to the earthquake was found. Geophysical investigation included deep seismic sounding (see Chapter 33); the Moho was located at a depth of about 40 kilometers. Four temporary seismological stations, equipped with instruments of moderate sensitivity (static magnifications 1000–3000) operated from June through October, 1949. About 1500 local shocks were recorded, 150 of them well enough for determination of epicenter and depth. The procedure depended entirely on S-P intervals, since no precise standard times were available at the temporary locations. All the epicenters were located within an area about 80 kilometers square, slightly elongated in the north-south direction, with the epicenter of the main shock not far from its eastern edge.

Geologists had anticipated that the center of disturbance would be found in the folded zone of the Kopet-Dagh range, which passes just south of Ashkhabad; instead, it was in the Kara-Kum desert area to the north, where a great depression in the basement rocks is filled up nearly level by Tertiary and Quaternary deposits. This depression thus behaves tectonically as a foredeep of the northward-fronting Kopet-Dagh arc.

A second expedition operated 7 temporary stations from June to September 1953, 4 with magnifications up to 10,000, and 3 with Kharin instruments magnifying from 30,000 to 35,000. Even at this high level of sensitivity, seismicity had decreased so far that only 100 weak shocks were recorded well enough for location. The distribution of epicenters agreed closely with that of the 1949 shocks. Depths, as before, were mostly determined as near 10 kilometers. An appearance of sharp recorded waves between P and S was attributed to transformation between P and S at discontinuities within the crust. Near each station very weak shocks were recorded with S-P intervals (or what were taken for them) of less than 2 seconds.

ACTIVE FAULTING IN ASIA MINOR

A series of earthquakes of great importance for the present purpose occurred in Anatolia (Asiatic Turkey), principally in 1939 and following years (Fig.

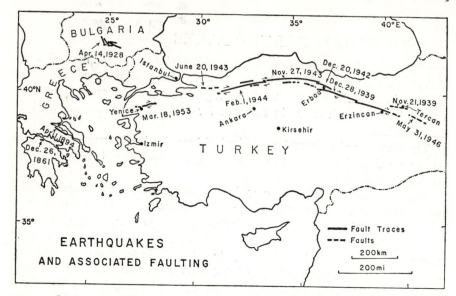

FIGURE 31-11   *Anatolia, showing epicenters and fault traces, 1938–1953. Greece and Bulgaria, indicating faulting of 1861, 1894, and 1928.*

31-11). On April 19, 1938, an earthquake occurred in the district of Kirsehir, originating near 39.5° N, 33.5° E; although the magnitude was only about 6¾, a fault trace was found in crystalline rock, trending N 75° W for 14 kilometers, with a throw of 30 to 60 centimeters and a right-hand strike-slip of 60 to 100 centimeters. This was within the central mass of Anatolia; the important earthquakes which followed occurred around its northern periphery.

Earthquakes of 1939

On November 21, 1939, an earthquake of magnitude near 6 caused much damage in the region of Tercan, in eastern Anatolia. No faulting was observed; the epicenter was located instrumentally near 40° N, 40° E. This may fairly be regarded as a foreshock of the great earthquake (magnitude 8.0) (Fig. 31-12) which followed on December 26, 1939 (at 23:57:16 G.C.T., and consequently on December 27, local date). The *International Seismological Summary* gives the same epicenter for both, but comparison of data shows a systematic difference. The best available epicenter as determined by Gutenberg for the great earthquake is near 39½° N, 39½° E,† near Erzincan.

Loss of life was between 20,000 and 30,000; over 30,000 dwellings were destroyed. The large homeless populations, exposed to snow and severe winter cold, suffered severely. The four largest communities where the majority of

_____
† This revises and replaces 39½° N, 38½° E, as given in *Seismicity of the Earth.*

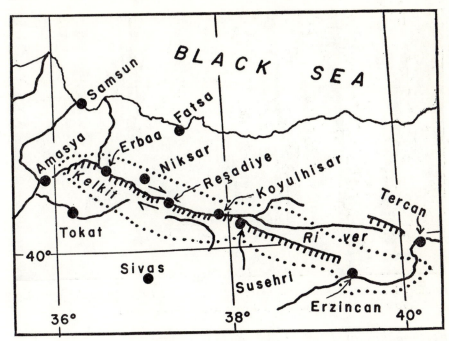

FIGURE 31-12   *Anatolian earthquake, December 26–27, 1939. Location map, showing meizoseismal area and faulting. [After Pamir and Ketin.]*

structures were destroyed were (from east to west) Erzincan, Susehri (Endires), Koyulhisar (Misas), Reşadiye, and Niksar (Neocaesarea). Faulting and high intensity extended from east of Erzincan to near Niksar, about 340 kilometers. The meizoseismal zone, following the fault line, was about 15 kilometers wide. The area over which the shock was felt was roughly elliptical, with major and minor axes of about 1300 and 600 kilometers.

The fault trace was mapped principally by looking for fissures (some 4 meters wide) and slides. At some points in and near towns there was clear evidence of strike-slip, as at Reşadiye, where a photograph (Fig. 31-13) shows right-hand offset, amounting to 3.7 meters (12 feet), of a row of poplars; at this location there was a downthrow of about 1 meter northward (away from the camera).

Since the epicenter of the principal earthquake was certainly on land, it is important that at Fatsa, on the Black Sea coast to the north, the sea retired 50 meters, then advanced about 20 meters beyond its usual mark, and finally returned to normal.†

---

† This earthquake provided one of the best tall stories, reported as follows: At Erbaa the local financial inspector and his family lived on an upper floor. The earthquake rent the structure from top to bottom, so that most of the family, rushing out of their bedrooms, were precipitated to the ground floor. The rent then closed, and the inspector himself, chivalrously bringing up the rear, was able to cross and descend by the stairs.

FIGURE 31-13   *Anatolian earthquake, December 27, 1939. Strike-slip offsetting row of trees at Reşadiye.* [*Parejas, Akyol, and Altinli.*]

Earthquakes of 1942, 1943, and 1944

Aftershocks in the narrow sense were frequent, and some of them were large; but in the following years generally high seismicity continued in Anatolia, and damaging earthquakes occurred well outside the main zone of disturbance. Those in this main zone, however, brought to light an extraordinary tectonic feature comparable with the San Andreas Rift. For the four principal events, instrumental results are as follows (magnitudes are revised).

Date	G.C.T.	Lat. N	Long. E	M
1939, Dec. 26	23:57:16	39½	39½	8.0
1942, Dec. 20	14:03:08	40½	36½	7.3
1943, Nov. 26	22:20:36	41	34	7.6
1944, Feb. 1	03:22:36	41½	32½	7.6

(For Turkish local time, 2 hours should be added to the G.C.T. times of occurrence; for the first and third shocks this advances the date.) These epicenters were determined by a rather rapid method and may easily be in error by a quarter degree in latitude or longitude. Some of the large earthquakes were apparently preceded by foreshocks, which confuse the recording at the nearer stations.

These earthquakes were accompanied by faulting progressing from east to west along the tectonic feature mentioned; this is a zone of faulting curving around the north margin of the Anatolian mass. In 1939 the faulting was developed chiefly along the valley of the Kelkit River, which is a typical rift valley. In some parts it is a deep canyon; for 96 kilometers it is almost straight, trending east-southeast–west-northwest. Southeastward the fault leaves the Kelkit and extends across the alluviated plain of Erzincan, where its topographic expression is less evident. The zone here is highly seismic; Erzincan has been destroyed by earthquakes repeatedly over many centuries.

The 1942 earthquake affected the region of Erbaa, near the west end of the meizoseismal zone of 1939. Faulting appears to have extended for about 35 kilometers, with two main traces. Downthrow was northward, and at two points right-hand strike-slip from ½ to 1 meter was observed.

In the 1943 earthquake, faulting continued westward for 280 kilometers more. There were northward downthrows of 90 to 100 centimeters, but no strike-slip was observed. From the end of this the faulting of the 1944 earthquake again continued west (and a little south), probably continuously for 180 kilometers, although part of the course was through mountainous country where the trace was not followed. The displacement was now predominantly strike-slip; gardens and walls in the town of Gerede, and a road between Bolu and Ilicia, showed right-hand strike-slip of 3 to 3.5 meters, the north side, which was shifted east, being downthrown 40 to 100 centimeters.

Tectonics

Geological description of this faulting altered as events and investigation progressed. In a small part of its course the Kelkit fault separates two divergent structures of the Alpide system, the Taurides on the south and the Anatolides on the north, and this was at first taken to be characteristic of the whole 1939 event. However, the eastward segment in the vicinity of Erzincan is an internal fracture in the Tauride system, while northwestward the zone first separates the Anatolides from the Pontides to the north of them and then passes completely into the Anatolide structure. The natural conclusions would be that, when there is stratigraphic contrast across the fault, it is an incident of the relative displacement, and that the fracture on the whole takes a course independent of older structures. Where the Kelkit zone separates the Anatolides and Pontides, the former, on the south side, are relatively downthrown about 100 meters; while along the whole fracture the new displacements are downthrown to the north. This contrasts with the usual behavior of strike-slip faults, where the vertical throw tends to reverse and scissor.

Strike-Slip

The workers on all these earthquakes seem to have established strike-slip only in towns or across roads. In the rural districts artificial markers are rare, and apparently evidence from offset natural features, tension cracks, and so on, either was not adequate or was not studied. In discussing this entire group of earthquakes the author is hampered by the incompleteness of accessible material. There seems to have been no inclusive report on the earthquake of 1939; this is due in part to war conditions. Much of the published material on all these shocks is in journals normally of limited circulation and still less well distributed during the war; some of it has appeared only in very condensed or summarized form.

Earthquake of 1953

On March 18, 1953, an earthquake of magnitude 7.2 originated in northwestern Anatolia near 40° N 27¼° E. The communities of Yenice and Gönen were violently shaken. There was a large fault trace about 50 kilometers long, not quite straight but independent of small topographic features, trending roughly east-west. No significant vertical displacement was seen, but there was consistent right-hand strike-slip; measured values were 1.5, 3.3, and 4.3 meters. This appears to be a continuously cultivated area, and offset roads were frequently found. Photographs of the trace show clear right-hand tension cracks (see Fig. 13-7), and one portrays a typical mole track. This trace belongs apparently to the same general group as the earthquakes of 1939–1944; but there is a gap of over 200 kilometers between the end of the 1944 trace and the beginning of the 1953 trace, so that it is not certain whether the active tectonic line is continuous or offset.†

Progressive development of faulting in successive earthquakes has been suggested on other occasions, usually on the basis of less convincing and chiefly macroseismic evidence, and almost never on such a scale as this. This faulting is cutting through the interior of the Alpide belt, between its north and south fronts. In situation it resembles the Baluchistan fault more nearly than any other we have discussed. The evident slight curvature of the main fault system invites explanation, which in the present state of knowledge is best postponed. We have here an example of the generalization already offered in Chapter 4 with regard to the Alpine structures; present tectonics do not usually represent continuation of the Alpine folding and thrusting, but slicing of the now relatively rigid masses along new fractures.

FAULTING IN GREECE

Earthquake of 1861

The earthquake in Greece on December 26, 1861, is a doubtful case in spite of competent field investigation. De Ballore discusses it in connection with that of Helice in 373 B.C. (Chapter 14), which may have had nearly

---

† On May 26, 1957, an earthquake of magnitude 7± originated near 41° N 31° E. It was accompanied by right-hand strike-slip faulting. (Personal communication from Dr. Kazim Ergin.)

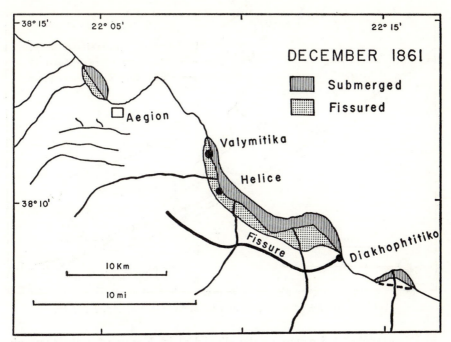

FIGURE 31-14   *Effects of the earthquake of December 1861 (Gulf of Corinth).*
[*After de Ballore, following Schmidt.*]

the same epicenter. In 1861 a great fissure appeared which followed a winding course near the base of hills extending north to the Gulf of Corinth
(Fig. 31-14); the area north of this fissure subsided about 2 meters. There
is no clear correlation with known faults, and the extent of the meizoseismal
area is that of an earthquake of only moderate magnitude. The whole description suggests a large slump.

Locris—1894

The earthquakes of April, 1894, near the coast of Locris, followed a pattern
in time like that described in Chapter 6 for Imperial Valley: on April 20
a strong and locally destructive shock, with almost no aftershocks; on April
27 a violent shock, followed 13 minutes later by a major earthquake, with
considerable development of faulting (Fig. 31-15), and a long normal series
of aftershocks, some of them quite strong.

The fault trace was nearly straight for a distance of 55 kilometers, trending east-southeast–west-northwest. The investigators are particularly attentive
to the gaping fissures on the trace, which were as wide as 4 meters. There
was regularly downthrow of the northeastern block, that toward the coast;
this amounted to only 30 centimeters where the trace cut sound Cretaceous

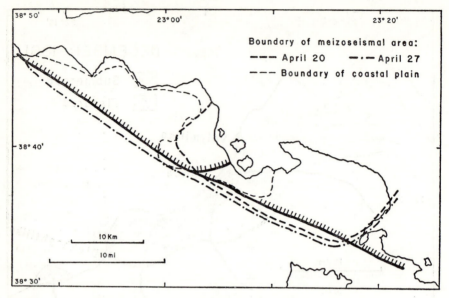

FIGURE 31-15   *Faulting in earthquakes of Locris, Greece, April 1894.*

rock but reached 1.5 meters on the alluvial plain. Not unnaturally, some observers considered this a secondary effect due to shaking; however, the trace was straight and independent of both relief and geology, so that, unlike the trace of 1861, it cannot be interpreted as slumping, but would need to be treated as the dropping of a block bounded by the fault.

At present, the chief interest attaches to the strike-slip, which was very imperfectly reported. The evidence consisted of tension cracks and, fortunately, offset of a dry stream bed. One has to combine the two principal sources; one author does not state the amount of horizontal displacement, the other calls it very slight. As usual, the direction is not well stated; it is given as "toward the northwest," and this appears to refer to the downthrown block, which would imply left-hand strike-slip.

AEGEAN DEEP-FOCUS EARTHQUAKES

In the Aegean region there are occasional large shocks at intermediate depth, some of which have caused much damage in Egypt, southern Greece, Crete, and other Aegean islands. The exceptional nature of these shocks, and the evidence of their unusual depth, were recognized as long ago as 1881 by Julius Schmidt. Their association with arc features, including the active volcano of Santorin (Thera), is at least roughly according to pattern. The latest of these large shocks occurred on June 26, 1926 (magnitude 8.2, revised; depth 100 kilometers). A large earthquake with epicenter near Santorin on July 9, 1956, appears to have been relatively shallow.

NOTES ON THE ITALIAN ARC

No surface faulting has been established in Italy; spectacular displacements in the Calabrian earthquakes of 1783 were due to enormous landslides. The geological description in Chapter 4, and the brief notice in Chapters 25 and 26, call for further discussion.

The Pacific arcs most similar to this are those like the Ryukyu and Banda Sea arcs, where the principal islands form an outer chain with no present volcanism, associated with gravity anomalies and large shallow earthquakes (feature B; this matches the seismic belt of the Apennines), while active volcanoes appear as islets on an inner arc (feature D; compare the Lipari Islands). Occurrence in the Italian region of shocks as deep as 300 or even 470 kilometers close to the volcanoes is not paralleled in the Ryukyu arc,† and only doubtfully in the Banda Sea. Although gravity anomalies are present, no strong negative belt follows the Apennines (it appears to run more easterly); and, while there is deep water in the southern Adriatic, there is nothing resembling the Pacific troughs.

Such differences are not great; and the vigorous program of seismological and geophysical investigation now going on in Italy, particularly at the Istituto Nazionale di Geofisica, should eventually make this arc one of the best understood in the world.

BULGARIA—1928

Details of the interesting observations now to be considered have been accessible to the writer only in the Bulgarian language; fortunately it has been possible to have the material translated.‡

On April 14 and 18, 1928, earthquakes of magnitude 6¾ originated in southern Bulgaria (see Fig. 31-11), resulting in what has been called the most serious earthquake calamity in the history of that country. The epicenter of the earlier shock was near Chirpan, where there was great destruction; that of the later was somewhat farther west near Popovitsa (Papazli). Faults were produced with maximum throw of 3 to 4 meters. (Fig. 31-16.) A precise leveling survey had been completed shortly before over lines extending through and well beyond the disturbed area; these lines were rerun one year

---

† A recent paper by Peterschmitt discusses the tectonic implications of the Italian deep-focus earthquakes.

‡ Translation prepared by Mr. J. M. Nordquist.

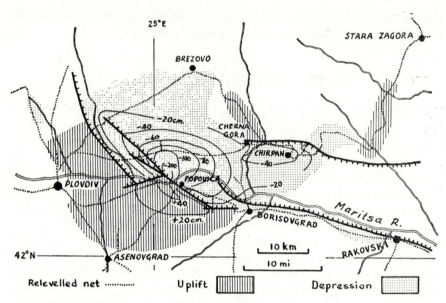

FIGURE 31-16   *Bulgarian earthquakes, April 1928. Faulting and changes of level.* [*After Jankof.*]

later, with results reported in an excellent paper by Jankof, who is careful not to go beyond his data. The meizoseismal area is in the Trakiiska plain. Jankof writes:

> Geologically, the shaken area is a depression in the mountainous region through which, from west to east, flows the river Maritsa. On the south the plain is bounded by the Rhodopian massif, made up chiefly of large granitic laccoliths, thick metamorphic strata, and locally of marble or of recent eruptive rocks. To the north the plain is bounded by the Sredna Gora, chiefly of granite and Senonian marls. The plain itself is moderately hilly only near the northern and southern borders. Its western part consists of diluvial and alluvial materials, while the eastern is chiefly Pliocene sediments with a little Eocene. Tectonically, the Trakiiska plain is a kettle basin . . .

Jankof presents a map showing the faults with contours of equal depression and uplift. In the Chirpan earthquake of April 14, two faults were formed. The northern, downthrown southward, extends from Cherna Gora eastward past Chirpan for 38 kilometers. The southern and longer fault, with northward downthrow, follows the general line of the Maritsa River, in places along its channel; beginning northwest of Borisovgrad, it extends east beyond Rakovski. The two faults, nearly parallel, are 13 to 16 kilometers apart.

In the earthquake on April 18, there was new faulting, downthrown north and northeast. Faulting extended about 10 kilometers west from Borisovgrad; thence, with a sharp change of direction, northwest past Popovitsa. Another northwest-trending trace appeared a few kilometers farther west. The over-all length of the southern group of traces formed in the two earthquakes was 105 kilometers.

The releveling indicated depression exceeding 50 centimeters in the vicinity of Chirpan, most of which must have occurred on April 14; but northwest of Popovitsa depression exceeded 300 centimeters. The contours of equal depression in this vicinity, which must largely represent displacements on April 18, enclose areas several times larger than the corresponding contours near Chirpan. Beyond the fault scarps, southwest of Popovitsa, uplift of over 40 centimeters was found which decreased southwestward away from the scarp. Depression also fell off rapidly in the other direction, corresponding to expectation of elastic rebound and to such observations as those for the Cutch earthquake of 1819. Less well established uplift was found beyond the scarp north of Chirpan.

Triangulation was also repeated over the area, but without any result indicating systematic horizontal displacement.

Jankof cites geological evidence that this entire area has sunk during most of Quaternary time, and he remarks on the discordance of this result with the occurrence of both uplift and downthrow in 1928.

This work places the observations in an entirely different category from those of other instances, otherwise apparently similar, in which we have to do with less adequate field observation and in which microseismic data clash with interpretation as faulting.

AFRICA—1928

Only one probably admissible instance of faulting during an earthquake has been reported for the African continent.† The data are largely unpublished, hence more space is accorded them here than might otherwise be proper.

This earthquake occurred on January 6, 1928, in the African rift region near the equator (Fig. 31-17). It was one of Turner's puzzling instances of "high focus" (see Chapter 19). To clarify this, Tillotson collected seismograms from many observatories and studied the readings in detail. The effect of apparent high focus seemed at first to be due mainly to late readings of the S phase, but Tillotson concluded that most of these times were correct as read. His later hypothesis, confirmed to some extent by close examination of the seismograms, was that the shock was double; the first shock had strong P waves with weak S, while the second, 10 seconds later, had weak P and

---

† For the destructive Algerian earthquake of 1954, Rothé and his collaborators have described spectacular fissures and scarps, all probably secondary, due either to shaking or to the readjustment of unconsolidated surface material after displacement of basement rock. They include a nearly semicircular system of scarps, 6 kilometers long, which appears to mark the head of a large landslide of earth-slump type.

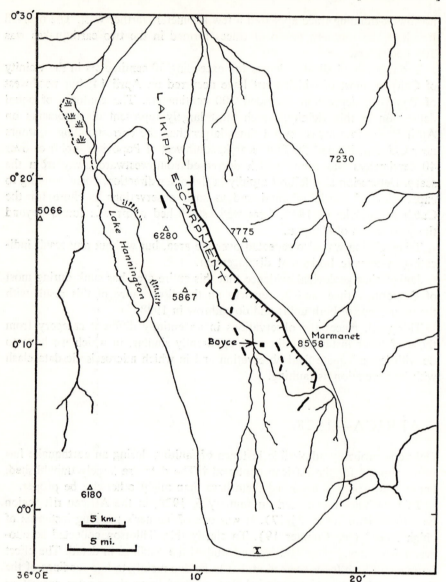

FIGURE 31-17   *Location map showing faulting in the Subukia earthquake, Africa, 1928.*

strong *S*. Both appeared to have originated at unusually shallow depth; however, the effect of the ellipticity of the earth may be involved, the equatorial elevation delaying the times at distant stations.

Tillotson's work was facilitated by checking against macroseismic data, for he was informed that the earthquake caused displacements in the Subukia (or Sabukia) Valley, chiefly along the Laikipia escarpment.† This was set forth in two principal reports, copies of which were sent to Tillotson, one signed by H. L. Sikes, Director of Public Works, Nairobi, the other by W. C. Simmons, petrologist. The former appeared in condensed form in the columns of the *London Times*; the latter has been published in a volume by Bailey Willis, but no map is given. Many photographs, and detailed field notes by Simmons, were also available, together with details of damage to structures, lists of aftershocks, and other information.

The Sikes report is reproduced in full as Appendix XV of this book. The Simmons report was originally accompanied by two maps; one, a location map on a small scale, is represented by our Figure 31-17. The other, a chart showing details of the observed trace, on a larger scale, was not supplied to Tillotson, nor apparently to Willis. Points referred to by lettering A, B, . . . in the Sikes report refer to this missing map.

The investigators clearly state that they consider the trace features due to normal faulting under gravity. Vertical displacement of a large block might lead to surface expression such as that described. On the other hand, many of the features, including open fissures, might have been due to lurching and sliding, and consequently secondary in character. Unfortunately insufficient detail for close analysis is given on the orientation of tension cracks at any one locality, but the occurrence of such cracks, with orientation in large part differing from that of the main trace, suggests strike-slip. This interpretation is also supported by the remark that where the trace changed direction it also changed character from open fissuring to a pressure ridge. This reads like a description of the two sides of a triangular facet below the main trace, indicating lateral displacement; or the pressure ridge may possibly have been a strike-slip mole-track feature.

It is doubtful whether the chief points could have been settled by the most carefully directed investigation on the ground; there is need for geological detail, precise leveling and triangulation, and accurate microseismic recording. Even where all these are available, as for the California earthquakes of 1952, interpretation is not easy.

Willis, in view of the known recent volcanism of the general region, considered possible volcanic origin for this shock. Its magnitude (7.1) alone renders this improbable. The obvious young faulting connected with the African rifts, and the continual moderate seismicity of the area, make us expect tectonic earthquakes as a matter of course; in spite of the cultural and climatic conditions, it is a little surprising that this imperfectly known occurrence is the only one of its kind thus far reported there.

---

† The present author is under the greatest obligations to Lt. Col. Ernest Tillotson for making the original documents and photographs available for study.

References

BIBLIOGRAPHY on individual shocks will be found in Appendix XVI; references on the controversial uplift of the Chilean coast in 1822 and 1835 appear there under those dates.

For stratigraphy and tectonics of Alpide structure refer to any of the large European geological treatises. Some of these generalize the particular circumstances of Alpine folding into universal laws, and ignore structures in other parts of the world which do not fit.

An important classic, for the Alpide belt particularly, is Suess, E., *Das Antlitz der Erde* (for detailed references see Chapter 1). The student in seismology will find it interesting to read his one-sided treatment of the Chilean changes of level and the Cutch earthquake of 1819 (Vol. 1., Part 1, Section 2); the errors of great men are highly instructive when studied with a view to understanding, and not merely to sneer or to justify one's own blunders.

A recent treatise of distinction is Goguel, J., *Traité de Tectonique*, Paris, 1952.

References given in Chapter 13 on rifts and faulting will supply background for some of the earthquakes discussed in the preceding chapter.

East Indies

Brouwer, H. A., *Geology of the Netherlands East Indies,* New York, 1925. A conservative statement.)

van Bemmelen, R. W., *The Geology of Indonesia,* Govt. Printing Office, The Hague, 1949, Vol. IA, Vol. IB (portfolio), Vol. II.

Umbgrove, J. H. F., "Geological history of the East Indies," *Bull. Am. Assoc. Petroleum Geologists,* vol. 22 (1938), pp. 1–70.

———, *Structural History of the East Indies,* Cambridge University Press, 1949.

Westerveld, J., *Phases of Mountain Building and Mineral Provinces in the East Indies, Report, 18th International Geological Congress, Great Britain 1948;* Part 13, London, 1952, pp. 245–255.

———, "Fasen van gebergtevorming en ertsprovincies in Nederlands Oost-Indië," *Ingenieur* (*Utrecht*), vol. 61, Mijnb. en Petroleumtechn., pp. M1–13, M15–27, M36. (More detailed in text and bibliography than the English version.)

Glaessner, M. F., "Geotectonic position of New Guinea," *Bull. Am. Assoc. Petroleum Geologists,* vol. 34 (1950), pp. 856–881.

Other references

Schmidt, J., *Studien über Vulkane und Erdbeben,* Leipzig, 1881. (Recognized some of the large deep-focus earthquakes of the eastern Mediterranean as such.)

Sieberg, A., Untersuchungen über Erdbeben und Bruchschollenbau im östlichen Mittelmeergebiet," *Medizinisch-naturw. Gesellschaft zu Jena, Denkschriften,* vol. 18 (1932), pp. 161–273. (Discusses the same group of earthquakes without recognizing their depth.)

Heck, N. H., "Earthquake history of the United States exclusive of the Pacific region," *U. S. Coast and Geodetic Survey, Spec. Publ.* 609 (1938). (Refer also to the serial, *United States Earthquakes.*)

Auden, J. B., "The bearing of geology on multipurpose projects," *Proc. 38th Indian Science Congress,* 1951, pp. 109–153. (Discusses the tectonics of Cutch.)

————, "A geological discussion on the Satpura hypothesis and Garo-Rajmahal gap," Proc. Nat. Inst. of Sciences of India, vol. 15 (1949), pp. 315–240. (Alignment of the Assam hills with the northern edge of the Peninsular shield. Sketch map of tectonic units.)

Peterschmitt, E., "Quelques données nouvelles sur les séismes profonds de la mer Tyrrhénienne," *Ann. di Geofisica* vol. 9 (1956), pp. 305–334.

Note added in proof: A. A. Treskov and N. A. Florensov (London *Times,* April 10, 1958) report faulting for the earthquake of December 4, 1957, 03:37 G.C.T., in the Gobi-Altai region, Mongolia. A straight line of fissuring extended east-west for 170 miles, with dip-slip up to 10 or 12 yards and strike-slip of 3 yards.

CHAPTER 32

Compressions and Dilatations

MOST of our geographical discussion has dealt with earthquakes for which there is macroseismic evidence of the nature of the displacements which produce them. As this is written (1956), several active research groups are developing microseismic evidence, of a type explained in Chapter 14, from a level of relatively isolated instances up to a major source of information.

This has been done largely by selecting individual earthquakes and assembling indications of first motion (*P* or *P'*) recorded at seismological stations all over the world. (The alternative, of investigating regular recordings of initial compression or dilatation at a given station, has also been used more and more; the results, though ultimately of great permanent value, are less readily interpreted in our present state of knowledge.) A valuable short summary has recently been published by Byerly, who pioneered this field as early as 1926 and established principles and methods now generally in use.

JAPAN

In Japan the foundation for this type of investigation was laid in papers by Nakano (1923) and Matuzawa (1925). The existence of over 100 recording stations in the relatively small area of the main Japanese islands makes the procedures more applicable to recording at short distances than anywhere else; however, the narrowness of Japan compared to its length seriously impairs the uniformity of distribution of stations in azimuth about a given epicenter.

In the majority of well-observed instances, for deep shocks as well as for shallow ones, azimuths in which recordings of first motion are primarily compression or primarily dilatation show the type of quadrantal distribution separated by nodal lines indicated in Figure 14-4 with the reversal characteristic of the Japanese region. That is, when the nodal lines are in the direction of the two main tectonic trends, which we may call northwest and northeast, initial compressions prevail in the north and south quadrants, initial dilatations in the east and west quadrants. As mentioned in Chapter 30, this would correspond to release of east-west compression, north-south tension, or both.

626

Even among shallow shocks there are local and individual deviations from this pattern, corresponding to the tectonic complexity of Japan; deep shocks show further complication, including the expected circle of reversal between short and long distances. Honda and Masatuka offer a wider generalization by stating that the direction of the indicated compressive stress in intermediate and deep shocks of the region is perpendicular to the trend of the zones of epicenters.

INVESTIGATION AT OTTAWA—
PREVAILING STRIKE-SLIP

The most extensive program of these investigations is now centered at the Dominion Observatory, Ottawa, where the principal workers are J. H. Hodgson and A. E. Scheidegger. The latter, in summarizing the evidence to date—which still appears fragmentary when mapped, as compared to the vast area of the unknown—points out that in the overwhelming majority of well-recorded instances the displacements are indicated as horizontal ("transcurrent," that is, strike-slip). Since many large earthquakes are too complex or for other reasons are not adequately recorded for this type of analysis, the meaning may simply be that strike-slip produces better recording for the purpose.

It is important, therefore, that Gutenberg on investigating compressions and dilatations for the major Kern County (California) earthquake of July 21, 1952, found the data satisfactorily in accord with thrust faulting with the known strike of the White Wolf fault, and with a dip of 63°. By analyzing the transverse waves he found indication of a left-hand strike-slip component, and so the resultant block displacements were nearly in a north-south line.

The use of recorded compressions and dilatations determines the strike of the fault only with ambiguity; the same distribution in azimuth will follow from right-hand strike-slip on a northwest-trending fault, or from left-hand strike-slip on a northeast-trending fault. In South America, where the major structures trend approximately northwest, the observations indicate that strike-slip, if parallel to these trends, is right-hand. As Byerly points out, this, if it represents the general condition, makes the continent move southeast relative to the Pacific Basin—in the same sense as the accepted displacement of North America, but contrary to those mechanical theories which call for drift of the continents away from the poles and toward the equator. One of the earthquakes appearing to show right-hand strike-slip is the Ancash earthquake (Peru, 1946) described in the previous chapter as showing conspicuous dip-slip, with no evidence of strike-slip (so far as the investigators could determine in open country). This conflict may not be so serious as it appears, judging from the numerous instances of strike-slip described in the preceding chapters as established by evidence which might easily have been missed, to say nothing of instances where unmistakable field evidence of strike-slip was overlooked in the first rapid reconnaissance work.

THE EAST INDIES

Important results have been published by Dr. A. R. Ritsema, at first at Utrecht but more recently at the Meteorological and Geophysical Service, Djakarta, Indonesia. Comparatively few earthquakes in the East Indies are well enough recorded at distant stations for this type of analysis. To results obtained by L. Koning, working at Amsterdam, for the deepest of all known large earthquakes (that of June 29, 1934, with depth of 720 kilometers under the Flores Sea), Ritsema added those for two others deeper than 600 kilometers, one under the Java Sea, the other under the Philippines. For all three the data are interpreted as indicating dip-slip along a plane dipping about 55° under the continent (northwesterly for the Flores Sea and Java Sea shocks, westerly for the Philippine shock). The sense of the dip-slip is opposite to that expected on the simplest theories of the tectonics of that region; it does not correspond to thrusting, but to normal faulting, giving tension at right angles to the structures instead of compression. Ritsema has also found a large proportion of apparent normal faulting among the deep earthquakes of the Tonga-Kermadec region. In more recent work Ritsema finds a high proportion of strike-slip for shallow shocks (but not for deep shocks) in the East Indies (preliminary report, Toronto, 1957).

CONTRIBUTIONS FROM THE USSR

At Moscow a vigorous program of investigating earthquake mechanism is being carried forward by V. I. Keilis-Borok and A. V. Vvedenskaya. The concept of a point source is abandoned, and the effect of the progression of faulting during the earthquake is considered. The method, now worked out and systematized to a high degree, involves observing not merely compressions and dilatations in P, but also the direction of initial displacement in S and the amplitude ratio of P to S. In theory this makes it possible to remove the ambiguities of interpretation which attend the use of compressions and dilatations only; in practice it demands extreme care in the interpretation of seismograms.[†] The method is applied to both local earthquakes and teleseisms.

As at Ottawa, prevailing strike-slip is found in a large majority of instances. Systematic study of earthquakes in the northwest Pacific arcs, from

---

† Recent model studies by Press at Pasadena show that diffraction effects due to finite length of the fault may significantly shift the azimuths in which nodes are observed for S waves.

Kamchatka south through the Kurile Islands and Japan, points to frequent strike-slip on fractures transverse to the arcs; this may be compared with the strike-slip on the great fault of the Mino-Owari earthquake, and possibly with the Fossa Magna.

CONCLUSIONS

There are no very definite conclusions at present. In spite of earnest effort, well-established results are still too few for world-wide generalization. Discussion of the validity of the method still arises, and there will probably be no general acceptance of any unexpected result until there are many more instances in which the microseismic inferences can be correlated with macroseismic data.

References

DISCUSSION and references in Chapter 14 cover the general method, and those investigations which analyze the seismograms at a single station. The following relate especially to the method which correlates initial motions at many stations for a single earthquake. For further references consult also the paper by Scheidegger cited in Chapter 14.

Kawasumi, H., "An historical sketch of the development of knowledge concerning the initial motion of an earthquake," *Publ. bureau central séismologique international*, Ser. A, vol. 15 (1937), pp. 258–330. (Very thorough; extensive bibliography.)

Honda, H., "On the mechanism and the types of the seismograms of shallow earthquakes," *Geophys. Mag.* (*Tokyo*), vol. 5 (1932), pp. 69–88.

————, "On the types of the seismograms and the mechanism of deep earthquakes, *ibid.*, pp. 301–326.

————, "Notes on the mechanism of deep earthquake," *ibid.*, vol. 7 (1933), pp. 257–267.

Honda, H., and Masatuka, A., "On the mechanism of the earthquakes and the stresses producing them in Japan and its vicinity," *Tôhoku Univ. Science Repts.*, (Sendai), Ser. 5, vol. 4 (1952), pp. 41–60; vol. 8 (1957), pp. 186–205.

Honda, H., "The mechanisms of the earthquakes," *ibid.*, vol. 9 (1957), Supplement, pp. 1–46.

Byerly, P., "The earthquake of July 6, 1934. Amplitudes and first motion," *B.S.S.A.*, vol. 28 (1938), pp. 1–13.

————, "Nature of faulting as deduced from seismograms," *Geol. Soc. Amer. Special Paper 62* (1955), pp. 75–86.

Gutenberg, B., "The first motion in longitudinal and transverse waves of the main shock and the direction of slip," Calif. *Dept. Nat. Res., Div. Mines, Bull.,* 171 (1955), pp. 165–170.

Many other valuable papers on this subject have been published by Kawasumi, Honda, Wadati, Yosiyama, and other Japanese authors; note those cited in Chapter 19.

Keilis-Borok, V. I., "Methods and results of the investigations of earthquake mechanism (a brief information)," *Publ. bureau central séismologique international,* Ser. A, vol. 19 (1956), pp. 383–394.

————— (ed.), "Issledovanie mekhanizma zemletryaseniy," *Trudy Geofiz. Inst. Akad. Nauk SSSR,* No. 40 (166), 1957.

Vvedenskaya, A. V., "Ob opredelnii dinamicheskikh parametrov ochagov zemletryaseniy po nablyudeniyam udalennykh stantsiy (Determination of dynamic parameters of the foci of earthquakes from the records of distant stations)," *Doklady Akad. Nauk SSSR,* vol. 80 (1951), pp. 591–594.

Malinovskaya, L. N., "Metodika opredeleniya mekhanisma zemlyetryaseniy," *Trudy geofiz. Inst. Akad. Nauk SSSR,* No. 22 (149), 1954.

Kogan, S. L., "K voprosu ob izuchenii mekhanisma glubokikh zemletryasenii," *Doklady Akad. Nauk SSSR,* vol. 99 (1954), pp. 385–388.

Hodgson, J. H., "Direction of faulting in some of the larger earthquakes of the Southwest Pacific, 1950–1954," Publ. Dominion Observatory, Ottawa, vol. 18 (1956), pp. 171–216.

————, "Direction of faulting in some of the larger earthquakes of the North Pacific, 1950–1953," *ibid.,* pp. 219–252.

————, "Nature of faulting in large earthquakes," *Bull. Geol. Soc. Amer.,* vol. 68 (1957), pp. 611–644.

McIntyre, D. B., and Christie, J. M., "Nature of the faulting in large earthquakes," *ibid.,* pp. 645–652. (Critical discussion of the preceding.)

Press, F., "Model study of elastic-wave radiation from faults," paper presented, Seismological Society of America, Los Angeles, April, 1957.

Ritsemá, A. R., "Earthquake-generating stresses in southeast Asia," *B.S.S.A.,* vol. 47 (1957), pp. 267–280.

CHAPTER 33

Seismology in the USSR

BETWEEN 1900 AND 1915 Russian seismology reached a high level of international recognition. This was largely due to Galitzin, who devised new and effective instruments, brought about their installation at many stations, and contributed in his publications to almost every feature of the interpretation of seismograms. His published course of lectures became an international standard manual.

Seismology was furthered under the Soviet regime. (See map, Fig. 33-1.) Much attention was drawn by a destructive earthquake in the Crimea in 1927. Nikiforov's seismographs for local earthquakes were installed at stations in the Crimea, the Caucasus, and central Asia. The latter group furnished data on tectonic conditions and earthquake risk which were used in routing the Turksib Railway, one of the first large Soviet engineering projects.

International relations interrupted during World War II were resumed from 1945 to 1947; publications and station bulletins were then reaching distant libraries and research centers. In 1948 this was largely broken off, with sporadic episodes of revival. Research results were published only in Russian, and in journals not always regularly available in other countries. Representatives ceased to attend international meetings. Current station bulletins and data no longer reached the international offices. During this interval there was a major reorganization.

In 1954 a strong Soviet delegation attended the international conference at Rome and presented an illuminating series of reports intended to bring the outside world up-to-date on the notable advances made since the war. These reports, in English, were published in the transactions of the conference, which appeared late in 1956. Meanwhile, publications in the Russian journals have been circulated freely, and the large manual on seismology and seismometry by Savarensky and Kirnos has been generously distributed. The following short summary is based on the Rome reports, with other details from the journals; *Geophysical Abstracts* has been a valuable guide.

Research has been greatly stimulated lately by the founding of geophysical and seismological institutes in many of the constituent republics of the USSR. The most active research center is still the Geophysical Institute (reorganized in 1956 and renamed Institute for Physics of the Earth) of the

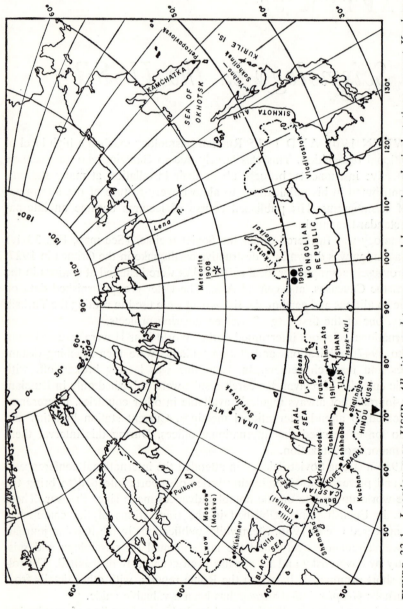

FIGURE 33-1   Location map, USSR. All cities and towns named have seismological stations, except Kuchan (in Iran) and Krasnovodsk.

Academy of Sciences at Moscow; the inclusive bulletin giving seismogram readings for all Soviet stations is edited there, and many publications come from its staff. Most of the contributions are in the journals of the Academy (*Akademiya Nauk SSSR; Doklady, Trudy, Izvestiya*). The regional institutions have their own publications and station bulletins. With 73 stations in operation (1955), over 10,000 earthquakes are recorded annually.

Galitzin and Nikiforov seismographs remain in operation, but most of the newer stations are equipped with instruments of two main types designed by Kirnos and Kharin, with electromagnetic photographic registration:

Type A, used generally for both near and distant earthquakes. Pendulum period 12.5 seconds, $h = 0.45$; galvanometer period 1 second, heavily over-damped ($h = 6$); effective magnification 1000, nearly constant for periods from 0.25 to 10 seconds.

Type B, for local earthquakes. Pendulum period 0.7 to 1 second, $h = 0.6$ to 0.9; galvanometer period 0.3 to 0.5 second, $h = 1.5$ to 3.0. Maximum magnification 20,000 to 30,000, with a sharp peak for periods between 0.3 and 0.6 second.

Low-magnification instruments with mechanical registration for recording strong motion are installed at selected stations.

In 1936 and 1939 E. A. Rozova determined the depth of the Moho in central Asia and the Caucasus as 50 and 60 kilometers. This is much greater than in European Russia. The result has lately been confirmed by refraction recording with large explosions.

Investigation of earthquake mechanism as inferred from displacements recorded by seismographs has been noted in Chapter 32.

Following the Ashkhabad earthquake of 1948, aftershocks were recorded at six closely spaced temporary stations (see Chapter 31).

Use of recorded times to calculate velocities in the interior of the earth requires knowledge of the angle of incidence of the seismic ray at the surface. This is usually determined from the time-distance curve (see Appendix VI); but, as Galitzin pointed out, it can be found directly from the ratio of vertical to horizontal recorded amplitudes, allowing for the effect of the free surface. Savarensky has refined the method, disposing of some theoretical objections, and has used it to obtain results in good agreement with those generally accepted.

Seismology and geophysics suffered a serious loss by the death in 1955 of G. A. Gamburtsev, Director of the Geophysical Institute at Moscow. His originality expressed itself in the development of many novel and fruitful techniques. Several of these were reported at Rome, as follows:

(1) An azimuth seismic station is a set of seismographs at a single location, mounted with their axes inclined and spaced around the surface of a cone. The instruments record in parallel on a single sheet, and phase relations among

the traces can be used to determine the azimuth from which a wave approaches, as well as its character as *P, S,* or surface wave.

(2) Extremely small local earthquakes are recorded with magnification up to several millions in the frequency range of 10 to 30 cycles per second; this is possible only in localities remote from all artificial disturbance.

(3) A development of the methods of geophysical prospecting is termed the correlation method of refracted waves (*korrelyatsionniy metod prelomlënnikh voln*†). It is a multi-channel system recording at high speed, with high sensitivity and special frequency characteristics, working with closely spaced geophones over a comparatively short profile. With large charges of explosive and under favorable conditions this method will penetrate sufficiently deep to record waves with apparent velocity near 8 kilometers per second (*Pn*), refracted below the Moho and yielding a measurement of crustal thickness.

(4) A far more successful development on the same basis is that of deep seismic sounding (*glubinnoe seysmicheskoe zondirovanie*†). Comparatively small charges are used, efficiency being increased by exploding in water at depths of 10 to 20 meters. Individual records are written at widely separated points on a long profile. This procedure has been thoroughly successful in the Tian Shan and adjacent parts of central Asia, in Turkmenia, and in the Caucasus. The level of the Moho is regularly determined, and P*, corresponding to the Conrad discontinuity (taken as separating granitic and basaltic layers) is observed. In many areas a thick upper sedimentary layer with low velocity is measured.

As in other countries, public and official attention tends to be attracted to the speculative possibilities of earthquake prediction. In 1954 a large special publication was issued which nominally dealt with that problem. Actually, the contributors made the sound point that any future hope of prediction rests on increasing knowledge of earthquakes and their causes; the contents ranged over nearly the whole of seismology.

Because of new developments introducing industrialization and urban population into areas previously almost unoccupied, geographic charting of seismicity, with a view to estimating future risk, has been an important factor in organized planning. Many pages of the 1954 volume, and many special publications, sometimes involving highly controversial discussion, have been devoted to problems of seismic zoning.‡

---

† These are often referred to in Russian papers by the Russian initial letters corresponding to KMPV and GSZ.

‡ *Seismicheskoe rayonirovannie.* A closer translation would be "seismic regionalization." The term "regionalization" appears with this meaning in dictionaries, but is seldom used.

SEISMIC ZONING

The seismic map of the USSR set up by Gorshkov and collaborators in 1941 and repeatedly revised, has official status. Specifications for buildings and other construction must conform to requirements for the degree of risk indicated on the map. To each area is attached a number representing the highest intensity† on the Mercalli (in Medvedev's version) scale to be expected there. These ratings range from V to IX. The large areas of the stable shields are left blank, indicating expected intensity not over IV. Seismic risk is shown as a serious problem only for marginal areas of the USSR, involving the northern front of the Alpide belt, the Pamir-Baikal zone, and a small part of the circum-Pacific belt. The maximum risk, corresponding to expected intensity IX or over, is shown only for parts of Turkmenia, of Central Asia, and of the Kurile-Kamchatka arc.

Imperfections in such mapping are manifest, and were recognized from the first. Even in districts which have been settled for centuries, historical data on earthquakes are inadequate. Strong shocks occasionally occur where no earlier ones are on record (see "Unusual Activity" in Chapter 6; a striking example in USSR territory was the Ashkhabad earthquake of 1948, Chapter 31). It is unwise to suppose that the highest intensity known to have affected a given locality will never be exceeded there in the future. Moreover, use of historical records tends to overemphasize seismicity near the older and larger centers of population.

History has been supplemented by statistics of small shocks recorded instrumentally for a few decades, especially in central Asia. The tacit assumption is that large earthquakes are most probable where small ones are frequent. Experience in California does not confirm this expectation, but suggests that local conditions for large and small shocks differ. Compare the remarks on the activity of the San Andreas fault in Chapter 27 with Petrushevsky's observation that in the Tian Shan and elsewhere there are large rigid crustal blocks which apparently resist minor stresses, fracture only when major deformations have accumulated, and consequently are subject chiefly to great earthquakes.

The need to consider geological as well as statistical evidence was emphasized by Gubin, in publications highly critical of the work of Gorshkov as well as of earlier writers. However, Gubin oversimplified the geological problem. He insisted that earthquake risk is concentrated near dislocations which stratigraphic and geomorphic evidence indicates as active in Pleistocene and Recent time.

---

† In Russian, intensity, as defined in this book, is usually denoted by *"ball'nost',"* meaning scale degree, or *"sila,"* meaning force or strength. *"Intensivnost'"* is used in the physical sense, as of the intensity of radiation; but with reference to earthquakes it often means magnitude.

Two principal factors have enabled Gubin's opponents to challenge many of his arguments. The first factor is the discordance between present seismicity and conspicuous young geologic structures noted in Chapter 27. Thus Petrushevsky remarks that epicenters in parts of the Tian Shan area are in localities of low relief, tens of kilometers distant from high young scarps. This may be compared with the relation of the Owens Valley earthquakes to the Sierra Nevada scarp.

The second factor is depth of focus. Earthquakes at depths of 500 to 600 kilometers in the Far East (Manchuria) are of little interest in seismic zoning, since they almost never attain damaging intensity at the surface. On the other hand, some of the larger earthquakes of the Hindu Kush series, originating at depths near 220 kilometers, have caused reported damage at points far distant within the USSR. In central Asia hypocenters range all the way from such depths up to shallow and almost superficial levels (as in the Garm district).

Tectonic displacement at depths of 30 kilometers or more need not manifest itself obviously at the surface. Evidence of former violent shaking, such as landslide scars or old fissures in alluvium, may be searched for, but the identification of such features as earthquake effects is usually uncertain. Actual deformation can be seen only if displacement extends upward to or near the surface, and its manifestation there may be complex (Chapters 13, 14).

From the Pamir plateau westward all the problems of arc tectonics arise. Activity is lower than that of the south-facing Alpide arcs, and still lower than in the Pacific arcs. Earthquakes are relatively few, and the other arc features are less clearly defined.

The Pamir-Baikal zone, where block structure prevails, includes the Tian Shan, which has been more thoroughly studied than any other part of central Asia. Here sharp differences in seismicity exist between mountain structures of nearly the same age and character. Petrushevsky's most successful correlation is with the age of the folding now superseded by block tectonics; the younger the last folding, the higher the seismicity, even though the block structures may be of the same age and size.

There has been search for some generally applicable criterion of geological evidence to be used in estimating seismicity. One cited by Petrushevsky is "contrast" in tectonic type between adjacent areas, with seismicity high at the contact where contrast is high. The concept of contrast proved difficult to define precisely, and there were puzzling exceptions. Gzovsky advocates using the horizontal gradient of the rate of stress change, determined from geomorphic, stratigraphic, and geodetic data; such a criterion will fit many facts, but still exceptions appear. Probably no one simply formulated criterion applies to all cases, or includes both arc and block tectonics. Evaluation of seismic risk ultimately will have to depend on thorough investigation of each individual area with respect to its local tectonic characteristics.

References

Savarensky, E. F., and Kirnos, D. P., *Elementi seysmologii i seysmometrii,* 2nd rev. ed., Moscow, 1955.

Gamburtsev, G. A., ed., *Problemi prognoza zemletryaseniy, Trudy Geofiz. Inst. Akad. Nauk SSSR,* No. 25 (152), 1954. A symposium on earthquake prediction, including:

Gubin, I. E., "O nekotorikh voprosakh seysmicheskogo rayonirovaniya," *ibid.,* pp. 36–73.

Gamburtsev, G. A., "Glubinnoe seysmicheskoe zondirovanie zemnoy kori," *ibid.,* pp. 124–133.

————, Riznichenko, Iu. V., and Berzon, I. S., *Korrelyationniy metod prelomlennikh voln,* Izdateltsvo Akad. Nauk SSSR, Moscow, 1952.

————, and Weizmann, P. S., "Sopostostavlenie dannikh glubinnogo seysmicheskogo zondirovaniya o stroenii zemloy kori v rayone severnogo Tyan-Shanya s dannimi seysmologii i gravitmetrii," *Izvestiya Akad. Nauk SSSR, ser. geofiz.,* 1956, pp. 1035–1043. (General report on crustal structure in the Tian Shan region).

Savarensky, E. F., "Development of seismic research and analysis of seismograms in the USSR," *Publ. Bureau central séismologique international,* Ser. A., vol. 19 (1956), pp. 249–256.

Gorshkov, G. P., "General survey of seismicity of the territory of the USSR," *ibid.,* pp. 257–259.

Kirnos. D. P., and Kharin, D. A., "Basic equipment of seismological stations in the USSR," *ibid.,* pp. 357–371.

Gamburtsev, G. A., "On some new methods of seismological research," *ibid.,* pp. 373–381.

Keylis-Borok, V. I., "Methods and results of the investigations of earthquake mechanism (a brief information)," *ibid.,* pp. 383–394.

Petrushevsky, B. A., "Znachenie geologicheskikh yavlyeniy pri seysmicheskom rayonirovanii," *Trudy Geofiz. Inst. Akad. Nauk SSSR,* 28 (155), 1955.

————, "O svyaz seysmicheskikh yavleniy na Uralo-Sibirskoy platforme i v Tyan-Shane s geologicheskoy obstanovkoy etikh territory," *Byull. Moskvskogo obshchestva ispitateley prirodi,* vol. 30 (6), 1955, pp. 31–53.

Gzovsky, M. V., "Tektonofizicheskoe obosnovanie geologicheskikh kriteriev seismichnosti," *Izvestiya Akad. Nauk SSSR, ser. geofiz.,* 1957, pp. 141–160.

Roberts, E. B., and Ulrich, F. P., "Seismological activities of the U. S. Coast and Geodetic Survey in 1948," *B.S.S.A.,* vol. 40 (1950), pp. 195–216. (Seismic probability map on p. 214.)

PART THREE

Appendixes

APPENDIX I

Chronology

ONLY EARTHQUAKES of exceptional interest are listed here. For greater details see the chronological bibliography in Appendix XVI.

373 B.C. Earthquake of Helice, Greece.

132 A.D. Chang Heng, seismoscope.

1556. Chinese earthquake; 830,000 deaths reported.

1707, October 28. Great Japanese earthquake.

1755, November 1. Lisbon earthquake.

1783, February 5 and following. Earthquakes of Calabria, Italy; scientific field investigation.

1819, June 16. Earthquake of Cutch, India; observed faulting (Allah Bund).

1830. Lyell, *Principles of Geology,* first edition.

1846. Robert Mallet; his first general paper on earthquakes.

1855. Kreil, early seismograph.
January 24. New Zealand earthquake; observed faulting.

1857, January 9. California (Fort Tejon) earthquake; observed strike-slip.
December 16. Earthquake in Italy (Kingdom of Naples); Mallet, field investigation.

1865. Seismological observations begun at Manila (with seismoscopes; seismographs installed 1881–1889).

1872, March 26. Earthquake in Owens Valley, California; observed scarps and strike-slip.

1879. Seismographs developed in Japan by the British group.

1880. Seismological Society of Japan organized.

1883, August 27. Explosion of Krakatoa.

1884. Rossi-Forel intensity scale set up. Seismological service established at Manila.

1885. Lord Rayleigh, paper on theory of "Rayleigh waves."

1887. Voigt, investigation of elasticity of crystals; definition and naming of tensors. Rudolph, first paper on seaquakes. Seismological stations established in California at Berkeley and Lick Observatory.

1888. Eduard Suess, *Das Antlitz der Erde.*

September 1. Amuri earthquake, New Zealand; strike-slip, offsetting fences.

1889, April 18. First seismogram of a teleseism identified.

1891, October 28. Mino-Owari earthquake, Japan; large strike-slip and dip-slip.

1892. Japanese Imperial Earthquake Investigation Committee established.

May 17. Sumatra earthquake; strike-slip shown by triangulation survey.

1894. New seismographs developed by Vicentini (Padua, Italy) and by Milne in England.

1896. Seismology Committee of the British Association for the Advancement of Science organized.

1897, June 12. Great earthquake of Assam, India; investigation by Oldham.

1898. Seismograph developed by Omori at Tokyo. Stations with Milne instruments initiated at Wellington (New Zealand) and Batavia (Java).

1899. Knott's equations for reflection and refraction coefficients of elastic waves.

September 3 and 10. Great Alaskan earthquakes; uplifts reaching 47 feet.

1901. Inverted pendulum seismograph developed by Wiechert. Geophysical Institute founded at Göttingen. *Gerlands Beiträge zur Geophysik* began publication.

1903. International Seismological Association organized.

1904. Seismological station initiated at Uppsala (Sweden).

1906. Galitzin electromagnetic seismograph developed (Russia). Vertical-component seismograph set up by Straubel and Eppenstein at Jena (Germany).

January 21. Major deep-focus earthquake under Japan.

January 31. Great earthquake, Colombia and Ecuador.

April 18. California earthquake; great extent of strike-slip faulting; survey retriangulations leading to elastic-rebound theory of earthquakes (H. F. Reid, Johns Hopkins University).

1907. Seismograph developed by Mainka at Strassburg. First paper of Göttingen series *Über Erdbebenwellen.*

1908, June 30. Great Siberian meteorite fall.

1909. Seismological station established at Riverview (near Sydney, Australia).

October 8. Earthquake near Zagreb (Croatia); discovery of the subcrustal discontinuity by A. Mohorovičić.

1910. New instruments at Berkeley.

1911. Seismological Society of America founded; its *Bulletin* began publication. New instruments at Lick Observatory station. Galitzin vertical-component seismograph developed (Russia). Theory of Love waves (A. E. H. Love, England).

November 16. South German earthquake.

1913. Station established at La Paz, Bolivia. Radius of the core determined by Gutenberg at Göttingen.

1915. Milne-Shaw seismograph developed by J. J. Shaw (England).

1918. First year covered by *International Seismological Summary*.

1919. Zoeppritz equations for coefficients of reflection and refraction of elastic waves published (posthumously).

1921. Carnegie Institution of Washington, Advisory Committee in Seismology appointed.

September 21. Explosion at Oppau, Germany.

1922. Changes in the variation of latitude and in other geophysical elements. *Geophysical Supplements to Monthly Notices* of the Royal Astronomical Society began publication. Turner (Oxford, England) discovered deep-focus earthquakes.

1923. A. Sieberg, *Erdbebenkunde*.

January. Wood-Anderson torsion seismometers began regular recording (Pasadena).

September 1. Great Japanese earthquake (Kwantō); destruction at Tokyo and Yokohama. Earthquake Research Institute (Tokyo) founded as a consequence.

1924. Jeffreys, *The Earth,* first edition. de Ballore, *Géologie sismologique*. Nikiforov torsion seismographs in service (USSR).

1925. Jesuit Seismological Association organized; headquarters at St. Louis. Seismological work transferred from U. S. Weather Bureau to Coast and Geodetic Survey.

June 29. Santa Barbara earthquake.

1926. *Geophysical Magazine* (Tokyo) began publication. Earthquake department of Pacific Board of Fire Underwriters established.

1927. Seismological Laboratory at Pasadena occupied. March 7, Tango earthquake (Japan); thoroughly investigated; two fault traces.

1928. K. Wadati, paper on shallow and deep earthquakes.

1929. *Handbuch der Geophysik* began publication. New electromagetic seismograph developed by Wenner (Bureau of Standards, Washington).

June 17. New Zealand earthquake; expansion of New Zealand seismology followed.

1931. Variable-reluctance seismometer developed by Benioff (Pasadena).

1932. Benioff strain seismometer developed. U. S. Coast and Geodetic Survey initiated a program of strong-motion recording in California.

1933, March 3 (Japanese date). Great Sanriku earthquake and tsunami.

March 10 (local date). Long Beach earthquake, California; Field Act regulating school construction passed by California legislature in consequence.

1934. *P'P'* and related waves discovered (at Pasadena).

1935. Magnitude scale published.

1936. Existence of the inner core suggested by Miss Lehmann (Copenhagen).

1937. Istituto Nazionale di Geofisica established; headquarters at Rome.

1939, December 27. Destructive earthquake in Turkey, first of a series with extended and connected faulting.

1940. T wave described and named by Linehan (Weston, Massachusetts).

May 18. Imperial Valley earthquake (California-Mexico); strike-slip.

1944. *Annales de Géophysique* began publication.

1946, April 1. Aleutian tsunami, destructive on Hawaii; seismic sea wave warning service organized in consequence.

July 24. Atomic bomb test at Bikini; *P* waves recorded at distant stations.

1947, April 18. Helgoland demolition.

1948. C. A. Whitten (U. S. Coast and Geodetic Survey), paper on continuing horizontal displacements in California.

1949. *Annali di Geofisica* began publication.

August 6. Large quarry blast at Corona, California.

1950, August 15. Great earthquake, Tibet and Assam.

1951. Strain-release curves developed by Benioff (Pasadena).

1952. *Lg* waves discovered by Ewing and Press (Columbia University, New York).

July 21. Major earthquake, Kern County, California.

1953. *Pa* and *Sa* waves discovered by Caloi at Rome (and independently by Ewing and Press).

1954, March 29. Earthquake 640 kilometers deep under Spain.

December 16. Major earthquake in Nevada; large and extensive faulting.

1957, July 1—1958, December 31. International Geophysical Year.

APPENDIX II

Earthquake-Resistant Construction

THESE NOTES may assist the engineering student, but he should by all means consult his professional authorities.

Design of large buildings to be earthquake-resistant is an important issue to professional engineers because of the high degree of responsibility involved. To the general public and the general seismologist the matter is of distinctly less consequence, because most past and present construction raises no such problems. Debate on such design has often been misused by interested groups to cast a smoke screen over the urgent need for adequate building codes and consistent enforcement to ensure the safety of ordinary structures; it is said, in effect, that engineers are not agreed what safe construction is, and therefore regulations are of no use. This is an outrageous misrepresentation; the simple rule-of-thumb provision against lateral accelerations of 0.1g or more has repeatedly demonstrated its practical efficiency (Chapter 8).

Controversy stems from the observations of John Milne, who during his residence in Japan was impressed by the flexible yielding and recovery of the ordinary light Japanese house during moderate earthquakes (though it often fails hopelessly in strong shaking). The idea of flexibility opposed to rigidity as a principle of safe design has persisted for more than 70 years. After the Tokyo earthquake of 1923 most engineers, Japanese and foreign, were satisfied with the results of the rigid-body principle in designing against lateral accelerations of 0.1g or more. Nevertheless, a few dissenters went so far as to advocate extreme flexibility, but neither theory nor unprejudiced evaluation of the facts of damage support their claims. On the other hand, high rigidity for tall buildings is unattainable. Even the most rigid of ten-story buildings (solid, reinforced concrete with no windows) yield sufficiently so that flexibility must be considered in careful design.

Flexibility and rigidity are obviously artificial extremes. Fundamentally, resistance to earthquakes is a problem in motion, not in static equilibrium. In theory, it calls for dynamic design; some writers emphasize this point as if it were a refutation of static design. However, direct application of dynamic analysis is difficult even in theory, and hardly feasible with present means. In actual design static forces are considered, with modification to allow for the dynamic

645

properties of the structure. Even large and tall buildings designed on this static basis have repeatedly proved to be earthquake-resistant; objection to the static design of small structures is either a quibble or a political argument. To quote G. W. Housner:

> During an earthquake a structure is excited into a more or less violent vibration, with resulting oscillatory stresses, which depend both upon the ground motion and the physical properties of the structure. This is such a complex dynamic problem that it does not appear feasible to make a precise dynamic stress analysis of the problem, particularly inasmuch as it is not possible to foretell the precise nature of future earthquake ground motion nor to compute precisely all of the physical properties of a structure before it is built. The present methods of design are based upon a static rather than a dynamic approach, the structure being designed to resist certain static lateral forces. The static lateral forces are intended to produce stresses of the same order of magnitude as the maximum dynamic stresses likely to be experienced during an earthquake. Because of the complexity of the vibration problem and the various factors influencing the dynamic behavior of a structure, it is not possible to state with certainty the correct static loads that should be used in all instances, so that the loads used in present design methods must be considered as approximations which will be improved as additional knowledge is gained.

Students well grounded in theoretical physics may occasionally be startled by the extreme simplification necessary in order to proceed in this direction. Much engineering discussion, many theoretical papers, and a great deal of valuable experimental work are based on nothing more sophisticated than our equation (10) of Chapter 15, representing the forced oscillations of a damped pendulum. Three main assumptions are involved:

(1) Distortions are small and within the elastic limit, so that Hooke's law applies.

(2) The structure may be treated as a system of one-dimensional elements which are considered as separate and responding independently to disturbance.

(3) Energy is transferred chiefly into the oscillation mode of lowest frequency of which the structure is capable; this is termed the fundamental frequency.

These assumptions are largely justifiable on both theoretical and experimental grounds; this is fortunate, because it would be difficult to operate without them. Critics have on occasion emphasized the limitations of such work. It can be said, however, that any design which is indicated as unsafe by such relatively simple methods should be rejected. Thus, if dangerously large oscillations are indicated (even though the validity of Hooke's law is assumed), it is pointless to remark that the elastic limit may be passed in an actual earthquake—as indeed it must if there is structural damage.

Assumption 2 becomes troublesome theoretically. It is easy to construct relatively simple mechanical systems which are not separable; that is, the motion cannot be treated as two or more independent oscillations. Even as an approximation, the assumption of separability may fail under high excitation, which may produce oscillations wholly different in character from the lowest or fundamental; this is seen clearly in experiments on the vibration of thick plates. Since the actual behavior of large engineering structures is usually tested only for small oscillations representing low excitation, this leaves a wide opening for adverse criticism.

Assumption 3 represents no general physical law. It is theoretically correct if the principal energy in the exciting vibration is concentrated in frequencies near or below the fundamental frequency of the structure, and if the conditions are proper to excite the fundamental mode of oscillation. Given the randomness of earthquake disturbance and its tendency to high energy in a relatively low range of frequencies, the assumption will generally be justified. However, design should consider the possibility of more or less accidental excitation of some higher mode to an unsafe degree; this higher mode, as just remarked, may differ radically in form from the fundamental.

Experimentation necessarily takes place to a great extent with small-scale models. Because of the principle of similitude, these must be built of different materials, or of similar materials connected by weaker elements, than in the large structures they represent. Calculations for this purpose necessarily assume Hooke's law and a simplified type of coupling involving separability into independent oscillators.

All these points are generally recognized and clearly stated by the men doing the work which is actually advancing our knowledge in this field. Others in discussion sometimes pick up the results and apply them beyond their proper limits, with a naïveté which may lead to conclusions diametrically opposite to those obvious in the data.

It occasionally happens that unexpected overload at one point of a structure causes local yielding which changes the mechanical characteristics of the whole. This implies faulty design, but the error may not be an obvious one. The point is of importance in judging the safety for further use of buildings which have been slightly damaged.

Valid experimentation with models requires close correlation with the recording of strong ground motion in earthquakes by instruments such as those operated by the U. S. Coast and Geodetic Survey. Simplified statements about prevailing periods and expected accelerations easily get into wide circulation, although strong-motion seismograms show periods spreading over a wide range, at least from 0.05 to 2 seconds. (See Chapters 3 and 8. Detailed analysis is due largely to Frank Neumann and to Housner and collaborators.) Purely seismological error results from the fact that most seismographs have a peak magnification near the period of the pendulum. Motion with this period is likely to produce the maximum of the recorded seismogram, and its amplitude may then be reported as that of the maximum ground motion, while the seismogram of an instrument with different characteristics may appear to yield a very different result for ground amplitude and acceleration.

Strong-motion recording has made it possible to verify the applicability of assumption 3 rather directly. One of the clearest early examples was that of a bank building in San Jose, California. During its construction, the fundamental periods of the building were determined repeatedly by a party of the U. S. Coast and Geodetic Survey, by means of a shaking machine. At the completion of construction, strong-motion recorders remained installed in the basement and at the top floor. Not long after (December 30, 1934), there was a locally strong earthquake; the recorded motion plainly showed excitation of the fundamental periods previously measured. There have been many later observations of the same kind.

Vibration tests with shaking machines are of value in determining the safety of existing structures as well as in design. Such tests are ordinarily limited to producing small and harmless building oscillations; a rare exception was a test carried out by Housner and associates on a reinforced-concrete warehouse about to be demolished. One of their interesting results was the low natural damping; this gave point to a proposal to introduce damping into the design of such structures to reduce the oscillation.

Models and small structures are now being set up with installed equipment to record the strains set up in actual earthquakes. Such a program would arrive at useful results in less time in a region of high seismicity, like Japan, than in one of moderate seismicity such as California.

Careful writers do not use the terms "earthquake-proof" or "fireproof" in reference to design or construction. Proper expressions are "earthquake-resistant" and "fire-resistant." It would be extremely difficult, in fact, to design a structure of any consequence which would not be destroyed by intensity XII. Fortunately, the maximum of known shaking probably is manifested only in the few greatest earthquakes under exceptional ground conditions. Good construction will withstand VIII with no damage, and IX with minor damage if any. Designing against X M.M., involving accelerations of the order of $0.7g$ and ground displacements (of long period) with amplitudes of the order of a foot, is a problem which has been approached only in the roughest way.

The crowding of metropolitan centers brings a demand for tall structures of the "skyscraper" class, twenty stories or more in height. *In the writer's personal opinion, such structures simply should not be built* in a region subject to severe earthquakes, unless they can be founded directly on the soundest type of rock, granitic preferably; certainly not on sand or sediments. Undoubtedly a structure of this class involving the best designs and materials, in which plans and specifications have been rigorously enforced, will not collapse into an incoherent heap of rubble like the masonry at Shillong in 1897 or the unreinforced brick structures at Compton and Long Beach in 1933; but it may be left in a weakened condition in which the only proper safety precaution will be to tear it down. Even under intensity IX, severe damage to contents and to non-structural elements such as partitions is to be looked for.

This touches on a point which has partly governed the attitude of insurance organizations toward the rigidity-flexibility controversy. A structure which distorts significantly under earthquake disturbance may possibly be so designed that internal stresses in its frame are less than those in a more rigid structure, but under distortion there may be damage to contents and to non-structural elements (such as partitions and plaster walls) representing a relatively large monetary loss. Compare the remarks quoted from the 1952 report in Chapter 8 on damage to buildings in Los Angeles.

Rigid construction seems the only logical procedure when the foundation is really soft. A solidly braced, box-like unit will survive much distortion and differential settling of its foundation; its worst losses will probably be due to tilting out of plumb. An example was the Procter and Gamble factory on extremely soft ground near Long Beach; its units survived the 1933 earthquake practically undamaged.

Whatever the general principle of design, all are agreed: that no construction should consist of units, connected or in contact, which respond incoherently to shaking, creating stresses through the connections which would not develop if the units were independent; and that no design should include minor units, such as parapets and architectural ornaments, inadequately attached and likely to fall to the ground in an earthquake.

References

THE LITERATURE cited at the end of Chapter 8 should be consulted. The quotation is from Housner, G. W., "The design of structures to resist earthquakes," pp. 271–277, *Bull.* 171, Calif. Dept. Nat. Resources, Div. Mines; this entire article is well worth reading. Refer also to:

Hudson, D. E., and Housner, G. W., "Structural vibrations produced by ground motion," *Proc. Am. Soc. Civil Engrs.*, vol. 81 (1955), Paper 816.

Alford, J. L., and Housner, G. W., "A dynamic test of a four-story reinforced concrete building," *B.S.S.A.*, vol. 43 (1953), pp. 7–16.

Engle, H. M., "Earthquake provisions in building codes," *ibid.*, vol. 43 (1953), pp. 233–237.

Duke, C. M., and Brisbane, R. A., "Earthquake strain measurements in a reinforced concrete building," *ibid.*, vol. 45 (1955), pp. 83–92.

Proc. World Conference on Earthquake Engineering, Berkeley, California, June 1956, San Francisco, 1956.

APPENDIX III

Intensity Scales—Notes and Addenda

THE ROSSI-FOREL SCALE

The most commonly used form of the Rossi-Forel (R.F.) scale reads as follows:

I. *Microseismic shock.* Recorded by a single seismograph or by seismographs of the same model, but not by several seismographs of different kinds: the shock felt by an experienced observer.

II. *Extremely feeble shock.* Recorded by several seismographs of different kinds; felt by a small number of persons at rest.

III. *Very feeble shock.* Felt by several persons at rest; strong enough for the direction or duration to be appreciable.

IV. *Feeble shock.* Felt by persons in motion; disturbance of movable objects, doors, windows, cracking of ceilings.

V. *Shock of moderate intensity.* Felt generally by everyone; disturbance of furniture, beds, etc., ringing of some bells.

VI. *Fairly strong shock.* General awakening of those asleep; general ringing of bells; oscillation of chandeliers; stopping of clocks; visible agitation of trees and shrubs; some startled persons leaving their dwellings.

VII. *Strong shock.* Overthrow of movable objects; fall of plaster; ringing of church bells; general panic, without damage to buildings.

VIII. *Very strong shock.* Fall of chimneys; cracks in the walls of buildings.

IX. *Extremely strong shock.* Partial or total destruction of some buildings.

X. *Shock of extreme intensity.* Great disaster; ruins; disturbance of the strata, fissures in the ground, rock falls from mountains.

Under I, the reference to seismographs of different types without specification of their characteristics is even more unsatisfactory than the references in the M.M. scale to damage in various types of construction without adequate description. The middle grades of the scale have been rearranged on the basis of later experience. R.F. X lumps together all effects of high intensity, which have later been separated. The correspondence between the two scales is as follows:

M.M.	I	II	III	IV	V	VI
R.F.	I	I-II	III	IV-V	V-VI	VI-VII

M.M.	VII	VIII	IX	X-XII
R.F.	VIII–	VIII+ to IX–	IX+	X

NOTES ON THE M.M. SCALE OF 1931

Chapter 11 includes critical comment on the simultaneous use of long-period and short-period effects and on the introduction of primary effects such as faulting along with the secondary effects of shaking.

Under IV the original abridged version reads "Sensation like heavy truck striking building." This gives an exaggerated impression which would properly correspond to VI.

Under IX the author has added "general damage to foundations." Such damage is characteristic; an example is the filled ground at San Francisco in 1906.

Under IX "Caused sea waves . . ." is omitted. This is an indication of magnitude rather than intensity, but all the circumstances are variable (Chapter 9).

Under XII "Waves seen on ground surfaces . . ." is omitted. This is a regrettable inclusion in the 1931 scale. As a criterion for high intensity such reports are worse than useless, since apparently reliable evidence of the kind comes from localities where the intensity is VI or even lower (Chapter 10).

Adjustment of the descriptions of masonry A, B, C, D to the scale has caused the author much concern. A serious stumbling block has been the entry in the 1931 scale at VIII and IX of "damage slight" and "damage considerable," respectively, for masonry structures "built especially to withstand earthquakes." If this were applied to masonry A, VIII and IX should be assigned to much stronger shaking than the other criteria for these grades indicate. The confusing wording originates in translation from the modified version of the Mercalli scale as given by Sieberg; under VIII he writes:

> Häuser europäischer Bauart erleiden, trotz solider Konstruktion, durchweg erhebliche Beschädigungen durch klaffenden Spalten im Mauerwerk . . . Bebenfest konstruierte (Japan usw). Ziegelbauten zeigen bereits leichte Beschädigungen, wie Risse, Abbröckeln von Bewurf usw. (vgl. Grad VII, bei europäischen Häusern) . . .†

> Even though solidly constructed, houses of European architecture generally suffer heavy damage by gaping fissures in the walls . . . Brick buildings constructed to be resistant to earthquakes (Japan etc.) already show slight damage, such as cracks, breaking off of (external) plaster, etc. (compare VII for European houses) . . .

From this, and from explicit references elsewhere in his writings, it is clear that Sieberg's idea of masonry "built specially to withstand earthquakes" is a

† A. Sieberg, *Erdbebenkunde,* Jena, 1923, p. 103.

distorted reproduction of Milne's first enthusiastic description of Japanese construction (which was formulated without experience of any really great earthquake, such as that which followed in 1891). Of such structures it may properly be written that they are damaged slightly at VIII and considerably at IX; nevertheless, they may be better than the European structures to which Sieberg refers at VII—also "solider Konstruktion," which the 1931 scale translates as "ordinary substantial buildings"; these are at best qualified as masonry C. It must be admitted that this does not seriously misrepresent the prevailing character of ordinary masonry in California in 1931; standards to be met by "ordinary substantial" construction have risen notably in 25 years, owing primarily to the efforts of engineers and insurance organizations. The gradual change in standards makes it necessary to define explicitly the types of construction to which the scale degrees refer.

QUESTIONNAIRE CARDS

Much useful information can be elicited by the proper use of questionnaire cards such as are circulated by the U. S. Coast and Geodetic Survey. These are designed for application of the M.M. scale. A recent form reads as follows:

An earthquake was felt, not felt, on

Date of shock . . . . Time . . . . $\begin{array}{l}\text{A.M.}\\\text{P.M.}\end{array}$

Place . . . .

Please *return the card* even if the shock was not felt, as such information is essential.

Please *underline* the words below which best describe the shock *AT YOUR LOCALITY* as given above. Reports from other places on separate cards are also desirable.

Motion rapid, slow, ... Shook how long . . . .

Felt by several, many, all; by observer.

In your home, in community, or . . . .

In building, wood, brick, ... , strongly, weakly built; on 1, 2, ... floor; lying down, sitting, active.

Outdoors, by observer, by others; quiet, active.

Direction of motion felt *outdoors;* N., NE., E., etc. . . . .

Ground underneath locality: Rock, soil, loose, compact, marshy, filled in, ... ; level, sloping steep.

Awakened no one, few, many, all; (in your home) (in community).

Frightened no one, few, many, all; (in your home) (in community).

Rattling of windows, doors, dishes, ...

Creaking of walls, frame, ...

Hanging objects, doors, etc., did, did not, swing, N., NE., etc.

Pendulum clocks did, did not stop; clocks faced N., NE., etc.

Trees, bushes shaken slightly, moderately, strongly . . . .

Shifted small objects, furnishings, ...

Overturned vases, etc., small objects, furniture, ...

Cracked plaster, windows, walls, chimneys, ground, ...
Fall of knickknacks, books, pictures, plaster, walls, ...
Broke dishes, windows, furniture, ...
Twisting, fall, of chimneys, columns, monuments, ...
Damage, none, slight, considerable, great, total in wood, brick, masonry, con-
    crete, ...
REMARKS:

Signature . . . .
Address . . . .

Any additional information will be appreciated. Clip additional sheet to this card.

APPENDIX IV

Elasticity and Wave Propagation—Proofs and Addenda

HERE FOLLOW further details on some theoretical points passed over in Chapter 16. Although not necessary for the understanding of that chapter, they may assist the conscientious student or instructor to fill in the gaps in treatment. The reader with solid mathematical training will notice a lack of rigor in the absence of such details as existence proofs and investigation of limits of applicability. These may be found in the handbooks referred to at the end of Chapter 16. Difficulties of serious concern to the working seismologist usually occur only when it is necessary to introduce discontinuities into the physical assumptions, for, although mathematical discontinuities do not exist in nature, they are sometimes essential to manageable theory.

SYMMETRY OF THE STRESS COMPONENTS

Consider a medium under uniform stress such that each component Xx, Xy, etc., has a value constant at all points. Mark off a cube with edges of any length L, center at the origin of coordinates, and faces perpendicular to the axes. The forces exerted across the faces of the cube toward the external medium are given by the stress components multiplied by L^2; these forces may be considered as applied at the centers of the six faces. The components of force across the face cutting the positive X axis are $L^2(Xx, Yx, Zx)$; those across the face cutting the negative X axis are $L^2(-Xx, -Yx, -Zx)$; and similarly for the components across each of the other cube faces.

Equilibrium requires that the sums of the moments about the axes shall vanish. Take moments about the axis of Z. The force components proportional to Zx, Zy, Zz are parallel to the axis, and their moments are zero; the components proportional to Xx and Yy have zero moments because their lines of action pass through the origin. The sum of the remaining moments is $L^3(Yx - Xy)$. Hence, for equilibrium, $Yx = Xy$, and by taking moments about the other axes we similarly find $Zx = Xz$ and $Zy = Yz$.

By considering the effect of taking L very small, these relations can be shown to be valid in equilibrium when the stresses vary continuously from point to point.

EQUATIONS OF MOTION UNDER ELASTIC FORCES

In the interior of an elastic solid which has been deformed, allowed to reach equilibrium under stress, and then released, consider a small cube of edge L. Since in equilibrium the forces have zero moment about the coordinate axes, this will continue to be the case, and $Xy = Yx$, etc., will hold (unless new forces are applied); the small cube will not be rotated, although it will generally be translated.

The forces acting to displace the cube may be represented as applied at the centers of its faces, and obtained by multiplying the stress components by L^2. Those acting parallel to X on the six faces can be written $-XxL^2$, $-XyL^2$, $-XzL^2$, $[Xx + L(\partial Xx/\partial x)]L^2$, $[Xy + L(\partial Xy/\partial y)]L^2$, $[Xz + L(\partial Xz/\partial z)]L^2$. Note that the two face centers at which we are comparing Xy have the same value of x and z, differing only in y; and similarly for Xz. Adding all these, the total force parallel to X is

$$\left(\frac{\partial Xx}{\partial x} + \frac{\partial Xy}{\partial y} + \frac{\partial Xz}{\partial z}\right)L^3$$

The corresponding acceleration is $\partial^2 u/\partial t^2$; if the mean density of the cube is ρ, we have

$$\rho\frac{\partial^2 u}{\partial t^2} = \frac{\partial Xx}{\partial x} + \frac{\partial Xy}{\partial y} + \frac{\partial Xz}{\partial z} \tag{1a}$$

with two similar equations for the Y and Z components.

The symmetry relations among the stress components may then be applied.

PRINCIPAL AXES OF STRESS

What happens to the stress components when we rotate the direction of the coordinate axes has already been illustrated in equations (1a) and (1b) of Chapter 16. These, and the others which are easily obtained from them by interchanging the coordinates, follow straightforwardly from the definitions of stress components. Handbooks such as Love's usually derive them by considering the equilibrium of a tetrahedron, three of whose faces are formed by the three coordinate planes meeting at the origin, while the fourth is perpendicular to the new axis of X'.

We reduce the nine equations to six by applying the symmetry of the stress components, and then ask for the condition that in the new system we have $X'y' = X'z' = Y'z' = 0$. This gives three equations to determine the direction cosines of the new coordinate axes with reference to the old. There are nine such direction cosines, but they are not independent. In fact, only three of them can be, since three numbers are sufficient to specify the orientation of the new axes with respect to the old. Hence there are actually three equations in three unknown quantities; solution is always possible. With given values of the stress components in the original system, the resulting orientation of the principal axes is in general unique (apparent multiplicity in solution is produced by the circumstance that any one of the principal axes can be called X' and either of the two remaining Y', giving six formally different choices). With physical symmetry in the stresses the uniqueness may disappear; thus, when the complete stress system is a uniform ("hydrostatic") compression, the stresses Xy, Xz, Yz vanish no matter what the choice of axes, and any set of perpendicular axes qualifies as principal axes. This requires that $Xx = Yy = Zz$, which is then true for any choice of axes, as well as $Xy = Xz = Yz = 0$.

PRINCIPAL AXES OF STRAIN

The effect of rotation of axes on the strain components is formally simple to derive. As in Chapter 16, take a, b, c as the cosines of the angles between the new axis of x' and the old axes of x, y, z. Then

$$x' = ax + by + cz \tag{1b}$$

and

$$u' = au + bv + cw \tag{1c}$$

whence

$$\frac{\partial u'}{\partial x} = a\frac{\partial u}{\partial x} + b\frac{\partial v}{\partial x} + c\frac{\partial w}{\partial x} \tag{1d}$$

with similar results for $\partial u'/\partial y$ and $\partial u'/\partial z$. From (1b),

$$\frac{\partial x'}{\partial x} = a \qquad \frac{\partial x'}{\partial y} = b \qquad \frac{\partial x'}{\partial z} = c \tag{1e}$$

Remembering that (1b) and corresponding equations represent a rotation without stretching, and considering the inverse transformation, "it easily follows" that

$$\frac{\partial x}{\partial x'} = a \qquad \frac{\partial y}{\partial x'} = b \qquad \frac{\partial z}{\partial x'} = c \tag{1f}$$

But, from the definition of $e_{x'x'}$ and the rules of partial derivatives,

$$e_{x'x'} = \frac{\partial u'}{\partial x'} = \frac{\partial u'}{\partial x}\frac{\partial x}{\partial x'} + \frac{\partial u'}{\partial y}\frac{\partial y}{\partial x'} + \frac{\partial u'}{\partial z}\frac{\partial z}{\partial x'} \tag{1g}$$

Into this we may substitute the result (1d) and those for the other two corresponding derivatives, as well as the expressions (1f). On collecting terms and applying the definitions of the strain components, it appears that

$$e_{x'x'} = a^2 e_{xx} + b^2 e_{yy} + c^2 e_{zz} + abe_{xy} + ace_{xz} + bce_{yz} \tag{1h}$$

Corresponding transformations for the other strain components are derived in the same manner. They differ from those for the stress components only in the absence of the factor 2 in certain terms; this could be adjusted by a change of definition.

Because of the similarity of these equations, the principal axes of strain can be found in exactly the same way as those of stress. For a homogeneous isotropic body obeying the generalized Hooke's Law, it follows from equations (4) of Chapter 16 that the principal axes of stress and strain are identical at any point. These relations, or the more generalized linear form applying to the non-isotropic (aeolotropic) case, can be used to derive the transformation equations for the stresses from those of strain; but this involves assuming the generalized Hooke's law, while the direct method of deriving the stress transformation from the equilibrium conditions does not.

REDUCED EQUATIONS OF MOTION

The equations of small motion for a homogeneous isotropic elastic body are usually given in a form which results from substituting into the first form of the equations of motion—represented by (2a) of Chapter 16 and (1a) of this appendix and the corresponding equations for the other two components—the stress-strain relations (4) of Chapter 16, in which the strain components are further expressed in terms of the derivatives of displacement. In simplifying it is only necessary to remember that the order of differentiation with respect to any two of x, y, z may be reversed without changing the value of the partial derivative. The elastic coefficients λ and μ are treated as constant (the medium being homogeneous) and therefore are not differentiated; if it is desired to allow directly for non-homogeneity there at once arises a whole group of additional terms involving the derivatives of the elastic parameters, and the result cannot be given in the relatively neat form which here follows.

It is desirable to introduce the abbreviation

$$\theta = e_{xx} + e_{yy} + e_{zz} = \frac{\partial u}{\partial x} + \frac{\partial v}{\partial y} + \frac{\partial w}{\partial z}$$

and also to use the Laplacian operator ∇^2. The student who is unaccustomed to vector analysis may simply consider this as an abbreviation, so that we write

$$\nabla^2 u = \frac{\partial^2 u}{\partial x^2} + \frac{\partial^2 u}{\partial y^2} + \frac{\partial^2 u}{\partial z^2} \tag{2}$$

and similarly for $\nabla^2 v$, $\nabla^2 w$, or $\nabla^2 \theta$.

The wave equation (6) of Chapter 16 then may be written

$$\nabla^2 A = \frac{1}{C^2} \frac{\partial^2 A}{\partial t^2} \tag{3}$$

The equations of motion then take the standard form.

$$\rho \frac{\partial^2 u}{\partial t^2} = (\lambda + \mu) \frac{\partial \theta}{\partial x} + \mu \nabla^2 u \tag{4a}$$

$$\rho \frac{\partial^2 v}{\partial t^2} = (\lambda + \mu) \frac{\partial \theta}{\partial y} + \mu \nabla^2 v \tag{4b}$$

$$\rho \frac{\partial^2 w}{\partial t^2} = (\lambda + \mu) \frac{\partial \theta}{\partial z} + \mu \nabla^2 w \tag{4c}$$

P AND S WAVES

If one takes the partial derivatives of equations (4a), (4b), (4c) with respect to x, y, z, respectively, takes advantage of the interchangeability of the order of differentiation, and again applies the definition of θ, the result is

$$\rho \frac{\partial^2 \theta}{\partial t^2} = (\lambda + 2\mu) \nabla^2 \theta \tag{5}$$

If one forms the derivative of (4a) with respect to y and subtracts from it the derivative of (4b) with respect to x, the first terms on the right disappear, giving

$$\rho \frac{\partial^2}{\partial t^2} \left(\frac{\partial u}{\partial y} - \frac{\partial v}{\partial x} \right) = \mu \nabla^2 \left(\frac{\partial u}{\partial y} - \frac{\partial v}{\partial x} \right) \tag{6a}$$

and, by the other two analogous operations on pairs of (4),

$$\rho \frac{\partial^2}{\partial t^2} \left(\frac{\partial u}{\partial z} - \frac{\partial w}{\partial x} \right) = \mu \nabla^2 \left(\frac{\partial u}{\partial z} - \frac{\partial w}{\partial x} \right) \tag{6b}$$

$$\rho \frac{\partial^2}{\partial t^2} \left(\frac{\partial v}{\partial z} - \frac{\partial w}{\partial y} \right) = \mu \nabla^2 \left(\frac{\partial v}{\partial z} - \frac{\partial w}{\partial y} \right) \tag{6c}$$

Equation (5) is a scalar wave equation; (6a), (6b), (6c) are the three components of a vector wave equation. In (5) the wave velocity is given by $C^2 = (\lambda + 2\mu)/\rho$, and in (6) by $C^2 = \mu/\rho$; these are the results given for P and S waves. Equation (5), in fact, represents wave propagation of the quantity θ, which is the cubical dilatation; it is a compressional-rarefactional wave.

The vector quantity which appears as the wave parameter in equations (6) is known as the *curl* of the displacement vector whose components are u, v, w.

This discussion proves at best only the possibility of the existence of two kinds of waves. It can be shown that any arbitrary initial displacement of components u, v, w, will separate into two parts, propagated with the two characteristic velocities, in one of which the displacement vector itself will satisfy (6), while in the other the individual displacement components will satisfy (5).

There is a relatively simple proof of the existence of P and S waves in a homogeneous isotropic medium, which also shows that these are then the only possible types of plane waves. Take a plane wave propagating with front perpendicular to the axis of X and with velocity C. This may be represented by

$$u = U(s) \qquad v = V(s) \qquad w = W(s) \qquad s = x - Ct \tag{7}$$

where U, V, W are any differentiable functions of s, their derivatives with respect to s being U', V', W', and the corresponding second derivatives U'', V'', W''. The components of strain may be calculated directly from their definitions, giving

$$e_{xx} = U' \qquad e_{xy} = V' \qquad e_{xz} = W' \qquad e_{yy} = e_{zz} = e_{yz} = 0 \tag{8}$$

From this,

$$\theta = e_{xx} + e_{yy} + e_{zz} = U' \tag{9}$$

We may apply the stress-strain equations (4) of Chapter 16 to these results and find

$$Xx = (\lambda + 2\mu)U' \qquad Yy = Zz = \lambda V' \qquad Xy = \mu V' \qquad Xz = \mu W' \qquad Yz = 0 \tag{10}$$

The derivatives of these enter into the equations of motion, (2) of Chapter 16, in which we need also the results that

$$\frac{\partial^2 u}{\partial t^2} = C^2 U'' \qquad \frac{\partial^2 v}{\partial t^2} = C^2 V'' \qquad \frac{\partial^2 w}{\partial t^2} = C^2 W'' \tag{11}$$

Finally we arrive at

$$(\lambda + 2\mu - \rho C^2)U'' = 0 \tag{12a}$$

$$(\mu - \rho C^2)V'' = 0 \tag{12b}$$

$$(\mu - \rho C^2)W'' = 0 \tag{12c}$$

If these three equations are to hold simultaneously for a given value of C, there are only two physical possibilities:

$$\text{a.} \quad \lambda + 2\mu - \rho C^2 = 0 \qquad V = W = 0 \tag{13a}$$

$$\text{b.} \quad \mu - \rho C^2 = 0 \qquad U = 0 \tag{13b}$$

Mathematically we may have $U'' = 0$ without $U = 0$, but this implies infinite linear increase of amplitude as the wave progresses and may be rejected; similarly for V'' or W''.

In possibility a, the velocity C has the value for P waves; since v and w vanish, the displacement is in the direction of propagation and the wave is longitudinal. In possibility b, the velocity is that of S waves; u vanishes and the displacements are transverse. The procedure has been perfectly general (there is no loss of generality in taking the X axis in the direction of propagation so long as the medium is isotropic and homogeneous), so that no other types of plane wave can exist in the given medium.

ELASTIC CONSTANTS

The stress-strain equations (4) of Chapter 16 are given in terms of the theoretically convenient Lamé parameters, λ and μ. Of these only μ, the coefficient of rigidity, is directly related to experiment. Other elastic constants may be expressed in terms of these by applying the equations to their definitions.

Place the medium under uniform hydrostatic compression so that

$$Xx = Yy = Zz = P \quad \text{and} \quad Xy = Xz = Yz = 0$$

The equations then become

$$P = \lambda\theta + 2\mu e_{xx} \tag{14a}$$

$$P = \lambda\theta + 2\mu e_{yy} \tag{14b}$$

$$P = \lambda\theta + 2\mu e_{zz} \tag{14c}$$

Adding, and again applying the definition of θ, we find:

$$k = \frac{P}{\theta} = \lambda + \frac{2\mu}{3} \tag{15}$$

We define k as the bulk modulus, or modulus of incompressibility.

Next take a rod of square cross section extending parallel to the X axis and subjected only to pressure or tension applied perpendicular to the ends. Assuming equilibrium, the stresses and strains are uniform throughout the rod. We may put $Xx = T$ and set the other stress components equal to zero. Then

$$T = \lambda\theta + 2\mu e_{xx} = (\lambda + 2\mu)e_{xx} + \lambda(e_{yy} + e_{zz}) \tag{16a}$$

$$0 = \lambda\theta + 2\mu e_{yy} = (\lambda + 2\mu)e_{yy} + \lambda(e_{xx} + e_{zz}) \tag{16b}$$

$$0 = \lambda\theta + 2\mu e_{zz} = (\lambda + 2\mu)e_{zz} + \lambda(e_{xx} + e_{yy}) \tag{16c}$$

and

$$0 = e_{xy} = e_{xz} = e_{yz} \tag{16d}$$

Equations (16b) and (16c) together require that $e_{yy} = e_{zz}$. From either we then find Poisson's ratio σ:

$$\sigma = -\frac{e_{yy}}{e_{xx}} = \frac{\lambda}{2\lambda + 2\mu} \tag{17}$$

Subtracting (16b) from (16a),

$$T = 2\mu(e_{xx} - e_{yy}) \tag{18}$$

Defining Young's modulus E as T/e_{xx}, we arrive at

$$E = 2\mu(1 + \sigma) \tag{19}$$

NUMBER OF ELASTIC PARAMETERS

With 6 different components of stress and 6 of strain, the general linear relationship involves 36 coefficients. These coefficients will also enter into the expression for potential energy of the deformed medium. If one then makes the nearly inevitable assumption that this energy is a single-valued function of the state of strain, symmetries arise which reduce the number of elastic parameters to 21. This theoretical maximum applies to the least symmetrical crystals, those of the trihedral system.

Seismology chiefly uses the result for an isotropic medium, when the 36 or 21 elastic parameters reduce to 2. This reduction can be made very simple, except for one step. Since there is now no physical distinction between the directions labeled X, Y, Z, the coefficients in the equations must remain the same when these labels are interchanged, because of the symmetry. We may write

$$Xx = Ae_{xx} + B(e_{yy} + e_{zz}) + C(e_{xy} + e_{xz}) + De_{yz}$$
$$Yy = Ae_{yy} + B(e_{xx} + e_{zz}) + C(e_{xy} + e_{yz}) + De_{xz}$$
$$Zz = Ae_{zz} + B(e_{xx} + e_{yy}) + C(e_{xz} + e_{yz}) + De_{xy} \qquad (20)$$
$$Xy = E(e_{xx} + e_{yy}) + Fe_{zz} + Ge_{xy} + H(e_{xz} + e_{yz})$$
$$Xz = E(e_{xx} + e_{zz}) + Fe_{yy} + Ge_{xz} + H(e_{xy} + e_{yz})$$
$$Yz = E(e_{yy} + e_{zz}) + Fe_{xx} + Ge_{yz} + H(e_{xy} + e_{xz})$$

In this form we have only 8 possibly different coefficients. Suppose now that, without changing the physical condition of stress or strain, we reverse the direction of the positive axis of Z. This will reverse the sign of Xz and Yz (but not of Zz or the other stress components) and the sign of e_{xz} and e_{yz}, but not of the other components of strain. The first equation above then must continue to hold when A, B, C, D are unaltered but the signs of e_{xz} and e_{yz} are reversed; for this to be generally true we must have $C = D = 0$. Similarly, for the fourth equation to remain valid we require $H = 0$. In the fifth equation X_z, e_{xz}, and e_{yz} change sign; reversing the sign of the whole equation and comparing it with the original form, we find that necessarily $E = F = 0$. The equations thus reduce to

$$Xx = Ae_{xx} + B(e_{yy} + e_{zz}) \qquad (21a)$$
$$Yy = Ae_{yy} + B(e_{xx} + e_{zz}) \qquad (21b)$$
$$Zz = Ae_{zz} + B(e_{xx} + e_{yy}) \qquad (21c)$$
$$Xy = Ge_{xy} \qquad Xz = Ge_{xz} \qquad Yz = Ge_{yz} \qquad (21d)$$

We have now reduced from 36 elastic coefficients to 3. This is the correct result for a cubic crystal, if we have taken X, Y, Z parallel to the three crystallographic axes, since in that case the three are equivalent, which is all we have assumed thus far. If we are dealing with an isotropic medium, our equations must hold no matter what our choice of axes; in other words, they must be unaffected by any rotation.

Keeping the axis of Z fixed, rotate the other axes by 45° so that the axis of X' falls between those of X and Y. Introducing the various cosines of the angles between the two sets of axes, equation (1a) of Chapter 16 becomes

$$X'x' = \tfrac{1}{2}(Xx + Yy) + Xy \qquad (22)$$

The transformations of the strain components are similar to those of the stress components, except that we have to write $2e_{xx}$ where we had X_x, etc. In the present case three of the resulting equations are:

$$2e_{x'x'} = e_{xx} + e_{yy} + e_{xy} \qquad (23a)$$
$$2e_{y'y'} = e_{xx} + e_{yy} - e_{xy} \qquad (23b)$$
$$e_{z'z'} = e_{zz} \qquad (23c)$$

We require that

$$X'x' = Ae_{x'x'} + B(e_{y'y'} + e_{z'z'}) \qquad (24)$$

Substituting the expressions of the new stress and strain components in terms of the old, and multiplying by 2,

$$Xx + Yy + 2Xy = (A + B)(e_{xx} + e_{yy}) + (A - B)e_{xy} + 2Be_{zz} \qquad (25)$$

Subtracting the expressions (21b), (21d) for Yy and Xy in terms of the strain components,

$$Xx = Ae_{xx} + B(e_{yy} + e_{zz}) + (A - B - 2G)e_{xy} \qquad (26)$$

This will be fulfilled if and only if $A - B - 2G = 0$, which gives the required reduction from 3 elastic parameters to 2. If we now put $B = \lambda$ and $G = \mu$, we derive the stress-strain equations in the exact form given as equation (4) of Chapter 16.

GROUP VELOCITY

The following form of the derivation of group velocity is due to Sommerfeld.[†]
Express a single wave in the form

$$u = A \exp i(kx - \omega t) \qquad (27)$$

so that

$$k = \frac{\omega}{C} = \frac{2\pi}{CT} = \frac{2\pi}{L} \qquad (28)$$

A wave group U including waves from $k_0 - \epsilon$ to $k_0 + \epsilon$ takes the form

$$U = \int_{k_0 - \epsilon}^{k_0 + \epsilon} A(k) \exp i [kx - \omega(k)t] \, dk \qquad (29)$$

Write

$$kx - \omega t = k_0 x - \omega_0 t + (k - k_0)x - (\omega - \omega_0)t \qquad (30)$$

and define

$$Q = \int_{k_0 - \epsilon}^{k_0 + \epsilon} A(k) \exp i [(k - k_0)x - (\omega - \omega_0)t] \, dk \qquad (31)$$

Then

$$U = Q \exp i(k_0 t - \omega_0 t) \qquad (32)$$

Now Q is constant if

$$(k - k_0)x - (\omega - \omega_0)t = \text{constant} \qquad (33)$$

Hence the amplitude Q appears to travel with a velocity b, and in the limit

$$b = \frac{d\omega}{dk} \qquad (34)$$

[†] A. Sommerfeld, *Atombau und Spektrallinien, Wellenmechanische Ergänzungsband,* Vieweg, Braunschweig, 1929, x + 352 pp. (The derivation appears on p. 47.)

But, since $\omega = Ck = 2\pi C/L$, we have

$$b = \frac{d\omega}{dk} = C + k\frac{dC}{dk} = C - L\frac{dC}{dL} \tag{35}$$

These are the usual formulas for the group velocity b in terms of the phase velocity C and its derivatives.

Table of Angles of Incidence

Angle of Incidence and Related Quantities for Shallow Shocks, Assuming Surface Velocity 6.34 km/sec $= v_0$. (As calculated by Gutenberg.)

$\theta°$	$1/\bar{v}$ (sec/deg)	$1/\bar{v}$ (sec/km)	$\bar{v}$ (km/sec)	$\sin i_0 = v_0/\bar{v}$	i^0	$di/d\theta$	$di/d\theta$ smoothed
20	11.00	0.0990	10.10	0.628	38.9		2.1
						2.05	
22	10.00	.0900	11.11	.571	34.8		0.70
						0.70	
24	9.65	.0868	11.52	.550	33.4		
						0.65	
26	9.20	.0838	11.93	.531	32.1		
						0.50	
28	9.05	815	12.26	.517	31.1		
						0.30	
30	8.90	801	12.48	.508	30.5		0.33
						0.30	
32	8.73	786	12.72	.498	29.9		
						0.30	
34	8.57	771	12.97	.489	29.3		
						0.25	
36	8.44	760	13.16	.482	28.8		0.19
						0.20	
38	8.32	749	13.35	.475	28.4		
						0.20	
40	8.23	741	13.50	.470	28.0		
						0.10	
42	8.18	736	13.59	.467	27.8		0.10
44	8.00	720	13.89	.456	27.1		0.80
						0.55	
46	7.80	692	14.45	.439	26.0		

Angle of Incidence and Related Quantities for Shallow Shocks, Assuming Surface Velocity 6.34 km/sec $= v_0$—*Continued*

$\theta°$	$1/\bar{v}$ sec/deg)	$1/\bar{v}$ (sec/km)	$\bar{v}$ (km/sec)	sin $i_0 =$ $v_0/\bar{v}$	i^0	di/di	$di/d\theta$ smoothed
						0.35	
48	7.50	675	14.81	.428	25.3		0.30
						0.25	
50	7.35	662	15.11	.420	24.8		
						0.20	
52	7.23	651	15.36	.413	24.4		
						0.15	
54	7.15	644	15.53	.408	24.1		0.15
						0.15	
56	7.07	636	15.72	.403	23.8		
						0.10	
58	7.02	632	15.82	.401	23.6		
						0.10	
60	6.95	626	15.97	.397	23.4		
						0.30	
62	6.80	612	16.34	.388	22.8		0.25
						0.25	
64	6.65	599	16.69	.380	22.3		
						0.25	
66	6.50	585	17.09	.371	21.8		
						0.25	
68	6.35	572	17.48	.363	21.3		
						0.30	
70	6.20	558	17.92	.354	20.7		0.25
						0.25	
72	6.05	545	18.35	.346	20.2		
						0.25	
74	5.90	531	18.38	.337	19.7		
						0.20	
76	5.78	520	19.23	.330	19.3		
						0.25	
78	5.65	509	19.65	.323	18.8		
						0.25	
80	5.50	495	20.20	.314	18.3		
						0.35	
82	5.29	476	21.01	.302	17.6		
						0.38	
84	5.08	457	21.88	.2897	16.84		0.36

Angle of Incidence and Related Quantities for Shallow Shocks, Assuming Surface Velocity 6.34 km/sec $= v_0$—*Continued*

$\theta°$	$1/\bar{v}$ (sec/deg)	$1/\bar{v}$ (sec/km)	$\bar{v}$ (km/sec)	$\sin i_0 = v_0/\bar{v}$	i^0	$di/d\theta$	$di/d\theta$ smoothed
						0.36	
86	4.87	438	22.83	.2777	16.12		
						0.32	
88	4.68	421	23.75	.2669	15.48		
						0.23	
90	4.60	414	24.15	.2625	15.22		0.18
						0.10	
92	4.54	409	24.45	.2593	15.03		
						0.06	
94	4.51	406	24.63	.2574	14.92		
						0.02	
96	4.50	405	24.69	.2568	14.88		0.02
						0.02	
98	4.49	404	24.75	.2561	14.84		
						0.02	
100	4.48	403	24.80	.2555	14.80		
						0.01	
102	4.47	402	24.85	.2550	14.77		0.01
						0.005	
104	4.47	402	24.83	.2547	14.76		

APPENDIX VI

Calculation of Velocities within the Earth

THE MATHEMATICAL PROCESS by which arrival times of seismic waves at the surface of the earth are used to compute velocities in its interior is not simple, and it is full of traps for the unwary. Fortunately, it leads to a procedure which is clear and easy to specify, though laborious in practice. What follows here is meant only to establish notation and lay down the outline of the method. Students thoroughly prepared in mathematics will do best to approach the matter from the standpoint of the theory of integral equations, particularly that first solved by Abel.

For simplification, the source of seismic rays is taken to be at the surface of the earth; it is relatively easy to correct for hypocentral depth. The earth and its properties are taken as spherically symmetrical, so that the required velocity V is a function of the radius vector r only. If i is the angle of incidence, then (Chapter 17)

$$\frac{r \sin i}{V} = p$$

where p is constant along any given ray.

We limit ourselves first to the special case in which velocity increases continuously with depth (dV/dr is negative). The rays are then convex downward, and each has a deepest level at $r = r_m$, with corresponding $V = V_m$. Values of all quantities at the surface are specified by a zero subscript. For a given ray,

$$p = \frac{r_0 \sin i_0}{V_0} = \frac{r_m}{V_m}$$

Here r_0 is the known radius of the earth. V_0 is also presumed to be known from the study of local earthquakes and artificial explosions (but this sometimes raises difficulties; see Chapter 18). A little geometry will show that

$$\frac{dt_0}{d\theta_0} = \frac{r_0 \sin i_0}{V_0} = p$$

where t represents the observed arrival times and θ is the central angle.

The ray parameter p thus being known, we have the ratio r_m/V_m. We need one more relation between these quantities to determine them individually, giving V_m

667

as a function of r_m, equivalent to V as a function of r. This is provided by integration along an individual ray (p constant); the resulting integral equation is then solved by integration which passes from ray to ray (p variable). This solution is a special case of a theorem due to Abel which involves a parameter usually denoted by n; we need only $n = \frac{1}{2}$. If

$$f(h) = K \int_0^h U'(y)(h - y)^{-1/2} \, dy \tag{A}$$

then

$$K\pi U(y) = \int_0^y f(h)(y - h)^{-1/2} \, dh \tag{B}$$

Here $U'(y)$ is the derivative of $U(y)$ with respect to y, and K is any constant.

To apply this we note that $\tan i = r \, (d\theta/dr)$, express $\tan i$ in terms of $\sin i$, and replace $\sin i$ by pV/r. This yields

$$\frac{d\theta}{dr} = \frac{pV}{r} (r^2 - p^2V^2)^{-1/2}$$

We integrate along a complete ray, taking note of its symmetry, so that

$$\theta = \int_{r_m}^{r_0} \frac{2p}{r} \left(\frac{r^2}{V^2} - p^2 \right)^{-1/2} dr$$

Next we introduce a new variable $x = r^2/V^2$. In writing the limits of integration, note that

$$x_m = \frac{r_m^2}{V_m^2} = p^2 \quad \text{and} \quad x_0 = \frac{r_0^2}{V_0^2}$$

then

$$\theta = \int_{p^2}^{x_0} 2p(x - p^2)^{-1/2} \left(\frac{d \log r}{dx} \right) dx$$

Now substitute $y = x - x_0$, $h = p^2 - x_0$, and reverse the order of integration. The result is

$$\theta = 2p \int_0^h (y - h)^{-1/2} \left(\frac{d \log r}{dy} \right) dy$$

Comparing this with (A), we write down the solution corresponding to (B), which is

$$\log r + C = \int_0^y \frac{\theta \, dh (h - y)^{-1/2}}{2p\pi}$$

$$= \int_{x_0}^x \frac{\theta \, dp^2 (p^2 - x)^{-1/2}}{2p\pi}$$

$$= \int_{r_0/V_0}^p \frac{\theta \, dp_1 (p_1^2 - x)^{-1/2}}{\pi}$$

The subscript of p_1 distinguishes the variable of integration from the limit p. To fix the constant of integration note that, when $p = r_0/V_0$, $r = r_0$, which means that $C = -\log r_0$, or $\log r + C = \log (r/r_0)$.

A substitution $p = x^{1/2} \cosh q$ reduces the integrand to $\theta \, dq/\pi$. This may be integrated by parts; the integrated term vanishes, since $\theta q = 0$ at both limits, and the ultimate result is

$$\pi \log \frac{r_0}{r} = \int_0^\theta q(\theta_1) \, d\theta_1$$

The value of r computed from this is actually r_m, corresponding to θ, which is the end point of integration. The value of $q(\theta_1)$ for smaller θ_1 follows from the definition of q:

$$\cosh q = p_1 x^{-1/2} = \frac{V_0 \, (dt/d\theta)}{r_0}$$

where $dt/d\theta$ is determined from the observed arrival times for $\theta = \theta_1$.

q may be plotted as a function of θ_1 on squared paper, and the integral evaluated by counting squares, mechanically, or with a desk calculating machine. One such integration is required for each value of $r = r_m$; to derive enough points for a detailed velocity-depth function becomes very laborious.

Slichter[†] and others have shown how to proceed in some important special cases when the velocity-depth function is discontinuous, or when velocity decreases with depth.

Abel's integral equation is discussed in many mathematical treatises; its application to the seismological problem (due to Wiechert, Herglotz, and Bateman) is to be found in most of the general handbooks on seismology cited as references to Chapter 1. The only readily accessible treatment giving detail for both is that by Macelwane (*Theoretical Seismology*). In their monographs and manuals Jeffreys and Bullen give a treatment of the seismological problem originally due to Rasch, which is shorter in statement and avoids the use of the properties of beta and gamma functions involved in the Abel solution.

[†] L. B. Slichter, "Theory of the interpretation of seismic travel-time curves in horizontal layers," *Physics*, vol. 3 (1932), pp. 273-295. H. Witte, "Beiträge zur Berechnung der Geschwindigkeit der Raumwellen im Erdinnern," *Göttinger Nachrichten*, 1932, pp. 199-241.

APPENDIX VII

Refraction and Reflection—the Zoeppritz Equations

FOR MANY APPLICATIONS it is necessary to calculate the theoretically expected ratios of the amplitudes of refracted and reflected elastic waves to those of incident waves at a plane interface between two elastic media. Equations for this purpose were developed by Zoeppritz in 1907; partly because of his death in 1908, they were not published until 1919. They have since been verified, used, and reprinted by many authors. Logical notation designed to indicate the meaning of the various quantities results in a complication of subscripts or different alphabets; this makes proofreading awkward, and has led to misprints in publication. The notation which here follows is easier to print, but will require a little more attention to decipher. The letters are assigned for this occasion only and should not be confused with any other notation in this book.

We work throughout with plane waves in the vicinity of a plane interface between two homogeneous media, which in general differ in both elastic constants and in density. There are then 6 possible waves, corresponding to rays at different angles with the normal to the interface:

	Velocity	Amplitude	Angle with Normal
Incident *P*	*U*	*A*	*a*
Incident *S*	*V*	*B*	*b*
Reflected *P*	*U*	*C*	*c*
Reflected *S*	*V*	*D*	*d*
Refracted *P*	*Y*	*E*	*e*
Refracted *S*	*Z*	*F*	*f*

The velocities will involve the elastic constants and densities of the two media. We also need the quantity K, which is the ratio of the density of the second medium to that of the first (in which the rays are incident).

All the angles and velocities conform to Snell's law:

$$\frac{\sin a}{U} = \frac{\sin b}{V} = \frac{\sin c}{U} = \frac{\sin d}{V} = \frac{\sin e}{Y} = \frac{\sin f}{Z} \tag{1}$$

The theory assumes that all 6 wave fronts represent pure sine waves of the same frequency n; this chosen frequency drops out of the equations. From the components of displacement in this form the stresses represented by each wave can be calculated [equations (4) of Chapter 16].

Two physical conditions are applied:

(1) The vector sum of the displacements due to all the waves along the boundary in one medium must equal the corresponding sum in the other medium; otherwise there would be relative slip.

(2) The corresponding components of stress must be equal on the two sides of the interface.

In most applications there is little interest in having both P and S waves incident together; moreover the equations immediately fall apart into the separate cases.

For an incident P wave,

$$(A - C) \sin a + D \cos b - E \sin e + F \cos f = 0$$

$$(A + C) \cos a + D \sin b - E \cos e - F \sin f = 0$$

$$-(A + C) \sin 2a + D \frac{U}{V} \cos 2b + EK \left(\frac{Z}{V}\right)^2 \frac{U}{Y} \sin 2e$$

$$- FK \left(\frac{Z}{V}\right)^2 \left(\frac{U}{Z}\right) \cos 2f = 0$$

$$-(A - C) \cos 2b + D \left(\frac{V}{U}\right) \sin 2b + EK \left(\frac{Y}{U}\right) \cos 2f$$

$$+ FK \left(\frac{Z}{U}\right) \sin 2f = 0$$

The case of an incident S wave separates automatically into the consideration of two possible polarizations. If, as in most seismological work, the interface is horizontal, these polarizations are in vertical and horizontal planes and are known respectively as SV and SH.

For SV:

$$(B + D) \sin b + C \cos a - E \cos e - F \sin f = 0$$

$$(B - D) \cos b + C \sin a + E \sin e - F \cos f = 0$$

$$(B + D) \cos 2b - C \left(\frac{V}{U}\right) \sin 2a + EK \left(\frac{Z^2}{VY}\right) \sin 2e$$

$$- FK \left(\frac{Z}{V}\right) \cos 2f = 0$$

$$-(B - D) \sin 2b + C \left(\frac{U}{V}\right) \cos 2b + EK \left(\frac{Y}{V}\right) \cos 2f$$

$$+ FK \left(\frac{Z}{V}\right) \sin 2f = 0$$

For SH many terms drop out, and the equations reduce to two:

$$B + D - F = 0$$

$$B - D - K \left(\frac{Z}{V}\right) \left(\frac{\cos f}{\cos b}\right) F = 0$$

There is an arbitrariness in the choice of positive directions which affects the signs of the amplitudes for transverse waves, particularly of C; this will account for some differences in the equations as reproduced by various authors.

The equations are solved algebraically for the ratio of the amplitudes of the derived waves to the incident amplitude A or B. As noted in Chapter 17, Snell's law (equation 1) may give values for sines and cosines of c, d, e, f which are greater than unity and correspond to imaginary angles; the corresponding cosines become imaginary. It is then necessary to introduce real and imaginary parts for the amplitudes. Each equation splits into two; one of the resulting systems can usually be solved quickly, and the labor lies chiefly in reducing the other. The physical amplitude corresponding to A is the square root of the sum of the squares of the real and imaginary parts.

Minor misprints in the equations have occurred in a number of standard reference works, among them those cited below. None, fortunately, appeared in Zoeppritz' original publication. Macelwane (1933) has no misprints. Macelwane (1936) gives the third equation for SV with a factor in the coefficient of E misprinted as Z/V (in our notation) instead of Y/V. Gutenberg (1932) misprints the third and fourth equations (for combined incident P and S) with f in the last terms instead of $2f$. Byerly (1942) gives our third equation for longitudinal waves with an additional factor K.

References

Zoeppritz' equations

Zoeppritz, K., "Über Erdbebenwellen VIIb," *Göttinger Nachrichten,* 1919, pp. 66–84.

Macelwane, J. B. (1933), *Nat. Research Council Bull. 90, Physics of the Earth, VI; Seismology,* pp. 116–120.

—————— (1936), *Introduction to Theoretical Seismology,* Part I, Chapter VII.

Byerly, P. (1942), *Seismology,* pp. 152–178.

Gutenberg, B. (1932), *Handbuch der Geophysik,* Vol. 4, pp. 42–57.

Knott's equations

Knott, C. G., "Reflection and refraction of elastic waves with seismological applications," *Phil. Mag.,* ser. 5, vol. 48 (1899), pp. 64–97. (Also given by Macelwane and in various handbooks.)

Discussion

Gutenberg, B., "Energy ratio of reflected and refracted seismic waves," *B.S.S.A.,* vol. 34 (1944), pp. 85–102. Abbreviated notation, references to the literature, numerous special results. Misprints: in equation (7b), for $M = \sqrt{\cos \eta/\cos \alpha}$, read $M = \sqrt{\cot \eta / \cot \alpha}$. In (13g), the numerator and denominator of f/e should begin with j instead of 1.

Slichter, L. B., and Gabriel, V. G., "Studies in reflected seismic waves, Part 1," *G. Beitr.*, vol. 38 (1933), pp. 228–238.

Slichter, L. B., "Studies in reflected seismic waves, Part II," *ibid.*, pp. 239–256. (This paper and the preceding one contain many computed results and much theoretical detail.)

Petrashen', G. I. (ed.), *Materiali kolichestvennogo izucheniya dinamiki seysmicheskikh voln* (Materials for quantitative study of the dynamics of seismic waves), Izdat. Leningrad. Univ., 2 vols., 1957. (Large tables. Coefficients of reflection and refraction, with much other numerical data and discussion.)

APPENDIX VIII

Transit-Time Tables, Magnitude Charts, etc.

TELESEISMS

Tables VIII-1 through VIII-10 (and Figs. VIII-1, VIII-2, and VIII-3) show representative data on transit times for shallow and deep teleseisms. More complete data will be found in the papers and handbooks referred to in Chapters 17, 18, 19. Times applying to intermediate and deep shocks are taken from Gutenberg and

Table VIII-1   Transit Times of *P* (min:sec)

$\Delta°$	Depth of Focus (km)							
	25	100	200	300	400	500	600	700
0	0:04	0:14	0:27	0:39	0:50	1:01	1:11	1:20
2	0:32	0:30	0:38	0:46	0:57	1:06	1:15	1:24
4	0:59	0:56	1:03	1:08	1:12	1:18	1:25	1:32
6	1:27	1:25	1:28	1:30	1:30	1:34	1:40	1:46
8	1:57	1:54	1:54	1:54	1:53	1:54	1:59	2:03
10	2:28	2:22	2:20	2:18	2:16	2:16	2:18	2:20
12	2:56	2:50	2:46	2:42	2:39	2:38	2:38	2:38
14	3:22	3:17	3:12	3:07	3:02	3:00	2:59	2:57
16	3:47	3:43	3:37	3:31	3:25	3:22	3:19	3:17
18	4:11	4:07	4:00	3:54	3:47	3:43	3:39	3:36
20	4:34	4:30	4:22	4:15	4:08	4:03	3:58	3:55
22	4:55	4:51	4:43	4:34	4:27	4:22	4:16	4:12
24	5:16	5:10	5:02	4:53	4:45	4:39	4:34	4:30
26	5:35	5:29	5:20	5:11	5:03	4:56	4:51	4:47
28	5:53	5:48	5:38	5:28	5:20	5:13	5:08	5:04
30	6:11	6:06	5:56	5:46	5:37	5:30	5:25	5:21

Table VIII-1 Continued

Δ°	Depth of Focus (km)							
	25	100	200	300	400	500	600	700
32	6:29	6:23	6:13	6:03	5:54	5:47	5:42	5:38
34	6:46	6:40	6:30	6:20	6:11	6:04	5:59	5:55
36	7:03	6:56	6:46	6:37	6:28	6:21	6:17	6:12
38	7:19	7:13	7:03	6:54	6:45	6:39	6:34	6:28
40	7:36	7:30	7:20	7:11	7:02	6:56	6:50	6:44
42	7:52	7:47	7:36	7:28	7.19	7:12	7:05	6:59
44	8:09	8:03	7:52	7:44	7:35	7:27	7:20	7:14
46	8:25	8:18	8:08	7:59	7:50	7:42	7:35	7:29
48	8:40	8:33	8:23	8:14	8:05	7:57	7:50	7:43
50	8:55	8:47	8:38	8:28	8:20	8.12	8:05	7:58
52	9:10	9:02	8:53	8:43	8:35	8:26	8:19	8:12
54	9:24	9:17	9:08	8:57	8:49	8:40	8:33	8:27
56	9:39	9:31	9:22	9:12	9:04	8:55	8:47	8:41
58	9:52	9:45	9:35	9:26	9:18	9:09	9:01	8:55
60	10:06	10:00	9:49	9:40	9:31	9:22	9:15	9:08
62	10:20	10:14	10:03	9:53	9:44	9:35	9:28	9:21
64	10:34	10:28	10:16	10:05	9:57	9:48	9:40	9:34
66	10:47	10:41	10:29	10:18	10:09	10:01	9:53	9:47
68	11:00	10:54	10:42	10:31	10:22	10:13	10:06	9:59
70	11:12	11:07	10:55	10:44	10:35	10:26	10:18	10:11
72	11:24	11:19	11:07	10:55	10:46	10:37	10:29	10:22
74	11:36	11:30	11:18	11:06	10:57	10:48	10:40	10:33
76	11:48	11:41	11:29	11:17	11:08	10:59	10:51	10:44
78	12:00	11:52	11:40	11:29	11:19	11:10	11:02	10:55
80	12:11	12:03	11:51	11:40	11:30	11:21	11:13	11:05
82	12:22	12:14	12:02	11:51	11:40	11:31	11:23	11:15
84	12:32	12:24	12:12	12:01	11:50	11:41	11:33	11:25
86	12:42	12:34	12:22	12:11	12:01	11:51	11:42	11:34
88	12:51	12:44	12:31	12:21	12:11	12:01	11:52	11:43
90	13:00	12:53	12:41	12:30	12:20	12:10	12:01	11:52
92	13:10	13:02	12:50	12:39	12:29	12:19	12:10	12:01
94	13:19	13:11	12:59	12:48	12:38	12:28	12:19	12:10
96	13:28	13:20	13:08	12:57	12:47	12:37	12:28	12:19

Table VIII-1 Continued

$\Delta°$	Depth of Focus (km)							
	25	100	200	300	400	500	600	700
98	13:37	13:28	13:17	13:06	12:56	12:46	12:37	12:28
100	13:46	13:37	13:25	13:14	13:04	12:55	12.46	12:37
102	13:55	13:46	13:34	13:23	13:13	13:04	12:55	12:46
104	14:06	13:55	13:43	13:32	13:22	13:13	13:04	12:55
106	14:15	14:04	13:52	13:41	13:31	13:22	13:13	13:04
108	14:24	14:13	14:01	13:50	13:40	13:30	13:21	13:13
110	14:33	14:22	14:10	13:59	13:49	13:39	13:30	13:22
115	14:52	14:44	14:32	14:21	14:11	14:01	13:52	13:43
120	15:14	15:06	14:54	14:43	14.33	14:23	14:14	14:05
125	15:36	15:28	15:16	15:05	14:55	14:45	14:36	14:27
130	15:59	15:51	15:39	15.28	15:18	15:08	14:59	14:50
140	16:44	16:36	16:24	16:13	16:03	15:53	15:44	15:35
150	17:29	17:21	17:09	16:58	16:48	16:38	16:29	16:20

Table VIII-2 Transit Times of S (min:sec)

$\Delta°$	Depth of Focus (km)							
	25	100	200	300	400	500	600	700
0	0:08	0:24	0:46	1:08	1:29	1:49	2:07	2:24
2	0:55	1:00	1:06	1:24	1:41	1:59	2:15	2:30
4	1:47	1:49	1:49	1:51	2:05	2:20	2:34	2:48
6	2:38	2:38	2:37	2:34	2:40	2:47	2:58	3:09
8	3:28	3:27	3:24	3:18	3:19	3:23	3:30	3:38
10	4:17	4:16	4:12	4:03	4:00	4:01	4:06	4:12
12	5:08	5:04	4:56	4:47	4:42	4:39	4:42	4:46
14	5:56	5:49	5:40	5:31	5:23	5:18	5:18	5:20
16	6:42	6:35	6:25	6:15	6:05	5:57	5:54	5:54
18	7:28	7:21	7:09	6:57	6:46	6:36	6:30	6:28
20	8:12	8:04	7:52	7:39	7:26	7:15	7:06	7:02
22	8:51	8:44	8:32	8:19	8:06	7:54	7:42	7:36
24	9:27	9:21	9:05	8:51	8:37	8:26	8:18	8:09

Table VIII-2 Continued

	Depth of Focus (km)							
Δ°	25	100	200	300	400	500	600	700
26	10:01	9:54	9:37	9:22	9:07	8:56	8:47	8:39
28	10:33	10:27	10:09	9:52	9:37	9:26	9:16	9:08
30	11:05	10:57	10:39	10:23	10:08	9:56	9:46	9:37
32	11:36	11:28	11:10	10:53	10:38	10:26	10:15	10:06
34	12:07	11:58	11:40	11:23	11:08	10:56	10:44	10:35
36	12:38	12:28	12:10	11:53	11:38	11:25	11:13	11:03
38	13:09	12:58	12:40	12:23	12:08	11:54	11:42	11:32
40	13:39	13:27	13:09	12:52	12:38	12:23	12:11	12:01
42	14:09	13:56	13:37	13:21	13:07	12:52	12:39	12:30
44	14:39	14:24	14:06	13:50	13:35	13:20	13:07	12:58
46	15:08	14:51	14:34	14:18	14:03	13:48	13:35	13:25
48	15:35	15:19	15:01	14:45	14:30	14:16	14:02	13:51
50	16:03	15:46	15:28	15:12	14:57	14:43	14:29	14:17
52	16:32	16:14	15:55	15:38	15:23	15:10	14:57	14:44
54	16:59	16:41	16:22	16:05	15:50	15:37	15:24	15:11
56	17:26	17:08	16:48	16:32	16:16	16:03	15:49	15:37
58	17:53	17:35	17:15	16:58	16:42	16:28	16:14	16:02
60	18:19	18:02	17:42	17:24	17:07	16:52	16:38	16:26
62	18:45	18:28	18:08	17:49	17:31	17:16	17:02	16:50
64	19:09	18:52	18:32	18:12	17:55	17:40	17:25	17:13
66	19:34	19:15	18:55	18:35	18:18	18:03	17:48	17:35
68	19:58	19:38	19:18	18:58	18:40	18:25	18:10	17:57
70	20:21	20:01	19:40	19:21	19:03	18:48	18:33	18:20
72	20:44	20:23	20:03	19:43	19:26	19:10	18:56	18:43
74	21:06	20:45	20:24	20:05	19:48	19:33	19:19	19:06
76	21:28	21:07	20:46	20:27	20:10	19:55	19:41	19:28
78	21:50	21:29	21:08	20:49	20:32	20:17	20:02	19:49
80	22:11	21:51	21:30	21:11	20:54	20:39	20:23	20:10
82	22:31	22:13	21:52	21:33	21:16	21:01	20:46	20:31
84	22:52	22:35	22:14	21:55	21:38	21:23	21:08	20:53
86	23:11	22:56	22:35	22:16	21:59	21:43	21:27	21:13
88	23:30	23:16	22:55	22:36	22:18	22:02	21:44	21:30

Table VIII-2 Continued

	Depth of Focus (km)							
Δ°	25	100	200	300	400	500	600	700
90	23:49	23:35	23:14	22:55	22:37	22:20	22:03	21:48
92	24:08	23:53	23:32	23:12	22:55	22:37	22:20	22:05
94	24:24	24:11	23:50	23:31	23:12	22:54	22:37	22:23
96	24:41	24:28	24:08	23:48	23:30	23:12	22:55	22:40
98	24:58	24:46	24:25	24:05	23:47	23:29	23:12	22:57
100	25:14	25:03	24:42	24:23	24:04	23:46	23:29	23:14
105	25:57	25:45	25:24	25:04	24:46	24:28	24:11	23:56
110	26:37	26:26	26:05	25:45	25:27	25:09	24:52	24:37
115	27:19	27:08	26:47	26:27	26:09	25:51	25:34	25:19
120	28:01	27:50	27:29	27:09	26:51	26:33	26:16	26:01

Table VIII-3 Intervals *S-P* (min:sec), with Corresponding
Transit Times *P-O* and Distances Δ

	For Depth 25 km		For Depth 600 km	
S-P	P-O	Δ°	P-O	Δ°
1:00	1:13	5.0	1:18	2.7
1:30	1:56	7.9	1:56	7.6
2:00	2:42	10.9	2:34	11.6
2:30	3:18	13.7	3:13	15.3
3:00	3:52	16.4	3:49	19.1
3:30	4:26	19.3	4:21	22.5
4:00	5:00	22.5	4:58	26.8
4:30	5:40	26.6	5:38	31.5
5:00	6:21	31.1	6:24	36.9
5:30	6:58	35.4	7:00	41.4
6:00	7:33	39.6	7:35	46.0
6:30	8:09	44.0	8:12	51.1
7:00	8:45	48.7	8:46	55.7
7:30	9:18	53.1	9:24	61.3
8:00	9:51	57.9	9:58	66.8
8:30	10:29	63.3	10:32	72.5
9:00	11:02	68.3	11:03	78.2
9:30	11:36	74.0	11:31	83.7
10:00	12:11	80.0	11:59	89.5
10:30	12:43	86.2	12:32	96.9
11:00	13:12	92.5	13:05	104.4
11:30	13:48	100.5	13:38	111.9
12:00	14:23	108.2	14:11	119.2

Table VIII-4   Transit Times of P' (min:sec)

Δ°	Depth (km)							
	25	100	200	300	400	500	600	700
110	18:33	18:23	18:10	17:58	17:48	17:38	17:29	17:21
115	18:44	18:35	18:22	18:10	17:59	17:49	17:40	17:32
120	18:54	18:43	18:30	18:19	18:08	17:57	17:47	17:38
125	19:04	18:54	18:42	18:30	18:19	18:09	17:59	17:50
130	19:14	19:04	18:52	18:40	18:29	18:19	18:09	18:00
135	19:23	19:11	18:59	18:47	18:36	18:26	18:16	18:06
140	19:30[1]	19:17	19:04	18:52	18:41	18:31	18:21	18:11
142	19:33[1]	19:19	19:07	18:55	18:45	18:35	18:26	18:17
145	19:39	19:27	19:15	19:03	18:52	18:42	18:33	18:24
150	19:46	19:36	19:23	19:11	19:01	18:50	18:41	18:31
155	19:55	19:45	19:32	19:21	19:10	18:59	18:49	18:40
160	20:01	19:51	19:38	19:26	19:15	19:04	18:54	18:45
170	20:09	19:57	19:44	19:32	19:21	19:11	19:02	18:52
180	20:11	20:00	19:47	19:35	19:24	19:13	19:04	18:54

[1] Motion complicated and initial instant uncertain.

Table VIII-4A   Transit Times of P_2' (min:sec)

Δ°	Depth (km)							
	25	100	200	300	400	500	600	700
150	20:08	19:57	19:45	19:34	19:24	19:15	19:06	18:58
160	20:55	20:46	20:34	20:23	20:13	20:04	19:55	19:47
170	21:43[1]	21:34	21:22	21:11	21:01	20:52	20:43	20:35
180	22:31[1]	22:22	22:10	21:59	21:49	21:40	21:31	21:23
190	23:19[1]	23:10	22:58	22:47	22:37	22:28	22:19	22:11

[1] Extrapolated times. Observations often about 15 seconds earlier.

Table VIII-5   Transit Times of *SKS* (min:sec)

Δ°	Depth (km)							
	25	100	200	300	400	500	600	700
80	22:20	22:05	21:43	21:24	21:06	20:49	20:34	20:20
85	22:54	22:38	22:16	21:57	21:38	21:22	21:06	20:52
90	23:24	23:08	22:46	22:26	22:08	21:51	21:36	21:22
95	23:53	23:37	23:16	22:56	22:37	22:20	22:05	21:50
100	24:19	24:02	23:41	23:22	23:02	22:45	22:29	22:14
105	24:43	24:25	24:04	23:44	23:25	23:08	22:51	22:35
110	25:05	24:49	24:27	24:07	23:48	23:30	23:13	22:57
115	25:27	25:10	24:48	24:28	24:09	23:50	23:33	23:17
120	25:47	25:29	25:08	24:48	24:28	24:09	23:52	23:36
125	26:02	25:44	25:23	25:03	24:43	24:25	24:08	23:51
130	26:13	25:56	25:35	25:15	24:55	24:37	24:19	24:02
135	26:25	26:07	25:46	25:26	25:06	24:47	24:29	24:12
140	26:36	26:17	25:55	25:35	25:15	24:56	24:38	24:21
145	26:45	26:26	26:03	25:42	25:23	25:04	24:45	24:28
150	26:50	26:31	26:09	25:48	25:28	25:09	24:51	24:33

Table VIII-6   Transit Times of *ScS* (min:sec)

Δ°	Depth (km)							
	25	100	200	300	400	500	600	700
0	15:30	15:14	14:51	14:30	14:09	13:49	13:32	13:14
10	15:39	15:23	15:00	14:39	14:19	13:59	13:42	13:25
20	16:04	15:46	15:25	15:03	14:44	14:25	14:08	13:51
30	16:45	16:25	16:03	15:43	15:24	15:06	14:49	14:33
40	17:39	17:20	17:00	16:40	16:20	16:02	15:45	15:29
50	18:42	18:20	18:00	17:39	17:21	17:04	16:47	16:31
60	19:53	19:26	19:05	18:44	18:26	18:08	17:50	17:35
70	21:09	20:41	20:21	20:01	19:43	19:26	19:08	18:53
80	22:30	22:03	21:43	21:23	21:04	20:45	20:28	20:12

Table VIII-7   *PcP-P* (min:sec)

Δ°	Depth (km)		Δ°	Depth (km)	
	25	600		25	600
30	2:57	2:40	60	0:44	0:38
35	2:25	2:12	70	0:20	0:18
40	1:56	1:44	80	0:06	0:05
45	1:35	1:26	90	0:01	0:01
50	1:16	1:08			

Table VIII-8 *P'P'-P* and *SKPP'-P* (min:sec)

Δ°	P'P'-P (min:sec) Depth (km)		SKPP'-P (min:sec) Depth (km)	
	25	600	200	600
40	32:27	32:10	—	—
45	31:39	31:25	—	—
50	30:56	30:42	—	—
55	30:12	29:59	33:26	32:42
60	29:28	29:15	32:42	31:58
65	28:46	28:35	32:01	31:18
70	28:02	27:53	31:19	30:38
75	27:27	27:21	30:39	29:59
80	26:48	26:42	30:00	29:19
85	26:14	26:08	29:19	28:38
90	25:41	25:35	28:39	27:58
95	25:09	25:03		
100	24:36	24:30		
105	24:01	23:55		
110	23:19	23:13		

Table VIII-9 *pP-P* and *pP'-P'* (min:sec)

Δ°	Depth (km)							
	50	100	200	300	400	500	600	700
30	0:10	0:20	0:39	0:58	1:12	1:24	—	—
60	0:13	0:24	0:47	1:06	1:24	1:39	1:54	2:08
90	0:14	0:26	0:49	1:11	1:33	1:52	2:10	2:26
142	0:14	0:28	0:52	1:14	1:36	1:55	2:14	2:32
180	0:14	0:30	0:54	1:18	1:40	2:01	2:22	2:39

Table VIII-10 *sP-P* and *sP'-P'* (min:sec)

Δ°	Depth (km)							
	50	100	200	300	400	500	600	700
30	0:16	0:32	1:02	1:32	1:57	2:22	2:44	—
60	0:18	0:36	1:07	1:38	2:05	2:31	2:57	3:22
90	0:18	0:38	1:10	1:44	2:14	2:42	3:09	3:36
142	0:18	0:38	1:12	1:46	2:18	2:48	3:17	3:44
180	0:18	0:38	1:12	1:46	2:18	2:48	3:17	3:44

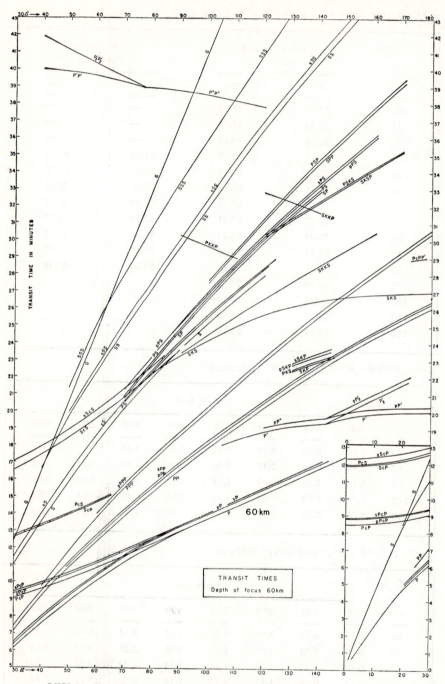

FIGURE VIII-1 *Transit times for focal depth 60 kilometers.*

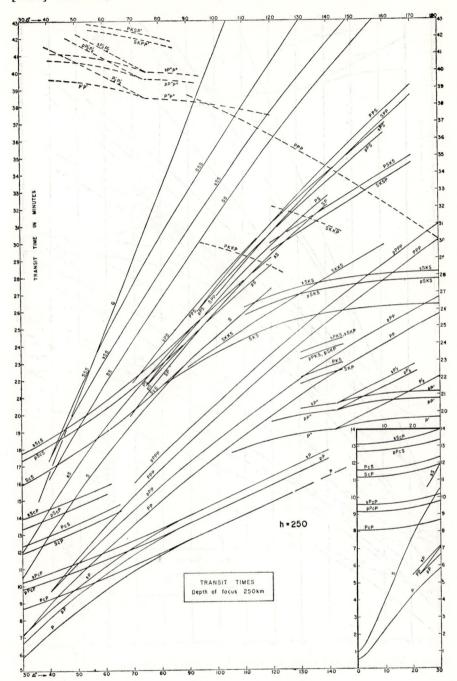

FIGURE VIII-2   *Transit times for focal depth 250 kilometers.*

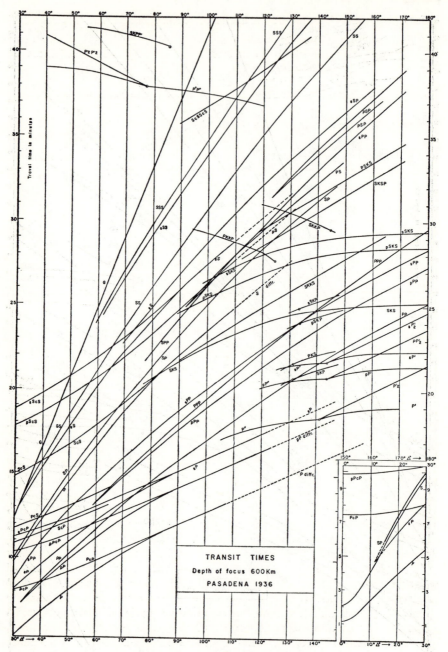

FIGURE VIII-3   *Transit times for focal depth 600 kilometers.*

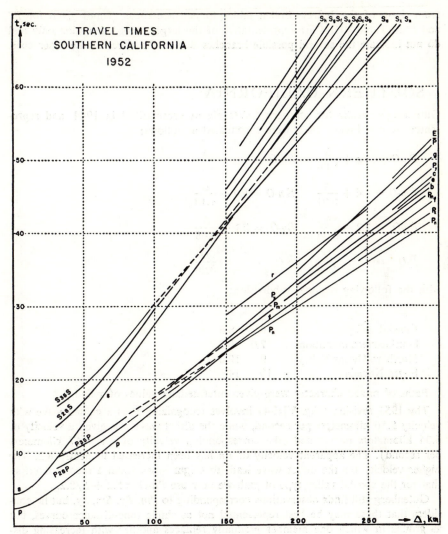

FIGURE VIII-4   *Transit times, Southern California.*

Richter, *Materials* . . . , 1936; these results should properly be revised slightly to bring them into harmony with those for shallow shocks, which are of later date.

The most important data, those for *P* with assumed hypocentral depth of 25 kilometers, are from Gutenberg's 1953 revision (*Calif. Dept. Nat. Resources, Div. Mines, Bull.* 171, pp. 162–163). The corresponding times for *S* are interpolated from the Jeffreys-Bullen tables of 1951, taking their depth 0.00 as corresponding to 33 kilometers, with corrections from residuals plotted by Gutenberg (*loc. cit.,* p. 161, Fig. 3).

Other times for shallow teleseisms are from Gutenberg and Richter, *On Seismic Waves,* except that those given for *ScS* are obtained by subtracting 6 seconds from those of the Jeffreys-Bullen tables for surface focus. Times given for P'' and P' in Table VIII-4 are representative of the larger recorded motion only and do not indicate the several possible branches due to the effect of the inner core.

SOUTHERN CALIFORNIA

Gutenberg's results for southern California as summarized in 1944, and represented by our Figure 18-2, were formulated as follows:

$$Pn\text{-}O = x + \frac{\Delta}{8.06} \qquad Sn\text{-}O = y + \frac{\Delta}{4.45}$$

$$Pm\text{-}O = 4.4 + \frac{\Delta}{6.94} \qquad Sm\text{-}O = 6.9 + \frac{\Delta}{4.10}$$

$$Py\text{-}O = 1.2 + \frac{\Delta}{6.05} \qquad Sy\text{-}O = 2.1 + \frac{\Delta}{3.65}$$

$$\bar{P}\text{-}O = \frac{D}{5.58} \qquad \bar{S}\text{-}O = \frac{D}{3.26}$$

with the following regional subdivision:

	x	y
Coastal valleys	6.2	8.5
Southeastern mountains	7.0	9.5
Northern Owens Valley	9	12.5
Sierra Nevada	10	14

Paths of mixed character were given intermediate values of x and y.

The 1951 revision (Fig. VIII-4) involves recognizing $\bar{P}$ as a guided wave with velocity 5.56 kilometers per second, while the direct wave p is given a velocity of 6.34 kilometers per second (the corresponding velocity of s is 3.67 kilometers per second). The apparent velocity of Pn has been increased to nearly 8.2. The higher velocity for the direct wave leads to origin times about 1 second later, so that for the coastal valley type of path we now use $Pn\text{-}O = 5.2 + \Delta/8.2$.

Gutenberg still finds observations corresponding to Pm, Py, Sm, Sy, but he considers that these may be best represented not as single time-distance curves, but as groups in which one member gradually replaces another with increasing distance, which would suit the idea that these are guided waves. Pm, however, shows sufficient sharpness for a refracted wave through a thin layer with velocity near 7 kilometers per second immediately above the Moho; there appear to be observed reflections from the upper surface of such a layer. The time of the largest Pm wave is given by $Pm\text{-}O = 3.5 + \Delta/7.01$. Similarly $Py\text{-}O = 1.2 + \Delta/6.21$; this is preceded by a wave Gutenberg letters c, for which the transit time is $0.2 + \Delta/6.22$. The nearly zero constant term for the last makes it practically continuous with the direct p.

AUXILIARY CHART FOR DEEP-FOCUS TELESEISMS

As indicated in Chapter 19, identification of phases for a deep-focus teleseism usually proceeds by trial and error, fitting the readings to charts drawn for various hypocentral depths. This work can be shortened if a few characteristic phases can be identified and used to find approximate depth and distance. On many seismograms S is immediately identifiable with little doubt, and often pP is so definite that question remains only as to whether it may be sP.

In such cases effective use may be made of the type of chart illustrated in Figure VIII-5. Similar charts have been constructed by many seismologists. One devised by Dr. Båth† has been used regularly and successfully by him and his assistants at Uppsala. In his original publication† he applied the method of least squares to reduce the results and calculate probable errors for distance, depth, and origin time.

Our figure has the same coordinate system as Båth's; distance Δ and depth H are taken as abscissa and ordinate, while curves are drawn for constant time intervals between phases. Those for constant pP-P and S-P were also drawn by Båth; the others differ from his. For example, his chart shows pPP-PP for large Δ where ours shows pP'-P'; numerically these differ slightly. To facilitate detecting misidentifications, curves for sP-P and PcP-P are given. Note that SKS and S coincide at about $\Delta = 84°$, independently of H; at greater distances SKS may be mistaken for S. $P'P'$-P is added as an adjunct to S-P, and P_2'-P' is drawn for distances over $142°$.

FIGURE VIII-5 *Chart for finding distance and depth of teleseisms from time intervals (in min:sec) between phases.*

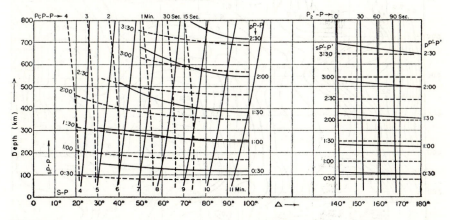

---

† Båth, M., "Sur une méthode pour calculer les tremblements de terre à foyer profonde à l'aide des phases d'une seule station séismographique," *Kgl. Svenska Vetenskapsakad. Handl.*, Ser. 3, vol. 20 No. 4 (1943).

REVISED VALUES OF A FOR PZ, 1955

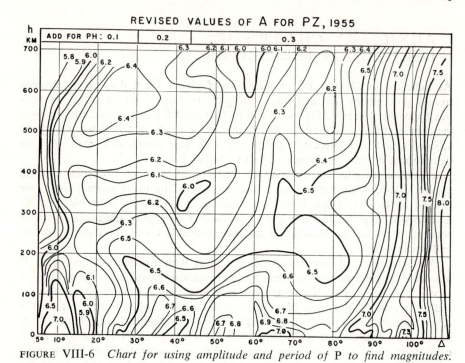

FIGURE VIII-6 *Chart for using amplitude and period of P to find magnitudes.*

FIGURE VIII-7 *Chart for using amplitude and period of S to find magnitudes.*

REVISED VALUES OF A FOR SH, 1955

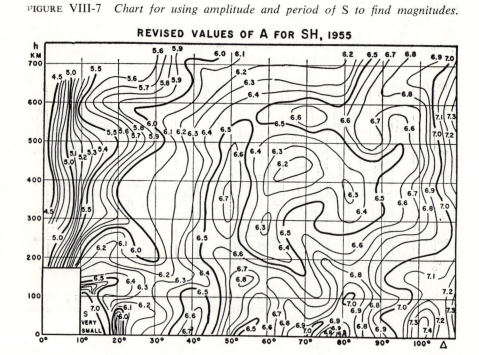

MAGNITUDE CHARTS

Figures VIII-6, VIII-7, and VIII-8 are charts for finding magnitude from the amplitude/period ratio in the maximum of the P, S, or PP group as recorded for teleseisms. The reduced maximum amplitude in microns is divided by the corresponding period in seconds, the logarithm is taken, and the station correction added. This result is then added to the quantity A read from the contours on the chart for the given phase, distance, and depth. The chart for S refers to the combined horizontal components; those for P and PP refer directly only to the vertical components (PZ, PPZ); to apply to the combined horizontal components (PH, PPH) add the corrections at the tops of the charts.

Example: Station, Pasadena. $\Delta = 60°$, $H = 300$ kilometers. Reduced ground amplitude, 3 microns; period, 1.5 seconds. Quotient amplitude/period, 2.0.

FIGURE VIII-8   *Chart for using amplitude and period of PP to find magnitudes.*

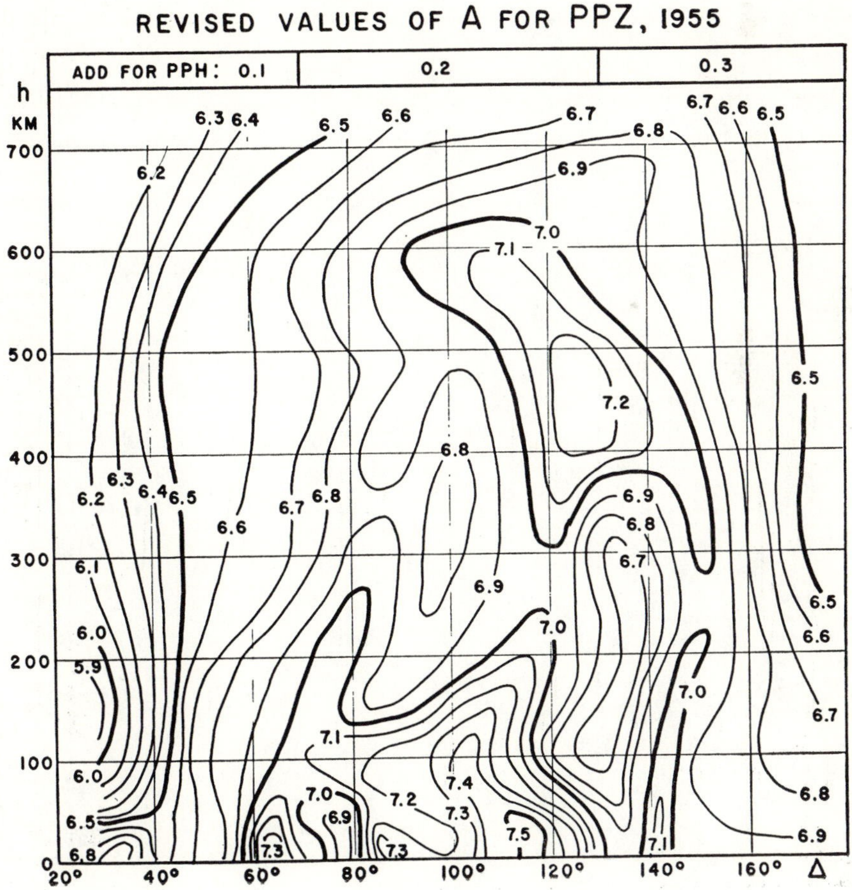

REVISED VALUES OF A FOR PPZ, 1955

Log quotient	+0.3
Station correction	+0.2
A from PZ chart	+6.3
Correction for PH	+0.3
Magnitude m	7.1

The station correction, and the log quotient for smaller shocks, may be negative. Station corrections seldom exceed ±0.3; lists will be found in Gutenberg's publications.

The calculation yields the body wave magnitude, equivalent to a single determination for Gutenberg's m. The corresponding M may be read from Table 22-4 or computed from equation (2) of Chapter 22:

$$M = 1.59m - 3.97$$

APPENDIX IX

Calculation of Times for Pn

LET US ASSUME that the behavior of *Pn* follows the simple assumption first made by Mohorovičić and illustrated in Figure 18-1. Call the thickness of the crustal layer d, and the depth of a hypocenter below the surface of this layer H. Let the velocity in the crust be V, and below the crust W.

The ray will be refracted horizontally in the lower medium if the angle of incidence in the upper medium is i such that

$$\sin i = V/W \tag{1}$$

The time along the complete ray to a point on the surface at distance Δ from the epicenter is

$$t = \frac{\Delta \sin i + A \cos i}{V} \tag{2a}$$

$$= \frac{\Delta + A \cot i}{W} \tag{2b}$$

where

$$A = 2d - H \tag{2c}$$

The result (2b) is often used in the form

$$t = \frac{\Delta}{W} + K \tag{3}$$

in which

$$K = \frac{2d - H}{W} \cot i \tag{4}$$

The form (3) also applies when there are several horizontal layers with constant velocity in each; K is then a constant depending on H, the velocities, and the thicknesses of the layers. Note that the time t is affected appreciably if the layering is not horizontal, so that any dip of the Moho or other crustal discontinuities alters the interpretation of observed times.

Times along the direct and the refracted ray will coincide for $\Delta = X$ such that

$$(X^2 + H^2)^{1/2} = X \sin i + KV \tag{5}$$

The solution of (5) is

$$X = KV \tan i \sec i + \sec^2 i \, (K^2 V^2 - H^2 \cos^2 i)^{1/2} \tag{6}$$

The trigonometric functions in (6) are easily computed by using (1). If $H < KV$, a good first approximation is

$$X = \frac{KV}{1 - \sin i} - \frac{H^2}{2KV} \tag{7}$$

Applying such formulas to the actual earth neglects variation of velocity within the crust but gives a fair representation of the observations.

For the effect of dip of the Moho, and for treatment of several horizontal layers, refer to works on seismic prospecting (Chapter 18).

APPENDIX X

Location of a Shallow Teleseism

THIS PROCESS requires an illustrative example, which will also show some of the preliminary procedure. The material is taken, with slight rearrangement, from one of Professor Gutenberg's notebooks. The study is based entirely on preliminary readings circulated by the stations in their own provisional bulletins or by way of the Strasbourg office. The chosen earthquake, of magnitude 7, was highly destructive on the Ionian Islands off the west coast of Greece.

Table X-1 shows a preliminary determination of the origin time. The columns list: the recording stations which reported times for P and S (times of S are not used when noted as doubtful); the time of recorded P, in minutes and seconds after 9h, August 12, 1953, G.C.T.; the time interval S-P in seconds up to 10 minutes (larger S-P intervals were not used because of possible confusion with SKS); the transit time P-O corresponding to S-P, from Table VIII-3; and the resulting origin time in seconds after 9h 23m. The results spread over about half a minute.

These are unselected data, and a fair sample of what is available for preliminary purposes. The reporting stations are of all sorts, including some with the best modern equipment, others with old and insensitive instruments. The readings are subject to all kinds of errors—of communication and misprinting as well as of measurement or interpretation. The mean of the 61 values for the origin time is 09:23:48.0. Generally the best approximate origin time is later than such a mean value, since for a shock of this magnitude it is more likely that S will be read late than that P will be mistimed, and a late S then gives an early origin time. For further work, represented in Table X-2, 09:23:50 was used as a first approximation. Preliminary work with the globe showed that the epicenter is not far from 38° N, 20° E. Examination of the *International Summary* led to using for comparison epicenter 38.8° N, 20.6° E; a shock on March 11, 1938, had been referred to this location, so that distances and azimuths for many stations were calculated for and tabulated in the *Summary*. A selection from these, representative of different azimuths, was made (using only stations for which preliminary times for the 1953 shock were available). For a few other stations distances were calculated from the approximate epicenter, using direction cosines. All these distances and azimuths appear in Table X-2, followed by the reported time of arrival of P in minutes and seconds following 09h. The column P-O gives the

Table X-1 Calculation of Approximate Origin Time—Earthquake
of August 12, 1953, 09h G.C.T.

Station	Arrival Time of P, 09h +		Interval S-P		Calculated P-O		Origin Time 09h 23m + sec
	m	s	m	s	m	s	
Reggio	24	54	00	49	00	57	57
Messina	24	55	00	49	00	57	58
Belgrade	25	32	01	25	01	46	46
Rome	25	42	01	28	01	53	49
Florence	26	04	01	42	02	15	49
Trieste	26	04	01	37	02	02	62
Siena	26	05	01	45	02	18	47
Prato	26	07	01	53	02	30	37
Hurbanovo	26	21	01	53	02	30	51
Tolmezzo	26	21	01	50	02	26	55
Pavia	26	33	02	00	02	42	51
Chur	26	45	02	12	02	56	49
Oropa	26	46	02	06	02	49	57
Praha	26	55	02	16	03	02	53
Zurich	26	56	02	30	03	18	38
Ksara	26	56	02	24	03	11	45
Ravensburg	27	00	02	32	03	20	40
Cheb	27	01	02	27	03	15	46
Stuttgart	27	04	02	21	03	08	56
Karlsruhe	27	10	02	32	03	21	49
Strasbourg	27	13	02	34	03	23	50
Algiers	27	14	02	44	03	34	40
Clermont	27	28	03	05	03	58	30
Tortosa	27	30	02	53	03	44	46
Paris	27	47	03	10	04	03	44
Alicante	27	47	03	06	03	59	48
Hamburg	27	53	03	23	04	18	35
De Bilt	28	04	03	28	04	24	40
Copenhagen	28	10	03	34	04	31	39
Cartuja	28	17	03	39	04	36	41
Toledo	28	19	03	34	04	31	48
Jersey	28	25	03	42	04	40	45
Kew	28	26	03	33	04	29	57
Uppsala	28	46	03	56	04	55	51
Coimbra	28	54	04	01	05	01	53
Lisbon	29	00	04	10	05	13	47

Table X-1 Continued

Station	Arrival Time of P, 09h +		Interval S-P		Calculated P-O		Origin Time 09h 23m + sec
	m	s	m	s	m	s	
Averroes	29	04	04	16	05	21	43
Rathfarnham	29	08	04	19	05	26	42
Aberdeen	29	11	04	14	05	19	52
Edinburgh	29	12	04	19	05	25	47
Azores	31	13	05	51	07	22	51
Quetta	31	16	05	57	07	30	46
New Delhi	32	27	06	54	08	38	49
Bombay	32	43	07	12	08	58	45
Poona	32	49	07	07	08	53	56
Kodaikanal	33	43	08	00	09	51	52
Calcutta	34	01	07	56	09	47	74
Shillong	34	04	08	12	10	06	58
Harvard	34	48	09	00	11	02	46
Bermuda	34	53	09	01	11	03	50
Ottawa	34	54	09	02	11	04	50
Palisades	35	02	09	05	11	08	54
Fordham	35	04	09	10	11	13	51
Washington	35	22	09	29	11	35	47
Morgantown	35	31	09	44	11	52	39
Martinique	35	35	09	44	11	52	43
Cherry Point	35	36	09	42	11	50	46
College	35	46	09	54	12	04	42
San Juan	35	47	09	45	11	53	54
Cincinnati	35	47	09	59	12	10	37

result of subtracting 09:23:50 from each *P* time. The column headed Δ (*P-O*) gives epicentral distances derived from *P-O* by using Table VIII-3. These are diminished by the calculated distances in the first column to give residuals in distance; the residuals are plotted as ordinates with azimuths as abscissas (Fig. X-1).

If the residuals are due only to the difference between the true and approximate epicenters, they should fall on a sine curve when plotted in this way. With a little attention to the required properties of such a curve (such as having crest and trough 180° apart in azimuth), it can be sketched through the points with fair certainty, as shown in the figure. In this example the curve is not symmetrical about the zero line, but lies higher on the plot. This indicates that the best origin time is later than that first assumed, and the corrected time is chosen as 09:23:53, which will provide the small shift in the distances needed. (The effect of changed origin time differs somewhat for the nearer and the more distant stations; in case

Table X-2 Data for Epicenter Revision—Earthquake of August 12, 1953.
Approximate origin time: 09:23:50.
Approximate epicenter: 38.8° N, 20.6° E.

	Δ° Calc	Az.	P = 09h + m	s	P-O m	s	Δ° (P-O)	Δ° Residual
Athens	2.6	120	24	34	00	44	2.7	+0.1
Belgrade	6.0	359	25	32	01	42	7.0	+1.0
Trieste	8.5	326	26	05	02	15	9.2	+0.7
Chur	11.4	318	26	45	02	55	11.9	+0.5
Praha	12.1	314	26	55	03	05	12.7	+0.6
Zurich	12.3	316	26	56	03	06	12.8	+0.5
Cheb	12.7	335	27	01	03	11	13.2	+0.5
Stuttgart	12.9	324	27	04	03	14	13.4	+0.5
Ksara	13.3	107	26	56	03	06	12.8	−0.5
Strasbourg	13.5	321	27	13	03	23	14.1	+0.6
Algiers	14.0	267	27	14	03	24	14.2	+0.2
De Bilt	17.1	326	28	04	04	14	18.3	+1.2
Copenhagen	17.8	344	28	10	04	20	18.8	+1.0
Toledo	19.1	283	28	19	04	29	19.6	+0.5
Cartuja	19.2	274	28	17	04	27	19.4	+0.2
Kew	19.4	319	28	26	04	36	20.2	+0.8
Uppsala	21.2	355	28	46	04	56	22.1	+0.9
Harvard	66.5	307	34	48	10	58	67.7	+1.2
Fordham	68.9	306	35	04	11	14	70.3	+1.4
Calcutta	59.3	85	34	01	10	11	60.7	+1.4
Tamanrasset	20.5	210	28	29	04	39	20.5	0.0
Bombay	49.2	110	32	43	08	53	49.7	+0.5
Tananarive	62.7	155	34	15	10	25	62.7	0.0
Kimberley	67.3	175	34	43	10	53	67.0	−0.3
Kochi	84.9	45	36	35	12	45	86.6	+1.7
Sendai	85.7	50	36	38	12	48	87.2	+1.5

of doubt, recalculate the residuals and plot the new curve.) The amount by which the epicenter has to be shifted is the amplitude of the sine curve (measured from the adjusted zero line); here it is nearly a degree (of a great circle). The minimum of the curve occurs just to the right of 180°, so that the direction of shift is south

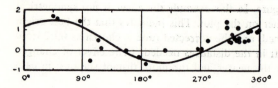

FIGURE X-1

Distance residuals plotted against station azimuth. Ionian Islands earthquake, August 12, 1953.

and a little west. The components of the shift in the cardinal directions may be read directly from the intercepts on the abscissas of 0°, 90°, 180°, 270° (this happens to be mathematically exact). When the new zero line is drawn at about +0.7, the intercept at 180° is about $-\frac{1}{2}$, and that at 270° slightly negative. The latitude is then decreased by about half a degree, and the longitude slightly decreased (moving westward), with the result 38¼° N, 20¼° E. Where there are larger shifts in longitude, it must be noted that the plotted curve gives arcs along great circles; the indicated east or west shift must be divided by the cosine of the latitude to give the actual change in longitude.

Stations recording this shock clearly at great distances were few. Riverview and Wellington reported P' at 09:43:28 and 09:43:54, respectively, which with origin time 09:23:53 gives transit times of 19m 35s and 20m 01s. The calculated distances of the two stations from the trial epicenter are 139.2° and 159.1°; the tabulated times for P' at these distances are 19m 29s and 20m 00s. These data thus give no evidence of other than shallow focal depth, a result to be expected from the effects of the shock and the large recorded surface waves. Chihuahua, at a calculated distance of 96.5°, reports SKS at 09:47:51, which corresponds closely to the tabulated transit time of 24m 01s for that distance.

APPENDIX XI

Example of Location from Near Stations

AN EXAMPLE of the possibilities of local-earthquake registration can be taken from the data of a small shock (magnitude 3.2) recorded as part of a special program during the aftershock period in Kern County, 1952 (Table XI-1).

Part A of the table shows calculation of origin time from the readings of P and S for the seven nearest stations. Each time interval S-P is multiplied by 1.37 to give P-O (this is equivalent to assuming Poisson's ratio to be ¼), and the result is subtracted from the time of P to give O. The results are all close to 15:14:57.9, except for Clear Creek Ranch and Chuchupate. The former station, which is the nearest, clearly has one of the deceptive early arrivals which obscure the true S at short distances; using this gives a late value for the origin time. The opposite discrepancy at Chuchupate is due to a late reading for S; actually there is no sharp S phase on this particular seismogram, and a wrong choice was made.

Various approximate procedures led to a trial epicenter at 35° 19′ N, 118° 30′ W; a few tests showed that a depth of 10 kilometers would suit the data better than the routine assumption of 16 kilometers. The results of calculation on this basis appear in Part B of the table. Tabulated D is derived from calculated Δ and $h = 10$ kilometers. The next column lists the arrival times for P; at Pasadena 19.1 corresponds to the small first arrival, 20.6 to a later sharp impulse. P-O is calculated by dividing D by the mean velocity, 6.34 kilometers per second established in earlier work for southern California. Subtracting this from P, there result the values of O. Most of these are close to the values 15:14:57.9 derived from S-P. Clear Creek Ranch is a little early; this may be due to timing error. The late origin time derived for Haiwee probably indicates only a late reading for the first motion; the station has a relatively high level of background disturbance, and the first P may easily be obscured in recording so small a shock. Note that the time at Tinemaha fits reasonably well, in spite of the fact that this station is 194 kilometers distant and therefore far beyond the distance at which Pn normally precedes the direct p; this is attributable to the cutting off or delay of Pn by the Sierra Nevada structure (and not to a lost first motion, since an equivalent transit time is found at Tinemaha for the sharp first arrivals recorded for much larger earthquakes in the same area).

The transit time of Pn is $\Delta/8.2$ plus a quantity which should vary with hy-

Table XI-1 Data for Earthquake of September 4, 1952, 15h 14m

A. Calculation of origin time from S-P

Station	P 15h 15m + (sec)	S-P (sec)	P-O (sec)	O 15h 14m + (sec)
Clear Creek Ranch	00.1	01.1	01.5	58.6
Piute Ranch	00.4	02.0	02.7	57.7
Parker Creek	02.1	03.1	04.2	57.9
Havilah	01.8	02.8	03.8	58.0
Woody	06.4	06.1	08.4	58.0
Chuchupate	10.0	10.1	17.8	56.2
China Lake	14.3	11.9	16.3	58.0

B. Calculation of origin time from P-O, assuming epicenter 35° 19′ N, 118° 30′ W, and $h = 10$ km

Station	Δ (km)	D (km)	P 15h 15m + (sec)	P-O (sec)	O 15h 14m + (sec)	$\Delta/8.2$	$P-\Delta/8.2$ 15h 15m + (sec)
Clear Creek Ranch	12.5	16.0	00.1	02.5	57.6		
Piute Ranch	12.1	15.7	00.4	02.5	57.9		
Havilah	21.5	23.7	01.8	03.7	58.1		
Parker Creek	25.1	27.0	02.1	04.3	57.8		
Woody	53.1	54.0	06.4	08.5	57.9		
Chuchupate	74.0	74.6	10.0	11.8	58.2		
China Lake	98.9	99.4	14.3	15.7	58.6		
Haiwee	103.1	103.5	15.3	16.3	59.0		
Mt. Wilson	127.7	128.1	18.6	20.3	58.3		
Pasadena	133.1	133.5	19.1	21.1	58.0		
			20.6			16.2	04.4
Dalton	142.0	142.3	21.3	22.5	58.8	17.3	04.0
Santa Barbara	147.5	147.8	22.8	23.3		18.0	04.8
Riverside	179.4	179.7	25.8			21.9	03.9
Big Bear	188.2	188.5	27.0			23.0	04.0
Tinemaha	194.4	194.6	29.4	30.9	58.5	23.7	
Barrett	337.7	337.8	45.9			41.2	04.7

pocentral depth but should be roughly constant for a given earthquake, variations being attributable to different structure along the path, including the effect of dip of the Moho. The last two columns show $\Delta/8.2$ and $P - \Delta/8.2$ for the more distant stations. The readings at Santa Barbara and Barrett are probably both slightly late—the former because of high background, the latter because of relatively great distance and small amplitude. The others give $P - \Delta/8.2$ near 4 seconds; this includes the second arrival at Pasadena, which may be read slightly late. For earthquakes in the same area which appear to be nearer the standard depth (about 16 kilometers), $P - \Delta/8.2$ ranges from about 5.5 to 6 at the same group of stations.

Slight improvement in the consistency of the solution may be had by making small changes in epicenter and depth, but this would certainly be driving the work well past the true limit of accuracy, no matter what small value may be calculated for the "probable error." No amount of statistical juggling will remove the obvious systematic errors involved in assuming constant velocities and uniform structures along all paths.

The limitations on these results represent the most favorable conditions. Even so, much uncertainty still remains as to depth and crustal structure.

A P P E N D I X   X I I

Calculation of Short Distances

IN WORKING with local earthquakes, distances are usually wanted from an assumed epicenter to the recording stations, with error not exceeding 1 kilometer. Especially when analytical methods of location, involving computation with given coordinates and distances, are used, it is desirable to determine distances to the tenth of a kilometer, thus avoiding rounding-off errors.

To measure distances of the order of 200 kilometers from a map, with an accuracy of 0.1 kilometer, requires time-consuming precautions in mapping and measurement; consequently such distances are usually computed from the given latitudes and longitudes.

The effect of the curvature of the earth is small enough so that distances up to about 500 kilometers are given with error not exceeding 0.1 kilometer by

$$\Delta^2 = \Delta x^2 + \Delta y^2 \qquad \Delta x = A \Delta \lambda \qquad \Delta y = B \Delta \phi$$

If Δ is the required distance in kilometers, and $\Delta\lambda$, $\Delta\phi$ are differences in longitude and latitude between the two given points (usually epicenter and station), expressed in minutes of arc (so that $1° 23' 36''$ becomes $83.6'$), sufficient accuracy is provided by taking A and B as the length in kilometers and decimals of one minute of arc of the parallel and meridian, respectively, centered at the mean latitude between the two given points. It remains to construct tables giving the values of A and B at intervals of 1 minute of the mean latitude (using the ordinary geodetic latitude as entered on maps).

The tables presented here were derived from data based on the Clarke spheroid (in the Smithsonian Geographical Tables), but the deviation from the internationally adopted spheroid of later date is a minor and negligible correction.

Table XII-1 has a wide range of applicability. It presents the coefficients A and B for every even degree of the mean latitude from 0° to 70°. The change in B from degree to degree is so small and regular that interpolation to any desired minute of latitude is easy. The corresponding differences in A are much less regular; but, as the third column of the table shows, this is due to a trigonometric factor, and the quantities $A/\cos \phi$ show even more regularity than B. For accuracy it is merely necessary to interpolate $A/\cos \phi$ for the required minute of latitude and then multiply by the cosine of that latitude.

Accuracy to four decimal places in A and B is usually sufficient. For B this

701

Table XII-1   Arcs of Meridian and Parallel

Mean Latitude ϕ (deg)	Arc of 1' of Meridian, B (km)	Arc of 1' of Parallel, A (km)	A/cos ϕ (km)
0	1.842808	1.855365	1.855365
1	1.842813	1.855087	1.855369
2	1.842830	1.854243	1.855374
3	1.842858	1.852840	1.855383
4	1.842898	1.850877	1.855396
5	1.842950	1.848353	1.855414
6	1.843011	1.845270	1.855434
7	1.843085	1.841628	1.855458
8	1.843170	1.837430	1.855487
9	1.843265	1.832675	1.855520
10	1.843372	1.827365	1.855555
11	1.843488	1.821502	1.855595
12	1.843617	1.815087	1.855638
13	1.843755	1.808122	1.855683
14	1.843903	1.800610	1.855733
15	1.844062	1.792552	1.855786
16	1.844230	1.783950	1.855842
17	1.844408	1.774808	1.855902
18	1.844595	1.765128	1.855966
19	1.844792	1.754912	1.856031
20	1.844998	1.744163	1.856100
21	1.845213	1.732887	1.856173
22	1.845437	1.721083	1.856248
23	1.845668	1.708757	1.856325
24	1.845907	1.695910	1.856404
25	1.846153	1.682550	1.856488
26	1.846408	1.668677	1.856573
27	1.846670	1.654297	1.856661
28	1.846938	1.639413	1.856750
29	1.847213	1.624032	1.856843
30	1.847495	1.608155	1.856937
31	1.847781	1.591788	1.857033
32	1.848073	1.574937	1.857132
33	1.848372	1.557605	1.857231
34	1.848673	1.539798	1.857331

Table XII-1   Continued

Mean Latitude ϕ (deg)	Arc of $1'$ of Meridian, B (km)	Arc of $1'$ of Parallel, A (km)	$A/\cos\phi$ (km)
35	1.848980	1.521522	1.857435
36	1.849290	1.502780	1.857538
37	1.849605	1.483580	1.857643
38	1.849922	1.463927	1.857750
39	1.850242	1.443827	1.857858
40	1.850565	1.423283	1.857964
41	1.850890	1.402307	1.858074
42	1.851217	1.380900	1.858184
43	1.851543	1.359070	1.858294
44	1.851873	1.336823	1.858403
45	1.852202	1.314167	1.858512
46	1.852531	1.291108	1.858623
47	1.852860	1.267653	1.858734
48	1.853188	1.243808	1.858842
49	1.853515	1.219582	1.858951
50	1.853842	1.194982	1.859061
51	1.854165	1.170013	1.859170
52	1.854487	1.144685	1.859276
53	1.854805	1.119005	1.859384
54	1.855122	1.092980	1.859488
55	1.855433	1.066618	1.859592
56	1.855742	1.039928	1.859695
57	1.856045	1.012918	1.859798
58	1.856345	0.985595	1.859896
59	1.856640	0.957968	1.859995
60	1.856928	0.930047	1.860094
61	1.857212	0.901837	1.860187
62	1.857490	0.873348	1.860279
63	1.857762	0.844590	1.860369
64	1.858025	0.815572	1.860459
65	1.858283	0.786300	1.860544
66	1.858533	0.756785	1.860627
67	1.858775	0.727037	1.860709
68	1.859008	0.697063	1.860787
69	1.859235	0.666873	1.860861
70	1.859452	0.636477	1.860934

Table XII-2   Coefficients B (lengths in kilometers of $1'$ of the meridian)

Mean Latitude ϕ	B	Mean Latitude ϕ	B
29° 51' to 30° 11'	1.8475	35° 14' to 35° 33'	1.8491
30 12 to 32	76	34 to 52	92
33 to 53	77	35 53 to 36 11	93
54 to 31 14	78	36 12 to 30	94
31 15 to 34	79	31 to 49	95
35 to 54	80	50 to 37 08	96
55 to 32 14	81	37 09 to 27	97
32 15 to 35	82	28 to 46	98
36 to 55	83	47 to 38 06	1.8499
56 to 33 15	84	38 07 to 24	1.8500
33 16 to 35	85	25 to 43	01
36 to 55	86	44 to 39 01	02
56 to 34 15	87	39 02 to 20	03
34 16 to 34	88	21 to 39	04
35 to 54	89	40 to 57	05
55 to 35 13	1.8490	58 to 40 16	1.8506

makes it possible to construct a simple auxiliary table, Table XII-2, from which the required B can be read rapidly. For A a much larger table is needed in order to eliminate the time otherwise needed to apply the cosine factor. These auxiliary tables for mean latitude 30° to 40° (Tables XII-2 and XII-3) are reproduced from an earlier publication.†

Where the same group of stations is repeatedly in use, other special tables can be constructed quickly. For example, A and B can be tabulated for a given station as functions, not of the mean latitude, but of the latitude of the other point (epicenter usually) to which distance is required. An alternative to the use of B is to compute (or determine from geographical tables) and tabulate the meridian distance in kilometers and decimals from the equator, 30°, or any fixed latitude, to every even minute of given latitude, and to the exact latitude of each station used; one subtraction then yields $B\Delta\phi$.

Example of calculation: Distance of Pasadena, 34° 08.9' N, 118° 10.3' W, from a trial epicenter at 32° 00' N, 119° 00' W. $\Delta\phi = 128.9'$, $\Delta\lambda = 49.7'$; mean latitude, $\phi = 33°$ 04.45'. From the tables, $B = 1.8484$, $A = 1.5562$. Hence $\Delta x = 77.3$ km, $\Delta y = 238.3$ km, and $\Delta = 250.5$ km.

---

† Richter, C. F., "Calculation of small distances," *B.S.S.A.*, vol. 33 (1943), pp. 243–250.

Table XII-3 Coefficients *A* for Given Mean Latitude Φ
(lengths in kilometers of 1′ of the parallel)

	30°	31°	32°	33°	34°	35°	36°	37°	38°	39°
00′	1.6082	1.5918	1.5749	1.5576	1.5398	1.5215	1.5028	1.4836	1.4639	1.4438
01	79	15	47	73	95	12	25	33	36	35
02	76	12	44	70	92	09	21	29	33	32
03	74	10	41	67	89	06	18	26	29	28
04	71	07	38	64	86	03	15	23	26	25
05	1.6068	1.5904	1.5735	1.5561	1.5383	1.5200	1.5012	1.4820	1.4623	1.4421
06	65	1.5901	32	58	80	1.5197	09	16	19	18
07	63	1.5898	29	56	77	94	06	13	16	15
08	60	96	27	53	74	90	1.5002	10	13	11
09	57	93	24	50	71	87	1.4999	07	09	08
10	1.6055	1.5890	1.5721	1.5547	1.5368	1.5184	1.4996	1.4803	1.4606	1.4404
11	52	87	18	44	65	81	93	1.4800	1.4603	1.4401
12	49	85	15	41	62	78	90	1.4797	1.4599	1.4398
13	46	82	12	38	59	75	87	94	96	94
14	44	79	09	35	56	72	83	90	93	91
15	1.6041	1.5876	1.5706	1.5532	1.5353	1.5169	1.4980	1.4787	1.4589	1.4387
16	38	73	04	29	50	66	77	84	86	84
17	36	71	1.5701	26	47	63	74	81	83	81
18	33	68	1.5698	23	44	60	71	78	79	77
19	30	65	95	20	41	56	67	74	76	74
20	1.6028	1.5862	1.5692	1.5517	1.5338	1.5153	1.4964	1.4771	1.4573	1.4370
21	25	59	89	14	35	50	61	68	69	67
22	22	57	86	11	32	47	58	64	66	63
23	19	54	83	08	28	44	55	61	63	60
24	17	51	81	05	25	41	52	58	59	56
25	1.6014	1.5848	1.5678	1.5502	1.5322	1.5138	1.4948	1.4754	1.4556	1.4353
26	11	45	75	1.5499	19	35	45	51	53	50
27	08	43	72	96	16	31	42	48	49	46
28	06	40	69	93	13	28	39	45	46	43
29	03	37	66	91	10	25	36	41	43	40
30	1.6000	1.5834	1.5663	1.5488	1.5307	1.5122	1.4932	1.4738	1.4539	1.4336
31	1.5998	31	60	85	04	19	29	35	36	33
32	95	29	58	82	1.5301	16	26	32	33	29
33	92	26	55	79	1.5298	13	23	28	29	26
34	89	23	52	76	95	10	20	25	26	22
35	1.5987	1.5820	1.5649	1.5473	1.5292	1.5106	1.4916	1.4722	1.4523	1.4319
36	84	17	46	70	89	03	13	18	19	16
37	81	15	43	67	86	1.5100	10	15	16	12
38	78	11	40	64	83	1.5097	07	12	13	09
39	76	09	37	61	80	94	04	09	09	05
40	1.5973	1.5806	1.5634	1.5458	1.5277	1.5091	1.4900	1.4705	1.4506	1.4302
41	70	03	31	55	74	88	1.4897	1.4702	1.4502	1.4298
42	68	1.5800	29	52	71	85	94	1.4699	1.4499	95
43	65	1.5798	26	49	67	81	91	95	96	92
44	62	95	23	46	64	78	87	92	92	88
45	1.5959	1.5792	1.5620	1.5443	1.5261	1.5075	1.4884	1.4689	1.4489	1.4285
46	57	89	17	40	58	72	81	86	86	81
47	54	86	14	37	55	69	78	82	82	78
48	51	83	11	34	52	66	75	79	79	74
49	48	81	08	31	49	63	71	76	75	71
50	1.5946	1.5778	1.5605	1.5428	1.5246	1.5059	1.4868	1.4672	1.4472	1.4267
51	43	75	1.5602	24	43	56	65	69	69	64
52	40	72	1.5599	22	40	53	62	66	65	60
53	37	69	97	19	37	50	58	62	62	57
54	34	66	94	16	34	47	55	59	59	54
55	1.5932	1.5764	1.5591	1.5413	1.5231	1.5044	1.4852	1.4656	1.4455	1.4250
56	29	61	88	10	28	40	49	52	52	47
57	26	58	85	07	24	37	45	49	48	43
58	23	55	82	04	21	34	42	46	45	40
59	20	52	79	1.5401	18	31	39	43	42	36
60	1.5918	1.5749	1.5576	1.5398	1.5215	1.5028	1.4836	1.4639	1.4438	1.4233

APPENDIX XIII

Selected List of Seismological Stations

Station—Country or State	Latitude (Geodetic)	Longitude	Geocentric Cosines a	b	c
Alger, Algeria	36° 46′ N	3° 03′ E	+8018	+ 427	+5960
Antofagasta, Chile	23 39 S	70 25 W	+3074	−8639	−3988
Apia, Samoa	13 48 S	171 47 W	−9615	−1390	−2371
Baguio, Philippines	16 25 N	120 35 E	−4883	+8262	+2809
Baku, USSR	40 23 N	49 54 E	+4920	+5843	+6454
Berkeley, California	37 52 N	122 16 W	−4224	−6692	+6113
Bogotá, Colombia	4 37 N	74 04 W	+2736	−9585	+0800
Bombay, India	18 54 N	72 49 E	+2797	+9045	+3219
Boulder, Colorado	40 00 N	105 16 W	−2023	−7410	+6403
Boulder City, Nevada	35 59 N	114 50 W	−3407	−7362	+5847
Brisbane,[1] Queensland	27 30 S	153 01 E	−7916	+4030	−4593
Bucharest, Rumania	44 25 N	26 06 E	+6436	+3153	+6974
Budapest, Hungary	47 29 N	19 04 E	+6410	+2215	+7349
Buenos Aires, Argentina	34 36 S	58 29 W	+4313	−7033	−5651
Cartuja, Granada, Spain	37 11 N	3 36 W	+7970	− 501	+6019
Cheb, Czechoslovakia	50 05 N	12 23 E	+6293	+1381	+7648
Chicago, Illinois	41 47 N	87 36 W	+ 313	−7471	+6639
Christchurch, New Zealand	43 32 S	172 37 E	−7213	+ 935	−6863
College, Alaska	64 52 N	147 50 W	−3615	−2274	+9042
Columbia, South Carolina	34 00 N	81 02 W	+1295	−8206	+5566
Djakarta, Indonesia	6 11 S	106 50 E	−2879	+9517	−1070
Eureka, Nevada	39 30 N	116 00 W	−3392	−6954	+6335
Fayetteville, Arkansas	36 06 N	94 12 W	−0593	−8078	+5865
Hamburg, Germany	53 28 N	9 55 E	+5889	+1030	+8016
Helwan, Egypt	29 51 N	31 20 E	+7419	+4518	+4953
Hong Kong	22 18 N	114 10 E	−3792	+8449	+3773

Station—Country or State	Latitude (Geodetic)	Longitude	Geocentric Cosines		
			a	b	c
Honolulu,[1] Hawaii	21 18 N	158 06 W	−8652	−3478	+3612
Huancayo, Peru	12 03 S	75 20 W	+2476	−9464	−2073
Hungry Horse, Montana	48 21 N	114 02 W	−2716	−6093	+7450
Irkutsk, USSR	52 16 N	104 19 E	−1519	+5954	+7889
Jena, Germany	50 56 N	11 35 E	+6199	+1271	+7743
Kew, England	51 28 N	0 19 W	+6255	− 34	+7802
Kimberley, South Africa	28 45 S	24 47 E	+7972	+3680	−4785
Kiruna, Sweden	67 50 N	20 26 E	+3556	+1325	+9252
Ksara, Lebanon	33 49 N	35 53 E	+6745	+4880	+5540
La Paz, Bolivia	16 30 S	68 08 W	+3573	−8903	−2822
La Plata, Argentina	34 55 S	57 56 W	+4364	−6965	−5697
Lwiro, Belgian Congo	02° 15′ S	28° 48′ E	+8757	+4814	−0391
Macquarie Island	54 30 S	158 57 E	−5444	+2093	−8123
Manila,[1] Philippines	14 40 N	121 05 E	−4997	+8288	+2516
Matsushiro, Japan	36 33 N	138 13 E	−6005	+5366	+5929
Melbourne, Victoria	37 50 S	144 58 E	−6483	+4545	−6108
Mérida, Yucatan, Mexico	20 57 N	89 37 W	+ 62	−9347	+3554
Moscow,[1] USSR	55 44 N	37 38 E	+4482	+3456	+8246
Mt. Hamilton, California	37 20 N	121 39 W	−4181	−6785	+6040
Nanking, China	32 03 N	118 48 E	−4091	+7442	+5280
New Delhi, India	28 35 N	77 12 E	+1949	+8576	+4760
Nouméa, New Caledonia	22 18 S	166 27 E	−9003	+2169	−3773
O'Higgins, Chilean Antarctica	63 20 S	57 54 W	+2398	−3822	−8924
Ottawa, Canada	45 24 N	75 43 W	+1739	−6828	+7096
Palisades, New York	41 00 N	73 54 W	+2098	−7272	+6536
Pasadena, California	34 09 N	118 10 W	−3916	−7311	+5587
Perth, W. Australia	31 57 S	115 50 E	−3705	+7651	−5266
Poona, India	18 32 N	73 51 E	+2639	+9114	+3159
Praha, Czechoslovakia	50 04 N	14 26 E	+6241	+1606	+7647
Pretoria, South Africa	25 45 S	28 11 E	+7949	+4260	−4322
Punta Arenas, Chile	53 09 S	70 54 W	+1971	−5691	−7983
Quetta, Pakistan	30 12 N	67 02 E	+3378	+7970	+5006
Rabaul, New Britain	4 12 S	152 10 E	−8820	+4657	− 727
Resolute, Canadian Arctic	74 41 N	94 54 W	−0227	−2648	+9640
Reykjavik, Iceland	64 08 N	21 54 W	+4069	−1636	+8987
Riverview, New South Wales	33 50 S	151 09 E	−7292	+4015	−5541
Rome,[1] Italy	41 54 N	12 31 E	+7287	+1618	+6653
St. Louis, Missouri	38 38 N	90 14 W	− 32	−7831	+6218
San Juan, Puerto Rico	18 23 N	66 07 W	+3844	−8683	+3136

Station—Country or State	Latitude (Geodetic)		Longitude		Geocentric Cosines		
					a	b	c
Santa Lucia,[1] Chile	33 26	S	70 39	W	+2770	−7890	−5483
Sapporo,[1] Japan	43 03	N	141 20	E	−5723	+4579	+6803
Scott Base, Antarctica	77 51	S	166 45	E	−2063	+0486	−9773
Scoresby-Sund, Greenland	70 29	N	21 57	W	+3117	−1256	+9418
Sendai, Japan	38 16	N	140 54	E	−6108	+4964	+6168
Shasta, California	40 42	N	122 23	W	−4073	−6420	+6495
Shillong, India	25 34	N	91 53	E	+0297	+9027	+4292
Strasbourg, France	48 35	N	7 46	E	+6579	+ 897	+7477
Stuttgart, Germany	48 46	N	9 12	E	+6531	+1057	+7499
Sverdlovsk, USSR	56° 50'	N	60 38	E	+2696	+4791	+8353
Tacubaya, Mexico	19 24	N	99 12	W	−1508	−9318	+3302
Taipei, Formosa	25 02	N	121 31	E	−4742	+7733	+4209
Tamanrasset, Algeria	22 47	N	5 31	E	+9186	+ 887	+3852
Tananarive, Madagascar	18 55	S	47 33	E	+6389	+6985	−3222
Tashkent,[1] USSR	41 20	N	69 18	E	+2662	+7044	+6579
Tokyo, Japan	35 41	N	139 46	E	−6214	+5259	+5807
Toledo, Spain	39 53	N	4 03	W	+7676	− 543	+6387
Trieste	45 39	N	13 45	E	+6814	+1668	+7126
Trinidad, West Indies	10 45	N	61 34	W	+4678	−8641	+1853
Tucson, Arizona	32 15	N	110 50	W	−3014	−7920	+5310
Uccle, Belgium	50 48	N	4 21	E	+6328	+ 482	+7728
Uppsala, Sweden	59 51	N	17 38	E	+4810	+1528	+8633
Victoria, British Columbia	48 31	N	123 25	W	−3662	−5550	+7469
Vladivostok, USSR	43 07	N	131 54	E	−4889	+5450	+6811
Warsaw, Poland	52 14	N	21 02	E	+5741	+2207	+7885
Wellington, New Zealand	41 17	S	174 46	E	−7505	+ 687	−6573
Zurich, Switzerland	47 22	N	8 35	E	+6721	+1014	+7334

[1] Values of a, b, c are taken when possible from the list compiled by the staff of the *International Summary* (Kew Observatory, 1951). Data have been added for new stations at Boulder (Colorado), Eureka (Nevada), Kimberley, Lwiro, Macquarie Island, Nouméa, O'Higgins, Quetta, Rabaul, Shillong. For the stations indicated with a superscript minor changes in coordinates have necessitated recalculating a, b, c. Some of these represent new locations, so that the data given in the 1951 tables apply to earlier years. The largest such change is at Honolulu; this station operated from 1903 to 1921 at Ewa, from 1921 to 1946 at the University of Hawaii (21° 18.1' N, 157° 49.3' W) and returned in 1946 to the Magnetic Observatory (21° 18' 13'' N, 158° 05' 44'' W) near its first location. The only station at Alger is now at the University; the old station at Alger-Bouzaréah (the Observatory) was discontinued in 1954. The Mt. Hamilton station is at the Lick Observatory, and it continues to be listed in the *International Summary* as "Lick." Santa Lucia, Chile, is the headquarters station in the city of Santiago. In 1956 the official designation of Resolute Bay was changed to Resolute. The spelling Uppsala is now official, replacing the older form, Upsala.

APPENDIX XIV

Lists of Large Earthquakes

Table XIV-1   Great Shallow Earthquakes—1896–1903*

Date			G.C.T.	Lat.	Long.	M
1897	Feb.	7	07:36±	40 N	140 E	8.3±
	Feb.	19	20:48±	38 N	142 E	8.3±
	Feb.	19	23:48±	38 N	142 E	8.3±
	May	13	12:30±	12 N	124 E	7.9±
	June	12	11:06±	26 N	91 E	8.7±
	Aug.	5	00:12±	38 N	143 E	8.7±
	Aug.	15	12:	18 N	120 E	7.9?
	Aug.	16	07:54±	39 N	143 E	7.9±
	Sept.	20	19:06:	6 N	122 E	8.6±
	Sept.	21	05:12±	6 N	122 E	8.7±
	Oct.	18	23:48±	12 N	126 E	8.1
	Oct.	20	14:24±	12 N	126 E	7.9
1898	Jan.	24	23:	?	?	7.9?
	Apr.	22	23:36±	39 N	142 E	8.3±
	Apr.	29	16:18±	12 N	86 W	7.9±
	June	29	18:36±	?	?	8.3±
	Aug.	31	19:54±	?	?	7.9±
	Nov.	17	12:48±	?	?	7.8±
1899	Jan.	24	23:43±	17 N	98 W	8.4±
	June	14	11:09±	18 N	77 W	7.8
	July	14	13:32?	?	?	7.8±
	Aug.	24	15:09±	?	?	7.8
	Sept.	4	00:22	60 N	142 W	8.3
	Sept.	10	17:04	60 N	140 W	7.8
	Sept.	10	21:41	60 N	140 W	8.6
	Sept.	29	17:03	3 S	128½ E	7.8
	Nov.	23	09:49±	53 N	159 E	7.9
	Nov.	24	18:42	32 N	131 E	7.8
	Nov.	24	18:55	32 N	131 E	7.8?

Table XIV-1   Continued

Date			G.C.T.	Lat.	Long.	M
1900	Jan.	11	09:07	?	?	7.8?
	Jan.	20	06:33	20 N	105 W	8.3 ±
	May	16	20:12	20 N	105 W	7.8 ±
	June	21	20:52	20 N	80 W	7.9
	July	29	06:59	10 S	165 E	8.1
	Oct.	7	21:04	4 S	140 E	7.8
	Oct.	9	12:28	60 N	142 W	8.3
	Oct.	29	09:11	11 N	66 W	8.4
	Dec.	25	05:04	43 N	146 E	7.8
1901	Jan.	7	00:29	2 S	82 W	7.8
	Apr.	5	23:30	45 N	148 E	7.9
	June	24	07:02	27 N	130 E	7.9
	Aug.	9	09:23	40 N	144 E	7.9
	Aug.	9	13:01	22 S	170 E	8.4
	Aug.	9	18:33	40 N	144 E	8.3
	Dec.	14	22:57	14 N	122 E	7.8
	Dec.	31	09:02	52 N	177 W	7.8
1902	Jan.	1	05:20	55 N	165 W	7.8
	Jan.	24	23:27	8 S	150 E	7.8
	Feb.	9	07:35	20 S	174 W	7.8
	Apr.	19	02:23	14 N	91 W	8.3
	Aug.	22	03:00	40 N	77 E	8.6
	Sept.	22	01:46	18 N	146 E	8.1
	Sept.	23	20:18	16 N	93 W	8.4
	Dec.	12	23:10 ±	29 N	114 W	7.8
1903	Jan.	14	01:47	15 N	98 W	8.3
	Feb.	1	09:34	48 N	98 E	7.8
	Feb.	27	00.43	8 S	106 E	8.1
	May	13	06:34	17 S	168 E	7.9 ±
	Dec.	28	02:56	7 N	127 E	7.8

* As listed by B. Gutenberg, *Trans. Am. Geophys. Union,* vol. 37 (1956), pp. 608–614; *m* replaced by *M* according to Table 22-4. Where ± follows the minute it may be in error up to 5 minutes.

Table XIV-2   Great Shallow Earthquakes ($M = 7.9$ or over), 1904–1956*

Date			G.C.T.	Lat.	Long.	M
1904	Jan.	20	14:52.1	7 N	79 W	7.9
	June	25	14:45.6	52 N	159 E	8.3
	June	25	21:00.5	52 N	159 E	8.1
	June	27	00:09.0	52 N	159 E	7.9
	Aug.	24	20:59.9	30 N	130 E	7.9
	Aug.	27	21:56.1	64 N	151 W	8.3
	Dec.	20	05:44.3	8½ N	83 W	8.3
1905	Feb.	14	08:46.6	53 N	178 W	7.9
	Apr.	4	00:50.0	33 N	76 E	8.6
	July	6	16:21.0	39½ N	142½ E	7.9
	July	9	09:40.4	49 N	99 E	8.4
	July	23	02:46.2	49 N	98 E	8.7
1906	Jan.	31	15:36.0	1 N	81½ W	8.9
	Apr.	18	13:12.0	38 N	123 W	8.3
	Aug.	17	00:10.7	51 N	179 E	8.3
	Aug.	17	00:40.0	33 S	72 W	8.6
	Sept.	14	16:04.3	7 S	149 E	8.4
	Dec.	22	18:21.0	43½ N	85 E	8.3
1907	Apr.	15	06:08.1	17 N	100 W	8.3
	Oct.	21	04:23.6	38 N	69 E	8.1
1911	Jan.	3	23:25:45	43½ N	77½ E	8.7
	June	7	11:02.7	17½ N	102½ W	7.9
	Aug.	16	22:41.3	7 N	137 E	8.1
1912	May	23	02:24.1	21 N	97 E	7.9
1913	Mar.	14	08:45:00	4½ N	126½ E	8.3
	Aug.	6	22:14.4	17 S	74 W	7.9
1915	May	1	05:00.0	47 N	155 E	8.1
1916	Jan.	1	13:20.6	4 S	154 E	7.9
	Jan.	13	08:20.8	3 S	135½ E	8.1
1917	Jan.	30	02:45.6	56½ N	163 E	8.1
	May	1	18:26.5	29 S	177 W	8.6 ± [1]
	June	26	05:49.7	15½ S	173 W	8.7
1918	Aug.	15	12:18.2	5½ N	123 E	8.3
	Sept.	7	17:16:13	45½ N	151½ E	8.3
	Nov.	8	04:38.0	44½ N	151½ E	7.9
1919	Apr.	30	07:17:05	19 S	172½ W	8.4
	May	6	19:41:12	5 S	154 E	8.1
1920	June	5	04:21:28	23½ N	122 E	8.3
	Sept.	20	14:39:00	20 S	168 E	8.3

Table XIV-2   Continued

Date			G.C.T.	Lat.	Long.	M
1920	Dec.	16	12:05:48	36 N	105 E	8.6
1922	Nov.	11	04:32.6	28½ S	70 W	8.4
1923	Feb.	3	16:01:41	54 N	161 E	8.4
	Sept.	1	02:58:36	35¼ N	139½ E	8.3
1924	Apr.	14	16:20:23	6½ N	126½ E	8.3
	June	26	01:37:34	56 S	157½ E	8.3
1926	Oct.	3	19:38:01	49 S	161 E	7.9[1]
	Oct.	26	03:44:41	3¼ S	138½ E	7.9
1927	Mar.	7	09:27:36	35¾ N	134¾ E	7.9
	May	22	22:32:42	36¾ N	102 E	8.3
1928	Mar.	9	18:05:27	2½ S	88½ E	8.1
	June	17	03:19:27	16¼ N	98 W	7.9
	Dec.	1	04:06:10	35 S	72 W	8.3
1929	Mar.	7	01:34:39	51 N	170 W	8.6[1]
	June	27	12:47:05	54 S	29½ W	8.3
1931	Jan.	15	01:50:41	16 N	96¾ W	7.9
	Feb.	2	22:46:42	39½ S	177 E	7.9
	Aug.	10	21:18:40	47 N	90 E	7.9
	Oct.	3	19:13:13	10½ S	161¾ E	8.1
1932	May	14	13:11:00	½ N	126 E	8.3
	June	3	10:36:50	19½ N	104¼ W	8.1
	June	18	10:12:10	19½ N	103½ W	7.9
1933	Mar.	2	17:30:54	39¼ N	144½ E	8.9
1934	Jan.	15	08:43:18	26½ N	86½ E	8.4
	Feb.	14	03:59:34	17½ N	119 E	7.9
	July	18	19:40:15	11¾ S	166½ E	8.1
1935	Sept.	20	01:46:33	3½ S	141¾ E	7.9
	Dec.	28	02:35:22	0	98¼ E	8.1
1938	Feb.	1	19:04:18	5¼ S	130½ E	8.6
	May	19	17:08:21	1 S	120 E	7.9
	Nov.	10	20:18:43	55½ N	158 W	8.7
1939	Jan.	25	03:32:14	36¼ S	72¼ W	8.3[1]
	Jan.	30	02:18:27	6½ S	155½ E	7.9
	Apr.	30	02:55:30	10½ S	158½ E	8.1
	Dec.	26	23:57:21	39½ N	39½ E	7.9
1940	May	24	16:33:57	10½ S	77 W	8.4[1]
1941	June	26	11:52:03	12½ N	92½ E	8.7[1]
	Nov.	18	16:46:22	32 N	132 E	7.9
	Nov.	25	18:03:55	37½ N	18½ W	8.4

Table XIV-2 Continued

Date		G.C.T.	Lat.	Long.	M
1942	Apr. 8	15:40:24	13½ N	121 E	7.9
	May 14	02:13:18	¾ S	81½ W	8.3
	Aug. 6	23:36:59	14 N	91 W	8.3[1]
	Aug. 24	22:50:27	15 S	76 W	8.6[1]
	Nov. 10	11:41:27	49½ S	32 E	8.3
1943	Apr. 6	16:07:15	30¾ S	72 W	8.3[1]
	May 25	23:07:36	7½ N	128 E	8.1
	July 29	03:02:16	19¼ N	67½ W	7.9
	Sept. 6	03:41:30	53 S	159 E	7.9
1944	Dec. 7	04:35:42	33¾ N	136 E	8.3
1945	Nov. 27	21:56:50	24½ N	63 E	8.3
1946	Aug. 2	19:18:48	26½ S	70½ W	7.9[1]
	Aug. 4	17:51:05	19¼ N	69 W	8.1
	Aug. 8	13:28:28	19½ N	69½ W	7.9
	Dec. 20	19:19:05	32½ N	134½ E	8.4
1947	July 29	13:43:22	28½ N	94 E	7.9[1]
1948	Jan. 24	17:46:40	10½ N	122 E	8.3
	Mar. 1	01:12:28	3 S	127½ E	7.9[1]
	Sept. 8	15:09:11	21 S	174 W	7.9
1949	Aug. 22	04:01:11	53¾ N	133¼ W	8.1
1950	Aug. 15	14:09:30	28½ N	96½ E	8.7
	Nov. 2	15:27:56	6½ S	129½ E	8.1[1]
	Dec. 2	19:51:49	18¼ S	167½ E	8.1[1]
1951	Nov. 18	09:35:47	30½ N	91 E	7.9
	Dec. 8	04:14:12	34 S	57 E	7.9
1952	Mar. 4	01:22:43	42½ N	143 E	8.6
	Mar. 19	10:57:12	9½ N	127¼ E	7.9
	Nov. 4	16:58:26	52¾ N	159½ E	8.4
1953	Nov. 25	17:48:52	34 N	141½ E	8.3
1954	None				
1955	None				
1956	None[2]				

* From *Seismicity of the Earth* (magnitudes as revised, 1956) by Gutenberg; *m* replaced by *M* according to Table 22-4.

[1] Shocks originating at depths of the order of 50–60 kilometers.

[2] The next shock in continuation of this list would be that of 1957 March 9, at 51° N 176° W (Aleutian Islands).

Table XIV-3   Known Intermediate Shocks ($M = 7.9$ or over), 1903–1956*

Date			G.C.T.	Lat.	Long.	Depth (km)	M
1903	June	2	13:17:	57 N	156 W	100?	8.3±
	Aug.	11	04:32.9	36 N	23 E	100+	8.3
1905	Jan.	22	02:43.9	1 N	123 E	90	8.4
	June	2	05:39.7	34 N	132 E	100±	7.9
1906	Sept.	28	15:24.9	2 S	79 W	150±	7.9
1907	June	25	17:54.6	1 N	127 E	200	7.9
1908	Mar.	26	23:03.5	18 N	99 W	80±	8.1
1909	Mar.	13	14:29.0	31½ N	142½ E	80	8.3
	July	7	21:37:50	36½ N	70½ E	230±	8.1
	Nov.	10	06:13.5	32 N	131 E	190	7.9
1910	Apr.	12	00:22:13	25½ N	122½ E	200	8.3
	June	16	06:30.7	19 S	169½ E	100	8.6
	Nov.	9	06:02.0	16 S	166 E	70±	7.9
1911	June	15	14:26.0	29 N	129 E	160±	8.7
1913	Oct.	14	08:08.8	19½ S	169 E	230	8.1
1914	Nov.	24	11:53:30	22 N	143 E	110±	8.7
1915	Sept.	7	01:20.8	14 N	89 W	80	7.9
1918	May	20	17:55:10	28½ S	71½ W	80±	7.9±
	Nov.	18	18:41:55	7 S	129 E	190	8.1
1919	Jan.	1	02:59:57	19½ S	176½ W	180	8.3
1921	Nov.	15	20:36:38	36½ N	70½ E	215	8.1
1926	June	26	19:46:34	36½ N	27½ E	100	8.3
1939	Dec.	21	21:00:40	0	123 E	150±	8.6
1943	July	23	14:53:09	9½ S	110 E	90	8.1
1950	Dec.	9	21:38:48	23½ S	67½ W	100	8.3
	Dec.	14	01:52:49	19¼ S	175¾ W	200	7.9
1951–1956	None						

* From *Seismicity of the Earth,* with revisions and additions by Gutenberg; *m* replaced by *M.*

Table XIV-4   Known Deep Shocks ($M = 7.9$ or over), 1903–1956*

Date	G.C.T.	Lat.	Long.	Depth (km)	M
1903 Jan. 4	05:07	20 S	175 W	400?[1]	8±
1904 June 7	08.17.9	40 N	134 E	350±	7.9±
1906 Jan. 21	13:49:35	34 N	138 E	340	8.4
1907 May 25	14:02:08	51½ N	147 E	600	7.9
1909 Feb. 22	09:21.7	18 S	179 W	550	7.9
1921 Dec. 18	15:29:35	2½ S	71 W	650	7.9
1932 May 26	16:09:40	25½ S	179¼ E	600	7.9
1937 Apr. 16	03:01:37	21½ S	177 W	400	8.1
1950 Feb. 28	10:20:57	46 N	144 E	340	7.9

* Revised by B. Gutenberg; *m* replaced by *M.*

[1] May possibly have been at intermediate depth.

APPENDIX XV

African Earthquake, 1928

Notes on earthquake of 6–1–28. On Saturday, 4–2–28, I proceeded by car to Farm No. 3230 (Major Boyce) for the purpose of making a reconnaissance during the week end of the movements which gave rise to the earthquake at about 10 hrs. 3 mins. P.M. on 6–1–28. Reports from the Ag. Executive Engineer, Nakuru, dated 18–1–28, 27–1–28 and 31–1–28 had already demonstrated the existence in the Sabukia Valley of a crack of the nature of a gravity fault running for some miles along the Laikipia Escarpment near its base and of small cracks on the floor of the valley on and in the neighborhood of Farm 3230. The damage done to buildings had also been shown to be greater in the Sabukia and Solai Valleys than elsewhere in the Colony. Disturbances of various descriptions had also been reported on native information from Lake Hannington. On Saturday afternoon I started for that locality on foot. On the way, natives who lived in the region of Lake Hannington assured me that there was no visible evidence of the effect of the earthquake other than landslides, the chief one being the falling of a large quantity of boulders from the steep scarp overlooking Lake Hannington. Nothing of the nature of cracks had been seen, and the lake had not been disturbed. This information was subsequently confirmed by other natives. I, therefore, did not visit Lake Hannington but diverged towards the lower part of the Sabukia Valley as all the evidence pointed to the epicentral area being situated along the Laikipia Escarpment overlooking that valley. The ensuing day (Sunday 5–2–28) was spent in examining the cleft which was found to extend along the Laikipia Escarpment for 10 miles, approximately from long. 36°–12′ E, lat. 0°–17′ N, to long. 36°–16′ E, lat. 0°–10′ N as shown roughly on the accompanying map as indicated by the line *AD*. [Letters *A, B, C, D* refer to a map not now available—C.F.R.] At the former point (*A* on the map) the cleft was found to have become so small in magnitude that it was difficult to follow, and at the latter point (*D* on map) it was lost in the steep southwestern slope of Marmanet to which it had diverged. Between *B* and *C* it showed the maximum movement. Unconfirmed native reports allege that it continues northward along the Laikipia Escarpment past Lake Baringo, and it is also stated on reliable authority that it reappears on the escarpment southeast of Marmanet and continues for five or more miles in that direction. It is also stated that another similar cleft appears on the flank of a valley parallel to the Sabukia Valley a few miles northeastward of it. Time was not available to investigate these statements. I returned to Nakuru on Sunday night.

2. The Laikipia Scarp may be regarded as forming the chief eastern wall of the Rift Valley in this area. The Sabukia River, which was carrying about ¾ cusec at the lowest point reached by me, flows parallel to its foot in a gorge cut by it through the valley. The scarp is much denuded and dissected by dry ravines which carry the flood drainage from it to the Sabukia River. In height it varies from 2000 to 2500 feet and has an average slope of 10° to 25°. It is an old fault scarp which is now much worn away, the lower part of the slope being formed mostly of screes, clays and decomposed volcanic rocks. It has

been regarded as Pliocene in age, but the evidence is not conclusive. Its formation was probably very slow and proceeded by a series of jolts occurring at intervals. Evidence exists at one locality of vulcanicity after the scarp had assumed a form approximating to its present outlines but probably not in recent times.

3. On the lower slopes of this scarp there has been a recurrence of gravity faulting along a plane which is so situated that one would expect it to be in the locality of one of the original main planes of faulting. Along this plane the strata has fractured and has subsided on the southwest or down hill side. The cracking of the underlying strata doubtless produced the earthquake. This cracking was probably caused primarily by the tension in the mass of volcanic rock below resulting from the loss of heat over a long period. The tension in the rock again reached the intensity of breaking point and caused adjustment of stresses by fracture and subsidence. It has manifested itself at the surface in a crack or cleft or series of adjacent ones following an average bearing of 138° along the slope of the Laikipia Escarpment for a distance of about 10 miles, or possibly much more. It varies a good deal from this average direction, bearings of 106° and 168° being noted at particular points. Transverse valleys are generally crossed by it at right angles, but the presence of hills in its course appears to have caused it to diverge round their flanks in some cases. Normally the down-throw is to the southwest, the southwestern side having sunk, so forming a step in the hill slope and the ground having opened in a trench. Occasionally, however, where the plane of movement diverges considerably from its average direction, the movement has been of the nature of reversed faulting, one side having been slightly thrust up and often overlapping the surface of the other side. The movement in those parts appears to have been due to pressure or shear through conjugate stresses being set up in particular directions as a result of the dominant tensional stress. The line of the fault appears to follow, on the average, the 6,000 ft. contour along the flank of the escarpment, but, owing to the presence of ravines and hills on the flank, its level varies in elevation by some 500 ft. in different parts. The amount of movement, both vertical and horizontal, which is registered by surface movement varies greatly and the form which it takes is also various. It increases gradually in magnitude from A to B; from B to C it is at its maximum, and from C to D it gradually becomes reduced. Ordinarily it takes the form of a cleft varying from a few inches in width to a maximum of 10 ft. with one or more small trenches roughly parallel to it within 50 ft. on either side. The width at the surface is generally much greater than at depth owing to disturbance of the soil and screes which have fallen into the cleft. Usually the northeastern or uphill edge of the cleft is higher than the southwestern or down hill side by an amount which varies from a few inches to a maximum of 11 ft. and shows the amount of vertical displacement at that particular point of the cleft. The small subsidiary clefts show similar but smaller differences of level between the two edges. In other cases, though rarely, the subsidence at the surface takes the form of a small trough fault, a strip of ground having subsided between parallel planes to a depth of 3 or 4 feet below the former surface of the ground. As previously mentioned, when the line diverges much from its average direction, the fracture sometimes takes the form of a reversed fault, the soil of one side with trees and other vegetation having sometimes been thrust over the other side to distances varying from a few inches to a few feet.

4. Small clefts showing vertical and horizontal displacement up to 1 foot in each direction are met with at intervals on the foothills below the main cleft and on the valley floor within a couple of miles of the cleft. These minor clefts vary greatly in direction, some being roughly parallel to the general direction of the main cleft on the flank of the escarpment; others are almost at right angles to it, while others again are intermediate in direction. As one would expect, the northeastern side of the valley below the main cleft seems to have subsided, the maximum movement being along the plane of the main cleft,

but causing subsidiary fractures in other parts as the subjacent strata subsided and adjusted itself.

5. The earth tremors resulting from the fracture of the strata have produced landslides on steep slopes, especially in the region of the Sabukia and Solai Valleys. Many large boulders have collected at the feet of such slopes, often overturning trees in their courses down the hillsides.

6. Damage has been done to many houses and other structures in the Sabukia and Solai Valleys and to a lesser degree in other parts, especially at Ravine, Fort Hall and Nyeri. A statement of the damage done to property in the Sabukia and Solai Valleys, as noted by the Ag. Executive Engineer, Nakuru, is attached to these notes with a rough estimate of the damage.

7. After-shocks, which are common after an earthquake and sometimes last for months at frequent intervals, have been occurring since the main shock of January 6th. They were still noticeable in the Sabukia Valley last week. These are due to the strata continuing to adjust itself locally by fracture or other movement to the altered stresses in the effort to achieve equilibrium. It is to be observed that the cleft intercepts all the drainage from the upper part of the Laikipia Escarpment for a distance of at least 10 miles, and it is not improbable that during the rains water in quantity will find its way into the cleft and may cause further adjusting movements. Such movements may be considerable or trivial, but their possibility should not be disregarded. One small stream, known as the Little Sabukia, has been intercepted by one of the clefts and will have to be flumed across it. It has been noted in many cases that streams formerly clear have become turbid. This is due to earth tremors disintegrating the soil so that particles formerly adherent to adjacent particles here become disconnected from them and easily removable by water. It is stated that some streams have become reduced in flow and others have increased. This cannot be determined quantitatively owing to the absence of measurements. It is reasonable to suppose that it has taken place. Reduction of flow would obviously be due to fracture or greater absorption owing to disintegration. Temporary increase of flow would be caused by disintegration on account of the earth tremors of water-bearing strata from which the stream is fed, so enabling that strata to discharge its water more rapidly than before.

8. It is to be observed that the earth movements which have taken place are not inconsiderable and are not disproportionate to some of those which have caused very destructive earthquakes. If a town had been situated in the Sabukia or Solai Valleys, the loss of life and property would have been likely to be considerable. The absence of earthquakes of importance in recent times in the Rift Valley is remarkable, for one would expect it to be a seismic zone where earth movements causing rejuvenation of fault scarps, as in this case, continued to occur at intervals.

9. It was intended to place a cement tie between the sides of the cleft to show any further movement. In no place could there be found sufficiently sound rock in the sides of the cleft to render this possible. Arrangements were therefore made for concrete pegs to be established at certain spots and their relative levels ascertained with precision. It can subsequently be determined by levelling whether relative movement has taken place or not.

10. It seems very desirable that 3 seismological stations should be established in Kenya. It is suggested that they might be at Nairobi, Nakuru and Nyeri. An estimate of the cost and advice regarding the type could be obtained from the Director of Geological Survey, Uganda.

Signed: H. L. Sikes
           Director of Public Works
           10th February, 1928

APPENDIX XVI

Chronological Bibliography

THIS BIBLIOGRAPHY lists publications on individual earthquakes or groups of earthquakes (generally excluding purely seismometric studies). References of other types are appended to the appropriate chapters of the text and may be located from the general index. For seismometric studies see Chapters 17, 18, 19, 21; for tsunamis see Chapter 9. Lists of large earthquakes are given in Appendix XIV and Chapters 22, 28, 29, 30.

Earthquakes are given here in chronological order, by local date of the country of origin. Page references to this book appear on the first line of each entry.

Abbreviations are:

B	F. de Montessus de Ballore, *Géologie sismologique* (1924)
D	C. Davison, *Great Earthquakes* (1936)
F	J. R. Freeman, *Earthquake Damage and Earthquake Insurance* (1931)
IM	A. Imamura, *Theoretical and Applied Seismology* (1937)
B.S.S.A.	*Bulletin of the Seismological Society of America*
C.R.	*Comptes rendus hebdomadaires de l'Académie des Sciences* (Paris)
E.R.I.	*Bulletin of the Earthquake Research Institute* (Tokyo)
Geophys. Mag.	*Geophysical Magazine* (Tokyo)
N. Z. Journ.	*New Zealand Journal of Science and Technology*, Series B
N. Z. Trans.	*Transactions of the Royal Society of New Zealand*

Abbreviations vol. and pp. are omitted from periodical references.

373 B.C. (Helice)   206, 616
    B, pp. 153–156.
    Pausanias, *Description of Greece,* vii, 24.
    Strabo, *Geography,* viii, 7.

818 A.D., A.D. 1703 (Japan)   206, 571
    IM, pp. 179–182.

1556 (China)   641
    IM, p. 142. (Date given as February 2.)
    Kuo, T. C., "On the Shensi earthquake of January 23, 1556," *Acta Geophysica Sinica* 6 (1957), 49–68. (Chinese, with English abstract.)

1707, October 28 (Nankaidō, Japan)   563
    IM, pp. 184–187.

1755, November 1 (Lisbon)   104–105, 110, 113–114
    D, Chapter 1 (pp. 1–28).
    Letters in *Trans. Roy. Soc. (London)* 49 (1755) 351–398.
    Reid, H. F., "The Lisbon earthquake of November 1, 1755," *B.S.S.A.*
        4 (1914) 53–80.
    Sousa, F. L. Pereira de, "Sur les effets en Portugal du megaséisme
        du 1er novembre 1755," *C.R.* 158 (1914) 2033–2035.
    ————, *O terremoto do 1er do novembro de 1755 em Portugal e um
        estudo demográfico*, Lisboa, 1919, 2 vols.
    Kendrick, T. D., *The Lisbon earthquake*. American edition, Lippin-
        cott, Philadelphia and New York, 1957.

1762, April 2 (Arakan, Burma)   607
    Mallet, F. R., "The mud volcanoes of Ramri and Cheduba," *Records
        Geol. Survey India* 11 (1878) 188–207.
    Halsted, E. P., "Report on the island of Chedooba," *Journ. Asiatic
        Society, Calcutta*, 10 (1841) 433 ff.
    Baird Smith, R., "Historical survey of Indian earthquakes . . . ,"
        ibid. 12 (1843) 1029 ff., especially p. 1050.

1811–1812 Dec. 16, etc. (New Madrid)   593–594
    B, pp. 16–33.
    D, Chapter III, pp. 54–67.
    Lyell, C., *A Second Visit to the United States of North America*,
        London, 1849, vol. 2, pp. 228–239.
    ————, *Principles of Geology*, 12th ed., London, 1875, Vol. 1, pp.
        452–453; Vol. 2, pp. 106–110.
    Shepard, E. M., "The New Madrid earthquake," *Journ. Geology* 13
        (1905) 45–62.
    Fuller, M. L., "The New Madrid earthquake," *U. S. Geol. Survey,
        Bull.* 494 (1912).
    Morse, W. C., "New Madrid earthquake craters," *B.S.S.A.* 31 (1941)
        309–319. (Describes evidence remaining in 1936 and 1940.)

1812, December 8 and December 21 (California)   113, 452, 466, 472
    Carpenter, F. A., "Early records of earthquakes in southern Califor-
        nia," *B.S.S.A.* 11 (1921) 1–3. Refer also to the catalogues cited
        in Chapter 28.

1819, June 16 (Cutch, India)   *47*, 189, 190, 192, 607–608
    B, pp. 137–148.
    D, Chapter IV, pp. 68–76.
    Oldham, R. D., "A note on the Allah Bund in the north-west of the
        Rann of Kucch," *Mem. Geol. Survey India* 28 (1898) 27–30.
    ————, "The Cutch (Kacch) earthquake of 16th June 1819, with a
        revision of the great earthquake of 12th June 1897," *ibid.* 46
        (1928) 71–147.

1822, November 19; 1835, February 20 (Chile)   190, 600
    B, pp. 68–72.

D, Chapters V, pp. 77–82; VI, pp. 89–94.

Graham, Maria, "An account of some effects of the late earthquake in Chili," *Trans. Geol. Soc. (London)*, Ser. 2, 1 (1824) 413–415.

"Earthquake in Chili, Feb. 20, 1835," *Am. Journ. Science* 28 (1835) 336–340.

Caldcleugh, A., "An account of the great earthquake experienced in Chile on the 20th of February, 1835," *Trans. Royal Soc. (London)* 1836, 21–26.

Fitzroy, R., "Sketch of the surveying voyages of his Majesty's ships Adventure and Beagle, 1825–1836," *Geographical Journ.* 6 (1836) 311–343.

Lyell, C., *Principles of Geology*, 12th ed., 1875, Vol. 2, pp. 94–97.

Suess, E., *Das Antlitz der Erde*, Vol. 1, Part 1, Section 2.

Brüggen, J., *Contribución a la geología sísmica de Chile*, Imprensa Universitaria, Santiago, 1943. (See especially pp. 100–110.)

1836, 1838 (California)   472, 473, 476
Louderback, G. D., "California earthquakes of the 1830's," *B.S.S.A.* 37 (1947) 33–74.

1848, October 16, 17, 19 (Awatere, New Zealand)   455, 538, 540–541
B, p. 208.

D, p. 266.

Lyell, C., *Bull. Soc. géol. française*, Ser. 2, 13 (1857) 661–667.

———, *Principles of Geology*, 12th ed., 1875, Vol. 2, pp. 82–89.

Cotton, C. A., "Submergence in the lower Wairau valley," *N. Z. Journ.* 35 (1954) 364–369.

1855, January 23 (New Zealand)   189, 538, 541–542
B, pp. 206–209.

D, pp. 266–267.

Lyell, C., *Principles of Geology*, 12th ed., 1875, Vol. 2, pp. 82–89.

Ongley, M., "Surface trace of the 1855 earthquake," *N. Z. Trans.* 73 (1943) 84–89.

1857, January 9 (Fort Tejon, California)   67, 472, 473, 475, 481–482, 533
B, pp. 118–119.

D, p. 178.

Lawson, A. C., *et al. The California Earthquake of 1906* (see reference under 1906), Vol. 1, pp. 449–451.

Wood, H. O., "The 1857 earthquake in California," *B.S.S.A.* 45 (1955) 47–67.

1857, December 16   66, 30–37, *35*, 45
Mallet, R., *Great Neapolitan earthquake of 1857, etc.*, London, 1862, 2 vols. (See Chapter 4 and references.)

1861, December 26 (Greece)   *612*, 616–617
B, pp. 149–153.

Schmidt, J., *Studien über Vulkanen und Erdbeben*, Leipzig, Vol. 2, 1881, p. 68.

1868, April 2 (Hawaii)   161–162, 166
B, pp. 239–248,

Wood, H. O., "On the earthquakes of 1868 in Hawaii," *B.S.S.A.* 4 (1914) 169–203.

——, "Volcanic earthquakes," Chapter 3, *Natl. Research Council Bull.* 90, *Seismology*, 1933.

1868, October 21 (Haywards, California)   472, 476
F, pp. 173–186.

Lawson, A. C., *et al., The California Earthquake of 1906*, Vol. 1, pp. 434–448.

1872, March 26 (Owens Valley)   67, 174, 183, 472, 499–503
B, pp. 76–81.

D, Chapter VII, pp. 96–104.

F, pp. 227–230.

Whitney, J. D., "The Owens Valley earthquake," *Overland Monthly* 9 (1872) 130–140, 266–278. Reprinted: *8th Annual Report of the State Mineralogist*, California State Mining Bureau, Sacramento, 1888, pp. 288–309.

Hobbs, W. H., "The earthquake of 1872 in the Owens Valley, California," *G. Beitr.* 10 (1910) 352–385.

Knopf, A., "A geologic reconnaissance of the Inyo Range and the eastern slope of the southern Sierra Nevada, California," *U. S. Geol. Survey, Prof. Paper* 110 (1918).

Muir, John, *The Yosemite*, New York, 1912, "Earthquake storms," pp. 76–86.

Mulholland, C., "The Owens Valley earthquake of 1872," *Ann. Publ. Historical Soc. Southern California*, 1894, pp. 27–34. (Right-hand offset near Lone Pine.)

"Geology of the Owens Valley region," *California Dept. Nat. Research, Div. Mines, Bull.* 170 (1954), map sheet No. 11.

1875, January 24? (Eastern California)   515–516
Turner, H. W., "Further contributions to the geology of the Sierra Nevada," *17th annual report, U. S. Geol. Survey*, 1896, pt. 1, 521–740 (see p. 593).

"Downieville folio," *U. S. Geol. Survey folio No. 37*, 1896; with map, 1/125,000.

1883, July 28 (Ischia)   159–161
B, pp. 255–262.

Johnston-Lavis, H. J., *Monograph of the Earthquakes of Ischia*, London and Naples, 1886.

Mercalli, G., *L'isola d'Ischia ed il terremoto del 28 luglio 1883*, Milan, 1884.

Davison, C., *A Study of Recent Earthquakes*, 1905. Chapter III.

1884, April 22 (Colchester)   406
Meldola, R., and White, W., *Report on the East Anglian Earthquake of April 22, 1884*, Macmillan, London, 1885.

Davison, C., *A History of British Earthquakes*, Cambridge University Press, 1924, pp. 337–344.

1886, August 31 (Charleston) 73, 130
F., pp. 283–316.
Dutton, C. E., "The Charleston earthquake of August 31, 1886,"
U. S. Geol. Survey, 9th Annual Rept., 1887–1888, pp. 203–528,
pls. VII–XXI.

1887, May 3 (Sonora, Mexico) 147, 594–595
B, pp. 81–86.
Goodfellow, G. E., "The Sonora earthquake," *Science* 11 (1888)
162–168. (Extracts from communications by Goodfellow to Dut-
ton.)
Aguilera, J. G., "Estudios de los fenómenos sísmicos del 3 de mayo
1887," *Anal. Ministerio Fomento República Mexicana,* 10 (1888).
————, "The Sonora earthquake of 1887," *B.S.S.A.* 10 (1920) 31–
44. (A translation, prepared for A. C. Lawson, of parts of the
preceding reference.)
Dutton, C. E., *Earthquakes in the Light of the New Seismology,* New
York, 1904, pp. 53 ff.
Staunton, W. F., "Effects of an earthquake in a mine at Tombstone,
Arizona," *B.S.S.A.* 8 (1918) 25–27.
MacDonald, B., "Remarks on the Sonora earthquake, *ibid.,* 74–78.

1887, June 9 (Vyernyi) 610
Mushketov, I. V., "Vernenskoe zemletryasenie 28 Maya (9 Juniya)
1887 g.," *Trudy geol. komitata* 10 No. 1, 1890–1891.

1888, September (Amuri) 459, 538, 543
B, pp. 209–210.
McKay, A., "On the earthquakes of September, 1888, in the Amuri
and Marlborough districts of the South Island," *Colonial Museum
and Geol. Survey N. Z.,* Rept. No. 20, pp. 1–16.
Cotton, C. A., "The Hanmer plain and the Hope fault," *N. Z. Journ.*
29 (1947) 10–17.

1891, October 28 (Mino-Owari) 68, 180, *556,* 557, 563–566
B, pp. 165–174.
D, Chapter VIII, pp. 105–129.
IM, pp. 50–52, 193–195.
Kotō, B. On the cause of the great earthquake in central Japan, 1891.
Journ. Coll. Science, Imp. Univ. Japan 5 (1893), part 4, 296–353,
pls. XXVIII–XXXV.

1892, May 17 (Sumatra) 606
B, pp. 131–134.
Muller, J. J. A., "De verplaatsing van eenige triangulatie-pilaren in
de Residentie Tapanoeli (Sumatra) tengevolge van de aardbeving
van 17 mei 1892," *Verhandel. Koninkl. Akad. Wetenschap. Amsterdam,* Section 1, 1895.
Reid, H. F., "Sudden earth movements in Sumatra in 1892," *B.S.S.A.*
3 (1913) 72–79.

1892, December 20 (Baluchistan)   170–171, 608–609
B, pp. 86–88.
Egerton, R. W., "Effect of earthquakes on the Northwest Railway,
India," *Engineering* 55 (1893) 698 ff.   ,
Griesbach, C. L., "Notes on the earthquake in Baluchistan on the 20th
December 1892, *Records Geol. Survey India* 26, Part 2 (1893)
57–64.
Davison, C., "Note on the Quetta earthquake of Dec. 20, 1892,"
Geol. Mag. 10 (1893) 356–360.
McMahon, A. H., "The southern borderlands of Afghanistan," *Geo-graphical Journ.* 9 (1897) 393–415, with map.

1894, April (Greece)   617–618
B, pp. 156–164.
Papasviliou, S.-A., "Sur le tremblement de terre de Locride (Grèce)
du mois d'avril 1894," *C.R.* 119 (1894) 112–114.
———, "Sur la nature de la grande crevasse produite a la suite du
dernier tremblement du terre de Locride," *ibid.* 380–381; errata,
p. 480.
Skuphos, Th. G., "Die zwei grosse Erdbeben in Lokris am 8/20 und
15/27 April 1894," *Zeitschr. ges. Erdkunde zu Berlin,* 29 (1894)
409 ff.

1894, October 23 (Sakata, Japan). See 1896.   See 1896

1896, August 31 (Riku-Ugo, Japan)   72, 566
B, pp. 88–91.
IM, pp. 82–83 (foreshocks).
Yamasaki, N., "Das grosse japanische Erdbeben im nördlichen Hon-shu am 31. August 1896," *Petermanns Mitteilungen* (1900).

1897, June 12 (Assam)   28, 47–56, 66, 100, 129–130, 190, 607
B, pp. 36–53.
D, Chapter X, pp. 138–157.
Oldham, R. D., "Report on the great earthquake of 12th June 1897,"
Mem. Geol. Survey India 29 (1899). (See our Chapter 5 and refer-ences.)

1899, September (Alaska)   596–600
B, pp. 174–190.
D, Chapter XI, pp. 158–174.
Tarr, R. S., and Martin, L., "The earthquakes at Yakutat Bay, Alaska,
in September, 1899," *U. S. Geol. Survey, Prof. Paper* 69 (1912).

1899, December 25 (San Jacinto, California)   496
Daneš, J. V., "Das Erdbeben von San Jacinto am 25. Dezember
1899," *Mitteilungen K.K. Geogr. ges. in Wien* No. 6–7 (1907)
339–347.
Claypole, E. W., "The earthquake at San Jacinto," *Am. Geologist* 25
(1900) 106–108.
See also 1918, April 21.

1901, November 16 (Cheviot, New Zealand) 538–544
B, pp. 210–211.
McKay, A., *Report on the Recent Seismic Disturbance within Cheviot County in Northern Canterbury and the Amuri District of Nelson, November and December, 1901,* Wellington, 1902.

1902, February 13, new style (Shemakha) 610
Bogdanovich, K. I., "Neskol'ko zamechaniy o zemletryaseniy v Shemakhe 30 yanvara 1902 goda (s kartoyu)," *Izv. postoyannoy tsentral'noy seysmicheskoy komissii* 1 (1903) 282–290.

1905, April 4 (Kangra, India) *47, 63*
B, pp. 53–60.
Middlemiss, C. S., "The Kangra earthquake of 4th April 1905," *Mem. Geol. Survey India* 38 (1910).

1906, January 31 (great earthquake, Colombia-Ecuador) 350, 351
Rudolph, E., and Szirtes, S., "Das kolumbianische Erdbeben am 31. Januar 1906," *G. Beitr.* 11 (1911) 132–199, 207–275.

1906, March 17 (Kagi, Formosa) 567
B, pp. 91–96.
Omori, F., "Preliminary note on the Formosa earthquake of March 17, 1906," *Imp. Earthquake Inves. Comm., Bull.* 1 (1907) 53 ff.

1906, April 18 (San Francisco) 67, 100, 107, 123–124, 139, 190–191, 205, 365, 476–487
B, pp. 96–118.
D, Chapter XII, pp. 175–200.
F, pp. 230–241, 319–367.
Lawson, A. C., *et al., The California Earthquake of 1906,* 2 vols., Washington, 1908 and 1910. (See references in Chapters 1 and 13, and the following reference.)
Jordan, D. S., *et al., The California Earthquake of 1906,* A. M. Robertson, San Francisco, 1907. (Readable, with some details not in the preceding references.)
Wood, H. O., " 'Apparent' intensity and surface geology," *Nat. Research Council Bull.* 90 (1933), 67–82. (Effects in San Francisco.)
Louderback, G. D., "Characteristics of active faults in the central Coast Ranges of California, with application to the safety of dams," *B.S.S.A.* 27 (1937) 1–28. (Describes offset of old dam in reservoir.)
Eckart, N. A., "Development of San Francisco's water supply to care for emergencies," *ibid.* 185–204. (Describes damage in 1906.)

1906, August 17 (Valparaiso) *272,* 600
B, pp. 415–418.
D, Chapter V, pp. 82–88.
Brüggen (reference for 1822).
Rudolph, E., and Tams, E., *Seismogramme des nordpazifischen und südamerikanischen Erdbebens am 16. August 1906, Begleitworte und Erläuterungen,* Strassburg, 1907.

1908, June 30 (Siberian meteorite)   157, 166
Tams, E., "Das grosse sibirische Meteor vom 30. Juni 1908 und die
bei seinem Niedergang hervorgerufenen Erd- und Luftwellen,"
Zeitschr. Geophysik 7 (1931) 34–37. (For other references see
Chapter 12.)

1911, January 4 (Tien Shan)   610
Bogdanovitch, Ch., Kark, I., Korolkow, B., and Mouchketow, D.,
"Tremblement de terre du 22 décembre 1910 (4 janvier 1911)
dans les districts septentrionaux du Thian-chan," *Comité géol.,
Mém.* 89 N.S. (1914). (In Russian.)
Gamburtzev, G. A., ed., "Problemi prognoza zemlyetryaseniy,"
Akad. Nauk, Trudy geofizicheskogo Instituta, No. 25 (152), 28,
29, 50.
Galitzin, B., "Das Erdbeben vom 3–4 Januar 1911," *Izvestiya Akad.
Imp. Nauk,* 1911, 127–136.

1911, February 18 (Pamir).   For references and discussion see Chapter 12.

1911, November 16 (South Germany)   142
Sieberg, A., and Lais, R., "Das mitteleuropäische Erdbeben vom 16.
November 1911, Bearbeitung der makroseismischen Beobachtungen," *Reichsanstalt für Erdbebenforschung, Jena, Veröff.* No. 4,
1925.
Gutenberg, B., *Die mitteleuropäischen Beben vom 16. November 1911
und vom 20. Juli 1913, Bearbeitung der instrumentellen Aufzeichnungen,* Zentralbureau der Internationalen seismologischen Assoziation, Strassburg, 1915.

1912, November 19 (Mexico)   595–596
B, pp. 215–219.
Urbina, F., and Camacho, H., "La zona megaseismica Acambay-
Tixmadeje, estado de Mexico, commovida el 19 de noviembre de
1912," *Inst. geol. Mexico, Bol.* 32 (1913).
de Ballore, F. de Montessus, "The Mexican earthquake of November
12, 1912," *B.S.S.A.* 7 (1917) 31–33. (November 19 is the correct
date.)

1915, June 22 (Imperial Valley)   533
F, pp. 257–259.
Beal, C. H., "The earthquake in the Imperial Valley, California,
June 22, 1915," *ibid.* 5 (1915) 130–149.

1915, October 2 (Pleasant Valley, Nevada)   67, 106, 124, 472, 502–506
Jones, J. C., "The earthquake of October 2, 1915, in Pleasant Valley,
Nevada," *ibid.* 190–205.
Page, B. M., "Basin range faulting of 1915 in Pleasant Valley, Ne-
vada," *Journ. Geology* 43 (1935) 690–707.
Muller, S. W., *et al.,* "Mt. Tobin quadrangle, Nevada," *U. S. Geol.
Survey,* 1951.
Ferguson, H. G., *et al.,* "Golconda quadrangle, Nevada," *ibid.* 1952.

1918, April 21 (San Jacinto, California)   496
  Townley, S. D., "The San Jacinto earthquake of April 21, 1918,"
  B.S.S.A. 8 (1918) 45–62.
  Rolfe, F., and Strong, A. M., "The earthquake of April 21, 1918,
  in the San Jacinto Mountains," *ibid.* 63–67.
  Arnold, R., "Topography and fault system of the region of the San
  Jacinto earthquake," *ibid.* 68–73.

1920, June 21 (Inglewood, California)   67, 533
  F, 381–384.
  Taber, S., "The Inglewood earthquake in Southern California, June
  21, 1920," *B.S.S.A.* 10 (1920) 129–145.
  ———, "The Inglewood fault zone," *ibid.* 14 (1924) 197–199.
  Kew, W. S. W., "Geologic evidence bearing on the Inglewood earth-
  quake of June 21, 1920," *ibid.* 13 (1923) 155–158.

1920, December 16 (Kansu, China)   350
  Dammann, Mlle. Y., "Le tremblement de terre du Kan-Sou du 16.
  Decembre 1920," *Publ. bureau central sismologique international,*
  Ser. B., 1924.
  Close, U., and McCormick, Elsie, "Where the mountains walked,"
  Nat. Geographic Mag. 41 (1922) 445–464.

1921, June 29 (New Zealand)   552
  Bullen, K. E., "The Hawke's Bay earthquake of 1921 June 29,"
  N. Z. Journ. 19 (1937) 199–205.

1922 (Taupo, New Zealand)   552
  Grange, L. I., "Taupo earthquakes, 1922," *N. Z. Journ.* 14 (1932)
  139–141.

1922, November 10 (Chile)   111
  Willis, B., *Studies in Comparative Seismology. Earthquake Conditions
  in Chile,* Carnegie Institution, Washington, D. C., 1929. (Includes
  seismometric study by Macelwane and Byerly.)

1923, September 1 (Kwantō, Japan)   107–108, 562, 567–571
  IM (on many pages).
  F, pp. 447–512, etc.
  Bureau of Social Affairs, Tokyo, *The Great Earthquake of 1923 in
  Japan,* Tokyo, 1926. (The principal general report.)
  Davison, C., *The Japanese Earthquake of 1923,* London, 1931. (To
  be used with caution.)
  Imamura, A., "A diary on the great earthquake," *B.S.S.A.* 14 (1924)
  1–5.
  Yamasaki, N., "Physiographic studies of the great earthquake,"
  Journ. Faculty Science, Tokyo Imp. Univ., Sect. 2, 2 (1926) 77–
  119.
  Shepard, F. P., "Depth changes in Sagami Bay during the great
  earthquake," *Journ. Geology* 41 (1933) 527–536. (There are
  many special papers, in Japanese journals especially, some of them
  in English and other foreign languages.)

Muto, K., "A study of displacements of triangulation points," *E.R.I.*
10 (1932) 384–392.

1925, May 23 (Tajima, Japan)   302, *563*, 571–572
D, pp. 213–214.

Imamura, A., "The Tazima earthquake of 1925," *Bull. Imperial
Earthquake Investigation Committee* 10, No. 3 (1928) 71–107.

Yamasaki, N., "On the cause of the Tajima earthquake of 1925,"
ibid. 109–113.

Kotō, B., "The Tazima earthquake of 1925," *Journ. Faculty Science,
Tokyo Imp. Univ.*, Sect. 2, vol. 2, pt. 1 (1926) 1–75, 8 pl.

1925, June 27 (local date) (Montana)   281, 365
Willson, F. F., "The Montana earthquake of June 27, 1925—damage
in Gallatin County," *B.S.S.A.* 16 (1926) 165–169.

1925, June 29 (Santa Barbara)   70, 93, 99, 101, 388, 534
F, pp. 391–420.

"The Santa Barbara earthquake," Symposium, *B.S.S.A.* 15 (1925)
251–333.

Kirkbride, W. H., "The earthquake at Santa Barbara, June 29, 1925,
as it affected the railroad of the Southern Pacific Company, *ibid.*
17 (1927) 1–7.

1926, January 28 (Thuringia)   154
Sieberg, A., and Krumbach, G., *Das Einsturzbeben in Thüringen
vom 28. Januar 1926,* Reichsanstalt für Erdbebenforschung, Jena,
Part 6, 1927.

1927, March 7 (Tango, Japan)   281, *563*, 573–578
D, Chapter XIV, pp. 212–245.

IM, pp. 207–215, etc.

The literature is voluminous; the following references are representative:

Yamasaki, N., and Tada, F., "The Oku-Tango earthquake of 1927,"
E.R.I. 4 (1928) 159–177.

Imamura, A., "On the destructive Tango earthquake of March 7,
1927," *ibid.* 179–202. (In Japanese with English summary.)

Kunitomi, S. I., "Note on the North Tango earthquake of March 7,
1927," *Geophys. Mag.* (Tokyo) 2 (1929) 65–89.

Takahasi, R. A., "A graphical determination of the position of the
hypocenter of an earthquake and the velocity of the propagation
of the seismic waves," *E.R.I.* 6 (1929) 231–244.

Nasu, N., "On the aftershocks of the Tango earthquakes," *ibid.*
246–332 and pl. XVIII. (In Japanese with English summary.)

———, "Further study of the aftershocks of the Tango earthquake,"
ibid. 7 (1929) 133–152. (In Japanese with English summary.)

———, "Supplementary study on the stereometrical distribution
of the aftershocks of the great Tango earthquake of 1927," *ibid.*
13 (1935) 335–399.

Tsuboi, C., "An interpretation of the results of the repeated precise levelling in the Tango district after the Tango earthquake of 1927," *ibid*. 6 (1929) 71–83.

————, "Investigation on the deformation of the earth's crust in the Tango district connected with the Tango earthquake of 1927," *ibid*. 8 (1930) 153–221, 338–345; 9 (1931) 423–434; 10 (1932) 411–434.

1927, November 4 (off Point Arguello, California)   534

Byerly, P., "The California earthquake of November 4, 1927," *B.S.S.A.* 20 (1930) 53–66.

1928, January 6 (Africa)   313, 621–623, 716–718

See Appendix XV.

Willis, B., *East African Plateaus and Rift Valleys*, Carnegie Institution, Washington, D. C., 1936; see pp. 321–328.

Tillotson, E., "The African Rift Valley earthquake of 1928 January 6," *M.N.R.A.S. Geophys. Suppl.* 4 72–93 (1937). Further note, *ibid*. p. 315.

1928, April (Bulgaria)   *612*, 619–621

Bonchev, St., and Bakalov, P., "Les tremblements de terre dans la Bulgarie du Sud les 14 et 18 avril 1928," *Rev. soc. géol. bulgare*, 1928. (As given by Sieberg, A., in *Handbuch der Geophysik*, IV 3, *Erdbebengeographie*, p. 758; see *ibid*. IV 2, *Geologie der Erdbeben*, pp. 581–582.)

Jankof, K., "Changes in ground level produced by the earthquakes of April 14 and 18, 1928, in southern Bulgaria," *Tremblements de terre en Bulgarie*, Nos. 29–31, Institut météorologique central de Bulgarie, Sofia, 1945, pp. 131–136. (In Bulgarian.) (The same publication includes other descriptive articles and photographs on these earthquakes.)

1929, March 9 (New Zealand)   552–553

Speight, R., "The Arthur's Pass earthquake of 9th March, 1929," *N. Z. Journ.* 15 (1933) 173–182.

1929, June 17 (local date) (West Nelson, New Zealand)   538, 544–547

Fyfe, H. E., "Movement on White Creek fault, New Zealand, during the Murchison earthquake of 17th June, 1929," *N. Z. Journ.* 11 (1929) 192–197.

Henderson, J., "The West Nelson earthquake of 1929," *ibid*. 19 (1937) 65–144.

Thomson, A., "Earthquake sounds heard at great distances," *Nature* 124 (1929) 686–688.

1929, July 8 (Whittier, California)   37–44

F, pp. 435–440.

Wood, H. O., and Richter, C. F., "Recent earthquakes near Whittier, California," *B.S.S.A.* 21 (1931) 183–203. (See Chapter 4.)

1929, November 18 (Grand Banks, Atlantic)   125, 339

Doxsee, W. W., "The Grand Banks earthquake of November 18, 1929, *Publ. Dominion Observatory Ottawa*, 7 (1948) 323–335.

Heezen, B. C., and Ewing, M., "Turbidity currents and submarine slumps, and the 1929 Grand Banks earthquake," *Am. Journ. Science* 250 (1952) 849–873.

1930, November 26 (Izu, Japan)   132, 578–582
D, Chapter XV, pp. 246–265.

Nasu, N., Kishinouye, F., and Kodaira, T., "Recent seismic activities in the Idu Peninsula, Part 1, *E.R.I.* 9 (1931) 22–35; Part II (by N. Nasu), *ibid.* 13 (1935) 400–416.

Ōtuka, Y., "The geomorphology of the Kano-gawa alluvial plain, the earthquake fissures of Nov. 26, 1930, and the pre- and post-seismic crust deformations," *ibid.* 10 (1932) 235–246, pls. XX-XXV.

———, "The geomorphology and geology of northern Idu Peninsula, the earthquake fissures of Nov. 26, 1930, and the pre- and post-seismic crust deformations," *ibid.* 11 (1933) 530–574, pls. XXIV–XXXVIII.

Yamaguti, S., "Deformation of the earth's crust in Idu Peninsula in connection with the destructive Idu earthquake of Nov. 26, 1930," *ibid.* 15 (1937) 899–934.

Takahasi, R., "Results of the precise levellings executed in the Tanna railway tunnel and the movement along the slickenside that appeared in the tunnel," *Bull. E.R.I.* 9 (1931) 435–453.

Nasu, N., "Comparative studies of earthquake motion above ground and in a tunnel (Part I)," *ibid.* 454–472.

Imamura, A., "A seismometrical study of the destructive North Idu earthquake of November 26, 1930," *Japanese Journ. Astronomy Geophysics* 8 (1931) 51–65, pls. XII-XVII.

Kunitomi, S. I., "Notes on the North Idu earthquake of Nov. 26, 1930," *Geophys. Mag. (Tokyo)* 4 (1931) 73–102.

Terada, T., "On luminous phenomena accompanying earthquakes," *Bull. E.R.I.* 9 (1931) 225–255.

1931 (Hawke's Bay, New Zealand)   71, 76, 538, 547–551
D, Chapter XVI, pp. 269–276.

Callaghan, F. R., Henderson, J., *et al.*, "The Hawke's Bay earthquake of 3rd February 1931," *N. Z. Journ.* 15 (1933) 1–116.

Bullen, K. E., "An analysis of the Hawke's Bay earthquakes during February, 1931," *ibid.* 19 (1938) 497–519.

1932, September 16 (Wairoa, New Zealand)   71, 551–552
Ongley, M., Walshe, A. E., Henderson, J., and Hayes, R. C., "The Wairoa earthquake of 16th September, 1932," *N. Z. Journ.* 18 (1937) 845–865.

1932, December 20 (Nevada)   472, 504, 507–508
Gianella, V. P., and Callaghan, E., "The earthquake of December 20, 1932, at Cedar Mountain, Nevada, and its bearing on the genesis of Basin Range structure," *Journ. Geology* 42 (1934) 1–22.

———, "The Cedar Mountain, Nevada, earthquake of December 20, 1932," *B.S.S.A.* 24 (1934) 345–384, pls. 21–29.

Wilson, James T., "Foreshocks and aftershocks of the Nevada earthquakes of December 20, 1932, and the Parkfield earthquake of June 7, 1934," *ibid.* 26 (1936) 189–194.

1933 (Long Beach, Calif.) 23, 25, 34, 50, 67, 76, 91, 129, 383, 497–499

Wood, H. O., "Preliminary report on the Long Beach earthquake," *B.S.S.A.* 23 (1933) 43–56.

Chick, A. C., "The Long Beach earthquake of March 10, 1933 and its effect on industrial structures," *Trans. Am. Geophys. Union,* 1933, 273–284.

Martel, R. R., "A report on earthquake damage to type III buildings in Long Beach," *U. S. Coast Geodetic Survey, Spec. Publ.* No. 201 (1936); *Earthquake Investigations in Southern California 1934– 1935,* Chapter 8, pp. 143–162.

Philbrick, F. P., "The effect of earthquakes on fire-alarm systems," *B.S.S.A.* 31 (1941) 1–8.

Du Ree, A. C., "Fire-department operations during the Long Beach earthquake of 1933," *ibid.* 9–12.

1934, January 15 (India) 56–62, 64, 67, 106–107

"The Bihar-Nepal earthquake of 1934," *Mem. Geol. Survey India,* 73 (1939). See Chapter 5.

1934, January 30 (Nevada) 508–509

Callaghan, E., and Gianella, V. P., "The earthquake of January 30, 1934, at Excelsior Mountains, Nevada," *B.S.S.A.* 25 (1935) 161– 168.

1934, March 5 (Pahiatua, New Zealand) 540, 553

Hayes, R. C., "The Pahiatua earthquake of 1934, March 5," *N. Z. Journ.* 19 (1937) 382–388.

Bullen, K. E., "On the epicentre of the 1934 Pahiatua earthquake," *ibid.* 20 (1938) 61–66.

1934, March 12 (Utah) 509–510

Neumann, F., "The Utah earthquake of March 12, 1934," *United States Earthquakes, 1934, U. S. Coast Geodetic Survey Ser.* No. 593 (1936), pp. 43–48.

1934, June 7 (Parkfield, California) 205, 534

Byerly, P., and Wilson, James T., "The central California earthquakes of May 16, 1933 and June 7, 1934," *B.S.S.A.* 25 (1935) 223–246. (See Wilson under 1932, and *U. S. Earthquakes 1934,* 48–50.)

1934, July 6 (off California) 471

Byerly, P., "Earthquakes off the coast of northern California," *B.S.S.A.* 27 (1937) 73–96.

————, "The Earthquake of July 6, 1934; amplitudes and first motion," *ibid.* 28 (1938) 1–13.

1935, April 21 (Formosa) 582–585

E.R.I. Supplementary Volume III, 1936, *Papers and reports on the Formosa earthquake of 1935.* (In Japanese, with some English summaries.) See especially Paper 3: Ōtuka, Y., "The earthquake of central Taiwan (Formosa), April 21, 1935, and earthquake faults," pp. 22–69, pls. I–IV. (English summary, pp. 70–74.)

Nishimura, S., *Report on the April 21, 1935 earthquake in Shinchiku and Taichu,* Taihoku, 1937. (In Japanese.)

1935, May 31 (Quetta, Baluchistan)   63

West, W. D., "Preliminary geological report on the Baluchistan (Quetta) earthquake of May 31st, 1935," *Records Geol. Survey India,* 69 (1936) 203–241.

———, "Geological account of the Quetta earthquake," *Trans. Min. Geol. Inst. India,* 30 (1936) 138–142.

Ramanathan, K. R., and Mukherji, S. M., "A seismological study of the Baluchistan (Quetta) earthquake of May 31, 1935," *Records Geol. Survey India,* 73 (1938) 483–513.

1935, October (Helena, Montana)   73–79

Neumann, F., "The Helena earthquakes of October and November, 1935," in *U. S. Earthquakes 1935, U. S. Coast Geodetic Survey,* Ser. No. 600, pp. 42–56.

Engle, H. M., "The Montana earthquakes of October, 1935; structural lessons," *B.S.S.A.* 26 (1936) 99–109.

Ulrich, F. P., "Helena earthquakes," *ibid.* pp. 323–339.

1939, December 27 (Turkey)   111, 135, 611–616

Pamir, H. N., and Ketin, I., "Das anatolische Erdbeben Ende 1939," *Geol. Rundschau* 32 (1941) 279–287.

Ketin, I., "Über die tektonisch-mechanischen Folgerungen aus den grossen anatolischen Erdbeben des letzten Dezenniums," *ibid.* 36 (1948) 77–83.

Parejas, E., Akyol, I. H., and Altinli, E., "Le tremblement de terre d'Erzincan du 27. decembre 1939 (secteur occidental)," *Rev. Faculté Sciences Univ. Istanbul,* Ser. B, 6 (1941) 187–222.

Pamir, H. N., "Les séismes en Asie Mineure entre 1939 et 1944. La cicatrice nord-anatolienne," *Proc. Internatl. Geol. Congress, Great Britain 1948,* London, 1950, Part XIII, pp. 214–218.

1940 (Imperial Valley)   72, 74–76, 100, 108, 178, 180, 181, 472, 487–495

Ulrich, F. P., "The Imperial Valley earthquakes of 1940," *B.S.S.A.* 31 (1941) 13–31.

Buwalda, J. P., and Richter, C. F., "Imperial Valley earthquake of May 18, 1940," *Bull. Geol. Soc. Amer.* 52 (1941) 1944. (Abstract.)

1942, 1943, 1944. Anatolia. See 1939.

1942, June 24 (Wairarapa, New Zealand)   542–552

Ongley, M., "Wairarapa earthquake of 24th June, 1942, together with map showing surface traces of faults recently active," *N. Z. Journ.* 25 (1943) 67–78.

1943, March 4 and September 10 (Tottori, Japan)   26, 563, 585–586

Omote, S., "The Tottori earthquake of March 4, 1943," *E.R.I.* 21 (1943) 435–457. (In Japanese with English summary.)

Matuzawa, T., "Über die Verschiebung von Komainu bei dem Tottori-Grossbeben," *ibid.* 22 (1944) 60–65. (In Japanese with German summary.)

Tsuya, H., "Geological observations on the earthquake faults (Sikano and Yosioka) of 1943 in Tottori prefecture," *ibid.* 22 (1944) 1–32. (In Japanese.)

Miyamura, S., "Die zwei Verwerfungen beim Tottoribeben vom 10. Sept. 1943," *ibid.* pp. 49–59. In Japanese with German summary.)

Omote, S., "A preliminary report on the aftershocks of the Tottori earthquake," *ibid.* pp. 33–41. (In Japanese.)

————, Aftershocks that accompanied the Tottori earthquake of Sept. 10, 1943," *ibid.* 33 (1955) 641–661. (2nd paper.)

1944, January 15 (San Juan, Argentina)   600–601
Castellanos, A., "El terremoto de San Juan," *Asoc. cultural de conferencias de Rosario, Ciclo de caracter general, 1944,* Publ. No. 6. Part B, pp. 76–242.

1945, January 13 (Mikawa, Japan)   563, 586–587
Tsuya, H., "The Fukōzu fault. A remarkable earthquake fault formed during the Mikawa earthquake of January 13, 1945," *E.R.I.* 24 (1946) 59–75. (In Japanese, with English summary.)

1946, March 15 (Walker Pass, California)   518, 519
Chakrabarty, S. K., and Richter, C. F., "The Walker Pass earthquakes and structure of the southern Sierra Nevada," *B.S.S.A.* 39 (1949) 93–107.

1946, November 10 (Ancash, Peru)   602–604, 627
Silgado, E., "The Ancash, Peru, earthquake of November 10, 1946," *ibid.* 41 (1951) 83–100.

1947, April 10 (Manix, California)   67, 516–518
Richter, C. F., "The Manix (California) earthquake of April 10, 1947," *B.S.S.A.* 37 (1947) 172–179.

Richter, C. F., and Nordquist, J. M., "Instrumental study of the Manix earthquakes," *ibid.* 41 (1951) 347–388.

1948, June 28 (Fukui, Japan)   587–588
Collins, J. J., and Foster, Helen, *Geology,* Vol. 1, *The Fukui earthquake, Hokuriku region, Japan, 28 June, 1948.*

Butler, D. W., Muto, K., and Minami, K., *Engineering,* Vol. II, Office of the Engineer, General Headquarters, Far East Command, Tokyo, 1949.

Tsuya, H., ed., *The Fukui Earthquake of June 28, 1948, Report of the Special Committee for the Study of the Fukui Earthquake,* Tokyo, 1950.

Omote, S., "On the aftershocks of the Fukui earthquake, Part 2," *E.R.I.* 28 (1950) 311–319, pl. VII.

1948, October 5 (Ashkhabad)   611
Savarensky, E. F., Linden, N. A., and Masarsky, S. I., "Zemletryaseniya Turkmenii i Ashkhabadskoe zemletryaseniye 1948 goda," *Izv. Akad Nauk S.S.S.R. ser. geofiz.* 1953, 1–26.

Rustanovich, D. N., "Nekotorie voprosi izucheniya seysmichnosti Ashkhabadskogo rayona," *ibid.* 1957, 10–20.

1948, December 4 (Desert Hot Springs, California)   495–496
Murphy, L. M., and Ulrich, F. P., *United States Earthquakes 1948,* pp. 19–23.
Richter, C. F., Allen, C. R., and Nordquist, J. M., "The Desert Hot Springs earthquake of December 4, 1948" (abstract), *Bull. Geol. Soc. Amer.* 67 (1956) 1780.

1949, December 26 (Imaichi, Japan)   588–590
Symposium, *E.R.I.* 28 (1950) 355–472, pls. VIII–XXXI.
Ikegami, R., and Kishinouye, F. "The acceleration of earthquake motion deduced from the overturning of the gravestones in case of the Imaichi earthquake on Dec. 26, 1949," *ibid.* 121–128.

1950, January 8, etc. (Cook Strait, New Zealand)   456
Hayes, R. C., "The Cook Strait earthquakes: 1950 Jan.–Feb., "*N. Z. Journ.* 33 (1952) 309–318.

1950, August 15 (Assam-Tibet)   *47,* 63–64, 128, 134, 351
Rao, M. B. R., ed., *A Compilation of Papers on the Assam Earthquake of August 15, 1950,* The Central Board of Geophysics, Calcutta, 1953.
For other references see Chapter 5.

1950, December 14 (Fort Sage Mountains)   515–516
Gianella, V. P., "Earthquake and faulting, Fort Sage Mountains, California, December, 1950," *B.S.S.A.* 47 (1957) 173–177.

1952, (Kern Co.) 69–71, 77, 83–84, 89–101, 142, 183, 338–339, 370, 519–531
Båth, M., and Richter, C. F., "Mechanisms of the aftershocks of the Kern County, California, earthquake of 1952." (In press.)
"Earthquakes in Kern County, California, during 1952," *Calif. Dept. Nat. Resources, Div. Mines, Bull.* 171 (1955).
Steinbrugge, K. V., and Moran, D. F., "An engineering study of the Southern California earthquake of July 21, 1952, and its aftershocks," *B.S.S.A.* 44 (1954) 199–462.

1953, March 18 (Anatolia)   180, 616
Ketin, I., and Roesli, F., "Makroseismische Untersuchungen über das nordwestanatolische Beben vom 18. März 1953," *Eclogae Geol. Helvetiae* 46 (1953) 187–208.
Dilgan, H., and Hagiwara, T., "Le tremblement de terre de Yenice (18 mars 1953)," *Publ. bureau central séismologique international,* Ser. A, 19 (1956) 287–295.

1954, July 6 and August 23 (Fallon, Nevada)   73, 510–511
"The Fallon-Stillwater earthquakes of July 6, 1954 and August 23, 1954," *B.S.S.A.* 46 (1956): Byerly, P., Historic introduction, pp. 1–3; Slemmons, D. B., Geologic setting, pp. 4–9; Tocher, D., Movement on the Rainbow fault, pp. 10–14; Steinbrugge, K. V., and Moran, D. F., Damage, pp. 15–33; Cloud, W. K., Intensity distribution and strong-motion seismograph results, pp. 34–40.

1954, September 9 (Algeria) 125, 133, 621
Rothé, J. P., "Le tremblement de terre d'Orléansville et la séismicité de l'Algérie," *La Nature,* No. 3237, Janvier 1955, 1–9.

1954, December 16 73, 179, 504, 511–515
Gianella, V. P., "Faulting and the Nevada earthquakes of 1915, 1932, 1954" (abstract), *Bull. Geol. Soc. Amer.* 66 (1955) 1650.
Whitten, C. A., "Crustal movement in California and Nevada," *Trans. Amer. Geophys. Union,* 37 (1956) 393–398.
Tocher, D., Romney, C., Whitten, C. A., Cloud, W. K., Steinbrugge, K. V., Moran, D. F., Reil, O. E., Slemmons, D. B., Larson, E. R., and Zones, C. P., "The Dixie Valley-Fairview Peak, Nevada, earthquakes of December 16, 1954," *B.S.S.A.* 47 (1957), 299–396.

1955, April 1 (Philippines) 124
Kintanar, R. L., Quema, J. C., and Alcaraz, A. P., *The Lanao earthquake, Philippines, April 1, 1955,* Weather Bureau, Manila, 1955 (mimeographed).

1956, February 9 and 14 (Baja California) 531–532
See summarized reports in "Seismological Notes," *B.S.S.A.* 46 (1956) 161–163.
Shor, G. G., and Roberts, E., "San Miguel, Baja California Norte, earthquakes of February, 1956," *B.S.S.A.* (in press).

1957, December 4 (Mongolia) 625

APPENDIX XVII

Notes from Toronto—1957

THE ELEVENTH General Assembly of the International Union of Geodesy and Geophysics convened at Toronto, Canada, September 3–14, 1957. This was a highly effective and productive meeting; its success was in no small part due to the excellent facilities provided by the University of Toronto and Canadian authorities.

As usual, few new lines of research were opened; most of the papers presented additional data and further developments in established fields. Also as usual, the principal benefits came not from the formal presentation of papers, but from the opportunity for discussion among those working on similar problems in different countries.

Developments of most interest to readers of this book were in connection with meetings of the International Association of Seismology and Physics of the Earth's Interior. Dr. K. E. Bullen chose for his opening presidential address the title "Seismology in our Atomic Age." Briefly reviewing the history of seismology, he pointed out that improvement of instruments and development of theory since World War II makes possible a new and more precise attack on the problem of the nature of the interior of the Earth. Data from large artificial explosions, correlated with those for natural earthquakes, are excellent for the application of these new methods. More or less by accident, some highly valuable data have come from the testing of atomic weapons; but official inertia has usually blocked any effective use of these tests for seismological purposes, and has interfered with publication of data. Dr. Bullen reported on results obtained, largely by applying detective methods to published bulletins of seismological stations, from recordings of the hydrogen bombs of 1954. In 1955 Dr. Bullen, as chairman of a committee, circulated a proposal that one or more bombs be exploded specifically for seismological purposes. In general, this was not successful; but with the cooperation of British and Australian governments, four bombs exploded at Maralinga in Central Australia in 1956 were timed and recorded. These gave the first reliable information as to the depth of the Moho (35 kilometers or over) and the velocity of Pn ($8.23\pm$ kilometers per second) in Australia.

Later during the meetings, Dr. D. S. Carder of the U. S. Coast and Geodetic Survey presented transit time data based on atomic bomb tests in Nevada, and on hydrogen bombs fired on Pacific Islands (including those studied inferentially by Bullen and Burke-Gaffney). These data were presented and are expected to be published in a form suitable for accurate seismological applications.

736

Shortly after Dr. Bullen's address, the U. S. Atomic Energy Commision tele-graphed advance notice of an underground atomic detonation at the Nevada proving area. Preparations were made to record this at many points in the United States and Canada. The firing, after some postponements, took place on September 19, 1957; the energy radiated in elastic waves was rather small for an atomic explosion, and distant observers were disappointed. However, good records were obtained at short distances, and because of the excellent timing valuable results are anticipated.

At a symposium on magnitude and energy, Båth reported his latest result, $\log E = 12.24 + 1.44M = 6.5 + 2.3m$.

Solov'ev, from USSR data, reported $\log E = 11.5 + 1.5M$.

Values of the relation between m and M close to those determined by Guten-berg and by Båth were reported by di Filippo and Marcelli (Rome), and by Zátopek, Kárník, and Vaněk (Prague).

A new paper by Shebalin reported extension of his earlier work on energy, intensity, and focal depth, now including data for 225 earthquakes.

The loss of energy in passing through a level near 80 kilometers is confirmed; Shebalin concludes that Gutenberg's values of m for depths greater than 80 kilometers require a correction of about -0.7. However, this result depends in part on using surface waves to determine M for deep shocks.

Several of the Antarctic stations set up for the International Geophysical Year have gone into operation; in addition to Scott Base, regular recording is going on, and occasional readings are available, for the South Pole, Wilkes Land (Knox Coast), and a Soviet station (Mirny) in East Antarctica.

During discussion of the Rocky Mountain Trench, a rift structure in Canada and the northern United States, it was suggested that it may be a strike-slip feature.

A paper investigating the anomalous distribution of intensity observed for deep-focus shocks in Japan, relating it to geological and geophysical data, was presented by Wadati and Hirono.

A highly successful symposium, organized by Dr. J. H. Hodgson, dealt with the mechanism of earthquakes as derived from seismograms by the methods dis-cussed in Chapter 32. Dr. Keylis-Borok reported theoretical calculations with the disconcerting result that the form of the group of P waves recorded at distant stations may be seriously distorted if the hypocenter is within a few wavelengths of a discontinuity. Questions of statistical interpretation of groups of shocks, with the object of removing ambiguities in determination for single events, were mooted between Dr. Hodgson and Dr. McIntyre (their earlier publications on this point are cited in Chapter 32). The proceedings of the symposium will appear in full in the publications of the Dominion Observatory, Ottawa.

Many papers dealt with crustal structure. Caloi presented evidence that at least part of the root of the Alps is due to thickening of the intermediate (Con-rad) layer. Reports from the USSR indicated that, while the same is true for the Tian Shan, the root under the Pamir plateau, where the Moho descends to depths of 60–70 kilometers, is constituted by thickening of the whole crust. New work by Press and Ewing, based on phase velocities for Rayleigh waves from a South Pacific earthquake on April 14, 1957, provided a survey of crustal thick-ness across the North American continent.

The USSR was represented by a large delegation. The writer was fortunate in the opportunity for discussion with Professor I. E. Gubin. The account of seismic zoning in Chapter 33 does him somewhat less than justice. He deserves great credit for strongly emphasizing the necessity of applying geological evidence to supplement historical and geophysical data. This point is conceded by his opponents, who stress failure to correlate identifiable surface features with present seismicity. Geological and geophysical reasons for this have been noted in Chapter 33; a purely seismological factor is overestimation of accuracy of instrumental determinations of epicenters and depths.†

Gubin has vigorously continued his field investigations; at many points his latest interpretations closely approach those of this book, though based mostly on different data. Seismic zoning progresses in the USSR; a new edition of the fundamental map, on a scale of 1:500,000, is in course of publication.

---

† Controversy has made much of an earthquake in 1895, damaging at Krasnovodsk on the east coast of the Caspian Sea, where the surface topography presents a featureless plain. The locality is on the prolongation of surface tectonic features; geophysical evidence indicates an active subsurface structure. Compare the occurrence of earthquakes in the level Panhandle area of northern Texas; the active subsurface structures there are well known as a result of oil exploitation.

INDEX

Names of authors cited in the text are capitalized. Page references to text figures are italicized; for geographical names this generally indicates a map. Individual earthquakes are indexed geographically; a cross-reference by date, as, "*see* 1929," indicates an entry in Appendix XVI.

ABEL, N. H., 667, 669
Abnormal earthquake patterns in time, 72–73, 385, 617; *see also* Foreshocks, swarms of
Acambay (Mexico), 726
Acceleration, correlated with intensity, 87, 136, 140, 148; damaging, 26, 88, 140, 381; exceeding g, 25–26, 50–51, 53–54, 89; minimum perceptible, 26, 127, 140; momentarily high, 25–26, 589
Accelerometer, 215
Acobamba (Peru), 602, *603*
Acts of God, 608
Adelie Land, 334, 423
Adobe structures, damage to, 86, 93, 380, 475, 494, 501, 504
Adriatic Sea, *31,* 434, 619
Aegean intermediate earthquakes, 618, 624–625
Aeolotropic elastic solid, 657
Afterworking, 74
Africa, 298, *431,* 434; earthquakes, *see* 1928, 1954; rifts, 169, 184, 418, 621, 716; stable shield, 397, *398;* South Africa, 157, 165, 289, 298, 333, 435
Aftershocks, 68–72, 204; alarm caused by, 69, 385; Assam, 1950, 63, 205; California, 1952, *see* Kern County, 1952; Imperial Valley, 489–490; late and large, 70–71, 75, 497, 551; nature and mechanics of, 74–77; Omori's rule for frequency of, 69; recording, *see* Temporary stations; related to risk, 71, 385; second-order, 69–70, 76–77; Whittier (1929), 44
Afton Canyon (California), 517
AGAMENNONE, G., 223, 387
Agnews (California), *478,* 479
Agram (Croatia), *see* Zagreb
Agricultural losses, 60, 90, 91, 108, 384, 488

Agua Blanca fault, *441,* 463, *531, 532, 533*
Agua Caliente fault, *441,* 445, 461, *474*
AGUILERA, J. G., 594, 723
Air reconnaissance, use of, 173, 181, 454
Air waves, 128, 134, 152, 163–164
Ajiro (Japan), *578, 579,* 580
Akaroa (New Zealand), *442,* 453
AKYOL, I. H., 614, 732
Alabama Hills (California), 404, 461, *500,* 501
Alamo Canal (Baja California), *487,* 491
Alamo River (California), 181, *487,* 491
Alaska, 187, 209, 289, 299, 331, *419,* 433; Denali fault rift, 173, 187; earthquake, *see* 1899
Aleutian: arc, *419,* 432; Islands, *419;* Trench, 111, 112, 115, *419;* earthquake of 1957, 115, 116, 369–370, 713
ALFORD, J. L., 103, 649
Algerian earthquake, *see* 1954
Algiers or Alger, *105,* 333, 706, 708
All Saints' Day, 104
Allah Bund, 607, *608,* 720
All-American Canal (California), *487,* 491–493
ALLEN, C. R., vi, 173, 185, 194, 195, 443, 444, 445, 463, 481, 494, 495, 532, 734
ALLEN, M. W., 467, 536
Alma-Ata (USSR), 610, *632;* earthquakes, *see* 1887, 1911
Alpide arcs, 414; compared with Pacific, 48, 402, 414, 434, 558, 607
Alpide belt, 357, 377, 397, 402, *429,* 434, 604, 624, 635, 636; structures, 32, 48, 406, 615, 616, 624; earthquakes, 32, 47–65, 357, 607–621
Alpine fault (New Zealand), 173, 186, 409, 416, 441, *442,* 446–448, 454, 455–456, 465, 553
Alpine orogeny, 31–32, 48, 406